Biomechanical Study and Analysis for Cardiovascular/Skeletal Materials and Devices

Biomechanical Study and Analysis for Cardiovascular/Skeletal Materials and Devices

Editors

Aike Qiao
Haisheng Yang
Yongliang Mu

Basel • Beijing • Wuhan • Barcelona • Belgrade • Novi Sad • Cluj • Manchester

Editors

Aike Qiao
Beijing University of Technology
Beijing
China

Haisheng Yang
Beijing University of Technology
Beijing
China

Yongliang Mu
Northeastern University
Shenyang
China

Editorial Office
MDPI AG
Grosspeteranlage 5
4052 Basel, Switzerland

This is a reprint of articles from the Special Issue published online in the open access journal *Journal of Functional Biomaterials* (ISSN 2079-4983) (available at: https://www.mdpi.com/journal/jfb/special_issues/biomechanical_mat).

For citation purposes, cite each article independently as indicated on the article page online and as indicated below:

Lastname, A.A.; Lastname, B.B. Article Title. *Journal Name* **Year**, *Volume Number*, Page Range.

ISBN 978-3-7258-1669-9 (Hbk)
ISBN 978-3-7258-1670-5 (PDF)
doi.org/10.3390/books978-3-7258-1670-5

© 2024 by the authors. Articles in this book are Open Access and distributed under the Creative Commons Attribution (CC BY) license. The book as a whole is distributed by MDPI under the terms and conditions of the Creative Commons Attribution-NonCommercial-NoDerivs (CC BY-NC-ND) license.

Contents

About the Editors ix

Preface xi

Aike Qiao, Tianming Du, Haisheng Yang and Yongliang Mu
Biomechanical Study and Analysis for Cardiovascular/Skeletal Materials and Devices
Reprinted from: *J. Funct. Biomater.* **2023**, *14*, 398, doi:10.3390/jfb14080398 1

Kohei Mitsuzuka, Yujie Li, Toshio Nakayama, Hitomi Anzai, Daisuke Goanno, Simon Tupin, et al.
A Parametric Study of Flushing Conditions for Improvement of Angioscopy Visibility
Reprinted from: *J. Funct. Biomater.* **2022**, *13*, 69, doi:10.3390/jfb13020069 4

Tao Li, Zhuo Zhang, Wenyuan Wang, Aijia Mao, Yu Chen, Yan Xiong and Fei Gao
Simulation and Experimental Investigation of Balloon Folding and Inserting Performance for Angioplasty: A Comparison of Two Materials, Polyamide-12 and Pebax
Reprinted from: *J. Funct. Biomater.* **2023**, *14*, 312, doi:10.3390/jfb14060312 16

Yiya Kong, Xiaobin Yu, Gang Peng, Fang Wang and Yajun Yin
Interstitial Fluid Flows along Perivascular and Adventitial Clearances around Neurovascular Bundles
Reprinted from: *J. Funct. Biomater.* **2022**, *13*, 172, doi:10.3390/jfb13040172 31

Makoto Ohta, Naoya Sakamoto, Kenichi Funamoto, Zi Wang, Yukiko Kojima and Hitomi Anzai
A Review of Functional Analysis of Endothelial Cells in Flow Chambers
Reprinted from: *J. Funct. Biomater.* **2022**, *13*, 92, doi:10.3390/jfb13030092 44

Yang Yang, Yijing Li, Chen Liu, Jingyuan Zhou, Tao Li, Yan Xiong and Ling Zhang
Hemodynamic Analysis of the Geometric Features of Side Holes Based on GDK Catheter
Reprinted from: *J. Funct. Biomater.* **2022**, *13*, 236, doi:10.3390/jfb13040236 63

Liang Wang, Akiko Maehara, Rui Lv, Xiaoya Guo, Jie Zheng, Kisten L. Billiar, et al.
Image-Based Finite Element Modeling Approach for Characterizing In Vivo Mechanical Properties of Human Arteries
Reprinted from: *J. Funct. Biomater.* **2022**, *13*, 147, doi:10.3390/jfb13030147 79

Rui Lv, Liang Wang, Akiko Maehara, Mitsuaki Matsumura, Xiaoya Guo, Habib Samady, et al.
Combining IVUS + OCT Data, Biomechanical Models and Machine Learning Method for Accurate Coronary Plaque Morphology Quantification and Cap Thickness and Stress/Strain Index Predictions
Reprinted from: *J. Funct. Biomater.* **2023**, *14*, 41, doi:10.3390/jfb14010041 96

Mengde Huang, Akiko Maehara, Dalin Tang, Jian Zhu, Liang Wang, Rui Lv, et al.
Human Coronary Plaque Optical Coherence Tomography Image Repairing, Multilayer Segmentation and Impact on Plaque Stress/Strain Calculations
Reprinted from: *J. Funct. Biomater.* **2022**, *13*, 213, doi:10.3390/jfb13040213 113

Qing-Zhuo Chi, Yang-Yang Ge, Zhen Cao, Li-Li Long, Li-Zhong Mu, Ying He and Yong Luan
Experimental Study of the Propagation Process of Dissection Using an Aortic Silicone Phantom
Reprinted from: *J. Funct. Biomater.* **2022**, *13*, 290, doi:10.3390/jfb13040290 127

Lili Long, Huimin Chen, Ying He, Lizhong Mu and Yong Luan
Lingering Dynamics of Type 2 Diabetes Mellitus Red Blood Cells in Retinal Arteriolar Bifurcations
Reprinted from: *J. Funct. Biomater.* **2022**, *13*, 205, doi:10.3390/jfb13040205 144

Peng Su, Chao Yue, Likun Cui, Qinjian Zhang, Baoguo Liu and Tian Liu
Quasi-Static Mechanical Properties and Continuum Constitutive Model of the Thyroid Gland
Reprinted from: *J. Funct. Biomater.* **2022**, *13*, 283, doi:10.3390/jfb13040283 159

Jingyuan Zhou, Yijing Li, Tao Li, Xiaobao Tian, Yan Xiong and Yu Chen
Analysis of the Effect of Thickness on the Performance of Polymeric Heart Valves
Reprinted from: *J. Funct. Biomater.* **2023**, *14*, 309, doi:10.3390/jfb14060309 174

Saneth Gavishka Sellahewa, Jojo Yijiao Li and Qingzhong Xiao
Updated Perspectives on Direct Vascular Cellular Reprogramming and Their Potential Applications in Tissue Engineered Vascular Grafts
Reprinted from: *J. Funct. Biomater.* **2022**, *14*, 21, doi:10.3390/jfb14010021 188

Hanbing Zhang, Tianming Du, Shiliang Chen, Yang Liu, Yujia Yang, Qianwen Hou and Aike Qiao
Finite Element Analysis of the Non-Uniform Degradation of Biodegradable Vascular Stents
Reprinted from: *J. Funct. Biomater.* **2022**, *13*, 152, doi:10.3390/jfb13030152 217

Zhuo Zhang, Yan Xiong, Jinpeng Hu, Xuying Guo, Xianchun Xu, Juan Chen, et al.
A Finite Element Investigation on Material and Design Parameters of Ventricular Septal Defect Occluder Devices
Reprinted from: *J. Funct. Biomater.* **2022**, *13*, 182, doi:10.3390/jfb13040182 232

Tianming Du, Yumiao Niu, Youjun Liu, Haisheng Yang, Aike Qiao and Xufeng Niu
Physical and Chemical Characterization of Biomineralized Collagen with Different Microstructures
Reprinted from: *J. Funct. Biomater.* **2022**, *13*, 57, doi:10.3390/jfb13020057 245

Wenjuan Yan, Fenghe Yang, Zhongning Liu, Quan Wen, Yike Gao, Xufeng Niu and Yuming Zhao
Anti-Inflammatory and Mineralization Effects of an ASP/PLGA-ASP/ACP/PLLA-PLGA Composite Membrane as a Dental Pulp Capping Agent
Reprinted from: *J. Funct. Biomater.* **2022**, *13*, 106, doi:10.3390/jfb13030106 256

Changxin Xiang, Xinyan Zhang, Jianan Zhang, Weiyi Chen, Xiaona Li, Xiaochun Wei and Pengcui Li
A Porous Hydrogel with High Mechanical Strength and Biocompatibility for Bone Tissue Engineering
Reprinted from: *J. Funct. Biomater.* **2022**, *13*, 140, doi:10.3390/jfb13030140 274

Yumiao Niu, Tianming Du and Youjun Liu
Biomechanical Characteristics and Analysis Approaches of Bone and Bone Substitute Materials
Reprinted from: *J. Funct. Biomater.* **2023**, *14*, 212, doi:10.3390/jfb14040212 288

Yang Liu, Tianming Du, Aike Qiao, Yongliang Mu and Haisheng Yang
Zinc-Based Biodegradable Materials for Orthopaedic Internal Fixation
Reprinted from: *J. Funct. Biomater.* **2022**, *13*, 164, doi:10.3390/jfb13040164 312

Qiqi Ge, Xiaoqian Liu, Aike Qiao and Yongliang Mu
Compressive Properties and Degradable Behavior of Biodegradable Porous Zinc Fabricated with the Protein Foaming Method
Reprinted from: *J. Funct. Biomater.* **2022**, *13*, 151, doi:10.3390/jfb13030151 332

Ayumi Kaneuji, Mingliang Chen, Eiji Takahashi, Noriyuki Takano, Makoto Fukui, Daisuke Soma, et al.
Collarless Polished Tapered Stems of Identical Shape Provide Differing Outcomes for Stainless Steel and Cobalt Chrome: A Biomechanical Study
Reprinted from: *J. Funct. Biomater.* **2023**, *14*, 262, doi:10.3390/jfb14050262 . . . . **348**

About the Editors

Aike Qiao

Aike Qiao, Ph.D., Professor of the Department of Biomedical Engineering, College of Life Science and Bioengineering, Beijing University of Technology. He received his Ph.D. degree in Fluid Mechanics from the Beijing University of Technology. Afterward, he worked as a Postdoctoral Research Associate in the Shantou University Medical College, China. He also conducted visiting research in the Division of Information Sciences, Japan Advanced Science and Technology, Japan. Dr. Qiao has over 30 years of research experience in cardiovascular biomechanics. His main research interests are numerical simulations of biomechanics involving endovascular intervention and biomedical devices.

Haisheng Yang

Haisheng Yang, Ph.D., Professor and Director of the Department of Biomedical Engineering, College of Life Science and Bioengineering, Beijing University of Technology. He received his Ph.D. degree in Mechanics/Biomechanics from the Harbin Institute of Technology (jointly with the University of California, Berkeley). Afterwards, he worked as a Postdoctoral Research Associate in the Purdue Musculoskeletal Biology and Mechanics Lab at the Department of Basic Medical Sciences, Purdue University, USA. He also conducted research at the Research Centre of Shriners Hospitals for Children-Canada at McGill University, Canada. Dr. Yang has over 10 years of research experience in orthopedic biomechanics and mechanobiology of bone adaptation and regeneration.

Yongliang Mu

Yongliang Mu, Ph.D., Associate professor, Institute of Advanced Materials Preparation and Application Technology, School of Metallurgy, Northeastern University. He received his Ph.D. on nonferrous metal metallurgy from Northeastern University. Afterward, he worked as a Postdoctoral Research Associate in the Light Alloy Net Forming National Engineering Research Center-LAF-NERC at the School of Materials Science and Engineering, Shanghai Jiao Tong University, China. He also conducted research at the Engineering Research Center for Advanced Material Preparation Technology of the Ministry of Education at Northeastern University, China. Dr. Mu has over 10 years of research experience in biomedical degradable alloys and porous Zn/Mg alloys for bone tissue repair and replacement.

Preface

In examining biological functional materials and devices closely concerned with cardiovascular diseases and orthopedic diseases, we edited the Special Issue "Biomechanical Study and Analysis for Cardiovascular/Skeletal Materials and Devices". We would like to thank the "Journal of Functional Biomaterials" for the opportunity to publish this work, and we would also like to thank all the associated researchers for their support.

This Special Issue includes a total of 22 articles. The range of interventional medical devices in cardiovascular and orthopedic fields is relatively broad. Therefore, the research content of this Special Issue involves the application of polymer biomaterials, metal biomaterials, organic biomaterials, composite biomimetic materials, etc. This Special Issue focuses on the research hotspots of the biomechanical properties of cardiovascular/skeletal materials and devices. By editing this Special Issue, we hope to help researchers better understand the fundamental problems and challenges in current research related to cardiovascular/skeletal materials and devices, and we also hope to provide theoretical support for solving clinical problems related to cardiovascular/skeletal diseases. In future work, more in-depth attention should be paid to cross-sectional fundamental and clinical issues related to bone-vascular cross-talk, and existing research should be utilized within clinical applications to effectively solve problems that plague human health.

Aike Qiao, Haisheng Yang, and Yongliang Mu
Editors

Editorial

Biomechanical Study and Analysis for Cardiovascular/Skeletal Materials and Devices

Aike Qiao [1,*,†], Tianming Du [1,†], Haisheng Yang [1] and Yongliang Mu [2]

[1] Faculty of Environment and Life, Beijing University of Technology, Beijing 100124, China; dutianming@bjut.edu.cn (T.D.); haisheng.yang@bjut.edu.cn (H.Y.)
[2] School of Metallurgy, Northeastern University, Shenyang 110819, China; muyl@smm.neu.edu.cn
* Correspondence: qak@bjut.edu.cn
† These authors contributed equally to this work.

The Special Issue entitled "Biomechanical Study and Analysis for Cardiovascular/Skeletal Materials and Devices" addresses biological functional materials and devices relevant to cardiovascular diseases and orthopedic conditions. We are grateful for the opportunity provided by the *Journal of Functional Biomaterials* and the strong support of the researchers involved.

This Special Issue comprises a total of 22 articles, covering a wide range of interventional medical devices in the cardiovascular and orthopedic fields. The research presented in this Special Issue covers various applications of polymer biomaterials, metal biomaterials, organic biomaterials, and composite biomimetic materials [1–6]. The primary focus of this Special Issue is to investigate the biomechanical properties of cardiovascular/skeletal materials and devices, reflecting current research interests. In the subsequent sections, we provide a concise overview of the key aspects and challenges in the study of commonly used biological functional materials in cardiovascular and orthopedic fields. We also discuss future research trends, drawing insights from the articles featured in this Special Issue.

1. Why Study the Biomechanical Properties of Biomaterials for Cardiovascular/Skeletal Applications?

Cardiovascular and orthopedic diseases are two global epidemic diseases with far-reaching medical and socio-economic consequences [7]. Understanding the mechanical characteristics of arterial walls, blood flow, and valves holds great importance for the diagnosis, management, and treatment of cardiovascular diseases [8–10]. At the same time, bone tissue serves as a crucial weight-bearing structure within the body [11]. Consequently, the mechanical properties of implant materials and devices play a pivotal role in addressing cardiovascular and orthopedic diseases.

2. Research Progress of "Biomechanical Study and Analysis for Cardiovascular/Skeletal Materials and Devices"

(1) **A new phenomenon of interstitial fluid (ISF) microflow in perivascular and adventitial spaces around neurovascular bundles was reported.** Within this Special Issue, Kong et al. presented novel observations regarding the microflow of interstitial fluid (ISF) within the perivascular and adventitial clearances (PAC) surrounding neurovascular bundles [12]. This study not only enhances our understanding of ISF circulation throughout the body but also provides insights into the fundamental architecture of PAC. It helps to lay the foundation for the kinematics and dynamics of the ISF flow along the PAC around neurovascular bundles. Consequently, it establishes a basis for investigating the kinematics and dynamics of ISF flow along the PAC surrounding neurovascular bundles.

(2) **Matching of mechanical properties and degradation performance of metal biomaterials is currently a focus of attention in both orthopedic and cardiovascular materials** [13]. In this Special Issue, Zhang et al. conducted numerical

simulations to investigate material degradation [14]. Their study revealed that the suggested non-uniform degradation model, incorporating multiple factors for biodegradable endovascular stents, exhibited distinct phenomena when compared to commonly employed models. Furthermore, the numerical simulation results were found to align more closely with real-world degradation scenarios. In addition, various biodegradable porous materials were developed and demonstrated favorable compatibility between degradation and mechanical properties. For instance, biodegradable porous zinc stents and high-strength porous hydrogels showed promising biocompatibility for bone tissue engineering [2,3].

(3) **Improving biomaterials through biomimetic mineralization.** The composite properties of cardiovascular and skeletal systems play a vital role in their remarkable functionality as human tissues. In recent studies, there has been considerable interest in the investigation of biomimetic materials. Du et al. highlighted the significance of collagen mineralization research, which not only provides insights into the formation mechanisms of physiological tissues in humans but also holds promise for the development of more suitable biological functional materials for treating orthopedic diseases [11,15].

(4) **Advanced imaging, detection equipment, and computer technology have also greatly promoted the development of this field** [16]. A balloon dilatation catheter plays a critical role in percutaneous transluminal angioplasty procedures. In this Special Issue, Li et al. aimed to enhance our understanding of the underlying patterns by employing a highly realistic simulation method for balloon folding. They compared the trackability of balloons constructed from different materials, seeking to provide more effective insights [1]. This simulation-based approach allows for the evaluation of balloon performance when navigating curved paths, offering more precise and detailed data feedback compared to traditional benchtop experiments. Additionally, Lv et al. conducted a meticulous quantification of coronary artery plaque morphology and predicted cap thickness and stress/strain index. They employed a combination of IVUS, OCT data, biomechanical models, and machine-learning techniques for accurate assessments [17]. Furthermore, Huang et al. developed an automatic multilayer segmentation and repair method to extract multilayer vessel geometries from OCT images, facilitating the construction of biomechanical models [18]. The proposed segmentation technique holds significant potential for wide-ranging applications in vulnerable plaque research.

3. **Summary**

Through curating this Special Issue, our intention is to enhance researchers' comprehension of the fundamental issues and challenges prevalent in current investigations concerning cardiovascular/skeletal materials and devices. Furthermore, we aim to offer theoretical foundations that contribute to resolving clinical problems associated with cardiovascular and skeletal diseases. In future endeavors, it is imperative to allocate more extensive attention to interdisciplinary investigations encompassing bone–vascular crosstalk, both at fundamental and clinical levels. Additionally, we emphasize the importance of translating existing research studies into practical clinical applications, thereby effectively addressing the pressing health concerns affecting humanity.

Author Contributions: Conceptualization, A.Q. and T.D.; investigation, T.D.; resources, A.Q. and T.D.; writing—original draft preparation, T.D. and A.Q.; writing—review and editing, A.Q. and T.D.; supervision, A.Q.; project administration, A.Q., T.D., H.Y. and Y.M.; funding acquisition, T.D. and A.Q. All authors have read and agreed to the published version of the manuscript.

Funding: This research was supported by the National Natural Science Foundation of China [grant numbers 12172018 and 12202023] and the Joint Program of Beijing Municipal—Beijing Natural Science Foundation [grant number KZ202110005004].

Data Availability Statement: The data presented in this paper are available on request from the corresponding author.

Conflicts of Interest: The authors declare no conflict of interest.

References

1. Li, T.; Zhang, Z.; Wang, W.; Mao, A.; Chen, Y.; Xiong, Y.; Gao, F. Simulation and Experimental Investigation of Balloon Folding and Inserting Performance for Angioplasty: A Comparison of Two Materials, Polyamide-12 and Pebax. *J. Funct. Biomater.* **2023**, *14*, 312. [CrossRef] [PubMed]
2. Ge, Q.; Liu, X.; Qiao, A.; Mu, Y. Compressive properties and degradable behavior of biodegradable porous zinc fabricated with the protein foaming method. *J. Funct. Biomater.* **2022**, *13*, 151. [CrossRef] [PubMed]
3. Xiang, C.; Zhang, X.; Zhang, J.; Chen, W.; Li, X.; Wei, X.; Li, P. A Porous Hydrogel with High Mechanical Strength and Biocompatibility for Bone Tissue Engineering. *J. Funct. Biomater.* **2022**, *13*, 140. [CrossRef] [PubMed]
4. Yan, W.; Yang, F.; Liu, Z.; Wen, Q.; Gao, Y.; Niu, X.; Zhao, Y. Anti-Inflammatory and Mineralization Effects of an ASP/PLGA-ASP/ACP/PLLA-PLGA Composite Membrane as a Dental Pulp Capping Agent. *J. Funct. Biomater.* **2022**, *13*, 106. [CrossRef] [PubMed]
5. Chi, Q.-Z.; Ge, Y.-Y.; Cao, Z.; Long, L.-L.; Mu, L.-Z.; He, Y.; Luan, Y. Experimental Study of the Propagation Process of Dissection Using an Aortic Silicone Phantom. *J. Funct. Biomater.* **2022**, *13*, 290. [CrossRef] [PubMed]
6. Kaneuji, A.; Chen, M.; Takahashi, E.; Takano, N.; Fukui, M.; Soma, D.; Tachi, Y.; Orita, Y.; Ichiseki, T.; Kawahara, N. Collarless Polished Tapered Stems of Identical Shape Provide Differing Outcomes for Stainless Steel and Cobalt Chrome: A Biomechanical Study. *J. Funct. Biomater.* **2023**, *14*, 262. [CrossRef] [PubMed]
7. Sellahewa, S.G.; Li, J.Y.; Xiao, Q. Updated Perspectives on Direct Vascular Cellular Reprogramming and Their Potential Applications in Tissue Engineered Vascular Grafts. *J. Funct. Biomater.* **2022**, *14*, 21. [CrossRef] [PubMed]
8. Wang, L.; Maehara, A.; Lv, R.; Guo, X.; Zheng, J.; Billiar, K.L.; Mintz, G.S.; Tang, D. Image-Based Finite Element Modeling Approach for Characterizing In Vivo Mechanical Properties of Human Arteries. *J. Funct. Biomater.* **2022**, *13*, 147. [CrossRef] [PubMed]
9. Ohta, M.; Sakamoto, N.; Funamoto, K.; Wang, Z.; Kojima, Y.; Anzai, H. A Review of Functional Analysis of Endothelial Cells in Flow Chambers. *J. Funct. Biomater.* **2022**, *13*, 92. [CrossRef] [PubMed]
10. Yang, Y.; Li, Y.; Liu, C.; Zhou, J.; Li, T.; Xiong, Y.; Zhang, L. Hemodynamic Analysis of the Geometric Features of Side Holes Based on GDK Catheter. *J. Funct. Biomater.* **2022**, *13*, 236. [CrossRef] [PubMed]
11. Niu, Y.; Du, T.; Liu, Y. Biomechanical characteristics and analysis approaches of bone and bone substitute materials. *J. Funct. Biomater.* **2023**, *14*, 212. [CrossRef] [PubMed]
12. Kong, Y.; Yu, X.; Peng, G.; Wang, F.; Yin, Y. Interstitial Fluid Flows Along Perivascular and Adventitial Clearances around Neurovascular Bundles. *J. Funct. Biomater.* **2022**, *13*, 172. [CrossRef] [PubMed]
13. Liu, Y.; Du, T.; Qiao, A.; Mu, Y.; Yang, H. Zinc-Based Biodegradable Materials for Orthopaedic Internal Fixation. *J. Funct. Biomater.* **2022**, *13*, 164. [CrossRef] [PubMed]
14. Zhang, H.; Du, T.; Chen, S.; Liu, Y.; Yang, Y.; Hou, Q.; Qiao, A. Finite Element Analysis of the Non-Uniform Degradation of Biodegradable Vascular Stents. *J. Funct. Biomater.* **2022**, *13*, 152. [CrossRef] [PubMed]
15. Du, T.; Niu, Y.; Liu, Y.; Yang, H.; Qiao, A.; Niu, X. Physical and Chemical Characterization of Biomineralized Collagen with Different Microstructures. *J. Funct. Biomater.* **2022**, *13*, 57. [CrossRef] [PubMed]
16. Zhang, Z.; Xiong, Y.; Hu, J.; Guo, X.; Xu, X.; Chen, J.; Wang, Y.; Chen, Y. A Finite Element Investigation on Material and Design Parameters of Ventricular Septal Defect Occluder Devices. *J. Funct. Biomater.* **2022**, *13*, 182. [CrossRef] [PubMed]
17. Lv, R.; Wang, L.; Maehara, A.; Matsumura, M.; Guo, X.; Samady, H.; Giddens, D.P.; Zheng, J.; Mintz, G.S.; Tang, D. Combining IVUS+ OCT Data, Biomechanical Models and Machine Learning Method for Accurate Coronary Plaque Morphology Quantification and Cap Thickness and Stress/Strain Index Predictions. *J. Funct. Biomater.* **2023**, *14*, 41. [CrossRef] [PubMed]
18. Huang, M.; Maehara, A.; Tang, D.; Zhu, J.; Wang, L.; Lv, R.; Zhu, Y.; Zhang, X.; Matsumura, M.; Chen, L.; et al. Human Coronary Plaque Optical Coherence Tomography Image Repairing, Multilayer Segmentation and Impact on Plaque Stress/Strain Calculations. *J. Funct. Biomater.* **2022**, *13*, 213. [CrossRef] [PubMed]

Disclaimer/Publisher's Note: The statements, opinions and data contained in all publications are solely those of the individual author(s) and contributor(s) and not of MDPI and/or the editor(s). MDPI and/or the editor(s) disclaim responsibility for any injury to people or property resulting from any ideas, methods, instructions or products referred to in the content.

Article

A Parametric Study of Flushing Conditions for Improvement of Angioscopy Visibility

Kohei Mitsuzuka [1,2], Yujie Li [1,3], Toshio Nakayama [4], Hitomi Anzai [1], Daisuke Goanno [1,2], Simon Tupin [1], Mingzi Zhang [1,5], Haoran Wang [1], Kazunori Horie [6] and Makoto Ohta [1,*]

1. Institute of Fluid Science, Tohoku University, 2-1-1 Katahira, Aoba-ku, Sendai 980-8577, Japan; k.mitsuzuka0326@gmail.com (K.M.); jessie.li@torrens.edu.au (Y.L.); hitomi.anzai.b5@tohoku.ac.jp (H.A.); daisuke.goanno.r1@dc.tohoku.ac.jp (D.G.); s.tupin@tohoku.ac.jp (S.T.); mingzi.zhang@mq.edu.au (M.Z.); victorytcwang@gmail.com (H.W.)
2. Graduate School of Biomedical Engineering, Tohoku University, 6-6 Azaaoba, Aramaki, Aoba-ku, Sendai 980-8579, Japan
3. Centre for Healthy Futures, Torrens University Australia, 1-51 Foveaux Street, Sydney, NSW 2010, Australia
4. National Institute of Technology, Nara College, 22 Yatacho, Yamatokoriyama 639-1080, Japan; tnakayama@ctrl.nara-k.ac.jp
5. Macquarie Medical School, Faculty of Medicine, Health, and Human Sciences, Macquarie University, 75 Talavera Rd., Sydney, NSW 2109, Australia
6. Department of Cardiology, Sendai Kousei Hospital, 4-15 Hirose, Aoba-ku, Sendai 980-0873, Japan; horihori1015@gmail.com
* Correspondence: makoto.ohta@tohoku.ac.jp; Tel.: +81-22-217-53-09

Abstract: During an angioscopy operation, a transparent liquid called dextran is sprayed out from a catheter to flush the blood away from the space between the camera and target. Medical doctors usually inject dextran at a constant flow rate. However, they often cannot obtain clear angioscopy visibility because the flushing out of the blood is insufficient. Good flushing conditions producing clear angioscopy visibility will increase the rate of success of angioscopy operations. This study aimed to determine a way to improve the clarity for angioscopy under different values for the parameters of the injection waveform, endoscope position, and catheter angle. We also determined the effect of a stepwise waveform for injecting the dextran only during systole while synchronizing the waveform to the cardiac cycle. To evaluate the visibility of the blood-vessel walls, we performed a computational fluid dynamics (CFD) simulation and calculated the visible area ratio (VAR), representing the ratio of the visible wall area to the total area of the wall at each point in time. Additionally, the normalized integration of the VAR called the area ratio (AR_{VAR}) represents the ratio of the visible wall area as a function of the dextran injection period. The results demonstrate that the AR_{VAR} with a stepped waveform, bottom endoscope, and three-degree-angle catheter results in the highest visibility, around 25 times larger than that under the control conditions: a constant waveform, a center endoscope, and 0 degrees. This set of conditions can improve angioscopy visibility.

Keywords: coronary angioscopy; flush conditions; CFD; two-phase flow; dextran injection

1. Introduction

Ischemic heart disease (IHD) (e.g., heart infarction and angina pectoris) has been a primary cause of death in the world [1]. The IHD is caused by stenosis inside the coronary artery [2]. With the chronic deposition of cholesterols and other substances in the blood, the arterial lumen becomes narrow, and the blood flow is partially or totally blocked. The lack of oxygen leads to the dysfunction of heart muscle. When medical doctors perform percutaneous coronary intervention (PCI) as a treatment method for IHD, they need to evaluate the condition of the thrombus and plaque before and after the procedure to perform the PCI procedure safely. The color of the thrombus and plaque is an essential marker of their condition [3,4].

Angiography, intravascular ultrasound (IVUS) [5–7] and optical coherence tomography [8,9] are frequently used to observe the blood-vessel wall. However, they provide only an indirect view of the plaque. Angioscopy provides sequential visual images through a camera and directly visualizes the wall and plaque. Medical doctors seek color information for the plaque on the wall, thrombus, and plaque protrusion, and the vessel's morphology [10,11]. There are several papers with statistical analyses describing the capabilities of angioscopy for plaque-color classification or stent-coverage investigations [12,13]. During angioscopy, for taking images, the blood in the artery needs to be removed by using a transparent liquid. Low-molecular-weight dextran is often used as the transparent liquid [14,15]. Previously, the coronary artery was completely blocked using a balloon to stop the blood flow. Then, dextran was sprayed out of the catheter to remove the blood. This method allowed doctors to record images easily [16]. However, balloon occlusion poses a risk of IHD because of the complete cessation of the blood supply [17], so the use of a balloon is gradually being phased out to avoid this risk [18]. The new method is performed without a balloon, and the dextran is sprayed out while the images are being taken. Komatsu et al. examined the possibility of increasing the visibility by increasing the flow rate [17]. However, there is still a risk of blood-flow blockage due to over-flushing.

To improve the visibility during angioscopy, several studies on shape improvements for the catheter have been conducted. Li et al. proposed several optimized designs for the catheter [19]. Yamakoshi et al. developed a catheter with a hood geometry at its tip for keeping the dextran in front of the camera [20]. Okayama et al. developed a catheter with holes on its side surface to spray the dextran to the wall, to make the flush flow reach the vessel wall [21]. Additionally, Faisal et al. investigated the flushing flow conditions and pressure inside the artery during angioscopy using a simplified model to optimize the hole shape on the catheter [22]. These studies and developments appear promising but realizing the use of these new devices in treatment may take time, as their effectivity is verified.

We hypothesize that the visibility can be improved not only with one condition, but with several conditions. Several candidate conditions that could affect the angioscopy visibility are shown in Table 1. Some of these may have more substantial effects on visibility. The improvement of these conditions will be applicable to not only the conventional devices, but also the next generation of devices. Therefore, this study sought to determine a method for realizing clearer angioscopy under different values for the parameters of the injection waveform, endoscope position, and catheter angle, using computational fluid dynamics (CFD).

Table 1. Flushing conditions considered to affect the angioscopy visibility. Underlined conditions were investigated in this study.

Flushing Conditions	
Geometry	Flow
Endoscope position in the catheter	Dextran injection waveform
Catheter angle	Dextran flow rate
Catheter position in the blood vessel	Dextran injection timing
Use of balloon	Blood flow rate

2. Materials and Methods
2.1. Model

Figure 1 shows coronary vascular, catheter, and endoscope models. These models were constructed using CAD software (SolidWorks, Dassault, France). The coronary vascular model was a straight 2.00 mm-diameter vessel. The angioscopy model consisted of a straight catheter and an endoscope. The straight catheter was set in the center of the blood vessel. The outer and inner diameters of the catheter were 1.70 and 1.42 mm, respectively. The total lengths of the blood vessel and catheter were 35.00 and 7.55 mm, respectively.

The endoscope model (Smart-iTM type S11, SURGE TEC Corp., Tokyo, Japan) consisted of the camera, 0.50 mm-diameter camera cord, and 0.36 mm-diameter guidewire model. The shape of the camera featured a lean on the camera lens. The camera lens was assumed to be placed at the orange point shown in Figure 2. The center axis of the camera lens was 0.30 mm away from the blood vessel's central axis in the y-direction. The viewing angle of the camera was set at 60 degrees based on the instructions for the endoscope. The endoscope was set in the center of the catheter, and this model was called the center model. In this study, the depth of the field was set as unlimited.

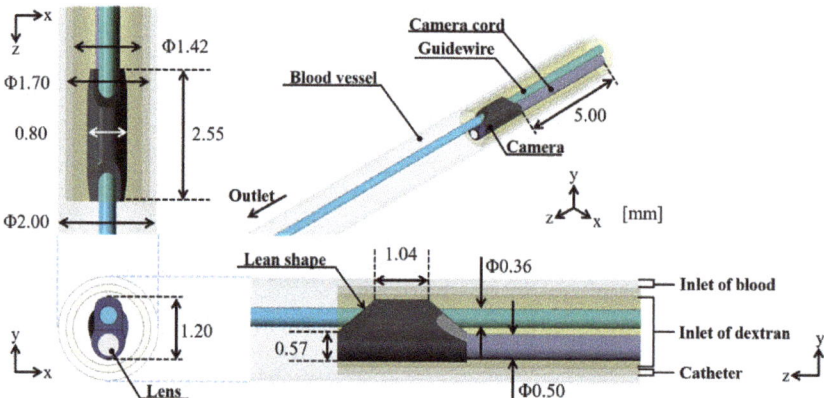

Figure 1. Basic model of blood vessel, catheter, and endoscope, called the center model. The straight catheter model was set in the center of the blood vessel. The endoscope was located at the center of the catheter.

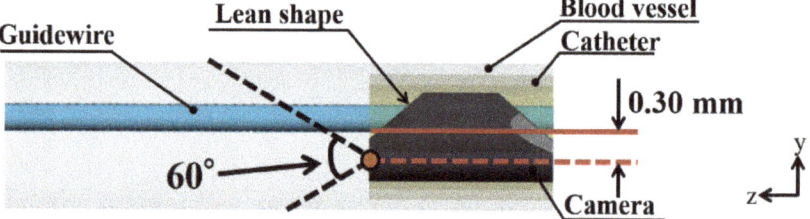

Figure 2. Camera viewing angle of the endoscope. The orange point and black dotted line represent the camera lens position and camera viewing angle, respectively. The red line denotes the center axis of the blood vessel. The red dashed line is passing through the camera lens position and perpendicular to the XY plane.

Here, we changed the endoscope position and catheter angle with respect to the angioscope geometry. Three models were prepared to change the endoscope position in the catheter. The endoscope was brought into contact with the top, side, and bottom wall of the catheter, respectively (Figure 3A). Additionally, the models with the top, side, and bottom endoscopes were called the top, side, and bottom (0°) models, respectively. Three models with different catheter angles were prepared. The 2.55 mm-length part of the catheter tip was angled at 1, 2, and 3 degrees in the bottom direction (Figure 3B). The endoscope in the angled catheter model was brought into contact with the catheter's bottom inner wall. The models with the 1-, 2-, and 3-degree catheters were called the bottom (1°), bottom (2°), and bottom (3°) models, respectively.

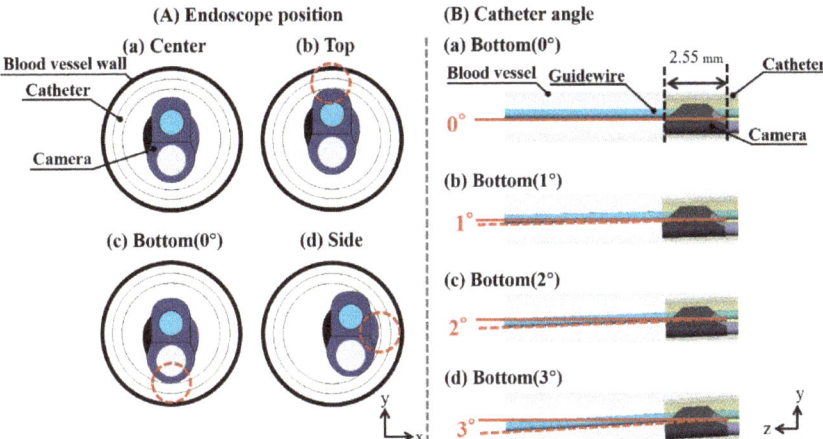

Figure 3. Models of different (**A**) endoscope positions in the catheter and (**B**) catheter angles. In (**A**), in addition to the model of (**a**) the center endoscope, the endoscope was brought into contact with the (**b**) top, (**c**) bottom, and (**d**) sidewall of the catheter. The red dashed circle of each position is the contact point between the endoscope and catheter wall. In (**B**), based on (**a**) the center model, the 2.55 mm-length part of the catheter tip was angled by 1, 2, and 3 degrees in the bottom direction. Red lines are the centerline of the blood vessel. Red dashed lines are the centerline of the angled part of the catheter.

The center model was used to study the effect of the injection waveform on the angioscopy visibility. Additionally, the stepped waveform was applied on all the other cases.

The fluid domains of the blood vessel and catheter were discretized into tetrahedron elements using ANSYS Meshing 2019 R1 (ANSYS Inc., Canonsburg, PA, USA), with finer elements of less than 0.0860 mm for the blood-vessel domain and 0.0220 and 0.0710 mm for the domain around and before the camera in the catheter, respectively. These mesh sizes were decided on based on the results of the mesh sensitivity analysis according to the grid convergence index (GCI) [23]. The mesh numbers of each model are shown in Table 2.

Table 2. The mesh numbers of each model.

Model	Mesh Number (Million)	Model	Mesh Number (Million)	Model	Mesh Number (Million)
center	3.3	Top	3.8	Bottom (1°)	6.0
		Bottom (0°)	5.6	Bottom (2°)	6.2
		Side	3.8	Bottom (3°)	5.7

2.2. Boundary Condition

As the blood flow rate at the inlet, the physiological pulsatile blood flow rate of the left anterior descending artery shown in Figure 4 was applied to a previous study [24]. The maximum and minimum blood flow rates were 54 and 9 mL/min, respectively. Two patterns of a dextran boundary condition with different injection waveforms were compared (Figure 4). Figure 4a shows the boundary condition for reproducing a situation in which a doctor injected dextran continuously under a constant-flow-rate condition within the dextran volume limitation. This injection waveform is called a constant waveform. Figure 4b shows the boundary condition for injecting dextran only during a 0.5 s systole. In this condition, it became possible to increase the maximum dextran flow rate up to 180 mL/min while the total volume of the dextran was maintained. This injection waveform is called the stepped waveform. The cardiac cycle of both the blood and dextran flow rate curve was 1.0 s in both injection waveform conditions. When the waveform was constant, after one cardiac cycle, only blood flowed from the inlet under the condition that the dextran flow rate was 0 during the 1.0 s cardiac cycle. The 2.0 s set of these two cycles was simulated for 4.0 s. The result of only the second injection was evaluated. When the waveform was stepped, all the cases were simulated for three cycles (3.0 s). The result of the only third cycle was evaluated. A zero-pascal relative pressure condition was set at the extended outlet. The artery, catheter, and endoscope wall were defined as stationary, with a no-slip condition. The CFD calculations were carried out using a finite volume method in ANSYS CFX 2019 R1 (ANSYS Inc., Canonsburg, PA, USA).

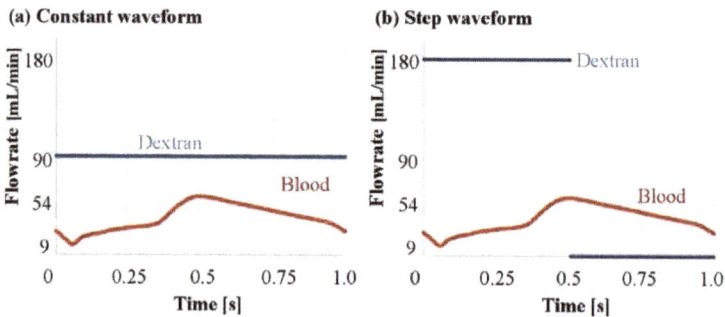

Figure 4. Inlet boundary condition of each injection waveform. The blood flow rate was the physiological pulsatile curve shown in both (**a**,**b**): (**a**) was continuous dextran injection under the constant-flow-rate condition; (**b**) was the dextran injection only during the 0.5 s systole. The injection when the blood flow rate was higher was stopped, and the maximum dextran flow rate was increased within the limitation of the dextran volume that could be injected.

2.3. Analysis Conditions

The density and viscosity of the blood were 1060 kg/m^3 and 3.50×10^{-3} Pa s, respectively [25]. Additionally, the dextran's density and viscosity were 1043 kg/m^3 and 4.99×10^{-3} Pa s, respectively, which Otsuka Pharmaceutical Co., Ltd. cited. The blood and dextran flow was assumed to be that of an incompressible Newtonian fluid within a turbulent regime under transient conditions. The k-ε model was selected to model the turbulent flow with wall functions to model the mass fluxes in the turbulent boundary layer based on molar fluxes and molar fractions of the non-condensable fluid. The timestep size for the simulation was 0.005 s. The root mean square normalized residuals of the momentum and continuity of the iterative calculation were set to 0.0001.

2.4. Evaluation Indices

To evaluate the angioscopy visibility, a movie of the inside of the blood vessel from a camera viewpoint was produced using EnSight v.10.2.3 (Computational Engineering International Inc., Apex, NC, USA). During the movie's production, the opaque volume due to the blood inside the blood vessel needed to be visualized. Then, the volume fraction of dextran (Vf) was used to define whether the volume was transparent or not. The Vf is the ratio of the volume of dextran to the total volume of the blood and dextran. The reason for choosing Vf as a parameter to define the opaque volume is that there is a relationship between the transparent mixture of the blood and dextran. The Vf was calculated by using Equation (1).

$$Vf = \frac{V_{dextran}}{V_{total}} \times 100 \qquad (1)$$

where $V_{dextran}$ and V_{total} represent the volume of the dextran and the total volume of the blood and dextran in one part, respectively. In this study, it was provisionally defined that the mixture of blood and dextran was transparent when Vf was more than 80%, referring to the optical density of fully oxygenated whole blood as a function of the hematocrit value in previous research [26].

From the movie of the inside of the blood vessel from the camera viewpoint, the visible area of the vessel wall for the camera within the camera view field ($CVA_{visible}$) and the total area of the vessel wall within the camera view field (CVA_{total}) could be calculated. When the function f(t) of the $CVA_{visible}$ percentage with respect to the CVA_{total} was defined as VAR at a certain time t, VAR could be expressed as shown in Equation (2).

$$VAR = \frac{CVA_{visible}}{CVA_{total}} \times 100 \quad (2)$$

Additionally, the area ratio of the VAR (AR_{VAR}) was defined to quantitatively evaluate the VAR. The AR_{VAR} was the normalized integral of VAR during one cycle. The integral of 100% VAR during one cycle was defined as A_{total} as expressed in Equation (3); AR_{VAR} was calculated by using Equation (4).

$$A_{total} = CVA_{total} \times \int_{one\ cycle} \Delta t \quad (3)$$

$$AR_{VAR} = \frac{\int_{one\ cycle} VAR\ \Delta t}{A_{total}} \times 100 \quad (4)$$

3. Results

Figure 5 shows the *Vf* distribution, streamline, and normalized yz and zx velocity vector of the yz and zx cross-section plane at 0.100 s after starting to flush when the endoscope position was at the center, top, or bottom in the catheter. Figure 6 displays the *Vf* distribution, streamline, and normalized yz and zx velocity vector of the yz and zx cross-section plane at 0.100 s after flushing when the endoscope position was at the center or side in the catheter, and the injection waveform was stepped. Figure 7 shows the camera view field of each flushing condition at 0.100 s after starting to flush. The deep-red, light-blue, and green parts represent the invisible volume due to low *Vf* (<80%), the guidewire, and the blood-vessel wall, respectively. Figures 8 and 9 indicate the VAR of each flushing condition as a function of time and AR_{VAR} calculated from Figure 7, respectively.

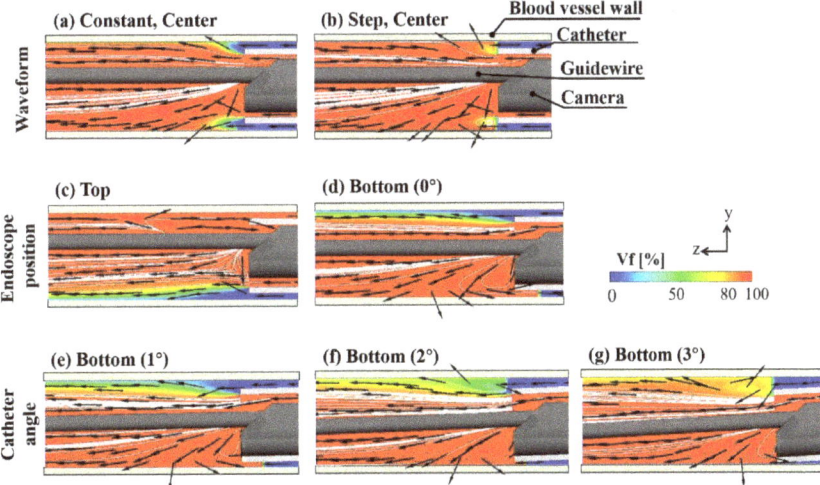

Figure 5. *Vf* distribution, streamline, and normalized yz velocity vector of yz cross-section plane at 0.100 s after starting to flush (x is the center of the blood vessel). Each row represents each flushing condition.

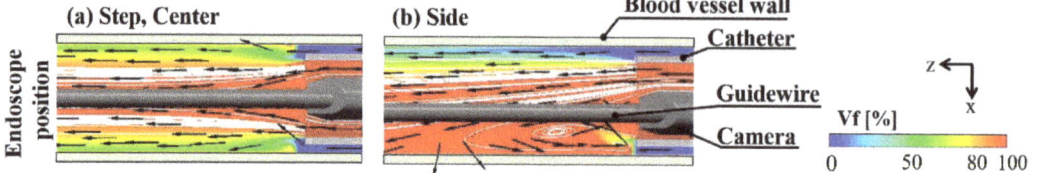

Figure 6. *Vf* distribution, streamline, and normalized yz velocity vector of zx cross-section at 0.100 s after starting to flush (y is the center of the blood vessel).

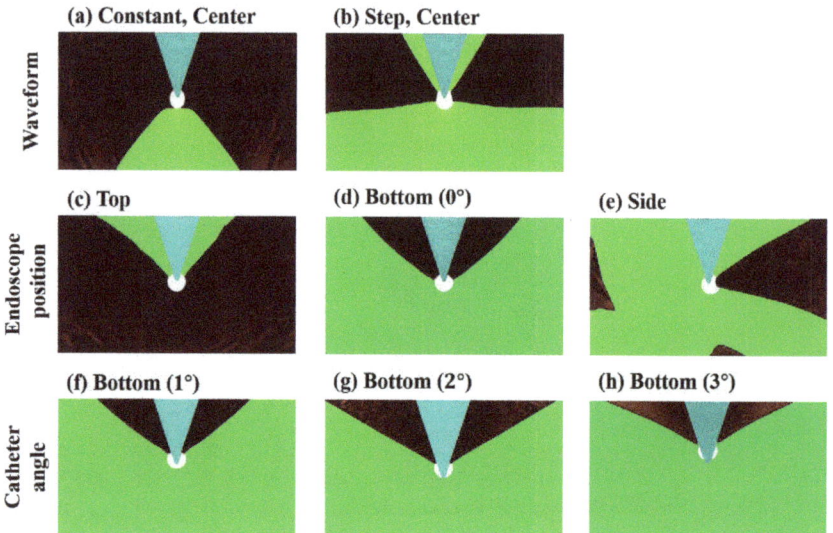

Figure 7. Camera view field of each flushing condition at 0.100 s after starting to flush. Each row represents each flushing condition. Deep red: the invisible volume due to low *Vf* (<80%). Light-blue: the guidewire. Green: the blood-vessel wall.

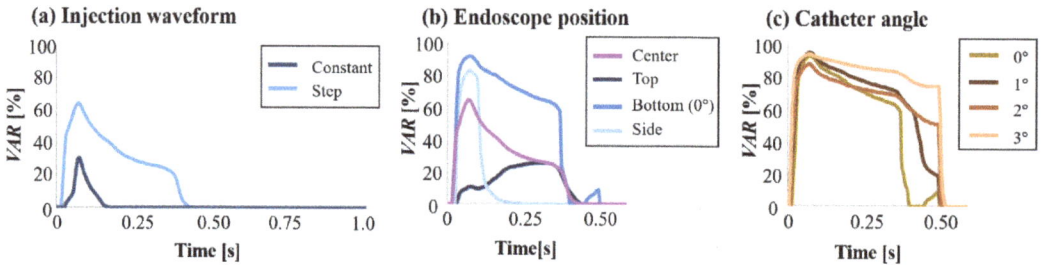

Figure 8. *VAR* evolution as a function of the injection waveform, the endoscope position, and the catheter angle.

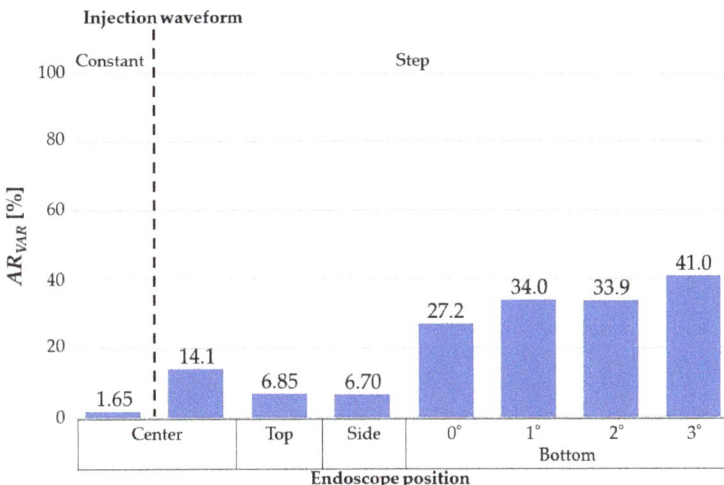

Figure 9. AR_{VAR}s for all the flushing conditions.

3.1. Effect of Injection Waveform

As shown in Figure 5a,b, the dextran flowed straight and to the blood vessel's top and bottom regions evenly. When the waveform was stepped, a dextran flow volume fraction at more than 80% covered the space between the camera and walls. Moreover, the flow reached the blood vessel's top and bottom walls rather than a constant waveform. As shown in Figure 7a,b, the camera could capture a larger area of the blood-vessel wall when the waveform was stepped rather than constant. As can be observed in Figures 8a and 9, the VAR for the stepped waveform was larger than that for the constant, and the AR_{VAR} for the stepped waveform was also 8.5 times as high as that for the constant waveform.

3.2. Effect of Endoscope Position

From Figure 5c,d and Figure 6, it can be observed that the dextran flow went straight and in the direction in which the endoscope was contacting the inner wall of the catheter and reached the blood-vessel wall. These phenomena can be observed in Figure 7c–e. The camera could capture the wall in the direction in which the endoscope was contacting the catheter's inner wall. Figure 8 shows that the VAR of the bottom (0°) model was larger than the VARs for the other positions, including the position of the center model. In Figure 9, the AR_{VAR} of the bottom (0°) model was larger than the AR_{VAR}s for the top, side, and center and around 1.9 times as high as that for the center.

3.3. Effect of Catheter Angle

As shown in Figure 5e–g, the dextran flow went straight and to the bottom region of the blood vessel. From Figure 7b,f–h, it can be observed that the camera could capture a larger area of the blood-vessel wall when the catheter angle was larger. In Figure 8c, it can be observed that the VAR of the bottom (3°) model was the largest and remained higher throughout the period of injection. From Figure 9, it can be observed that the AR_{VAR}s of the bottom (1°) and bottom (2°) models were higher than the AR_{VAR} of the bottom (0°) model, but there was almost no difference in the AR_{VAR} between the bottom (1°) and bottom (2°) models. Additionally, the AR_{VAR} of the bottom (3°) model was the largest and around 1.5 times as high as that of the bottom (0°) model.

4. Discussion

Angioscopy is frequently used to observe the blood-vessel wall, and the conditions under which angioscopy is performed affects the success of operations. The flow of dextran depends on various flushing conditions, and the effects of each condition on the visibility are unclear. This study found out a way to realize clearer angioscopy under different values for the parameters of the injection waveform, endoscope position, and catheter angle, using CFD.

4.1. The Effect of Injection

Regarding the effect of the injection waveform, stepped injection can increase the visibility by 8.5 times compared to that realized with constant injection. This shows that increasing the volumetric ratio of dextran to blood increases the visibility. Additionally, this finding shows that the injection method during the operation can be improved while keeping the total volume of dextran the same.

4.2. The Effect of Angioscopy Shape on Visibility Improvement

The angle of the catheter and the position of the endoscope were analyzed to improve the angioscopy visibility. The effects of these geometry conditions are not even, and the position of the endoscope had a greater effect (1.9 times) than the catheter angle (1.5 times).

Overall, a three-degree angle for the catheter with stepped flow may improve the angioscopy visibility by 25 times compared with a zero-degree angle with constant injection. The combination of good flushing conditions will considerably improve the angioscopy visibility.

4.3. The Effects of the Parameters in Clinical Situations

The angle and position of the catheter depend on the position of the disease or the geometry of the artery in a given clinical situation. Adjusting the blood flow rate or dextran flow rate may offer potential for improving the Vf; however, the clinical situations frequently require the blood flow rate to be maintained. Thus, adjusting the injection waveform or injection timing can be tried as a method for improving the visibility.

4.4. Evaluation Indices

Here, we show the VAR and AR_{VAR} for angioscopy visibility. The VAR represents the visibility at a given moment, and the AR_{VAR} represents the total visibility over the period of injection. The VAR may be used for evaluating the visibility if a medical doctor needs to objectively evaluate a lesion. The AR_{VAR} may be used for visibility evaluation if a medical doctor wants to visualize the whole region. This study proposes these quantitative evaluation indices as novel parameters for the assessment of angioscopic operations. These evaluation indices provide useful data that may be used to help clinicians in improving the efficacy of angioscopic procedures and to optimize device design through comparisons between models with different camera shapes, while their further application in clinical procedures needs to be investigated in more comprehensive studies.

4.5. Limitation

The k-ε model was selected to model the turbulent flow based on the preliminary experiment results and the previous studies. In the initial experiment, the flow during angioscopy was visualized using particle image velocimetry, and turbulent flow was observed around the endoscope despite the low Reynolds number. This turbulent flow is considered to have been caused by the complicated geometry of the camera. The k-ε model has been widely used to simulate turbulent blood flow [27–29]. The blood and dextran flows were assumed to represent those of an incompressible Newtonian fluid, which was an ideal condition. In addition, the blood vessel was assumed to be stationary in this study but possesses elasticity in a living body [30]. Therefore, comparison with experimental results will increase the practical relevance of this study; however, due to the complexity of the experimental study, we first performed this simulation to obtain preliminary results. In vitro experiments will be considered as further work on this topic. In clinical cases, some flushing conditions may not be used because of the patient's geometrical condition. In this case, medical doctors may need to choose the best combination from these flushing conditions. For determining the ideal combination, an optimization method can be useful [31,32].

Additionally, idealized catheter and guidewire conditions were used in this study. The shapes of the catheter and guidewire have curves that may affect the visibility. Takashima et al. developed a simulator for the motion of a catheter and guidewire inside a blood vessel [33–36]. Simulating the flow under conditions reproducing the shape and motion of the catheter and guidewire in the blood vessels at a clinical site, with reference to those studies, will be considered for future work.

Additionally, the previous paper states that inducing too much pressure inside the coronary artery by flushing is unsafe for a patient [22]. The increase in pressure inside the coronary artery induced by dextran injection needs to be considered in the future.

This study concludes that the visibility depends on the temporal flush volume displacing the blood volume. However, in current clinical situations, several factors such as the blood volume and catheter position are not known before treatment. Experimental trials using an endoscope are necessary to confirm the mechanism of the effect on visibility.

To translate this simulation study to patient care, there is a need for preclinical (3D models) as well as clinical testing. Moreover, diseased walls of vessels are fragile. Therefore, a study considering different vessel thicknesses and fragility needs to be conducted before proceeding to real-life applications. Injecting dextran at a high flow rate could potentially rupture fragile vessels.

5. Conclusions

The purpose of this study was to investigate the effect of the injection waveform, endoscope position, and catheter angle on the angioscopy visibility by CFD analysis.

Regarding the injection waveform, it was found that the AR_{VAR} for a stepped waveform was 8.5 times as high as that for a constant waveform. Regarding the endoscope position, the AR_{VAR} of the bottom (0°) model was higher than the AR_{VAR}s of the top, side, and center models and around 1.9 times as high as that of the center model. Regarding the catheter angle, it turned out that the AR_{VAR}s of the bottom (1°) and bottom (2°) models were higher than the AR_{VAR} of the bottom (0°) model, but there was almost no AR_{VAR} difference between the bottom (1°) and bottom (2°) models. The AR_{VAR} of the bottom (3°) model was the largest and around 1.5 times as high as that of the bottom (0°) model.

Therefore, it was found that the condition of the bottom (3°) model with a stepped waveform was the best for a higher AR_{VAR}. Additionally, the AR_{VAR} of the bottom (3°) model with a stepped waveform was around 25 times as large as that of the bottom (0°) model with a constant waveform.

Author Contributions: Conceptualization, M.O., T.N. and K.M.; methodology, K.M., T.N. and S.T.; software, K.M., M.Z., Y.L., H.W. and H.A.; validation, K.M., S.T., Y.L., M.Z., H.A., K.H. and H.W.; formal analysis, D.G., M.O. and K.M.; investigation, K.M.; resources, M.O.; data curation, K.M.; writing—original draft preparation, K.M.; writing—review and editing, M.O., S.T. and H.A.; visualization, K.M. and D.G.; supervision, M.O.; project administration, M.O.; funding acquisition, M.O. All authors have read and agreed to the published version of the manuscript.

Funding: This research was funded by Grants-in-Aid for Scientific Research, KAKENHI B (JP20H04557), Kakenhi C (22K12795), JP18K18355 Japan Society for the Promotion of Science. This research was also supported by the Collaborative Research Project 2020, Institute of Fluid Science, Tohoku University (J20R001, J21I074, J22I075, and J22I068).

Institutional Review Board Statement: Not applicable.

Informed Consent Statement: Not applicable.

Data Availability Statement: Not applicable.

Acknowledgments: This research was partially supported by the Creation of a Development Platform for Implantable/Wearable Medical Devices by a Novel Physiological Data Integration System project within the Program on Open Innovation Platform with Enterprises, Research Institute and Academia, of the Japan Science and Technology Agency.

Conflicts of Interest: The authors declare no conflict of interest.

References

1. Nowbar, A.N.; Gitto, M.; Howard, J.P.; Francis, D.P.; Al-Lamee, R. Mortality from Ischemic Heart Disease: Analysis of Data from the World Health Organization and Coronary Artery Disease Risk Factors from NCD Risk Factor Collaboration. *Circ. Cardiovasc. Qual. Outcomes* **2019**, *12*, e005375. [CrossRef] [PubMed]
2. Institute of Medicine (US) Committee on Social Security. *Cardiovascular Disability Criteria Cardiovascular Disability: Updating the Social Security Listings*; National Academies Press: Washington, DC, USA, 2010.
3. Nakamura, F. Clinical Application of Coronary Angioscopy for Diagnosing and Treating Coronary Artery Disease. *BME* **2005**, *43*, 8–11. [CrossRef]
4. Ueda, Y.; Ohtani, T.; Shimizu, M.; Hirayama, A.; Kodama, K. Assessment of Plaque Vulnerability by Angioscopic Classification of Plaque Color. *Am. Heart J.* **2004**, *148*, 333–335. [CrossRef] [PubMed]
5. Hitchner, E.; Zayed, M.A.; Lee, G.; Morrison, D.; Lane, B.; Zhou, W. Intravascular Ultrasound as a Clinical Adjunct for Carotid Plaque Characterization. *J. Vasc. Surg.* **2014**, *59*, 774–780. [CrossRef] [PubMed]
6. Kotsugi, M.; Takayama, K.; Myouchin, K.; Wada, T.; Nakagawa, I.; Nakagawa, H.; Taoka, T.; Kurokawa, S.; Nakase, H.; Kichikawa, K. Carotid Artery Stenting: Investigation of Plaque Protrusion Incidence and Prognosis. *JACC Cardiovasc. Interv.* **2017**, *10*, 824–831. [CrossRef] [PubMed]
7. Shinozaki, N.; Ogata, N.; Ikari, Y. Plaque Protrusion Detected by Intravascular Ultrasound during Carotid Artery Stenting. *J. Stroke Cerebrovasc. Dis.* **2014**, *23*, 2622–2625. [CrossRef] [PubMed]
8. De Donato, G.; Setacci, F.; Sirignano, P.; Galzerano, G.; Cappelli, A.; Setacci, C. Optical Coherence Tomography after Carotid Stenting: Rate of Stent Malapposition, Plaque Prolapse and Fibrous Cap Rupture According to Stent Design. *Eur. J. Vasc. Endovasc. Surg.* **2013**, *45*, 579–587. [CrossRef] [PubMed]
9. Yoshimura, S.; Kawasaki, M.; Yamada, K.; Hattori, A.; Nishigaki, K.; Minatoguchi, S.; Iwama, T. Optical Coherence Tomography (OCT): A New Imaging Tool during Carotid Artery Stenting. In *Optical Coherence Tomography*; IntechOpen: London, UK, 2013.
10. Kondo, H.; Kiura, Y.; Sakamoto, S.; Okazaki, T.; Yamasaki, F.; Iida, K.; Tominaga, A.; Kurisu, K. Comparative Evaluation of Angioscopy and Intravascular Ultrasound for Assessing Plaque Protrusion during Carotid Artery Stenting Procedures. *World Neurosurg.* **2019**, *125*, e448–e455. [CrossRef] [PubMed]
11. Mark, D.B.; Nelson, C.L.; Califf, R.M.; Harrell, F.E.; Lee, K.L.; Jones, R.H.; Fortin, D.F.; Stack, R.S.; Glower, D.D.; Smith, L.R.; et al. Continuing Evolution of Therapy for Coronary Artery Disease. Initial Results from the Era of Coronary Angioplasty. *Circulation* **1994**, *89*, 2015–2025. [CrossRef] [PubMed]
12. Kubo, T.; Imanishi, T.; Takarada, S.; Kuroi, A.; Ueno, S.; Yamano, T.; Tanimoto, T.; Matsuo, Y.; Masho, T.; Kitabata, H.; et al. Implication of Plaque Color Classification for Assessing Plaque Vulnerability: A Coronary Angioscopy and Optical Coherence Tomography Investigation. *JACC Cardiovasc. Interv.* **2008**, *1*, 74–80. [CrossRef] [PubMed]
13. Kotani, J.; Awata, M.; Nanto, S.; Uematsu, M.; Oshima, F.; Minamiguchi, H.; Mintz, G.S.; Nagata, S. Incomplete Neointimal Coverage of Sirolimus-Eluting Stents: Angioscopic Findings. *J. Am. Coll. Cardiol.* **2006**, *47*, 2108–2111. [CrossRef] [PubMed]
14. Komatsu, S.; Ohara, T.; Takahashi, S.; Takewa, M.; Minamiguchi, H.; Imai, A.; Kobayashi, Y.; Iwa, N.; Yutani, C.; Hirayama, A.; et al. Early Detection of Vulnerable Atherosclerotic Plaque for Risk Reduction of Acute Aortic Rupture and Thromboemboli and Atheroemboli Using Non-Obstructive Angioscopy. *Circ. J.* **2015**, *79*, 742–750. [CrossRef] [PubMed]
15. Ueda, Y.; Asakura, M.; Yamaguchi, O.; Hirayama, A.; Hori, M.; Kodama, K. The Healing Process of Infarct-Related Plaques: Insights from 18 Months of Serial Angioscopic Follow-Up. *J. Am. Coll. Cardiol.* **2001**, *38*, 1916–1922. [CrossRef]
16. Harken, D.E.; Glidden, E.M. Experiments in Intracardiac Surgery. *J. Thorac. Surg.* **1943**, *12*, 566–572. [CrossRef]
17. Komatsu, S.; Ohara, T.; Takahashi, S.; Takewa, M.; Yutani, C.; Kodama, K. Improving the Visual Field in Coronary Artery by with Non-Obstructive Angioscopy: Dual Infusion Method. *Int. J. Cardiovasc. Imaging* **2017**, *33*, 789–796. [CrossRef]
18. Mitsutake, Y.; Ueno, T. The Role of Coronary Angioscopy in the BRS Era. *J. Jpn. Coron. Assoc.* **2017**, *23*, 48–54. [CrossRef]
19. Li, Y.; Zhang, M.; Tupin, S.; Mitsuzuka, K.; Nakayama, T.; Anzai, H.; Ohta, M. Flush Flow Behaviour Affected by the Morphology of Intravascular Endoscope: A Numerical Simulation and Experimental Study. *Front. Physiol.* **2021**, *12*, 733767. [CrossRef]
20. Yamakoshi, K.; Tanaka, S. Endoscope and Blood Vessel Endoscope System. JP Patent P2011-87859A, 2011.
21. Okayama, K.; Nanto, S.; Sakata, Y. Vascular Endoscope Catheter and Vascular Endoscope. JP Patent P2019-115755A, 2019.
22. Faisal, S. Hemodynamics of Endoscopic Imaging of Chronic Total Occlusions. Master's Thesis, University of Washington, Seattle, WA, USA, 2018.
23. Saitta, S.; Pirola, S.; Piatti, F.; Votta, E.; Lucherini, F.; Pluchinotta, F.; Carminati, M.; Lombardi, M.; Geppert, C.; Cuomo, F.; et al. Evaluation of 4D Flow MRI-Based Non-Invasive Pressure Assessment in Aortic Coarctations. *J. Biomech.* **2019**, *94*, 13–21. [CrossRef]

24. Xie, X.; Wang, Y.; Zhu, H.; Zhou, H.; Zhou, J. Impact of Coronary Tortuosity on Coronary Blood Supply: A Patient-Specific Study. *PLoS ONE* **2013**, *8*, e64564. [CrossRef]
25. Hoi, Y.; Meng, H.; Woodward, S.H.; Bendok, B.R.; Hanel, R.A.; Guterman, L.R.; Hopkins, L.N. Effects of Arterial Geometry on Aneurysm Growth: Three-Dimensional Computational Fluid Dynamics Study. *J. Neurosurg.* **2004**, *101*, 676–681. [CrossRef]
26. Mroczka, J.; Szczepanowski, R. Modeling of Light Transmittance Measurement in a Finite Layer of Whole Blood-a Collimated Transmittance Problem in Monte Carlo Simulation and Diffusion Model. *Opt. Appl.* **2005**, *35*, 311–331.
27. Banks, J.; Bressloff, N.W. Turbulence Modeling in Three-Dimensional Stenosed Arterial Bifurcations. *J. Biomech. Eng.* **2007**, *129*, 40–50. [CrossRef] [PubMed]
28. Jahangiri, M.; Saghafian, M.; Sadeghi, M.R. Numerical Simulation of Hemodynamic Parameters of Turbulent and Pulsatile Blood Flow in Flexible Artery with Single and Double Stenoses. *J. Mech. Sci. Technol.* **2015**, *29*, 3549–3560. [CrossRef]
29. Saqr, K.M.; Tupin, S.; Rashad, S.; Endo, T.; Niizuma, K.; Tominaga, T.; Ohta, M. Physiologic Blood Flow Is Turbulent. *Sci. Rep.* **2020**, *10*, 15492. [CrossRef]
30. Tupin, S.; Saqr, K.M.; Ohta, M. Effects of Wall Compliance on Multiharmonic Pulsatile Flow in Idealized Cerebral Aneurysm Models: Comparative PIV Experiments. *Exp. Fluids* **2020**, *61*, 164. [CrossRef]
31. Anzai, H.; Ohta, M.; Falcone, J.L.; Chopard, B. Optimization of Flow Diverters for Cerebral Aneurysms. *J. Comput. Sci.* **2012**, *3*, 1–7. [CrossRef]
32. Putra, N.K.; Palar, P.S.; Anzai, H.; Shimoyama, K.; Ohta, M. Multiobjective Design Optimization of Stent Geometry with Wall Deformation for Triangular and Rectangular Struts. *Med. Biol. Eng. Comput.* **2019**, *57*, 15–26. [CrossRef]
33. Takashima, K.; Ota, S.; Ohta, M.; Yoshinaka, K.; Ikeuchi, K. Development of Computer-Based Simulator for Catheter Navigation in Blood Vessels (1st Report, Evaluation of Fundamental Parameters of Guidewire and Blood Vessel). *Trans. Jpn. Soc. Mech. Eng. Ser. C* **2006**, *72*, 2137–2145. [CrossRef]
34. Takashima, K.; Ota, S.; Ohta, M.; Yoshinaka, K.; Mukai, T. Development of Computer-Based Simulator for Catheter Navigation in Blood Vessels (2nd Report, Evaluation of Torquability of Guidewire). *Trans. Jpn. Soc. Mech. Eng. Ser. C* **2007**, *73*, 2988–2995. [CrossRef]
35. Takashima, K.; Ohta, M.; Yoshinaka, K.; Mukai, T.; Oota, S. Catheter and Guidewire Simulator for Intravascular Surgery (Comparison between Simulation Results and Medical Images). *IFMBE Proc.* **2009**, *25*, 128–131. [CrossRef]
36. Takashima, K.; Horie, S.; Takenaka, M.; Mukai, T.; Ishida, K.; Ueda, Y. Fundamental Study on Medical Tactile Sensor Composed of Organic Ferroelectrics. In Proceedings of the 2012 Fifth International Conference on Emerging Trends in Engineering and Technology, Himeji, Japan, 5–7 November 2012; IEEE: Piscataway, NJ, USA, 2012; pp. 132–136.

Article

Simulation and Experimental Investigation of Balloon Folding and Inserting Performance for Angioplasty: A Comparison of Two Materials, Polyamide-12 and Pebax

Tao Li [1], Zhuo Zhang [1], Wenyuan Wang [2], Aijia Mao [3], Yu Chen [3], Yan Xiong [1,*] and Fei Gao [2,*]

[1] College of Mechanical Engineering, Sichuan University, Chengdu 610065, China; 2021223025146@stu.scu.edu.cn (T.L.)
[2] Chengdu Neurotrans Medical Technology Co., Ltd., Chengdu 610065, China
[3] Department of Applied Mechanics, Sichuan University, Chengdu 610065, China
* Correspondence: xy@scu.edu.cn (Y.X.); phil.gao@neurotrans.com.cn (F.G.)

Abstract: Background: A balloon dilatation catheter is a vital tool in percutaneous transluminal angioplasty. Various factors, including the material used, influence the ability of different types of balloons to navigate through lesions during delivery. Objective: Thus far, numerical simulation studies comparing the impacts of different materials on the trackability of balloon catheters has been limited. This project seeks to unveil the underlying patterns more effectively by utilizing a highly realistic balloon-folding simulation method to compare the trackability of balloons made from different materials. Methods: Two materials, nylon-12 and Pebax, were examined for their insertion forces via a bench test and a numerical simulation. The simulation built a model identical to the bench test's groove and simulated the balloon's folding process prior to insertion to better replicate the experimental conditions. Results: In the bench test, nylon-12 demonstrated the highest insertion force, peaking at 0.866 N, significantly outstripping the 0.156 N force exhibited by the Pebax balloon. In the simulation, nylon-12 experienced a higher level of stress after folding, while Pebax had demonstrated a higher effective strain and surface energy density. In terms of insertion force, nylon-12 was higher than Pebax in specific areas. Conclusion: nylon-12 exerts greater pressure on the vessel wall in curved pathways when compared to Pebax. The simulated insertion forces of nylon-12 align with the experimental results. However, when using the same friction coefficient, the difference in insertion forces between the two materials is minimal. The numerical simulation method used in this study can be used for relevant research. This method can assess the performance of balloons made from diverse materials navigating curved paths and can yield more precise and detailed data feedback compared to benchtop experiments.

Keywords: angioplasty; balloon dilatation catheter; finite element analysis; bench test; insertion force; balloon pleating simulation

1. Introduction

Peripheral artery disease (PAD) is a long-term arterial disease caused by atherosclerosis in the peripheral vessels, leading to complications in limbs such as pain, ulcers, gangrene, and reduced function [1]. Percutaneous transluminal coronary angioplasty (PTCA) is an effective approach for treating symptomatic atherosclerotic peripheral artery disease [2]. This disease affects a wide patient population worldwide, with more than 200 million individuals afflicted [3]. The balloon-tipped catheter is a device delivered to the site of a lesion, and it functions to increase the lumen's diameter [4]. The roles played by different types of balloons in treatment vary greatly [5–7]. For instance, drug-coated balloons can directly deliver antiproliferative drugs to the local lesion in the blood vessel without the need for implanting a stent, while cutting balloons have metal blades attached to their surface that can cut through plaque during expansion. Additionally, different balloon structures, such as dual-wire and scoring balloons, cater to specific functional requirements. In terms of compliance, balloons can be categorized into compliant, semi-compliant, and non-compliant balloons. Compliant balloons adapt better to the morphology of the blood vessels, whereas non-compliant balloons have a superior ability to compress hard lesions. PTCA balloons are primarily composed of thermoplastic polymers, including polyethylene terephthalate (PET), nylon, polyethylene with additives, polyvinyl chloride (PVC), and polyurethane used along with nylon [8]. The materials used for PTCA balloons significantly affect their characteristics [9]. Currently, Polyamide 12 and Pebax are widely used in the market to manufacture PTCA balloons [10]. Trackability, which refers to the force needed to navigate a balloon catheter through a tortuous path to the target lesion [11], is a crucial indicator for evaluating the performance of PTCA balloons [12,13]. A lower value is preferred [14]. The trackability of a real balloon is tested in the folded state, and good folding properties, which determine the ease of folding and unfolding, also play a crucial role in maintaining the performance of the balloon after repeated inflations [12]. Therefore, the quality of the folded state directly affects the trackability of the balloon.

Finite element analysis (FEA) has become a powerful tool for the optimization process of coronary stents and balloon catheters. There have been many studies that have conducted experiments and simulations on balloons from various perspectives. Geith et al. [15] proposed a simple, microstructurally motivated constitutive model aimed at mimicking the pronounced anisotropic material response observed in the performed experiments. Helou et al. [16] presented a modeling method for simulating percutaneous transluminal angioplasty (PTA) endovascular treatment and evaluated the effects of balloon design, plaque composition, and balloon sizing on acute post-procedural outcomes after PTA. Dong et al. [17] investigated the efficacy of post-dilation balloon diameter and inflation pressure in improving the stent expansion in a calcified lesion. Rahinj et al. [18] analyzed a non-uniform balloon stent expansion pattern comprised of variations in the stent axial position on the balloon, balloon length, balloon folding pattern, and balloon wall thickness. De Beule et al. [19] proposed a trifolded balloon methodology, which was confirmed through experiments and agreed with manufacturers' data. Wiesent et al. [20] performed a stent life-cycle simulation including balloon folding, stent crimping, and the free expansion of the balloon–stent system. Bukala et al. [21] deployed the "kissing balloon" stenting technique applied to patients with bifurcation stenosis. Hamed et al. [22] verified through experiments that the cross-sectional scanning morphology of a balloon under different pressures and the diameter curve under applied pressure can be consistent with a 2D simulation, demonstrating the consistency between simulation and experimental results. Gajewski et al. [23] demonstrated that material stiffness has a significant impact on the degree of occlusion of the balloon in the artery. The clear advantage of a contrasting engineering simulation is that it provides a comprehension evaluation of a design, offering insightful feedback and thereby minimizing the need for complex and expensive experiments, which are often difficult to carry out.

However, to our knowledge, there have been no studies comparing simulated folding and tracking performances between two different materials. Although Sirivella et al. [11] successfully simulated the pushability and trackability of polyamide PTCA balloons in the folded state using finite element analysis and compared the results with experiments, they did not consider the pre-stress of the folded balloon. Instead, they directly established the initial model of the balloon in the folded configuration, which compromises some realism. On the other hand, Geith et al. [24] simulated a complete process of balloon folding and pleating which closely resembled the real scenario. Nonetheless, their study focused on stent expansion and did not investigate the trackability of the balloons.

In this article, we performed a simulation study using the FEA method to investigate the folding of nylon and Pebax material balloons, as well as the trackability of balloon catheters passing through a 90-degree ideal blood vessel under identical conditions. In addition, a benchtop experiment was conducted to further validate the simulation results.

2. Materials and Methods

2.1. Bench Tests

The bench test offered clear insights into the performance of the balloon dilatation catheter within tortuous anatomical structures. The test evaluated the insertion forces of folded balloon catheters made of different materials as they navigated grooves—a simulation of the transport conditions within actual curved vessels. The quality of trackability can be evaluated based on the measured insertion force. Therefore, bench tests can provide a reference for balloon design and clinical selection [13].

The tests were conducted using a test tracking fixture (Figure 1), as specified in ASTM F2394-07 [13]. It simulated the bending shapes of coronary arteries on a two-dimensional plane without considering the shapes of lesions. It was composed of two plates, one above and one below, with grooves representing the curved vessels. The grooves featured varying degrees of curvature at different positions. The folded balloon was pushed into the groove through the entrance. A multi-segment displacement load was applied to the balloon: the total push distance was set at 40 mm, the individual push distance at 5 mm, and the push speed was 2 mm/s. After each movement, the subsequent push was executed at regular 2.5 s intervals. A force-measuring device captured the push force exerted by the balloon as it navigated the bend, as depicted in Figure 1.

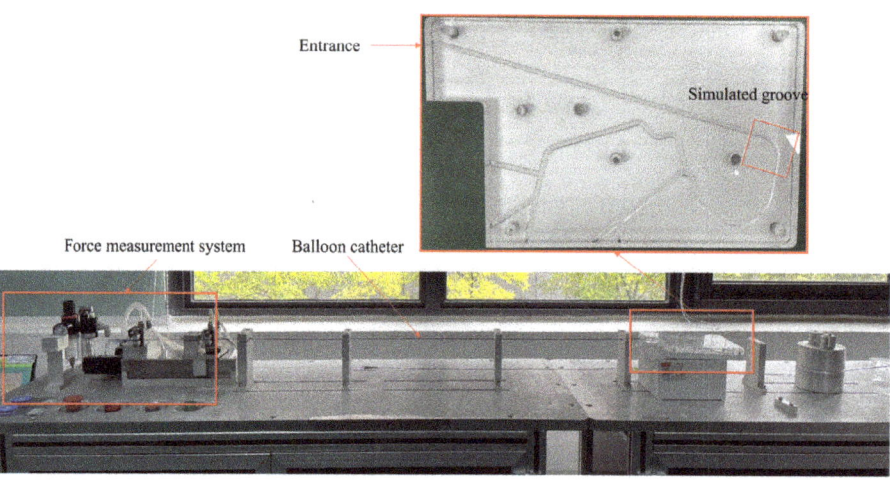

Figure 1. Bench test tracking fixture.

2.2. Materials

Nylon-12 (polyamide) and Pebax are the two materials most commonly used in balloons [15,25,26]. Two materials were used in this study: Grilamid L25 (nylon 12) from EMS and Pebax® 7033 SA 01 MED resin (a thermoplastic elastomer from Arkema). They are specially designed to meet the stringent requirements of medical applications such as minimally invasive devices, and they both exhibit good biocompatibility [26]. The nylon stress–strain curve was redrawn from a Grilamid L25 technical data sheet. The stress–strain relationship of Pebax is shown in [26]. An elasto-plastic material with an arbitrary stress as a function of strain curve and an arbitrary strain rate dependency can be defined using MAT_89 in LS-DYNA [27]; the model was selected, and the stress–strain curve was input to fit the material properties. The densities, moduli of elasticity, and Poisson's ratios of nylon and Pebax [15] are shown in Table 1. Clearly, nylon-12 is the stiffer of the two materials. The shaft is composed of high-density polyethylene (HDPE), which was modeled as linear elastic, as per Table 1 [28].

Table 1. Material properties.

Material	Density (g/cm^3)	Young's Modulus (MPa)	Poisson's Ratio
Nylon	1.01	1100	0.4
Pebax	1.01	414	0.4
HDPE	0.953	1000	0.46

*MAT_89 used the Cowper–Symonds constitutive model [29]. The equation of this model is:

$$\sigma_s = \sigma_0 \left(1 + \left(\frac{\dot{\varepsilon}}{C}\right)^{1/p}\right) \quad (1)$$

Taking the logarithm of both sides of the equation and simplifying it, we obtain:

$$\log_{10} \dot{\varepsilon} = \log_{10} C + P \log_{10}\left(\frac{\sigma_s}{\sigma_0} - 1\right) \quad (2)$$

where σ_s is the yield stress of the material, σ_0 is the quasistatic (0.001 s^{-1}) yield stress of the material, $\dot{\varepsilon}$ is the strain rate of the material, and P and C are material constants determined by test [30].

2.3. Geometric Models

During the process of manufacturing balloon catheters, cylindrical balloons are first compressed into a folded shape using a balloon-folding machine. Then, they are encapsulated into a smaller diameter using a balloon-crimping machine. Therefore, the balloons are in a folded state before inflation. In order to make the simulation more realistic, the crimping process, which involves the generation of pre-stress in the balloon, should be taken into account.

The balloon model is derived from the Euphora balloon dilation catheter (Medtronic, Minneapolis, MN, USA), featuring a principal diameter of 4 mm and a distal shaft diameter of 0.91 mm. The balloon's thickness is 0.02 mm [18] (Figure 2). The pleating tool, as shown in Figure 3, has an inner diameter of 2 mm. This tool functions by initially compressing the balloon into three wings and subsequently facilitating each wing's rotation around the axis, thereby inducing the regular folding of the balloon.

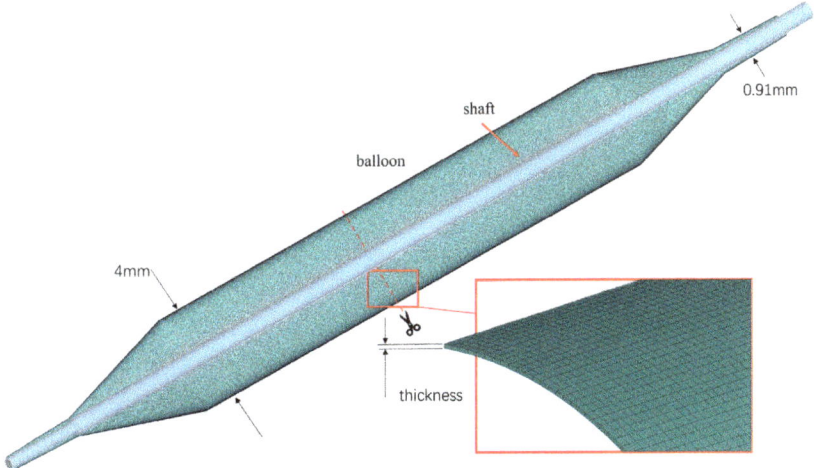

Figure 2. Balloon and shaft.

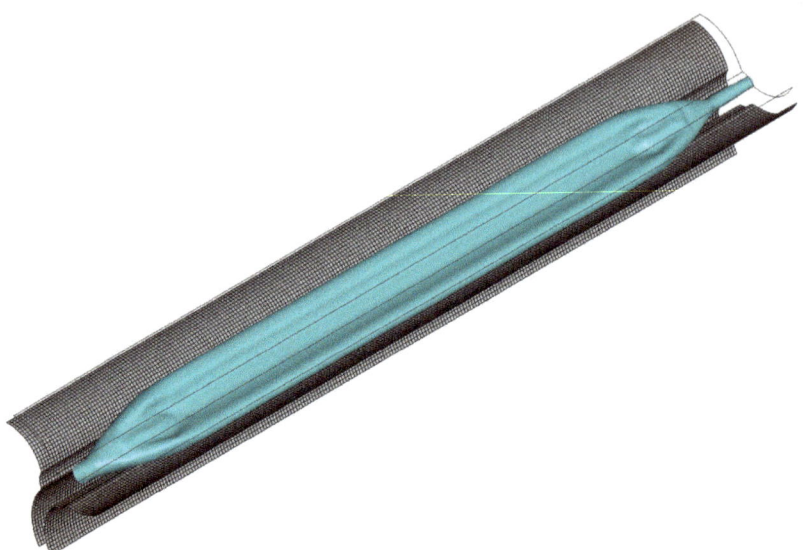

Figure 3. Pleating tools.

In order to more accurately simulate the working process of a balloon-crimping machine, the simplistic cylindrical model that was previously widely used was replaced with a novel design composed of 8 plates combined in a specific formation. They formed a chamber that could be enlarged or reduced (as shown in Figure 4). The three-winged balloons were folded and compressed in this chamber, and the encapsulated balloons' final diameters were then determined.

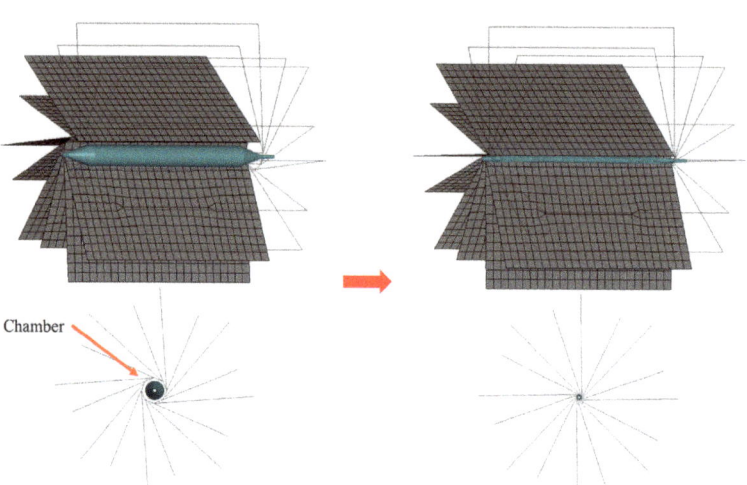

Figure 4. Balloon-crimping simulation.

A geometry model, referred to as the "pipeline", featuring a groove shape consistent with the ASTM F2394-07 bench test fixture, was designed with a diameter of 1.5 mm [13] (Figure 5). The model features a 90-degree angle between the inlet and outlet, replicating the grooves traversed by the balloon during the force-monitoring phase of the bench test (Figure 1). To prevent mesh distortion and convergence failure due to the pipeline's edges being cut by balloons, a transitional outward extension was created at the inlet (Figure 5).

The outlet was bent 90 degrees relative to the inlet, spanning a total length of 50 mm. The pipeline was modeled as a rigid body and fixed in space.

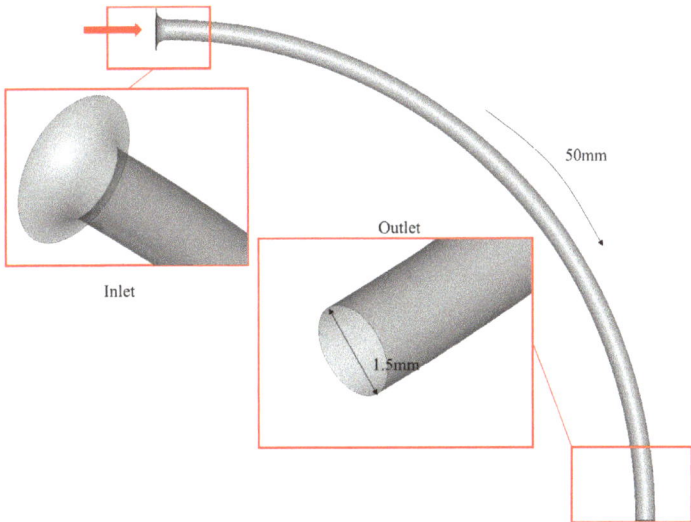

Figure 5. Pipeline. After folding, the balloon will be fed into the inlet.

2.4. Finit Element Modeling and Boundary Conditions

The pleating tool moved toward the axis by a distance of 2 mm, resulting in the balloon being compressed into a three-fold configuration (three-wing configuration). Then, it rotated along the axis, causing the three-fold to rotate counterclockwise by 120 degrees. Given that both ends of the balloon were secured to the shaft and a specific negative pressure was exerted on the balloon's inner surface at that moment, the balloon tightly adhered to and was fixed onto the shaft, while the three outward folds rotated with the pleating tool and attached to the shaft.

At the initial moment, both ends of the balloon were fixed to the shaft. After the pleating tool compressed the balloon, a negative pressure of 6.5 atm was applied to the inner surface of the balloon [23], causing it to contract. The degrees of freedom of the shaft in the x and y directions were fixed, and a continuous displacement load of 20 mm was applied to one end of the shaft. The reason for not employing multi-segment displacement is that when replicating the experimental conditions, the abrupt acceleration and deceleration of the shaft can lead to a significant dynamic effect [31], leading to a loss of the realism of the balloon's shape.

The contact between the balloon and the pipeline was set to have an ideal friction coefficient of 0.02 [32], while the shaft and the pipeline were set to have no friction. To make the balloon adhere more closely to the shaft, the friction coefficient between the balloon and the shaft was set to 0.2 so that a certain amount of friction force would be generated between them after the negative pressure was applied, preventing excessive distortion of the balloon at both ends during rotation and folding.

A mesh convergence study was conducted to ensure that the calculation results were not significantly affected by variations in the number of mesh elements [33]. In this study, a mesh convergence study was performed for the balloon. Four models with different mesh sizes were created with 54,177, 150,156, 213,539, and 336,542 elements, corresponding to mesh sizes of 0.1 mm, 0.06 mm, 0.05 mm, and 0.04 mm, respectively (Table 2). The models were assigned the material properties of Pebax and subjected to the same simulations of folding, crimping, and insertion into the pipeline. The maximum stress on the balloon surface at the minimum crimped diameter was recorded for each model (Figure 6). The results show that the maximum stress difference between the third and fourth models is less than 5%, indicating convergence and satisfying the criteria for mesh convergence. Therefore, the mesh size of the third model was chosen for subsequent simulations, ensuring computational accuracy without excessive computational time due to redundant mesh elements.

Table 2. Mesh convergence study.

Model	1	2	3	4
Mesh size (mm)	0.1	0.06	0.05	0.04
Number of elements	54,177	150,156	213,539	336,542
Maximum stress (MPa)	42.0	49.5	55.0	56.6

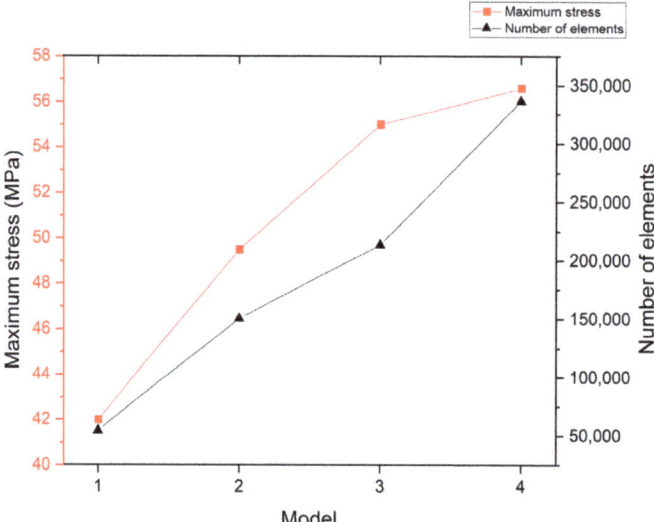

Figure 6. The variation in maximum stress with respect to the number of elements in the mesh convergence study.

Based on the mesh convergence study, the balloon model was discretized into four-node, fully integrated shell elements with five through-shell thickness integration points. It was discretized using 213,539 mixed shell elements with a thickness of 0.02 mm, and the shaft was discretized using 34,438 mixed solid elements. All other components were set as rigid bodies, and their mesh densities had little effect on the simulation results.

The model discretization was performed using ANSA v21.0 (BETA CAE Systems, Switzerland), and the simulations were performed on 16 CPUs of an AMD EPYC 7532 (GHz) workstation using LS-DYNA Release 13 (LSTC, Livermore, CA, USA).

3. Results

3.1. Bench Tests

Figure 7 presents the results of the insertion force tests for the nylon and Pebax balloons. The force–displacement curves, characterized by several peaks due to discontinuous loading, demonstrate that the insertion forces of both balloons initially increase and then decrease with greater displacement. The maximum insertion force for nylon is 0.866 N, substantially greater than that of Pebax, which is only 0.156 N.

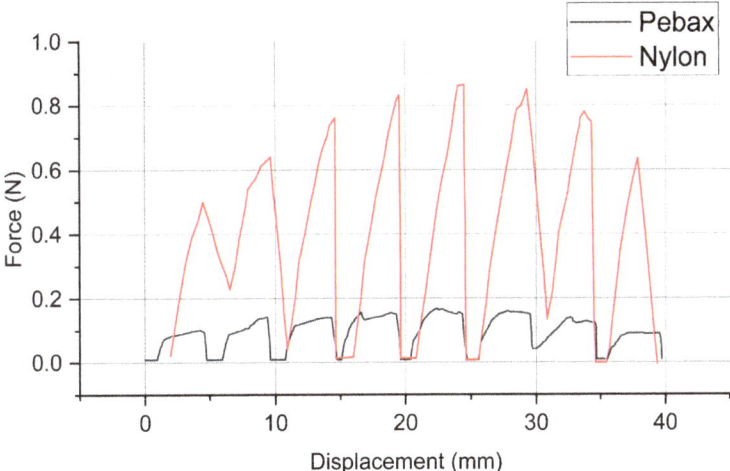

Figure 7. Test results of insertion forces for two types of balloon materials: Grilamid L25 (nylon 12) and Pebax® 7033 SA 01 MED.

3.2. Numerical Results

3.2.1. Dynamic Folding and Delivery Process of the Balloon

The pleating tool compressed the balloon from its initial form (Figure 8a) into a tri-wing shape (Figure 8c), with the three wings remaining in an inflated state. Upon the application of negative pressure to the balloon's inner surface, the three wings tightly adhered to each other (Figure 8d). As the pleating tool rotated counterclockwise, it dove the three wings (Figure 8e), which then sequentially attached to the shaft (Figure 8f). Subsequently, the simulated balloon-crimping machine continuously constricted the chamber, yielding a fixed balloon diameter of 1.5 mm. Following this, the balloon was axially inserted into the pipeline to test the insertion force (Figure 9). The balloon's shape within the pipeline demonstrated that the bent balloon exhibited evenly spaced wrinkles on the concave side Figure 10), causing the adjacent area to bulge beyond the original diameter.

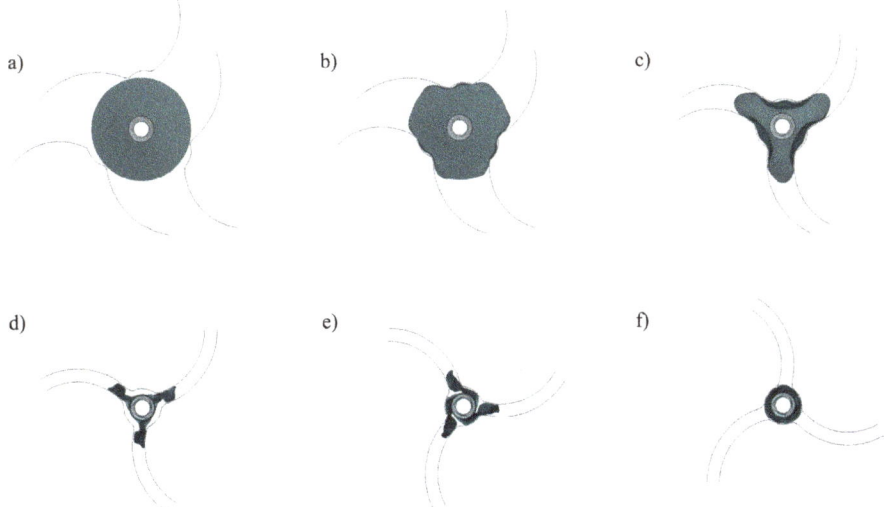

Figure 8. Pleating process. (**a**–**c**): Compression of the balloon by the pleating tool. (**d**–**f**): The three wings are folded onto the balloon.

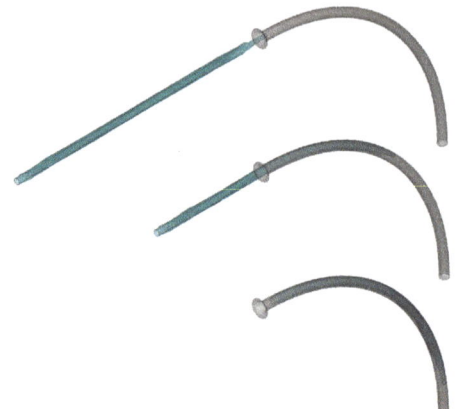

Figure 9. The process of inserting the folded balloon into the pipeline.

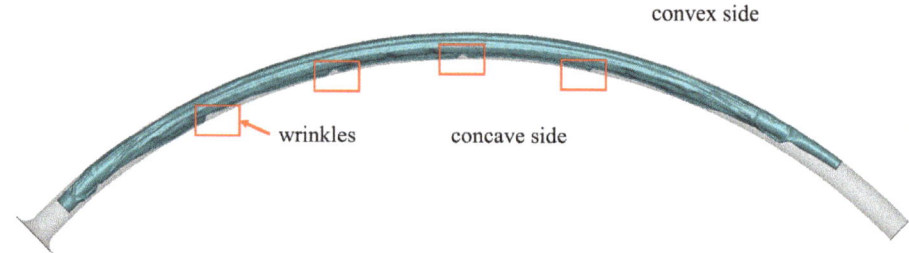

Figure 10. Equally spaced wrinkles in the concave side.

3.2.2. Balloon Stress–Strain Analysis

The stress distribution and effective plastic strain information can help us understand how the balloon behaves under different loading conditions, such as a folding state or bending state. This allows for an evaluation of the balloon's mechanical integrity and the identification of any areas prone to excessive stress or deformation. After radially crimping the balloons to achieve the same diameters, the stress levels of the nylon balloon were found to be higher than those of the Pebax balloon. The maximum stress was located at the proximal end, with 61.9 MPa for the nylon balloon and 55 MPa for the Pebax balloon, which is consistent with the results of previous studies [24]. The post-folding stress distribution in both balloons was more concentrated at the edges of the three folds and at the balloon ends (Figure 11), where the deformation exceeded that of the remaining areas. Figure 12 more clearly illustrates that higher effective plastic strains consistently manifested at the folds. Within the same fringe range, the Pebax balloon's middle showed a greater effective plastic strain than the nylon balloon, with the Pebax balloon reaching a maximum effective plastic strain of 0.948, which is higher than the nylon balloon's 0.879.

3.2.3. Trackability

The insertion force between the balloon and the pipeline was monitored during the process of pushing the balloon into the pipeline (Figure 13). The balloons of both materials showed a gradual increase in insertion force with the increase in pushing distance, and the insertion force reached its maximum value after the balloon was pushed into the pipeline completely at 2.16 ms. Nylon had a maximum insertion force of 1.016 N, while Pebax had a maximum insertion force of 1.021 N.

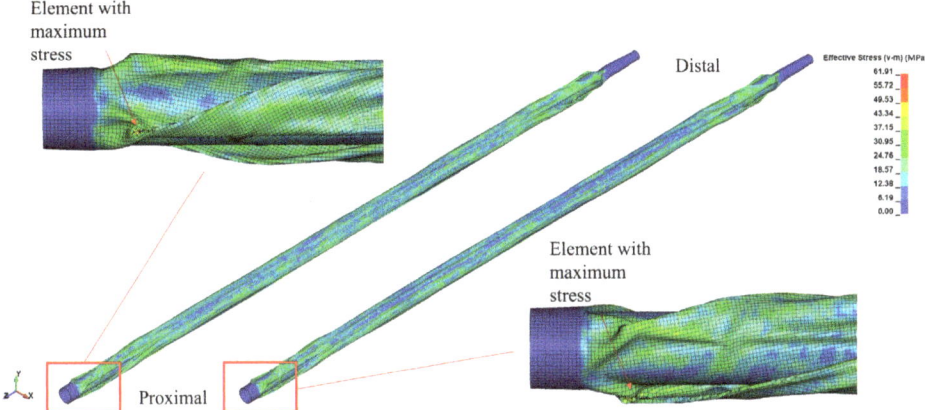

Figure 11. Stress distribution and location of elements with maximum stress after folding for the two types of balloon materials, nylon (**left**) and Pebax (**right**).

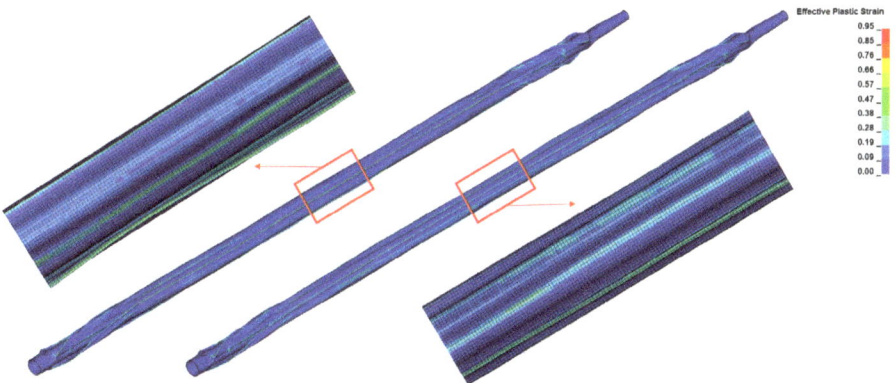

Figure 12. The effective plastic strain fields of the two types of balloon materials: Nylon (**left**) and Pebax (**right**).

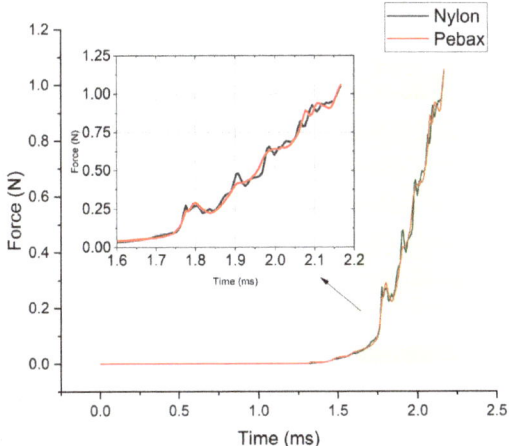

Figure 13. Temporal variation in contact force for two types of balloons.

Similar to the experimental results, the simulation results also exhibited a gradual increase in the force value for the initial 20 mm of the inserting displacement. In this study, the simulation was conducted specifically for the process of the balloon being inserted into the pipeline rather than the process of the balloon being pushed out of the pipeline. Therefore, unlike in the experiment, the contact force did not decrease after reaching its maximum value. In the experiment, the insertion force of the Pebax balloon was significantly lower than that of nylon. However, in the simulation, the insertion force of nylon was only slightly different and higher than Pebax's at certain moments. The maximum insertion force of the nylon balloon in the simulation was slightly larger than the experimental value of 0.866 N.

3.2.4. Surface Energy Density

The surface energy density provides valuable information about the interaction between the balloon material and the surrounding environment. It can affect the balloon's ability to navigate through blood vessels, interact with the vessel wall, and perform its intended functions effectively. The surface energy density of the pipeline with balloons fully inserted is shown in Figure 14. Throughout the insertion process, the surface energy density continued to increase and was primarily distributed on the convex surface of the pipeline and the folds of the balloon. The balloon's surface energy density is significantly higher than that of the pipeline. The Pebax-balloon-loaded pipeline had higher surface energy density than the nylon-balloon-loaded pipeline.

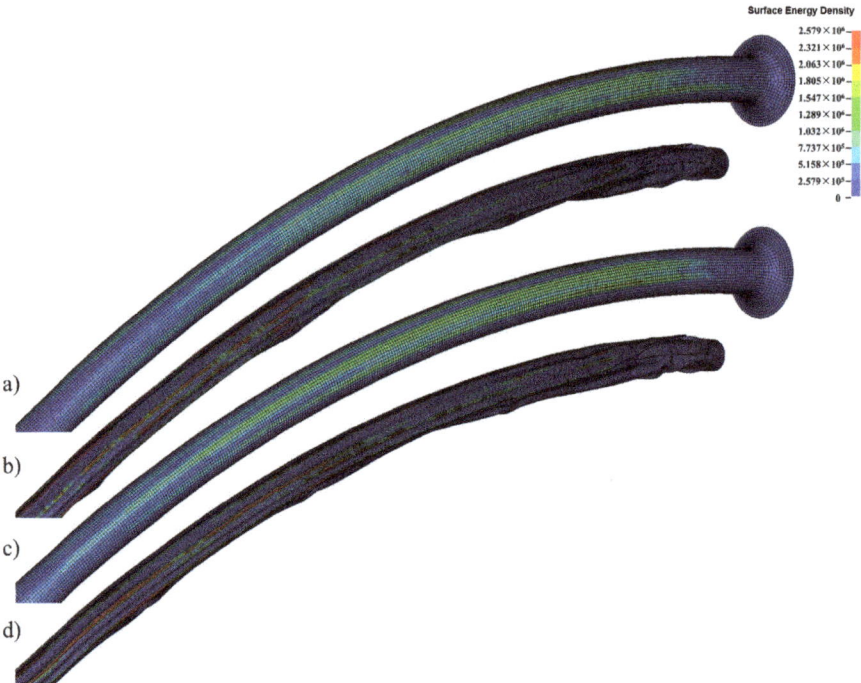

Figure 14. Surface energy densities of two balloons: (**a**) the pipeline loaded with a nylon balloon; (**b**) Nylon balloon. (**c**) The pipeline loaded with a Pebax balloon; (**d**) Pebax balloon.

4. Discussion

With the same shaft, the nylon balloon exhibited a greater insertion force than the Pebax balloon in the experiments. In the simulations, the coefficient of friction between the balloon and the pipeline was set to the same value for both cases for better variable control [32]. The simplification of the friction coefficient only affects the numerical values of the insertion force, while the comparison of stress and strain remains unaffected by it. The frictional force originated from both the balloon's tendency to expand outward in its folded state, exerting pressure on the pipeline wall, and from the pressure exerted on the wall as the balloon bent along the pipeline. Although the friction force is also related to velocity [34], the same boundary conditions are imposed on both so that the effect of velocity can also be neglected. The use of the same friction coefficient also results in a less significant difference in the insertion force between the two materials in the simulation compared to the experiment [35]. The difference between the experimental result of 0.866 N for the nylon balloon and the simulated result of 1.016 N can be attributed to the given friction coefficient. The friction coefficient used in the simulation was obtained from previous studies [32] and may not perfectly match the friction coefficient in our actual experimental environment. However, using the same friction coefficient, controlled for the variables, allows for a better assessment of the impact of material stiffness on the insertion force without the need to consider variations in surface characteristics such as roughness.

The experimental results seem to indicate that the stiffer balloon material has more friction during insertion and requires more insertion and retraction forces because it exerts more pressure on the wall than the softer material. However, numerical simulations revealed a negligible difference in the insertion forces between the two materials, contrary to the experimental results. This suggests that the influence of material stiffness on the balloon insertion force is minimal when surface characteristics are disregarded. Therefore, the significant disparity in the insertion force test results between the nylon and Pebax balloons in the experiment is attributed to the differences in their surface characteristics rather than nylon being harder and thus requiring greater force and Pebax being softer and thus requiring less force. This finding is counterintuitive.

Both experiments and simulations demonstrated that as the balloon progresses deeper into the pipeline and subsequently retreats, the frictional force initially increases and then decreases. This suggests that the friction force during the insertion and retraction processes correlates with the length of the balloon within the pipeline. Thus, selecting balloons of different lengths may be necessary to ensure that the friction force during insertion/retraction is not too high for different patients [36,37]. Our future research will explore the use of balloon models of the same material but different lengths to simulate the friction force through grooves, which can provide a reference for appropriate balloon selection.

After balloon folding, the deformation near the connection with the shaft is substantial. The element with the highest stress also occurs in this area, suggesting that this region dictates the overall balloon's stress concentration level [24]. This implies that a more thoughtful optimization of this area's structure could potentially reduce the maximum stress experienced after balloon folding. In this study, the proximal and distal folds exhibit slight differences. It is well documented that the elements with the maximum stress are at the distal end of both balloons. This is because the distal diameter of 0.91 mm is larger than the proximal diameter of 0.6 mm, and the transition from the balloon to the connection is steeper at the distal end, leading to the formation of deeper folds (Figure 15). Therefore, a smoother transition at the connections between balloons and shafts, particularly at the neck, could reduce the maximum stress of the balloons after folding.

Wrinkling arises when the balloon undergoes compression on its concave side while passing through the pipeline instead of stretching on the convex side. If the bend only stretches the balloon on the convex side, it only changes the length of the balloon along the axial direction without negatively affecting the balloon's deliverability. However, squeezing the balloon on the concave side will cause some folds to exceed the original diameter, resulting in an increased cross-sectional profile of the balloon during bending. This condition is unfavorable for navigating through tortuous anatomy [38] and can negatively impact the balloon insertion force, deliverability, and flexibility. To improve balloon delivery, the folded balloon should stretch on the convex side rather than being squeezed on the concave side when bending along the axial direction. This can be achieved by designing a fold structure to be less prone to wrinkling or by fabricating the balloon from a material characterized by superior compression resistance and ease of stretching.

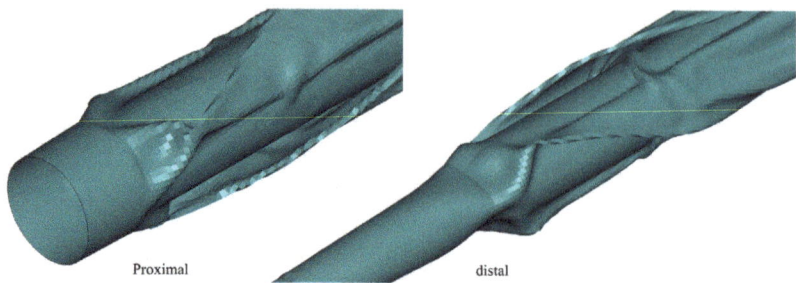

Figure 15. Comparison of balloon folds.

In this study, the assumption of using the same friction coefficient for both materials in contact with the pipeline is a simplified model. In reality, the contact properties between different materials may differ, which is a limitation of this study. In addition, fluid–structure interactions were neglected, as in previous studies [39–41] that simulated stent implantation in blood vessels with inflated balloons. These factors need to be considered in future work.

5. Conclusions

This study simulated the folding process of two distinct balloon materials and inserted the folded balloons into a simulated pipeline to monitor the insertion forces. Experimental tests were conducted on both types of balloons to measure the actual insertion forces. Through comparing the experimental data and numerical simulation results, it was found that under the assumption of the same friction coefficient in the numerical simulation, the difference in insertion force between the two materials was insignificant. This contrasted with the experimental results in which the harder nylon balloon demonstrated a significantly higher insertion force in comparison to the Pebax balloon. This indicates that the impact of material stiffness on the balloon insertion force is minimal when the surface characteristics of the materials are not considered. Therefore, in practical production and application, for balloons with poor trackability, emphasis should be placed on improving surface smoothness, such as by adding lubricious coatings. Additionally, the stress and effective plastic strain results after folding showed that the harder nylon balloon experienced greater stress after folding, while the softer Pebax balloon experienced a higher level of effective plastic strain. This suggests that nylon balloons are likely to exert greater pressure on the curved vessel wall. This simulation of balloon folding and insertion forces offers a generalized methodology for simulating the trackability of balloons made from different materials. This would be advantageous for selecting suitable balloon materials for various scenarios and may provide valuable insights for the design of balloon structures and folding methods.

Author Contributions: Conceptualization, T.L. and Y.C.; methodology, T.L. and Z.Z.; software, T.L.; validation, T.L. and Z.Z.; formal analysis, T.L.; investigation, W.W.; resources, W.W.; data curation, T.L. and A.M.; writing—original draft preparation, T.L.; writing—review and editing, T.L.; visualization, T.L.; supervision, Y.C. and Y.X.; project administration, Y.C.; funding acquisition, Y.C. and F.G. All authors have read and agreed to the published version of the manuscript.

Funding: Key Research Programs of Science & Technology Department of Sichuan Province (23ZDYF1576). National Natural Science Research Foundation of China (12172239).

Institutional Review Board Statement: Not applicable.

Informed Consent Statement: Not applicable.

Data Availability Statement: The data presented in this study are available on request from the corresponding author.

Conflicts of Interest: The authors declare no conflict of interest.

References

1. Bevan, G.H.; Solaru, K.T.W. Evidence-based medical management of peripheral artery disease. *Arterioscler. Thromb. Vasc. Biol.* **2020**, *40*, 541–553. [CrossRef] [PubMed]
2. Tepe, G.; Laird, J.; Schneider, P.; Brodmann, M.; Krishnan, P.; Micari, A.; Metzger, C.; Scheinert, D.; Zeller, T.; Cohen, D.; et al. Drug-coated balloon versus standard percutaneous transluminal angioplasty for the treatment of superficial femoral and popliteal peripheral artery disease: 12-month results from the IN.PACT SFA rando. *Circulation* **2015**, *131*, 495–502. [CrossRef]
3. Jun, S.; Gaetano, S. Update on peripheral artery disease: Epidemiology and evidence-based facts. *Atherosclerosis* **2018**, *275*, 379–381.
4. Caradu, C.; Lakhlifi, E.; Colacchio, E.C.; Midy, D.; Bérard, X.; Poirier, M.; Ducasse, E. Systematic review and updated meta-analysis of the use of drug-coated balloon angioplasty versus plain old balloon angioplasty for femoropopliteal arterial disease. *J. Vasc. Surg.* **2019**, *70*, 981–995. [CrossRef] [PubMed]
5. Fanelli, F.; Cannavale, A.; Boatta, E.; Corona, M.; Lucatelli, P.; Wlderk, A.; Cirelli, C.; Salvatori, F.M. Lower limb multilevel treatment with drug-eluting balloons: 6-month results from the DEBELLUM randomized trial. *J. Endovasc. Ther.* **2012**, *19*, 571–580. [CrossRef] [PubMed]
6. Scheinert, D.; Duda, S.; Zeller, T.; Krankenberg, H.; Ricke, J.; Bosiers, M.; Tepe, G.; Naisbitt, S.; Rosenfield, K. The LEVANT I (Lutonix paclitaxel-coated balloon for the prevention of femoropopliteal restenosis) trial for femoropopliteal revascularization: First-in-human randomized trial of low-dose drug-coated balloon versus uncoated balloon angioplasty. *JACC Cardiovasc. Interv.* **2014**, *7*, 10–19. [CrossRef]
7. Werk, M.; Albrecht, T.; Meyer, D.-R.; Ahmed, M.N.; Behne, A.; Dietz, U.; Eschenbach, G.; Hartmann, H.; Lange, C.; Schnorr, B.; et al. Paclitaxel-coated balloons reduce restenosis after femoro-popliteal angioplasty: Evidence from the randomized PACIFIER trial. *Circ. Cardiovasc. Interv.* **2012**, *5*, 831–840. [CrossRef]
8. Bukka, M.; Rednam, P.J.; Sinha, M. Drug-eluting balloon: Design, technology and clinical aspects. *Biomed. Mater.* **2018**, *13*, 032001. [CrossRef]
9. Mircea, P.; Sorin, B.; Florin, B.; Cleric, G.; Rita, A. Comparative Investigation on Polymer Drug-coating Balloons Used in Infrapopliteal Angioplasty Based on Angiosomes Concept. *Mater. Plast.* **2022**, *59*, 183–193.
10. Amstutz, C.; Weisse, B.; Haeberlin, A.; Burger, J.; Zurbuchen, A. Inverse Finite Element Approach to Identify the Post-Necking Hardening Behavior of Polyamide 12 under Uniaxial Tension. *Polymers* **2022**, *14*, 3476. [CrossRef]
11. Sirivella, M.L.; Rahinj, G.B.; Chauhan, H.S.; Satyanarayana, M.V.; Ramanan, L. Numerical Methodology to Evaluate Trackability and Pushability of PTCA Balloon Catheter. *Cardiovasc. Eng. Technol.* **2023**, *14*, 315–330. [CrossRef] [PubMed]
12. Schmitz, K.; Behrens, P.; Schmidt, W.; Behrend, D.; Urbaszek, W. Quality Determining Parameters of Balloon Angioplasty Catheters. *J. Invasive Cardiol.* **1996**, *8*, 144–152. [PubMed]
13. *ASTMF2394-07*; Standard Guide for Measuring Securement of Balloon Expandable Vascular Stent Mounted on Delivery System. ASTM: West Conshohocken, PA, USA, 2013.
14. Schmidt, W.; Lanzer, P.; Behrens, P.; Topoleski, L.D.T.; Schmitz, K.-P. A comparison of the mechanical performance characteristics of seven drug-eluting stent systems. *Catheter. Cardiovasc. Interv.* **2009**, *73*, 350–360. [CrossRef]
15. Geith, M.A.; Eckmann, J.D.; Haspinger, D.C.; Agrafiotis, E.; Maier, D.; Szabo, P.; Sommer, G.; Schratzenstaller, T.G.; Holzapfel, G.A. Experimental and mathematical characterization of coronary polyamide-12 balloon catheter membranes. *PLoS ONE* **2020**, *15*, e0234340. [CrossRef] [PubMed]
16. Helou, B.; Bel-Brunon, A.; Dupont, C.; Ye, W.; Silvestro, C.; Rochette, M.; Lucas, A.; Kaladji, A.; Haigron, P. Influence of balloon design, plaque material composition, and balloon sizing on acute post angioplasty outcomes: An implicit finite element analysis. *Int. J. Numer. Methods Biomed. Eng.* **2021**, *37*, e3499. [CrossRef]
17. Dong, P.; Mozafari, H.; Lee, J.; Gharaibeh, Y.; Zimin, V.; Dallan, L.; Bezerra, H.; Wilson, D.; Gu, L. Mechanical performances of balloon post-dilation for improving stent expansion in calcified coronary artery: Computational and experimental investigations. *J. Mech. Behav. Biomed. Mater.* **2021**, *121*, 104609. [CrossRef] [PubMed]
18. Rahinj, G.B.; Chauhan, H.S.; Sirivella, M.L.; Satyanarayana, M.V.; Ramanan, L. Numerical Analysis for Non-Uniformity of Balloon-Expandable Stent Deployment Driven by Dogboning and Foreshortening. *Cardiovasc. Eng. Technol.* **2022**, *13*, 247–264. [CrossRef] [PubMed]
19. De Beule, M.; Mortier, P.; Carlier, S.; Verhegghe, B.; Van Impe, R.; Verdonck, P. Realistic finite element-based stent design: The impact of balloon folding. *J. Biomech.* **2008**, *41*, 383–389. [CrossRef]
20. Wiesent, L.; Schultheiß, U.; Schmid, C.; Schratzenstaller, T.; Nonn, A. Experimentally validated simulation of coronary stents considering different dogboning ratios and asymmetric stent positioning. *PLoS ONE* **2019**, *14*, e0224026. [CrossRef]
21. Karol, B.J.; Jerzy, M.; Piotr, K. Finite element analysis of the percutaneous coronary intervention in a coronary bifurcation. *Acta Bioeng. Biomech.* **2014**, *16*, 23–31.

22. Azarnoush, H.; Pazos, V.; Vergnole, S.; Boulet, B.; Lamouche, G. Intravascular optical coherence tomography to validate finite-element simulation of angioplasty balloon inflation. *Phys. Med. Biol.* **2019**, *64*, 095011. [CrossRef] [PubMed]
23. Gajewski, T.; Szajek, K.; Stępak, H.; Łodygowski, T.; Oszkinis, G. The influence of the nylon balloon stiffness on the efficiency of the intra-aortic balloon occlusion. *Int. J. Numer. Methods Biomed. Eng.* **2018**, *35*, e3173. [CrossRef] [PubMed]
24. Geith, M.A.; Swidergal, K.; Hochholdinger, B.; Schratzenstaller, T.G.; Wagner, M.; Holzapfel, G.A. On the importance of modeling balloon folding, pleating, and stent crimping: An FE study comparing experimental inflation tests. I. *J. Numer. Methods Biomed. Eng.* **2019**, *35*, e3249. [CrossRef] [PubMed]
25. Warner, J.A.; Forsyth, B.; Zhou, F.; Myers, J.; Frethem, C.; Haugstad, G. Characterization of Pebax angioplasty balloon surfaces with AFM, SEM, TEM, and SAXS. *J. Biomed. Mater. Res. Part B Appl. Biomater.* **2015**, *104*, 470–475. [CrossRef] [PubMed]
26. Sadeghi, F.; Le, D. Characterization of polymeric biomedical balloon: Physical and mechanical properties. *J. Polym. Eng.* **2021**, *41*, 799–807. [CrossRef]
27. Alexandru, R.N.; Mihaela, O. Simulation of the Single Point Incremental Forming of Polyamide and Polyethylene Sheets. *MATEC Web Conf.* **2019**, *290*, 03014.
28. Stoppie, N.; Van Oosterwyck, H.; Jansen, J.; Wolke, J.; Wevers, M.; Naert, I. The influence of Young's modulus of loaded implants on bone remodeling: An experimental and numerical study in the goat knee. *J. Biomed. Mater. Res. Part A* **2008**, *90A*, 792–803. [CrossRef]
29. Hallquist, J.O. *LS-DYNA®Keyword User's Manual Volume II Material Models*; Livermore Software Technology Corporation: Livermore, CA, USA, 2013; p. 10.
30. Zhang, K.; Li, W.; Zheng, Y.; Yao, W.; Zhao, C. Compressive Properties and Constitutive Model of Semicrystalline Polyethylene. *Polymers* **2021**, *13*, 2895. [CrossRef]
31. Kurra, S.; Prakash, R.S. Effect of time scaling and mass scaling in numerical simulation of incremental forming. *Appl. Mech. Mater.* **2014**, *612*, 105–110.
32. Sharei, H.; Kieft, J.; Takashima, K.; Hayashida, N.; van den Dobbelsteen, J.J.; Dankelman, J. A Rigid Multibody Model to Study the Translational Motion of Guidewires Based on Their Mechanical Properties. *J. Comput. Nonlinear Dyn.* **2019**, *14*, 101010.
33. Salaha, Z.F.M.; Ammarullah, M.I.; Abdullah, N.N.A.A.; Aziz, A.U.A.; Gan, H.-S.; Abdullah, A.H.; Abdul Kadir, M.R.; Ramlee, M.H. Biomechanical Effects of the Porous Structure of Gyroid and Voronoi Hip Implants: A Finite Element Analysis Using an Experimentally Validated Model. *Materials* **2023**, *16*, 3298. [CrossRef] [PubMed]
34. Dong, S.; Sheldon, A.; Pydimarry, K.; Dapino, M. Friction in LS-DYNA®: Experimental Characterization and Modeling Application. In Proceedings of the 14th International LSDYNA Users Conference, Detroit, MI, USA, 12–14 June 2016; pp. 3–12.
35. Katarzyna, K.; Maciej, S.; Tomasz, C. Determination of urethral catheter surface lubricity. *J. Mater. Sci. Mater. Med.* **2008**, *19*, 2301–2306.
36. Tan, M.; Urasawa, K.; Koshida, R.; Haraguchi, T.; Kitani, S.; Igarashi, Y.; Sato, K. Comparison of angiographic dissection patterns caused by long vs short balloons during balloon angioplasty of chronic femoropopliteal occlusions. *J. Endovasc. Ther.* **2018**, *25*, 192–200. [CrossRef] [PubMed]
37. Hiromasa, T.; Kuramitsu, S.; Shinozaki, T.; Jinnouchi, H.; Morinaga, T.; Kobayashi, Y.; Domei, T.; Soga, Y.; Shirai, S.; Ando, K. Impact of total stent length after cobalt chromium everolimus-eluting stent implantation on 3-year clinical outcomes. *Catheter. Cardiovasc. Interv.* **2017**, *89*, 207–216. [CrossRef] [PubMed]
38. Byrne, R.; Neumann, F.-J.; Mehilli, J.; Pinieck, S.; Wolff, B.; Tiroch, K.; Schulz, S.; Fusaro, M.; Ott, I.; Ibrahim, T.; et al. Paclitaxel-eluting balloons, paclitaxel-eluting stents, and balloon angioplasty in patients with restenosis after implantation of a drug-eluting stent (ISAR-DESIRE 3): A randomised, open-label trial. *Lancet* **2013**, *381*, 461–467. [CrossRef]
39. Schiavone, A.; Abunassar, C.; Hossainy, S.; Zhao, L.G. Computational analysis of mechanical stress–strain interaction of a bioresorbable scaffold with blood vessel. *J. Biomech.* **2016**, *49*, 2677–2683. [CrossRef]
40. Xiaodong, Z.; Mitsuo, U.; Kiyotaka, I. Finite element analysis of cutting balloon expansion in a calcified artery model of circular angle 180°: Effects of balloon-to-diameter ratio and number of blades facing calcification on potential calcification fracturing and perforation reduction. *PLoS ONE* **2021**, *16*, e0251404.
41. Mortier, P.; Holzapfel, G.; De Beule, M.; Van Loo, D.; Taeymans, Y.; Segers, P.; Verdonck, P.; Verhegghe, B. A Novel Simulation Strategy for Stent Insertion and Deployment in Curved Coronary Bifurcations: Comparison of Three Drug-Eluting Stents. *Ann. Biomed. Eng.* **2009**, *38*, 88–99. [CrossRef]

Disclaimer/Publisher's Note: The statements, opinions and data contained in all publications are solely those of the individual author(s) and contributor(s) and not of MDPI and/or the editor(s). MDPI and/or the editor(s) disclaim responsibility for any injury to people or property resulting from any ideas, methods, instructions or products referred to in the content.

Article

Interstitial Fluid Flows along Perivascular and Adventitial Clearances around Neurovascular Bundles

Xiya Kong [1,2], Xiaobin Yu [3], Gang Peng [3], Fang Wang [1,2,*,†] and Yajun Yin [3,*,†]

1. Department of Cardiology, Beijing Hospital, National Center of Gerontology, Institute of Geriatric Medicine, Chinese Academy of Medical Sciences, Beijing 100730, China
2. Graduate School, Peking Union Medical College, Chinese Academy of Medical Sciences, Beijing 100730, China
3. Department of Engineering Mechanics, Tsinghua University, Beijing 100084, China
* Correspondence: bjh_wangfang@163.com (F.W.); yinyj@mail.tsinghua.edu.cn (Y.Y.)
† These authors contributed equally to this work.

Abstract: This study reports new phenomena of the interstitial fluid (ISF) microflow along perivascular and adventitial clearances (PAC) around neurovascular bundles. The fluorescent tracing was used to observe the ISF flow along the PAC of neurovascular bundles in 8–10 week old BALB/c mice. The new results include: (1) the topologic structure of the PAC around the neurovascular bundles is revealed; (2) the heart-orientated ISF flow along the PAC is observed; (3) the double-belt ISF flow along the venous adventitial clearance of the PAC is recorded; (4) the waterfall-like ISF flow induced by the small branching vessel or torn fascia along the PAC is discovered. Based on the above new phenomena, this paper approached the following objectives: (1) the kinematic laws of the ISF flow along the PAC around neurovascular bundles are set up; (2) the applicability of the hypothesis on the PAC and its subspaces by numerical simulations are examined. The findings of this paper not only enriched the image of the ISF flow through the body but also explained the kernel structure of the ISF flow (i.e., the PAC). It helps to lay the foundation for the kinematics and dynamics of the ISF flow along the PAC around neurovascular bundles.

Keywords: interstitial fluid (ISF); microflow; perivascular and adventitial clearance; neurovascular bundles; accompanying vein; accompanying artery; kinematic law

1. Introduction

The interstitial fluid (ISF) is a vital body fluid that connects with the blood, lymphatic fluid, and intracellular fluid. In the traditional physiological view, the majority of ISF is bound within the hydrogel-like extracellular matrix that cannot flow freely [1]. This study reports new phenomena of the interstitial fluid (ISF) microflow along perivascular and adventitial clearances (PAC) around neurovascular bundles.

Recently, the dynamic flow of the ISF along the perivascular space (PVS) has caused ongoing concern. The PVS, also known as the "Virchow-Robin space", is the space that surrounds the parenchymal vessels, with the glial membrane as the outer boundary and the vascular wall as the inner boundary. The PVS, together with the entry of penetrating arteries and the exit of draining veins from the cerebral cortex, is an ISF or cerebrospinal fluid (CSF)-filled, pial-lined structure [2,3]. Iliff et al. named the efficient convective exchange clearance system of CSF and ISF along the PVS as the "glymphatic system", which relies on astrocytes via aquaporin-4 in the central nervous system [4,5]. Its function has been demonstrated successively in human and animal models in association with various neurological disorders [6,7]. The diameter of cerebral PVS in mice was measured to be approximately 40 µm, and its integrated cross-section area was approximately 1.4 times that of the adjacent vessels found by two-photon microscopy [8,9]. The force to drive the cerebral ISF flow in the glymphatic system is considered to be significantly correlated with arterial pulsation but not with respiratory movement [4,10].

Carare et al. regarded the basement membranes of cerebral capillaries and arteries as the counterpart of cerebral lymphatic pathways. The drainage of the ISF occurs along the intramural periarterial drainage (IPAD). However, no empirical evidence of tracers coming into and out of the parenchyma was found on cortical veins [11–13]. The vasodilation movements produced by contraction and diastole of cerebrovascular smooth muscle cells are the main drivers of IPAD [14].

Until the recent studies of cerebral ISF flow, the available experimental evidence does not provide an adequate description of how the venous walls and the surrounding space drain the ISF, compared to the arteries. In any case, the functional role of venous drainage of the ISF appears to be minor [15]. Apart from the brain, Li et al. demonstrated that long-distance ISF flow exists in the arterial and venous adventitia and their surrounding fibrous connective tissues throughout the whole body, and possibly also in the epineurium and skin [16,17].

The aforementioned studies all focused on isolated blood vessels, which were not able to explain the experimental phenomenon (i.e., the ISF flow within the perivascular) in this study. Meanwhile, it should be emphasized that the majority of veins (excluding special superficial venous) are accompanied by an artery and a nerve (or nerves) to form neurovascular bundles [18]. The specificity and integrality of neurovascular bundles structure should be considered to provide a complete and less contradictory basis for ISF flow dynamics. Consequently, accompanying neurovascular bundles are observed as a holistic object in this study to locate the ISF flow space, and the frame with which the diverse ISF flow patterns can be interpreted consistently.

The paper is organized as follows. Section 4 introduces our experimental materials and methods. Section 2 shows the experimental results, including (a) the structure of the PAC, (b) the diverse ISF microflow patterns inside the PAC, and (c) the novel phenomenon of the waterfall-like ISF flow. Section 3 refined the kinematic laws of the ISF flow along the PAC around neurovascular bundles, based on the three kinematic laws of ISF flow over isolated arteries and veins [19].

These findings have not only enriched the image of the ISF flow through the body (except in the brain), but have also explained the kernel structure of the ISF flow (i.e., PAC). Only by clarifying the kernel structure of the ISF flow can we further study the pathways and properties of the ISF flow and elucidate the kinematic mechanism. It helps to develop the kinematics and dynamics of the ISF microflow along the PAC around neurovascular bundles.

2. Results

2.1. Fluorescence-Stained and Histological Analysis of the PAC Structure around Neurovascular Bundles

5.0 µL fluorescent tracer was supplied. The observed site is 1.5–2.0 cm away from the supply point (Figure 1A). The flow pathways have been fluorescently stained along the direction of the ISF flow. The ISF flows towards the heart in the same direction as venous blood flow, and opposite to arterial blood flow (Figure 1B). The fluorescent-stained neurovascular bundles were subjected to rapidly frozen sectioning and Verhoeff's Van Gieson (EVG) staining.

The observed ISF flows mainly occur in three areas, namely: the venous adventitia, the arterial adventitia, and the connective tissue surrounding the accompanying artery and vein. The entire flow space is termed perivascular and adventitial clearance (abbreviated as PAC). Furthermore, the PAC can be more finely divided into three subspaces, namely arterial adventitial clearance (PAC(a)), venous adventitial clearance (PAC(v)), and perivascular connective tissue flow space (PAC(p)) (Figure 2).

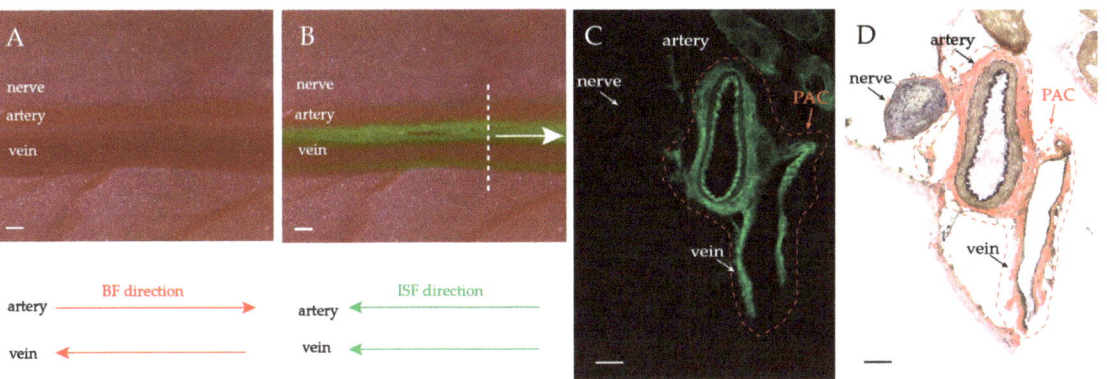

Figure 1. ISF flow along PAC around neurovascular bundles. (**A**) Brightfield observation before fluorescing. (**B**) Brightfield observation with fluorescent lamp opening after fluorescing. The heart direction is located left. Scale bar: 200 μm. (**C,D**) Frozen fluorescent sections and EVG staining show the structure of PAC. Magnification 20×. The arrows indicate nerve, artery, vein, and PAC. Scale bar: 50 μm.

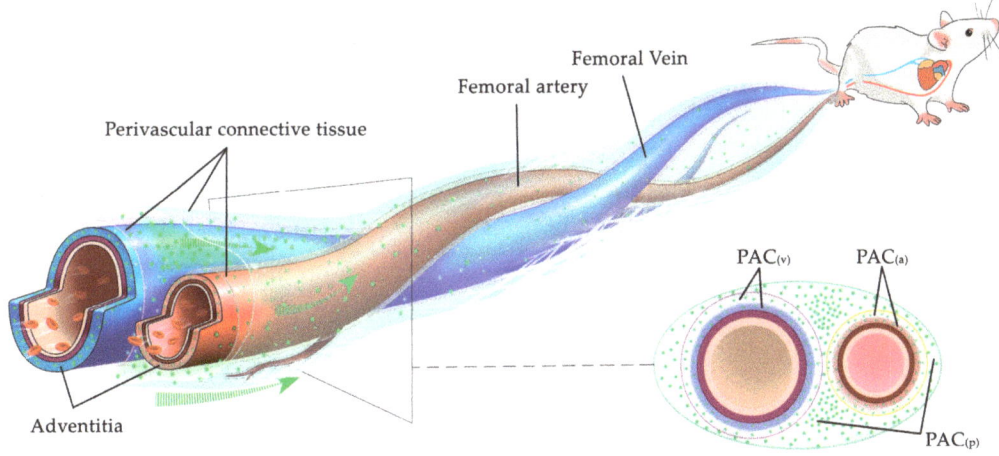

Figure 2. The observed space for ISF flow along PAC around neurovascular bundles. Perivascular and adventitial clearance (PAC) can be divided into arterial adventitial clearance (PAC(a)), venous adventitial clearance (PAC(v)) and perivascular connective tissue flow space (PAC(p)).

The histological results show that PAC(a), PAC(v), and PAC(p) were fluorescence-stained (Figure 1C). It is worth noting that soluble fluorescent tracers can penetrate and migrate into the medial and intima layers of the artery and vein. Therefore, the presence of fluorescent signals in the media smooth muscle and the intimal layer does not mean that they are all the natural flow pathways of the ISF. Corresponding to the EVG staining (Figure 1D), there is abundant loose connective tissue around the artery, the vein, and the nerve. The vascular adventitia itself is also loose connective tissue, with no clear boundary between them and the surrounding loose connective tissue. The EVG results show that the diameter of the PAC(a) is approximately 200 μm ($n \geq 5$), PAC(v) is approximately 400 μm ($n \geq 5$), the nerve is approximately 200 μm ($n \geq 5$), and the PAC is approximately 600–750 μm ($n \geq 5$).

The ISF flow pattern and the topological structure of the PAC shown in Figure 1 are universalities. To properly depict the ISF flow phenomena along the PAC in the following experiments, a general diagram of the PAC is abstracted from Figure 1 and displayed in Figure 2. The fluorescent spots in Figure 2 mark the ISF flow spaces. From Figure 2, we can predict that once the relative positions between the outer connective tissue membrane and adventitia change, both the ISF flow spaces and the flow patterns will change. This prediction is confirmed in the following experiment.

2.2. Heart-Orientated ISF Flow along PAC(p)

1.0 µL fluorescent tracer was supplied. The femoral artery and vein observed are 1.5–2.0 cm away from the supply point (Figure 3A,B). The fluorescent ISF flows along the connective tissue gap (i.e., PAC(p)) between the accompanying femoral artery and vein after a few seconds (Figure 3C and Supplementary Materials Video S1). Meanwhile, there is no fluorescent signal over PAC(a) or PAC(v) (Figure 3C). Then the fluorescent signal began to appear on the vein (Figure 3D). At the same time, in the gap between the accompanying artery and vein, the fluorescence is brighter, and the width of the fluorescence belt is wider (Video S1). The direction of the ISF flow is right-to-left and pointed to the heart. The flow direction towards the heart is always the same as the venous blood flow. In short, the PAC(p) is one of the ISF's directional flow channels. Because the ISF mainly flows inside the PAC(p) or the gap between the accompanying vein and artery, we may term the phenomenon "gap-flow". The gap width is about 300 µm, so the gap-flow is a continuous microflow.

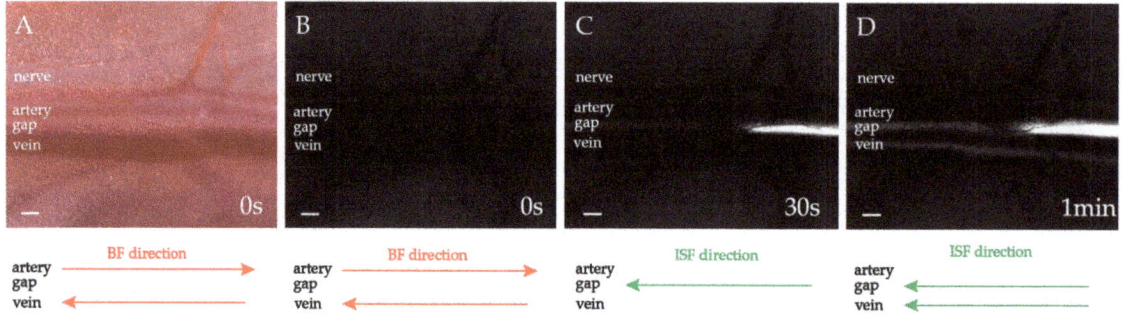

Figure 3. ISF flow along PAC(p). (**A**) Brightfield observation before fluorescing. (**B**) Fluorescence observation before fluorescing. (**C,D**) Fluorescence observation after fluorescing. The heart is located left. The ISF flow is pointed to the heart. Scale bar: 200 µm.

2.3. Double-Belt ISF Flow along PAC(v) around Neurovascular Bundles

1.0 µL fluorescent tracer was supplied. The femoral artery and vein observed are 1.5–2.0 cm away from the supply point (Figure 4A,B). After a few seconds, two parallel "fluorescence flow belts (FFB)" with equal width appeared simultaneously along the PAC(v) (Figure 4C). Each FFB is approximately 1/3 of the venous diameter (Figure 4C). As time went on, both FFBs widened, and the brightness increased. The two FFBs gradually fused into a single one after about 1 minute (Figure 4D and Video S2). The direction of the ISF flow is right-to-left, pointed to the heart.

The flow phenomenon in Figure 4 is termed "double-belt flow". Aside from the equal-width double-belt flows, unequal-width double-belt flows were also similarly observed.

Two parallel FFBs with unequal widths appear simultaneously along the PAC(v) a few seconds after the fluorescent tracer was supplied (Figure 4F). The two FFBs are approximately 1/4 and 1/2 width of the venous diameter, respectively. Both FFBs widen, and their brightness increases with time. After 1 minute, the two FBs gradually fused into one (Figure 4G and Video S3).

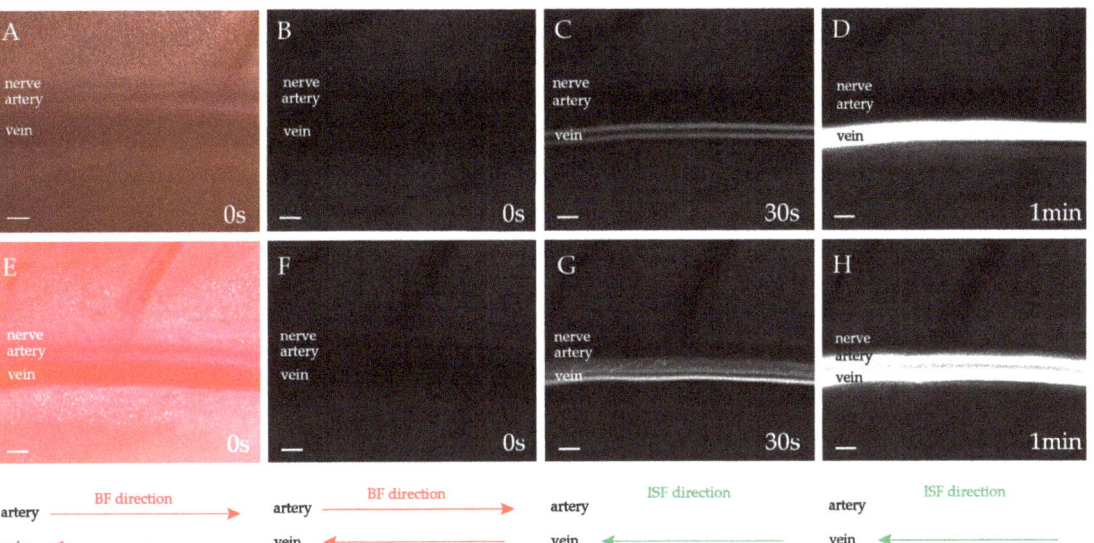

Figure 4. Double-belt ISF flow along PAC(v). (**A–D**) Equal-width double-belt ISF flow; (**E–H**) Unequal-width double-belt ISF flow. (**A,E**) Brightfield observation before fluorescing. (**B,F**) Fluorescence observation before fluorescing. (**C,D,G,H**) Fluorescence observation after fluorescing. The heart is located left. The ISF flow is pointed to the heart. Scale bar: 200 µm.

Both the equal and unequal double-belt flows along the PAC(v) around neurovascular bundles are rapid directional ISF flows.

How do we explain the double-belt ISF flow pattern? The relative locations of the adventitia and the multi layers of perivascular connective tissue may provide the answers. If the multi layers of connective tissue have adhered to the venue adventitia—such adhesion always occurs in neurovascular bundles—then an adhesion belt (i.e., the black no-flow belt) in Figure 4C may be formed. This adhesion belt will separate the flow space along the venue adventitia into two parts, where the two streams of ISF (i.e., the white flow belt) are observed.

2.4. Waterfall-like ISF Flow Induced by the Small Branching Vessel or Torn Fascia along PAC

The above experiments display the smooth ISF flows along the PAC. In our experiments, we also captured the various unsmooth ISF flows along the PAC. Next, the unsmooth flows caused by obstacles in the ISF flow pathway were reported.

1.0 µL fluorescent tracer was supplied. Using ophthalmic forceps, we gently tear the fascia formed by connected tissue covering the outermost layer of the femoral vein. A fascia layer is torn across the middle of the vein, 1.5–2.0 cm away from the supplying point (Figure 5A). Although the outer membrane of the multi-layered fascia was torn, the left layers can still ensure that the PAC structure is integral. When the fluorescent ISF flows, a fissure fascia is observed across the middle of the vein, and a waterfall-like ISF flow is induced along the PAC (Figure 5B and Video S4).

The same effect will also be created when a small nourishing vessel crosses over the main vein (Figure 5C). The waterfall-like ISF flow along the PAC is induced by the small branching vessel crossing the central part of the vein (Figure 5D). The ISF can be seen to overturn the small branching vessel to create an angled waterfall-like flow across the main vein (Video S5).

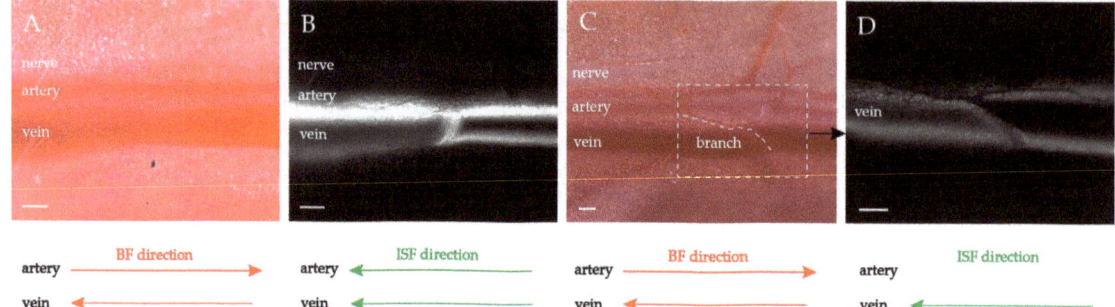

Figure 5. The small branching vessel or torn fascia induced waterfall-like ISF flow along PAC. (**A,B**) Torn fascia across the vein induced waterfall-like ISF flow. (**C,D**) The small vessel across the vein induced a waterfall-like ISF flow. (**A,C**) Brightfield observation before fluorescing. (**B,D**) Fluorescence observation after fluorescing. The heart is located in the left direction. The ISF flow is pointed to the heart. Scale bar: 200 μm.

The Supplementary Video shows that the partial ISF is modulated by the small vessel and flows along the branching. It can be deduced that the ISF flow is massive enough to form a local waterfall-like flow by any obstructive barrier in the PAC.

3. Discussion

3.1. The PAC Is the Kernel Structure for ISF Flow

Unlike the classical intraluminal flow (e.g., blood, lymphatic fluid), which is well known in traditional physiology, the ISF flow is a particular non-luminal flow. What is the flow space of this kind of non-luminal fluid is an essential and complicated key issue and the cornerstone of in-depth research. Only by clarifying the kernel structure of the ISF flow can we further study the pathways and properties of the ISF flow and elucidate the kinetic mechanism.

As a flow space for the ISF, the PAC in the body and the PVS in the brain are comparable. In the glymphatic system, the cerebral ISF flow space is believed to be the PVS. However, there are still many controversies. Prior to in vivo two-photon imaging, the existence of the PVS was doubted. There are significant morphological differences in the PVS when observed in vivo and dead states. The PVS may disappear due to dehydration, and the tracer was deposited into the adjacent collagen fibers when the animal died or during the tissue fixation process [9,20]. The research evidence on the drainage of the cerebral ISF along the IPAD pathway was controversial: (1) The results of the tissue sections showed that the tracer was deposited into the smooth muscle and basement membrane layers, but whether the region of tracer deposition marked the real ISF flow was inexplicable [21]. (2) Whether the pressure within the IPAD was able to drive a steady flow, and if the water resistance within the IPAD is so big that it can accommodate a large cerebral ISF, are all unclear [22]. These controversies are sufficient to show that revealing the kernel structure of the ISF flow is a very complex matter. However, it must be pointed out that when comparing the PAC and PVS, there is a need to note that the ISF flow environment in the lower limbs is quite different from that in the brain. For instance, (1) The brain has close to 80% of high water content [23]. The brain's extracellular fluids consist of ISF, blood plasma, and cerebrospinal fluid (CSF). CSF is also involved in the glymphatic system. (2) The blood–brain barrier (BBB) is a particular anatomic and physiologic barrier separating the circulating blood from the extracellular fluid in the central nervous system [24]. We have tried to elucidate the kernel structure of the ISF flow in the lower limbs, which is the most crucial purpose of this article. The cerebral ISF flow has its characteristics, but the ISF flow also occurs in the whole body. We chose to focus on blood vessels in the lower limbs because femoral arteries and veins are easily exposed, which facilitates real-time observation in vivo. (2) Unlike injection tracer into the cisterna magna, we could achieve a relatively quantitative and direct supply of fluorescent tracer to trace pathways in the lower limbs.

Our study revealed the new phenomena of ISF flow along vessel walls and in the surrounding space. The vascular adventitia is mainly composed of loose connective tissue containing extracellular matrix (ECM), fibroblasts and small perivascular nerves, in which loosely arranged, spiral or longitudinally-distributed collagen, elastic and reticular fibers form a stress-bearing viscoelastic skeletal structure filled with large amounts of amorphous material (e.g., hyaluronic acid) [25]. The perivascular space of both arteries and veins is also surrounded by abundant loose connective tissue. It is difficult to draw a clear boundary between the adventitia of blood vessels and their surrounding connective tissue (Figure 1D). The connective tissue becomes looser as it moves outwards, centered on the blood vessels [26,27]. The abundant loose connective tissue along the vascular walls provides a flowable space for a large amount of ISF flow within the PAC (Figure 2).

As mentioned above, the PAC includes three regions (i.e., PAC(a), PAC(v) and PAC(p), see Figure 2). The three subspaces are not three mutually exclusive subsets but are interconnected and complementary. Because of the diversity of biological structures and individual differences, when one subspace in the PAC is looser than another, its flow resistance will be smaller and the ISF will flow relatively faster. Therefore, when a very small amount of fluorescent tracer is supplied, sometimes it seemed that only one of the subspaces was observed to be brightened. However, the real flow should occur in all subspaces at the same time. In addition, for no-blood-vessel-accompanying nerve bundles or only the artery accompanying the vein, the three subspaces are not necessarily present simultaneously. They are combined by a subset of them, so the flow space on an isolated artery (vein) can be formed by the combination of PAC(a) or PAC(a) + PAC(p) (PAC (a) or PAC(v) + PAC(p)).

Up until now, most flows have been observed along the PAC. There is no clear answer as to whether there is an ISF flow along the neural epithelium. The reason might be that the dynamic images of fluorescent ISF flowing along the neural epithelium in vivo were not captured due to the limitation of the fluorescence stereo microscopes. However, the static images given in the previous study by Li et al. showed that the neural epithelium was fluorescently brightened in some situations. More experimental evidence is still required. Whatever the case, it is certain that the accompanying nerve provides a partial boundary for the ISF flow along the PAC. Therefore, we adopted a holistic perspective, using the neurovascular bundles as the object of study. It is possible to clarify the basic structure of the PAC to understand the diversity of the ISF flow patterns.

3.2. Spontaneous ISF Flow Instead of Pressure-Driven Flow or Tracer Diffusion

The shown ISF flow along the PAC is spontaneous. In other words, the ISF flow is not driven by external pressure such as injection, and it is not diffusion. The reason is as follows. In our all experiments, the fluorescent tracer was supplied with minimal amounts (1.0–5.0 µL), slowly and with no pressure in vivo (Figure 6A). This approach differs from pressure injection, which may cause the fluorescent tracer to flow simultaneously to the heart and far from the heart (Figure 6C).

Moreover, in the dead state, supplying the fluorescent tracer may produce diffusion in all directions (Figure 6B). Supplementary Experiment-1 was provided to repeat supplying in the dead state. In Supplementary Experiment-1, the 0.5 µL fluorescent tracer was supplied to the vascular adventitia in the middle of the exposed vessels in the dead state of mice, and we found that the fluorescent ISF was diffused (Figure S1).

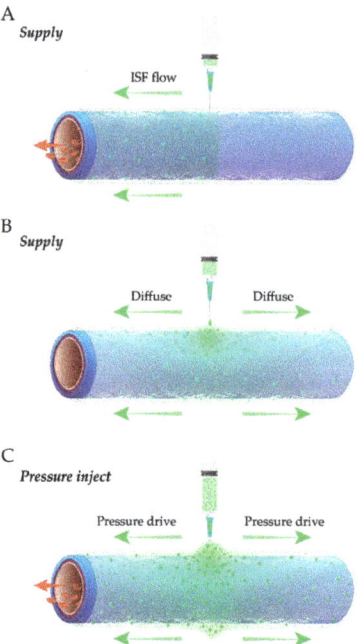

Figure 6. Three different fluorescent tracer approaches. (**A**) Supplying fluorescent tracer in vivo. (**B**) Supplying fluorescent tracers in the dead state. (**C**) Pressure injection in vivo.

3.3. The Diversity of ISF Flow Patterns within PAC

To investigate the effect of the geometry of the arteriovenous and their surrounding fascia on the flow distribution, simplified models are considered. For example, the cross-section shown in Figure 7, where the circle and the ellipse characterize the outer fascia and the arteriovenous membranes, respectively. At the small Reynolds number, the viscous laminar flow in the fully developed and eventually formed a constant flow. As this study focuses on the flow space rather than the dynamic pattern, the model ignores the changes in the longitudinal direction of the vessel due to pulsatile or systalic motion. The z-related terms in the nonlinear Navi-Stokes equations all degenerate to zero, and the equations degenerate to a two-dimension Poisson's equation. The flow velocity distribution problem within the cross-section becomes a Poisson equation boundary value problem under Dirichlet boundary conditions, with the governing equation given by

$$\frac{\partial^2 u_z}{\partial x^2} + \frac{\partial^2 u_z}{\partial y^2} = \frac{1}{\mu}\frac{dp}{dz}. \tag{1}$$

where: (x, y, z) represents the spatial position coordinate; u_z is the velocity in the z-direction; μ is the kinematic viscosity; $\frac{dp}{dz}$ is the pressure gradient along the z-axis.

The arterial radius r_a is selected as the characteristic length, with the characteristic velocity denoted as $\frac{r_a^2}{\mu}\frac{dp}{dz}$, then the following dimensionless parameters can be introduced

$$\xi = \frac{x}{r_a},\ \eta = \frac{y}{r_a},\ u = u_z \frac{\mu}{r_a^2}\frac{dz}{dp}. \tag{2}$$

The governing equation then becomes

$$\frac{\partial^2 u}{\partial \xi^2} + \frac{\partial^2 u}{\partial \eta^2} = 1 \tag{3}$$

On the outer boundary, the transmembrane permeation rate u_x and u_y are small compared with the longitudinal velocity. As a solid fibrous structure, the fascia could not flow itself, so the boundary condition can be considered as a Dirichlet boundary, i.e., $u = 0$ on the boundary.

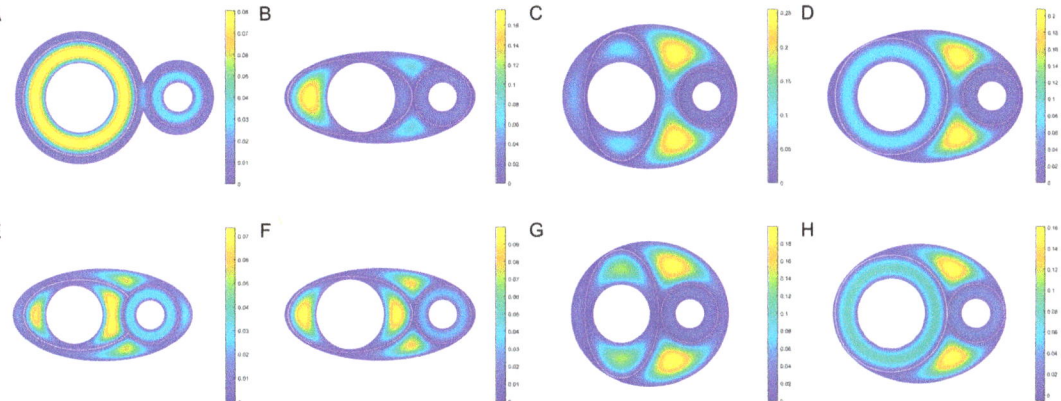

Figure 7. Simulated ISF flowing patterns. Three separated flowing spaces are PAC(a), PAC(v) and PAC(p) respectively. (**A–H**) Flowing velocity distribution on different cross-section geometry. On the boundaries, the velocity is zero without permeating. Flow velocity decreases from yellow to blue.

In realistic organisms, regular rounded fascia formed by loose connective tissues and regular rounded outer membranes of arteries and veins are scarce. The simplification here to a circular or elliptical shape is primarily because: (a) the effect of the relative relationship between the arteriovenous rather than the specific geometry on the flow distribution was discussed in a zero-order approximation sense; (b) the diverse flow patterns observed in the experiment were reproduced as much as possible based on the relative positions by limited shapes; (c) it avoided an introduction of subjective boundaries. The selected representative geometric parameters are shown in Figure 8 and Table 1.

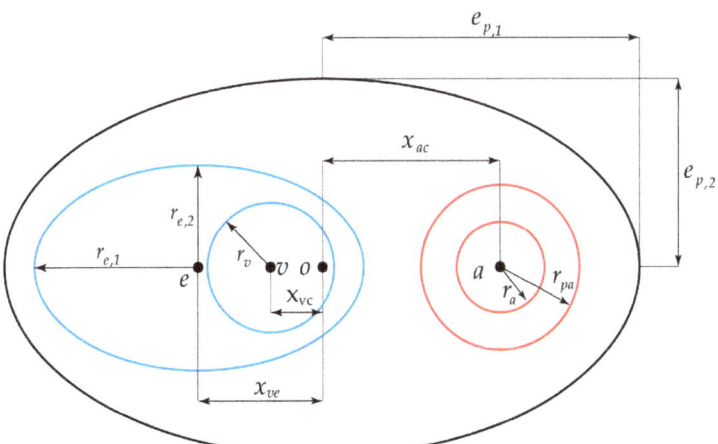

Figure 8. Schematic diagram of cross-sectional geometry. The marked geometric variables are the feature parameters when modelling, and all geometric lengths have been nondimensionalized using the radius of the artery r_a, and r_{pa} is fixed as 2. All the dimensionless value of the parameters are listed in Table 1.

Table 1. Simulation parameters.

No.	Vein				Artery		PVC	
	r_v	x_{vc}	$e_{v,1}$	$e_{v,2}$	x_{ve}	x_{ac}	$e_{p,1}$	$e_{p,2}$
(b)	2.4	1.2	4.4	2.6	2.0	4.6	6.8	3.2
(c)	2.4	2.0	2.6	4.4	2.0	2.0	6.2	5.0
(d)	2.4	2.4	4.0	4.0	2.4	4.8	7.0	4.6
(e)	2.0	0.4	4.0	2.4	1.6	5.0	6.4	3.2
(f)	2.4	2.0	4.4	2.6	2.0	4.6	6.8	3.2
(g)	2.0	2.0	2.6	4.4	2.0	4.8	5.6	5.0
(h)	2.4	2.0	4.0	4.0	2.0	4.0	6.2	4.8

The data in the table have been nondimensionalized with respect to the arterial radius.

In Figure 7, the circles and ellipses illustrate the cross-section of the neurovascular bundle and the topology of the PAC. Although circles cannot exactly represent the various complex patterns found in natural organisms, topological combinations of circles and ellipses have been able to functionally reproduce the experimental flow patterns, see Figure 7. For example, Figure 7A corresponds to the result that only veins are fluorescently stained, as shown in Figure 4; Figure 7C,G show that the fluorescent appears mainly in the arteriovenous space (i.e., PAC(p)) and slightly diffuses to the vein or artery, as shown in Figure 3; Figure 7D,H show that both the entire PAC(v) and PAC(p) are stained, as shown in Figure 1. It should be emphasized that the flow pattern and the various cross-section geometries do not correspond exactly to the real situation. In Figure 7A,D,H, although there are differences in the geometry and relative position relationships of the arterioles and fascia, the flow patterns presented are similar. It should be noted again, that the geometry in this simulation is only one of many possibilities, and the main purpose is to simulate the experimentally observed flow pattern, not to reproduce the exact shape of the organism species' fascia. Importantly, the simulation results elucidate that the three flows (i.e., the flows within PAC(a), PAC(v) and PAC(p)) are sufficient to reproduce the different flow patterns observed in various experiments.

3.4. Heart-Orientated ISF Flow

Although the ISF flow direction is a little bit contrary to intuitions, the evidence for heart orientation is solid enough (Videos S1–S5). To strengthen the reader's confidence, one supplementary experiment was provided (Figure S2). In the Supplementary Experiment-1, the 0.5 µL fluorescent tracer was supplied to the vascular adventitia in the middle of the exposed vessels in vivo. The image showed that the fluorescent ISF flows along the upstream venous adventitial (i.e., PAC(v) and PAC(p)), flowing toward the heart (Figure S2C,E). However, the downstream vessels of the fluorescent supplied point did not appear in the fluorescent signal (Figure S2D,F). The Supplementary Videos and Figures provide vivid and reliable evidence for the heart-orientated ISF flow.

3.5. The Kinematic Laws of ISF Flow along PAC around Neurovascular Bundles

In previous work [19], the three kinematic laws of ISF flow for isolated arteries and veins have been proposed. Based on the new phenomena, we took a holistic perspective of the accompanying artery, vein and nerve. The real topology structures and flow images of the PAC are highly variable in vivo. However, there is one commonality or invariance, i.e., the direction of the ISF flow remains the same in the whole process, which is supported by extensive empirical evidence. Based on such an invariance, the kinematic laws of the ISF along the PAC around neurovascular bundles were proposed.

The First law: There exists an ISF flow along the PAC around neurovascular bundles.

The Second law: Along the PAC around the neurovascular bundles, the direction of the ISF flow is the same as that of the accompanying venous blood flow and opposite to that of the accompanying arterial blood flow.

The second law shows that on the extremities, the ISF flow along the PAC is towards the heart, i.e., it is heart-orientated.

4. Materials and Methods

4.1. Experimental Animals

Thirty male BALB/c mice aged 8–10 weeks were purchased from Beijing HFK Bio-Technology Co., Ltd., (Beijing, China). The mice were anesthetized using a combination of 2.5% Avertin at 0.12–0.15 mL/10 g intraperitoneally and 0.02–0.04 mL/10 g Meloxicam subcutaneously. After the experiment, the remaining mice were executed using the overdose anesthesia method. All experiments involving animals conformed to ARRIVE (Animal Research: Reporting of In Vivo Experiments) guidelines. This study protocol was reviewed and approved by Institutional Animal Care and Use Committee, Tsinghua University (No.: 20-YYJ1).

4.2. Surgical Operation and Fluorescent Tracing

Surgical incisions on the mice's legs were made from foot to groin to fully expose the vein, artery and nerve. A litter of the outer connective tissue covering the vascular adventitia was removed by forceps. 1.0–5.0 µL fluorescent sodium (Guangzhou Baiyunshan Pharmaceutical Co., Ltd., Guangzhou, China) was supplied slowly to the vascular adventitia near the foot and ankle (Figure 6).

4.3. Fluorescing Imaging

Observe the ISF flow through Fluorescence Stereo Microscopes ZEISS Axio Zoom V16 (ZEISS, Jena, Germany). The dynamic videos were recorded with a time series of 4s/frame for continuous filming.

4.4. Histological Staining

The tissue specimens were obtained immediately after fluorescing imaging. The fixed tissues were embedded with OCT compound then 5 um slices were prepared, cutting by $-20\ °C$ freezing microtome. Immediate fluorescence images were acquired by a fluorescence microscope (Nikon, Tokyo, Japan) equipped with a charge-coupled device (CCD) (Nikon, Tokyo, Japan). The remaining sections were fixed and dehydrated according to the standard treatment of dewaxing, hydration and staining with Elastic Van Gieson (EVG). The images were acquired by an upright light microscope (Olympus, Tokyo, Japan).

4.5. Sample Size Calculation

Taking into account the unique design to confirm that the ISF flow along the PAC is universal in mice, traditional sample size calculations are not applicable. One thing that needs to be clarified is that such an experiment has been repeated hundreds of times in mice, and the phenomena of the ISF flow along the PAC are proved universal. In this article, to measure the diameter of PAC(a), PAC(v), and PAC(p), we used thirty 8–10 week old BALB/c mice in order to reduce errors.

5. Conclusions

In this study, the presence of the ISF flow was verified around the neurovascular bundles, and its flow along vascular walls was confirmed to be massive. Based on the new phenomena, the internal subspaces of PAC, i.e., PAC(a), PAC(v) and PAC(p), were proposed. Using numerical simulations, it was confirmed that different combinations of the three subspaces were sufficient to reproduce the various flow patterns observed in experiments, which indirectly validated the applicability of the PAC together with its subspaces. The kinematic laws of the ISF flow along the PAC around neurovascular bundles are proposed. We hope that the novel phenomena above may draw the attention of researchers and may be examined or checked by other laboratories.

Up until now, the physiological functions of the rapid ISF flows are still unclear. Even so, it is still reasonable to believe that such a long-range, continuous and heart-orientated

ISF flow is indispensable and vital to life systems. Of course, if the rapid ISF flows are circulated, then the physiological functions may be imaginable. The findings of this study lay the foundation and open the way for establishing and developing the kinematics and dynamics of the ISF flow along the PAC around neurovascular bundles.

6. Limitations

This study has not provided an accurate velocities range of the ISF flow, and has only given centimeters per second (cm/s) as the velocity magnitude. The ISF flow is high-speed compared to the lymphatic fluid flow rate (1.8–6 cm/min) [28,29]. We have tried various experimental methods to quantify the ISF flow. For example, tracking the fluorescent signal front-end through continuous images recorded by Fluorescence Stereo Microscopes has been performed. Nevertheless, the interference of uneven and variational background signal intensity to the actual fluorescence signal front-end cannot be avoided. It is also limited by the recording resolution and sensitivity to the fluorescent signal of the CCD. In addition, fluorescent microspheres used in Mestre et al. [9] and Bedussi et al. [30] were applied to track the ISF flow along the PAC without being able to form a continuous flow. We speculate that the possible reason for this is that there is a different environment for the ISF flow in the legs compared to the watery brain. Measuring the accurate velocity range of the ISF flow along the PAC is an issue for future research to explore.

Supplementary Materials: The following are available online at https://www.mdpi.com/article/10.3390/jfb13040172/s1, Figure S1: Diffusion of fluorescent tracers in the dead state, Figure S2: The fluorescent ISF flows towards the heart in vivo, Video S1: ISF flow along PAC(p), Video S2: Equal-width double-belt ISF flow along PAC(v), Video S3: Unequal-width double-belt ISF flow along PAC(v), Video S4: Torn fascia across the vein induced waterfall-like ISF flow, Video S5: The small vessel across the vein induced a waterfall-like ISF flow.

Author Contributions: Conceptualization, Y.Y., Y.K. and X.Y.; methodology, Y.K. and X.Y.; investigation, Y.K. and X.Y.; formal analysis, Y.Y., Y.K. and X.Y.; writing—original draft preparation, Y.K.; writing—review and editing, Y.Y., F.W., X.Y., and G.P.; supervision, F.W. and Y.Y. All authors have read and agreed to the published version of the manuscript.

Funding: This research was funded by National Natural Science Foundation of China (12050001, 11672150).

Institutional Review Board Statement: All experiments involving animals conformed to ARRIVE (Animal Research: Reporting of In Vivo Experiments) guidelines. This study protocol was reviewed and approved by Institutional Animal Care and Use Committee, Tsinghua University (No.: 20-YYJ1).

Informed Consent Statement: Not applicable.

Data Availability Statement: Not applicable.

Conflicts of Interest: The authors declare no conflict of interest.

References

1. Hall, J.E. *Guyton and Hall Textbook of Medical Physiology*; Saunders/Elsevier: Amsterdam, The Netherlands, 2011.
2. Krueger, M.; Bechmann, I. CNS pericytes: Concepts, misconceptions, and a way out. *Glia* **2010**, *58*, 1–10. [CrossRef]
3. Sepehrband, F.; Cabeen, R.P.; Choupan, J.; Barisano, G.; Law, M.; Toga, A.W.; Alzheimer's Disease Neuroimaging Initiative. Perivascular space fluid contributes to diffusion tensor imaging changes in white matter. *Neuroimage* **2019**, *197*, 243–254. [CrossRef]
4. Iliff, J.J.; Wang, M.; Zeppenfeld, D.M.; Venkataraman, A.; Plog, B.A.; Liao, Y.; Deane, R.; Nedergaard, M. Cerebral arterial pulsation drives paravascular CSF-interstitial fluid exchange in the murine brain. *J. Neurosci.* **2013**, *33*, 18190–18199. [CrossRef]
5. Iliff, J.; Wang, M.; Liao, Y.; Plogg, B.; Peng, W.; Gundersen, G.; Benveniste, H.; Vates, G.; Deane, R.; Goldman, S.; et al. A paravascular pathway facilitates CSF flow through the brain parenchyma and the clearance of interstitial solutes, including amyloid β. *Sci. Transl. Med.* **2012**, *4*, 147ra111. [CrossRef]
6. Ji, C.; Yu, X.; Xu, W.; Lenahan, C.; Tu, S.; Shao, A. The role of glymphatic system in the cerebral edema formation after ischemic stroke. *Exp. Neurol.* **2021**, *340*, 113685. [CrossRef] [PubMed]
7. Plog, B.; Nedergaard, M. The Glymphatic System in Central Nervous System Health and Disease: Past, Present, and Future. *Annu. Rev. Pathol.* **2018**, *13*, 379–394. [CrossRef]

1. Mestre, H.; Du, T.; Sweeney, A.M.; Liu, G.; Samson, A.J.; Peng, W.; Mortensen, K.N.; Staeger, F.F.; Bork, P.A.R.; Bashford, L.; et al. Cerebrospinal fluid influx drives acute ischemic tissue swelling. *Science* **2020**, *367*, eaax7171. [CrossRef]
2. Mestre, H.; Tithof, J.; Du, T.; Song, W.; Peng, W.; Sweeney, A.; Olveda, G.; Thomas, J.; Nedergaard, M.; Kelley, D. Flow of cerebrospinal fluid is driven by arterial pulsations and is reduced in hypertension. *Nat. Commun.* **2018**, *9*, 4878. [CrossRef]
3. Kedarasetti, R.; Drew, P.; Costanzo, F. Arterial pulsations drive oscillatory flow of CSF but not directional pumping. *Sci. Rep.* **2020**, *10*, 10102. [CrossRef] [PubMed]
4. Albargothy, N.; Johnston, D.; MacGregor-Sharp, M.; Weller, R.; Verma, A.; Hawkes, C.; Carare, R. Convective influx/glymphatic system: Tracers injected into the CSF enter and leave the brain along separate periarterial basement membrane pathways. *Acta Neuropathol.* **2018**, *136*, 139–152. [CrossRef] [PubMed]
5. Morris, A.; Sharp, M.; Albargothy, N.; Fernandes, R.; Hawkes, C.; Verma, A.; Weller, R.; Carare, R. Vascular basement membranes as pathways for the passage of fluid into and out of the brain. *Acta Neuropathol.* **2016**, *131*, 725–736. [CrossRef] [PubMed]
6. Carare, R.; Bernardes-Silva, M.; Newman, T.; Page, A.; Nicoll, J.; Perry, V.; Weller, R. Solutes, but not cells, drain from the brain parenchyma along basement membranes of capillaries and arteries: Significance for cerebral amyloid angiopathy and neuroimmunology. *Neuropathol. Appl. Neurobiol.* **2008**, *34*, 131–144. [CrossRef] [PubMed]
7. Aldea, R.; Weller, R.; Wilcock, D.; Carare, R.; Richardson, G. Cerebrovascular Smooth Muscle Cells as the Drivers of Intramural Periarterial Drainage of the Brain. *Front. Aging Neurosci.* **2019**, *11*, 1. [CrossRef] [PubMed]
8. Wardlaw, J.M.; Benveniste, H.; Nedergaard, M.; Zlokovic, B.V.; Mestre, H.; Lee, H.; Doubal, F.N.; Brown, R.; Ramirez, J.; MacIntosh, B.J.; et al. Perivascular spaces in the brain: Anatomy, physiology and pathology. *Nat. Rev. Neurol.* **2020**, *16*, 137–153. [CrossRef] [PubMed]
9. Li, H.; Yin, Y.; Yang, C.; Chen, M.; Wang, F.; Ma, C.; Li, H.; Kong, Y.; Ji, F.; Hu, J. Active interfacial dynamic transport of fluid in a network of fibrous connective tissues throughout the whole body. *Cell Prolif.* **2020**, *53*, e12760. [CrossRef] [PubMed]
10. Li, H.; Yang, C.; Yin, Y.; Wang, F.; Chen, M.; Xu, L.; Wang, N.; Zhang, D.; Wang, X.; Kong, Y.; et al. An extravascular fluid transport system based on structural framework of fibrous connective tissues in human body. *Cell Prolif.* **2019**, *52*, e12667. [CrossRef]
11. McGurk, S. Moore: Clinically Oriented Anatomy—Seventh international edition. *Nurs. Stand.* **2013**, *28*, 28. [CrossRef]
12. Yin, Y.; Li, H.; Peng, G.; Yu, X.; Kong, Y. Fundamental kinematics laws of interstitial fluid flows on vascular walls. *Theor. Appl. Mech. Lett.* **2021**, *11*, 100245. [CrossRef]
13. Benias, P.; Wells, R.; Sackey-Aboagye, B.; Klavan, H.; Reidy, J.; Buonocore, D.; Miranda, M.; Kornacki, S.; Wayne, M.; Carr-Locke, D.; et al. Structure and Distribution of an Unrecognized Interstitium in Human Tissues. *Sci. Rep.* **2018**, *8*, 4947. [CrossRef] [PubMed]
14. Mestre, H.; Mori, Y.; Nedergaard, M. The Brain's Glymphatic System: Current Controversies. *Trends Neurosci.* **2020**, *43*, 458–466. [CrossRef] [PubMed]
15. Faghih, M.; Sharp, M. Is bulk flow plausible in perivascular, paravascular and paravenous channels? *Fluids Barriers CNS* **2018**, *15*, 17. [CrossRef] [PubMed]
16. Natali, F.; Dolce, C.; Peters, J.; Gerelli, Y.; Stelletta, C.; Leduc, G. Water dynamics in neural tissue. *J. Phys. Soc. Jpn.* **2013**, *82* (Suppl. A), SA017. [CrossRef]
17. Wilhelm, I.; Krizbai, I.A. In vitro models of the blood–brain barrier for the study of drug delivery to the brain. *Mol. Pharm.* **2014**, *11*, 1949–1963. [CrossRef]
18. Tinajero, M.G.; Gotlieb, A.I. Recent Developments in Vascular Adventitial Pathobiology: The Dynamic Adventitia as a Complex Regulator of Vascular Disease. *Am. J. Pathol.* **2020**, *190*, 520–534. [CrossRef]
19. Witter, K.; Tonar, Z.; Schopper, H. How many Layers has the Adventitia?—Structure of the Arterial Tunica Externa Revisited. *Anat. Histol. Embryol.* **2017**, *46*, 110–120. [CrossRef]
20. Barallobre-Barreiro, J.; Loeys, B.; Mayr, M.; Rienks, M.; Verstraeten, A.; Kovacic, J. Extracellular Matrix in Vascular Disease, Part 2/4: JACC Focus Seminar. *J. Am. Coll. Cardiol.* **2020**, *75*, 2189–2203. [CrossRef] [PubMed]
21. Enzmann, D.R.; Pelc, N. Normal flow patterns of intracranial and spinal cerebrospinal fluid defined with phase-contrast cine MR imaging. *Radiology* **1991**, *178*, 467–474. [CrossRef]
22. Crescenzi, R.; Donahue, P.M.; Hartley, K.G.; Desai, A.A.; Scott, A.O.; Braxton, V.; Mahany, H.; Lants, D.K.; Donahue, M.J. Lymphedema evaluation using noninvasive 3T MR lymphangiography. *J. Magn. Reson. Imaging* **2017**, *46*, 1349–1360. [CrossRef] [PubMed]
23. Bedussi, B.; Almasian, M.; de Vos, J.; VanBavel, E.; Bakker, E. Paravascular spaces at the brain surface: Low resistance pathways for cerebrospinal fluid flow. *J. Cereb. Blood Flow Metab.* **2018**, *38*, 719–726. [CrossRef]

Review

A Review of Functional Analysis of Endothelial Cells in Flow Chambers

Makoto Ohta [1,*], Naoya Sakamoto [2,*], Kenichi Funamoto [1,*], Zi Wang [1,3], Yukiko Kojima [1,4] and Hitomi Anzai [1]

1. Institute of Fluid Science, Tohoku University, 2-1-1 Katahira, Aoba-ku, Sendai 980-8577, Japan; zi.wang.p3@dc.tohoku.ac.jp (Z.W.); kojima.yukiko.r8@dc.tohoku.ac.jp (Y.K.); hitomi.anzai.b5@tohoku.ac.jp (H.A.)
2. Graduate School of Systems Design, Tokyo Metropolitan University, 1-1 Minami-Osawa, Tokyo 192-0397, Japan
3. Graduate School of Biomedical Engineering, Tohoku University, 6-6-12, Aramaki Aza Aoba Aoba-ku, Sendai 980-8579, Japan
4. Graduate School of Engineering, Tohoku University, 6-6, Aramaki Aza Aoba Aoba-ku, Sendai 980-8579, Japan
* Correspondence: makoto.ohta@tohoku.ac.jp (M.O.); sakan@tmu.ac.jp (N.S.); funamoto@tohoku.ac.jp (K.F.); Tel.: +81-22-217-5309 (M.O.); +81-42-677-2709 (N.S.); +81-22-217-5878 (K.F.)

Abstract: The vascular endothelial cells constitute the innermost layer. The cells are exposed to mechanical stress by the flow, causing them to express their functions. To elucidate the functions, methods involving seeding endothelial cells as a layer in a chamber were studied. The chambers are known as parallel plate, T-chamber, step, cone plate, and stretch. The stimulated functions or signals from endothelial cells by flows are extensively connected to other outer layers of arteries or organs. The coculture layer was developed in a chamber to investigate the interaction between smooth muscle cells in the middle layer of the blood vessel wall in vascular physiology and pathology. Additionally, the microfabrication technology used to create a chamber for a microfluidic device involves both mechanical and chemical stimulation of cells to show their dynamics in in vivo microenvironments. The purpose of this study is to summarize the blood flow (flow inducing) for the functions connecting to endothelial cells and blood vessels, and to find directions for future chamber and device developments for further understanding and application of vascular functions. The relationship between chamber design flow, cell layers, and microfluidics was studied.

Keywords: flow chamber; endothelial cells; coculture techniques; microfluidics; lab-on-a-chip

Citation: Ohta, M.; Sakamoto, N.; Funamoto, K.; Wang, Z.; Kojima, Y.; Anzai, H. A Review of Functional Analysis of Endothelial Cells in Flow Chambers. *J. Funct. Biomater.* 2022, 13, 92. https://doi.org/10.3390/jfb13030092

Academic Editor: Xinping Zhang

Received: 30 April 2022
Accepted: 28 June 2022
Published: 12 July 2022

Publisher's Note: MDPI stays neutral with regard to jurisdictional claims in published maps and institutional affiliations.

Copyright: © 2022 by the authors. Licensee MDPI, Basel, Switzerland. This article is an open access article distributed under the terms and conditions of the Creative Commons Attribution (CC BY) license (https:// creativecommons.org/licenses/by/ 4.0/).

1. Introduction

Vascular tissue is composed of many layers; the innermost of which is called the endothelium, which is made of vascular endothelial cells (ECs). Therefore, ECs are constantly contacting the blood flow and are subjected to the flow's mechanical stress to express their functions. To elucidate this phenomenon, the morphology of vascular ECs was analyzed in vivo [1]. Later, a method for seeding ECs in a chamber was proposed to observe the various functions of ECs. Flow load was applied using the culture medium to simulate blood flow. This method has become indispensable for studying EC functions induced by flow. Recent advances in chamber design technology have led to the development of chambers with ECs and smooth muscle cells (SMCs) cocultured. The relationships between ECs and the function of SMCs and other deeper layers in blood vessels using the chambers have become clear. There is a cascading effect of the influence of blood flow through ECs into the deeper tissues of blood vessels. The chamber-based method has contributed to the elucidation of this cascade. In addition, the development of microfluidic devices through the advancement of microfabrication technology has enabled us to construct a three-dimensional microvascular network that reproduces the microenvironment in vivo and clarifies the dynamics of cell groups (cells).

As described above, the functional expression and application of ECs induced by flow are closely related to the development of many chambers and devices. The purpose of this study is to show the methods of inducing flow for functionalizing ECs by inducing flow and affecting blood vessels. Additionally, this study will find directions for future designs of chambers and device developments for further elucidation and application of vascular functions.

2. Materials and Methods

A keyword-based search of Pubmed (https://pubmed.ncbi.nlm.nih.gov) was performed. The search was performed between 23 June and 20 August 2021. The keywords used are shown in Figure 1, and the number of hits for each keyword is shown. The keywords were divided into the three primary categories as shown in Figure 1. The first category is called monolayer of EC. In the chamber, EC are seeded with monolayer cells, and the functions are induced by flow to examine the stimulation in ECs or with ECs. By using assembly techniques, the chambers are constructed as parallel plate, T-chamber, step, cone plate, and stretch, respectively dependent on the induced flow.

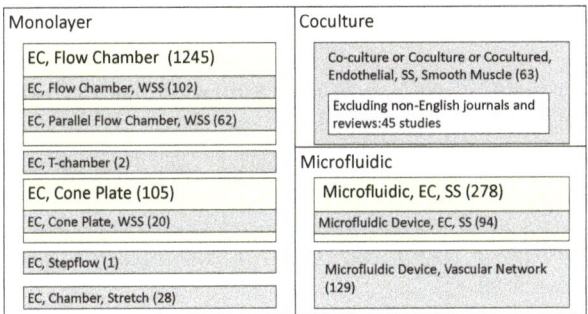

Figure 1. The main categories of paper searches using keywords in Pubmed.

The second category is called 'coculture', meaning that ECs induced by flows function by flow with SMCs. The third category is called 'microfluidic', meaning that the flow channels are fabricated more complexly and more precisely with patterns to use rare, diseased, or patient-specific cells with ECs stimulated by flows.

3. Results

3.1. Monolayer Chambers

3.1.1. Parallel Chamber

Parallel chambers introduce a unidirectional flow, and the cross-sectional area of the fluid domain is constant, as shown in Figure 2. A layer of EC is created at the bottom of the fluid domain in the chambers. Polystyrene [2], PDMS [3], ULTEM [4], glass for seeding cells in polycarbonate [5], gelatin-coated polyester sheet [6], and fibronectin-coat [7] are often used as the material of the chamber. A pulsating or steady flow was introduced into the chamber. The flow estimations in the chamber were accompanied mostly by computational fluid dynamics (CFD). On the other hands, LDV to experimentally measure the flow was performed and compared with some calculated values by Avari et al. [4], and microparticle image velocimetry was used by Lafaurie-Janvore et al. [8]. In the pulsatile flow, the actin microfilament and NO production of ECs were compared in wall shear stress (WSS) during exercise in vivo [9].

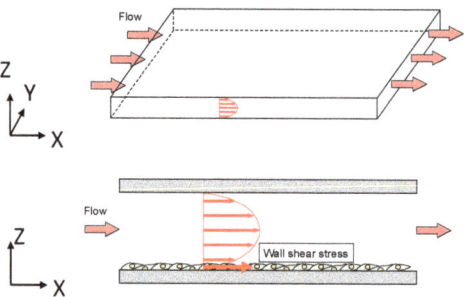

Figure 2. Flow domain in a parallel chamber. The arrows in the flow domain show the velocity distribution.

According to the purpose of these studies, we can categorize the EC's reactions or the activities of signals, including adhesion on the EC. Parallel chambers have been used to study not only the behavior of ECs themselves, but also the relationship between the adsorption of proteins on ECs and activation behavior.

ECs orient and elongate their shapes along with the flow. This has been quantitatively measured by Levesque et al. using the shape index [10]. The results showed that the shape index of bovine ECs almost converged after 24 h at 8.5 Pa WSS. Prasad et al. investigated the response of inositol phosphate levels [11]. Munn et al. investigated cell fluxes under several flow conditions such as steady or saw-tooth patterns [12]. Lafaureie-Janvore et al. showed the polarization of a cell on micron-size lines [8]. Anzai et al. proposed the disturbance effect by the placement of a stent strut on the cell culture in a parallel flow chamber [13]. Wang et al. investigated the cell density in the gap of two stent struts using a parallel flow chamber [14].

The relationship between EC walls, such as endothelium and fluid domain, were studied from signal activation to adhesion on the wall.

Cadroy et al. (1997) studied thrombogenicity using patient blood [15]. Gopalan et al. (1997) studied the emigration of neutrophil at 0.2 Pa [16], and Gosgnach et al. (2000) found that angiotensin converting enzyme is expressed by WSS [17]. Han et al. reported that shear stress induces mitochondrial reactive nitrogen species formation and inhibits the electron flux of the electron transport chain at multiple sites [18]. Popa et al. showed the formation of CD154-induced ultra-large von Willebrand factor (ULVWF) on ECs under 0.25–1 Pa WSS and transmigration of monocytes [19].

Lawrence et al. investigated the adhesion of polymorphnuclear leukocytes (PMNL) to endothelium (HUVEC) under 0.098 Pa [20]. Barabino et al. (1997) studied the adsorption of sickle red blood cells using parallel chambers. Viegas et al. (2011) investigated the adsorption of Methicillin-resistant Staphylococcus aureus on HUVECs in a 0–1.2 Pa parallel plate flow chamber [5]. Ozdemir et al. (2012) investigated the adsorption of soluble fibrinogen-mediated melanoma-polymorphonuclear neutrophils (PMNs) using a commercially available parallel plate flow chamber [21]. Kona et al. (2012) measured the poly (D, L-Lactic-co-glycolic acid) (PLGA) biodegradable nanoparticles uptake into ECs, which are considered to be injured arterial walls under fluid shear stress [22]. Rychak et al. investigated microbubbles adhesion to endothelium with deformation by WSS [23].

Previous studies have also revealed that WSS stimulates mechanosensitive molecules and intracellular signaling pathways, which can induce morphological and functional changes in ECs. For example, Takahashi and Berk showed that 1. 2 Pa WSS caused rapid activation of extracellular signal-regulated kinase (ERK1/2), which is one of the mitogen-activated protein (MAP) kinases and important for changes in EC gene expression [24]. Li et al. reported that phosphorylation of focal adhesion kinase (FAK), which plays a pivotal role in mechano-chemical transduction signaling pathways in ECs, is crucial in shear stress-induced activation of ERK1/2 [25]. Shear stress increases the influx of Ca^{2+}, known as a second messenger for the control of a variety of cell functions, in ECs mediated by Ca^{2+} ion channels [26]. Chachisvilis et al. demonstrated that conformational changes in G-protein-coupled receptors (GPCR) and the activation of G-protein, molecular switches changing the activity of GTP, are also induced by shear stress [27].

As summary, parallel plate chambers are widely used to find the cascade and mechanism of EC signals under dynamic response such as adhesion to ECs under uniaxial flow. As for chamber performance, WSS is used between 0.098 Pa and 8.5 Pa. In the future, a further development direction would be to conduct experiments under high WSS, where, for example, cardiovascular diseases are thought to develop, and next-generation chamber design using CFD methods is needed [3,28].

3.1.2. T-Chamber

The T-chamber is a chamber with a T-shaped cross section, as shown in Figure 3. The flow forms an impinging flow at the stagnation point in the center of the T-junction, and a high WSSG is loaded on the EC. This chamber assumes a bifurcation of an artery, which is the frequent site for a cerebral aneurysm. The T-chamber of Meng et al. (2008) has a WSS of 4 Pa and a WSSG of 30 Pa/cm. Sakamoto et al. (2010) investigated the orientation of ECs in chambers with a WSS of 2–10 Pa and a WSSG of 0–340 Pa/cm for 24–72 h under flow loading.

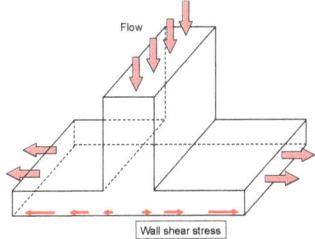

Figure 3. Flow domain in the T-chamber.

Additionally, Dunn's group revealed lymphatic ECs are sensitive to WSSG [29,30]. Several WSSGs are produced using a six-well impinging flow chamber and the migration direction of ECs induced by WSSG is changed by the densities of ECs.

3.1.3. Step Flow Chamber

A step flow chamber is a chamber that has a vertical step expansion at the entrance (Figure 4). This chamber could be used to generate disturbed flow including flow separation and reattachment following vortex. The gradient WSS is generated to mimic the flow condition of a vascular branch. A step flow chamber could study the stimuli of disturbed WSS on EC. Chiu et al. investigated the disturbed flow sustained by activated EC sterol regulatory element binding protein 1 (SREBP1) [31]. The integrins in the SREBP1 mechanoactivation play an important role in the modulation of EC lipid metabolism. Bao et al. found a temporal gradient in the shear effect of the NO expression process in EC [32]. Continually using a step flow chamber, Bao et al. also reported that the temporal gradient in shear induces ERK1/ERK2, c-fos, and Cx43 in the EC signaling pathway [33], and temporal gradient in shear effect EC proliferation [34]. Studies found the lipid bilayer in the EC membrane could sense the WSS and change membrane fluidity [35–37].

By using an EC-SMC coculture model together with a step flow chamber, Chen et al. found the disturbed flow could induce EC and SMC expressions of adhesion molecules and chemokines, which contributed to the increased white blood cell adhesion and transmigration [38].

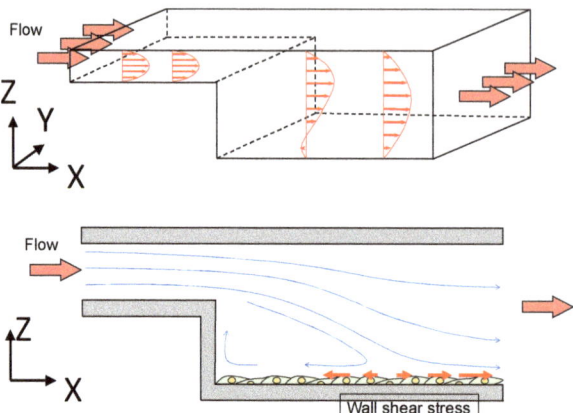

Figure 4. Flow domain in a step flow chamber. The red arrows in the flow domain show the velocity distribution, and the blue arrows show the flow patterns in the flow domain.

3.1.4. Cone Plate Flow Chamber

One form of flow chamber that provides hydrodynamic loading to the surface of cultured cells is the cone-plate type flow chamber, as shown in Figure 5. As a result of Pubmed survey with the keyword terms 'endothelial cell' and 'cone plate' and 'wall shear stress', 20 articles were found. Of these, 17 were flow exposure experiments using ECs.

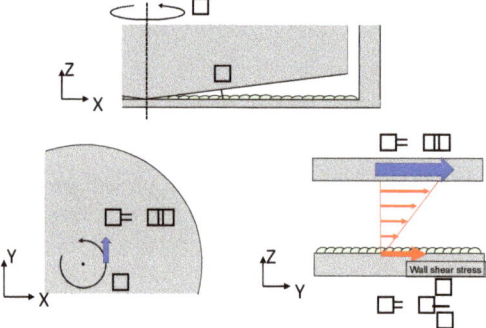

Figure 5. Flow domain in a cone plate flow chamber. The red arrows in the flow domain show the velocity distribution, and the blue arrows show the direction of the cone.

The cone plate flow chamber system was started in 1981 by Dewey, C.F. et al. [39]. It was developed based on a commercially available cone plate viscometer. In the cone plate type flow chamber, the cone is rotated to generate Couette flow in the small gap between the cone and the plate. Fluid shear stress occurs in a moving viscous fluid as a tractive force against a solid body. Newton's law of viscosity states that the shear stress τ is proportional to the strain rate: $\tau = \mu \frac{du}{dy}$, where u is the fluid velocity and y is the distance from the wall. On the rotational device, the equation can be approximated using angular velocity ω and cone angle α as $\tau = \mu \frac{\omega}{\alpha}$, under the assumption that the cone angle is sufficiently small [39]. While blood is a suspension and exhibits non-Newtonian properties, the culture medium is considered as a Newtonian fluid and is assumed to follow the above equations.

Then, the cell layer on the bottom plate was assumed to be exposed by uniform shear. For EC exposure, wall shear up to 120 dyne/cm^2 has been applied. By fixing the rotation speed and direction of the cone, a constant laminar flow is given [40–46].

One advantage of the cone plate type flow chamber is that transient flow can be applied using the same device as constant flow. Bongrazio et al. performed turbulent flow exposure (average WSS of 0.6 Pa (6 dyn/cm^2)) [43]. Franzoni et al. also performed transient flow exposure based on the sine waveform which ranges from 0.5 to 2.5 Pa (from 5 to 25 dyn/cm^2) of WSS [47]. O'Keeffe et al. performed transient flow exposure based on the waveform obtained by CFD on the right coronary artery [48]. Maroski et al. and Parker et al. also applied a transient waveform [44,49] based on arterial flow profiles: a so-called 'atheroprone' profile ranges from −8.9 to 3.7 dyn/cm^2 (−0.15 dyn/cm^2 mean) and 'atheroprotective' profile with the range from 13.3 to 43.7 dyn/cm^2 (20 dyn/cm^2 mean) introduced by Dai, G. et al. [50]. The same as these two articles, Franzoni et al. [51] applied waveform-WSS which involved the inflammatory phenotype of ECs provided by Feaver et al. [52].

Note that the nonlinearity of flow appears because of the secondary flow at the edge of the cone and can alter the shear stress on the plate surface, especially at high shear rate conditions with a larger cone angle. A numerical experiment by Shankaran et al. showed the increase of WSS magnitude with radial position away from the center according to the increase of Re number and cone angle [53]. In their experiment, 0.5° cone did not deviate from the primary flow value. However, 2° cones with a shear rate of 1000 s^{-1} caused the WSS to be ~5.1-fold larger than the primary flow value.

3.1.5. Stretch Chambers

ECs constantly undergo cyclic stretching due to blood pressure applied to the vessel wall. The purpose of using the stretch chamber shown in Figure 6 is to examine ECs loaded by the mechanical role of the wall in addition to being induced by flow. In 1990, Vedernikov et al. first demonstrated that female pigs' (8–10 weeks of age) left circumflex coronary arteries were directly stretched with a strain gage to examine the tension as a contractile [54]. They examined the effect of the existence of the ECs under stretch on the tension using the ring of the coronary arteries. However, in 1997, Rosaries et al. studied the response of EC collected from calf thoracic aorta by seeding them in a stretch chamber [55]. In this method, 4–25% stretching at 0.5 Hz or 1 Hz is applied to examine EC deformation [56,57], NO generation [58], and Ca^{2+} concentration. In 2008, Katanosaka et al. observed tyrosine phosphorylation and actin dynamics using fibronectin dots in a chamber with a cyclic and uniaxial strain [59]. Hashimoto-Komatsu et al. used the Boyden chamber method to investigate the function of Angiotensin II and its relationship to the function of microtubules in ECs involved in cell-cell adhesion [60].

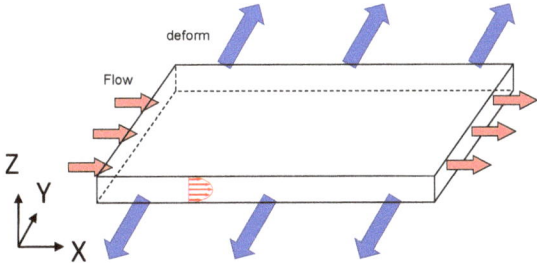

Figure 6. Flow domain in a stretch chamber. The red arrows show the velocity distribution in the flow domain, and the blue arrows show the direction of the wall of chamber.

3.2. Summary of Monolayer Chambers

Monolayer chambers can reveal the flow influence on the ECs with other substances. Using a parallel chamber, WSS induced by one-direction flow was applied and the WSS stimulated ECs are the cascade and mechanism of EC signals. T-chambers have WSSG, step flow chambers give vortex, cone chambers give different directions (time dependent on flow), and stretch chambers show the vessel wall deformation. These chambers relate to the representative of arterial geometry such as bifurcation using T-chamber, which is the frequent site for a cerebral aneurysm. The ECs in the chamber have a relation to the diseases. However, the ECs in the monolayer chamber do not have any response to/from/with other cells such as SMC or other organs. The monolayer chamber is missing the physiology of the wall including SMC and the extracellular matrix.

3.3. Application of WSS to EC-SMC Coculture

3.3.1. Interaction between ECs and SMCs

As described in the previous chapter, in vitro experiments with various types of flow chambers have revealed the effect of WSS on the morphology and function of vascular ECs. Diverse functions of ECs are responsive to WSS stimuli, such as vasoconstriction/dilation (vascular tone), wall permeability, leukocyte adhesion, thrombogenesis, and vascular wall remodeling. Endothelial responses to WSS are now widely recognized to play a critical role in vascular physiology and pathology such as arteriosclerosis and aneurysms.

In addition to the ECs lining the lumen of arterial walls, SMCs in the middle layer of blood vessel walls, 'tunica media', are also important in vascular physiology and pathology. In normal healthy arterial walls, SMCs reside in a quiescent and contractile state, referred to as the 'contractile type', and play a major role in vascular tone. However, in the medial degeneration site widely recognized as a pathologic feature of aortic diseases such as arteriosclerosis and aneurysms, SMCs dedifferentiate into a proliferative and synthetic state (synthetic phenotype) and exhibit low contractility and high proliferative and migration ability, with secretions of physiological activity factors and extracellular matrix (ECM) proteins. Since SMCs are not directly exposed to blood flow, the response of ECs to the WSS condition causes changes in the phenotype and function of SMCs through cell–cell interaction (cross-talk) [61]. Investigations focusing on cell–cell interactions between ECs and SMCs under WSS conditions have been performed to understand the role of EC response to WSS in vascular physiology and pathology in more detail.

3.3.2. EC and SMC Coculture Models for WSS Experiments

Coculture assays have been demonstrated to study the cellular interactions between ECs and SMCs in vitro. The coculture systems for investigating the effects of WSS on the EC-SMC interactions have also been developed. In this review, focusing on the interaction between ECs and SMCs in vascular physiology, we searched for the keyword terms 'coculture' or 'cocultured' in addition to 'endothelial', 'shear stress', and 'smooth muscle' in Pubmed (Figure 1). From the 63 identified articles from 1995 to July 2021, excluding non-English journals and reviews, we examined 45 studies that performed WSS experiments for coculture of ECs and SMCs.

The coculture methods used in the examined studies can be classified as follows (Figure 7).
1. Double-side type (flat and tubular types; Figure 7A,B): ECs and SMCs are cocultured on the opposite sides of a porous membrane (flat type, 20 cases (44.4%), tubular type, 6 cases (13.3%), total of 26 cases (57.8%));
2. Single-side type (Figure 7C): ECs and SMCs are cocultured on the same surface (3 cases (6.7%));
3. Direct culture type (Figure 7D): ECs are directly cultured above a pre-cultured SMC layer (7 cases (15.6%));
4. 3D culture type (Figure 7E): ECs are cultured on type I collagen gel or the other types of ECM gel containing embedded SMCs (8 cases (17.8%));
5. Another type: ECs are cultured on the inside of a culture insert, and SMCs are cocultured on the bottom of the well in which the culture insert was placed (1 case (2.2%)).

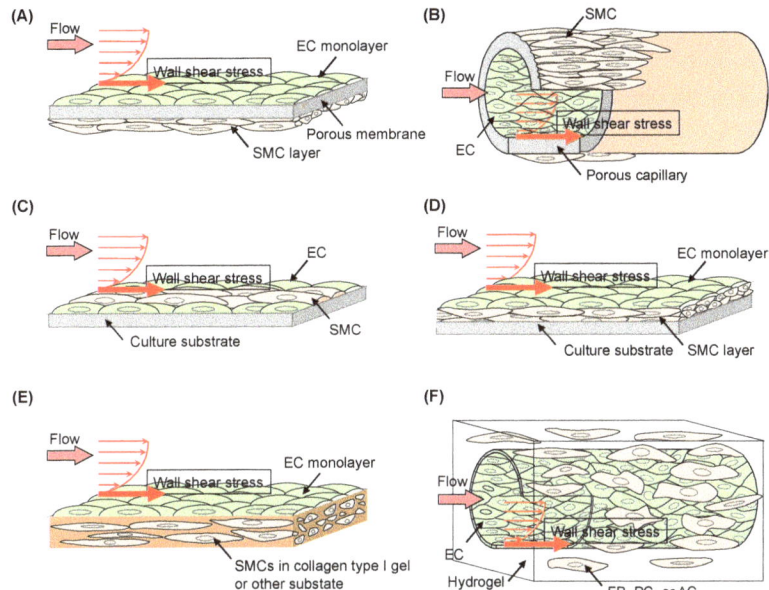

Figure 7. Schematic illustration of types of EC and SMC coculture models for WSS experiments found in the literature. (**A**) Coculture of ECs and SMCs on the opposite sides of a flat porous membrane. (**B**) Coculture of ECs and SMCs on the opposite sides of a porous tube. (**C**) Coculture of ECs and SMCs on the same surface (mixed or arranged). (**D**) Coculture of ECs directly above pre-cultured SMCs. (**E**) Coculture of ECs on the surface of type I collagen gel or the other types of ECM gel containing SMCs. (**F**) Coculture of ECs forming a capillary-like structure in a hydrogel with surrounding fibroblasts, pericytes, and/or astrocytes. In either of the models, except for C, WSS exerted only on ECs. FB, fibroblast; PC, pericyte; AC, astrocyte.

The 'double-side type' is the most common method for WSS experiments for coculture of ECs and SMCs among the studies examined, and many of these studies used originally-developed parallel plate flow chambers, which can incorporate commercially available cell culture inserts and apply WSS to ECs on the bottom of the culture inserts [62–72]. In some cases of the double-sided type, tubes or capillaries made of porous materials were also used [73–77]. There seems to be no essential difference in the EC-SMC cross-talk and WSS acting on ECs compared to the method using culture inserts, but in the tubular type, ECs are surrounded by SMCs, which is a similar environment to that of blood vessels. The 'direct culture type', '3D culture type', and 'single-side type' are generally formed in the flat plane shape of cocultures in the culture dish, and the parallel plate flow chambers are commonly adopted in these types of coculture methods. Experiments using a cone (disk) plate flow chamber [69,78] and perfusing culture medium into an original tubular substrate having a honeycomb cross-sectional shape [79] have also been reported as methods for application of WSS to EC-SMC cocultures.

3.3.3. Responses of ECs and SMCs to WSS under Coculture Conditions

Many studies used EC-SMC coculture experiments to understand the mechanism of arteriosclerosis. Therefore, these coculture studies also investigated the effects of WSS environments lower than the physiological levels of arteries, and have shown that WSS applied to ECs causes functional changes in SMCs not exposed to WSS. In addition, it has also been reported that the effects of WSS on ECs cocultured with SMCs are different from those on monocultured ECs, and factors that act as signal transmitters in the intercellular cross-talk in coculture environments. These studies have then revealed that the conditions of static culture and WSS lower than ~0.5 Pa have atheroprone effects on the behavior and functions of ECs and SMCs related to the formation and development of arteriosclerosis, and physiological levels of WSS induce atheroprotective responses in cells. Increased migration and proliferation of SMCs have been observed in arteriosclerotic lesions, and it has been reported that the application of 1 Pa of WSS on ECs suppresses the growth of SMCs not exposed to WSS compared to static conditions [66,69]. The WSS applied to ECs also suppresses the migration of SMCs [75,80,81], and it was revealed that nitric oxide (NO) produced by ECs in response to WSS plays a critical role in the suppression of SMC migration [80]. An increase in leukocyte adhesion and invasion of the blood vessel wall has also been observed in arteriosclerosis pathology, and the effects of coculture and WSS environment have been reported on this phenomenon. Coculturing with SMCs increases the expression of adhesion proteins such as intercellular adhesion molecule-1 (ICAM-1), vascular cell adhesion molecule-1 (VCAM-1), and E-selection on the surface of ECs, as well as the expression and secretion of chemokines and cytokines such as monocyte chemotactic protein-1 (MCP-1), growth-related oncogene-α (GRO-α), and interleukin-8 (IL-8) that promote leukocyte migration, and these expressions and secretions induced by the coculture are suppressed by a physiological level of 1.5 Pa WSS condition [62,63,69,82]. It has also been shown that WSS applied to ECs suppressed leukocyte invasion stimulated by an SMC coculture environment [83]. The WSS conditions of ECs also affect the phenotype of cocultured SMCs. The expression of contractile markers, smooth muscle α-actin (SMα-actin), SM-myosin heavy chain (SM-MHC), and calponin in SMCs was increased by 1.2 Pa of WSS applied to cocultured ECs [69]. Hastings et al. exposed ECs to atheroprone oscillatory WSS and pulsatile WSS, simulating physiological conditions, and evaluated the phenotypes of ECs and SMCs. As a result, oscillatory WSS decreased the expression level of the physiological state of quiescent phenotypic ECs' markers such as endothelial NO synthase (eNOS) and Tie2, and similarly decreased the contractile markers, SMα-actin, SM-MHC, and myocardin, of SMCs compared to the physiological conditions [63]. In cross-talk between ECs and SMCs under WSS conditions, the roles of prostacyclin (PGI$_2$) [69] secreted by ECs, platelet-derived growth factor (PDGF)-BB, and transforming growth factor-β1 (TGF-β1) [67] have also been shown.

3.3.4. Advanced Applications of EC and SMC Coculture Models

As stated above, coculture studies have revealed that EC-SMC interactions have crucial roles in vascular physiology and pathology under WSS conditions, especially in the pathogenesis of arteriosclerosis. Extending the work to include constructing tissue-engineered blood vessels and developing screening assay platforms for drug discovery as an alternative to animal testing, the improvement of EC-SMC coculture systems and the conduct of flow exposure experiments lasting longer than the typical conventional 24 h have also been performed. Cultured SMCs generally show the synthetic type while the normal state of SMCs in the arteries shows a contractile phenotype, and it has been pointed out that the effects of synthetic SMCs on cocultured ECs are different from those of contractile cells [84,85]. In addition to the phenotype of cocultured SMCs, they have focused on the effect of mechanical properties of culture inserts widely used in 'double-side type' coculture and the suppression of direct contacts between ECs and SMCs by the inserts [84,85]. They have conducted the 'direct culture type' coculture of ECs and SMCs. Some studies have demonstrated WSS experiments with EC-SMC coculture, which was constructed with SMCs differentiated into a contractile type by culturing with a serum-free culture medium in advance [86,87]. Longer-term WSS experiments using bioreactors have also been performed. For the purpose of investigating molecular mechanisms related to angiogenesis, vascular wall remodeling, and vascular disease, Janke et al. developed a bioreactor as an 'artificial artery', in which ECs were cultured on the inside of porous capillaries and SMCs on the outside, and conducted the application of WSS for 5 days [77]. To evaluate blood-brain barrier (BBB) characteristics in the cerebrovascular network, Cucullo et al. made an artificial vascular system mimicking cerebral capillary and venous segments by connecting a vein model in which ECs and SMCs were cocultured and a capillary model composed of endothelial and astrocyte coculture [76]. They applied WSS of 0.3 Pa to the vein model as well as WSS of 1.6 Pa to the capillary model for 3 weeks, and examined the relationship between the formation of cell–cell adhesion as assessed by transendothelial electronic resistance (TEER) and endothelial permeability in the cerebrovascular network and the experimental period. Since in vitro cell culture experiments have strong advantages in studying cell–cell cross-talk at the molecular level, coculture WSS experimental systems not only with ECs with SMCs but also with other types of cells such as pericytes, astrocytes, and valve interstitial cells will be more important for a detailed understanding of physiology and pathology, as well as for regenerative medicine and drug discovery.

Recently, it has been pointed out that high WSS has an effect on pathology such as aneurysms and arterial dissection [88,89], but physiological levels of up to 4 Pa WSS have been examined in coculture studies of ECs and SMCs because these WSS experiments have also been performed mainly for atherogenesis, and one study showed the effects of WSS up to about 10 Pa on EC-SMC coculture [90]. Since cross-talk between ECs and SMCs is considered to play an important role in pathogenesis associated with higher WSS conditions as well, high WSS experiments for coculture will also be required for elucidating these pathologies.

3.4. Cellular Experiments of ECs with Microfluidic Devices

3.4.1. Microfluidic Cellular Experiments

Due to the advancement of microfabrication technology, cellular experiments using microfluidic devices, which culture cells inside microchannels as shown in Figure 7, have been performed to observe cellular dynamics since the 2000s. We searched for keyword terms 'microfluidic' or 'microfluidic device' with 'endothelial cell' and 'shear stress' in Pubmed, resulting in the identification of 278 or 94 articles, respectively (Figure 1). In addition, in order to include state-of-the-art technology of microfluidic cellular experiments into this review, search for keyword terms 'microfluidic device' in addition to 'vascular network' identified 129 articles. Here, referring to the search results as appropriate, we provide an overview of cellular experiments using microfluidic devices.

In fabricating microfluidic devices, a convex channel pattern is first created on a silicon wafer using SU-8 photolithography or on a plastic polymer such as ABS resin by milling machine. The channel pattern is then transferred to polydimethylsiloxane (PDMS) [91] or a hydrogel [92,93] by soft lithography. Inlets and outlets are punched in the PDMS or hydrogel mold to access each channel, and a layer of the same material or a cover glass is bonded to the channel-patterned surface to form the channels. The devices consist of channels for cell culture, for hydrogels to mimic an ECM, and for loading various conditions on the cells (media and gas channels). Hereby, the control of environmental factors (mechanical and chemical stimuli) on cells is achieved, and cell dynamics under conditions that reproduce in vivo microenvironments can be evaluated. The usage of a microfluidic device for cellular experiments can save rare cells, such as patient-derived cells and stem cells, and expensive experimental reagents. Cell adhesion area and the fluid volume in a microfluidic device are smaller than those in cell culture dishes and wells, and small amounts of cells and reagents are sufficient for performing experiments. The coculture of multiple types of cells and their three-dimensional culture are feasible in the device. In addition, since the device is fabricated using transparent materials, it enables high-resolution and real-time observation of cell dynamics.

3.4.2. Microfluidic Devices for Shear Stress Applications

Many experiments have been performed with ECs in confluency in a microfluidic device to investigate their dynamics under flow exposure. Various shear stresses were applied to the ECs cultured in the channel by flowing the cell culture medium using pumps such as syringe pumps [94], roller pumps [95], or by applying hydraulic head pressure between the inlet and outlet of the channel [96]. To generate various levels of steady shear stress at the same time, other than directly controlling the shear stress by adjusting the flow rate of the cell culture medium, there are ways to change the geometry of the flow channels in the device [97,98]. The width or height of a single channel can be continuously [30,99–103] or stepwisely [104–107] enlarged or reduced. Otherwise, the channel size is changed by branching a channel into multiple channels [108–111] or by adjusting the circular diameter [112]. Even if the flow rate of the cell culture medium flowing in the entire channel is constant, the flow velocity varies with the cross-sectional area, and a large shear stress can be generated at a location with a narrow cross-sectional area. The shear stress can also be controlled by controlling the flow rate of the cell culture medium with the setting flow resistance of branching channels [113–116]. The effect of spatiotemporal gradients of shear stress on cells can also be studied by varying the shear stress within the flow channel [117]. In order to generate unsteady shear stress, it is common to combine a microfluidic device with the supply of cell culture medium by a pump [118,119]. Experiments have been conducted to load shear stress on the ECs by generating periodic flow, such as beating [120,121]. Pumpless microfluidic devices have also been proposed to periodically generate bidirectional flow [122]. These microfluidic methods enable us to simulate both physiologically healthy situations of shear stress and situations at the site of onset of arteriosclerosis and other diseases.

In experiments using microfluidic devices, it is possible to observe the dynamics of ECs under multiple stimuli in addition to flow exposure. By fabricating the device with elastic materials, the device itself can be deformed by stretching and shrinking to exert mechanical stimuli over the ECs [106]. Changes in the components such as adding glucose [100], vascular endothelial growth factor (VEGF) [123], tumor necrosis factor (TNF-α) [124,125], adenosine triphosphate (ATP) [117], or EDTA [126] in the cell culture medium yield to exert chemical stimuli over the cells. Experiments combined with micropatterning can also be conducted by modifying the surface properties of the substrate like hardness and hydrophilicity by coating with hydrogel or other materials [127,128] or by plasma treatment [8]. Furthermore, the effects of oxygen tension on ECs can be studied by manipulating the dissolved gas components in the cell culture medium [125,129,130]. As for a device structure, it has been proposed that a channel for culturing ECs and another separated channel be located sandwiching a membrane with nano-sized pores [131–133] or a hydrogel. This structure allows a coculture with different types of cells, such as pericytes or astrocytes that support blood vessels [134], cancer cells that promote angiogenesis [135], and epithelial cells that exist on the other side of the monolayer of ECs [136]. Consequently, it is possible to observe the dynamics of ECs under the interaction with other types of cells. Furthermore, observation of effects of blood cells (platelets, leukocytes, and red blood cells) [112,137] or parasitic protozoa and bacteria such as toxoplasma [138] are injected into the channel where ECs are cultured, considering it as a blood vessel.

3.4.3. Experiments with EC monolayer

With monolayers of ECs formed in the microfluidic device, morphological changes of the cells in response to environmental factors, including shear stress, have been observed. It has been shown that cells orientate in the flow direction by flow exposure, but orientate in a direction orthogonal to the flow direction when the shear stress is very high. The presence or absence of flow affects the differentiation and phenotype of ECs and alters their properties. The expression of intercellular adhesion molecules such as ICAM-1 and cell-substrate adhesion molecules, as well as cytoskeletal changes by actin filament, play important roles in the morphological changes of cells, and they vary according to the magnitude and period of the flow exposure. Although ECs forming a monolayer are distributed like paving stones, individual cells do not lose their motility, and random collective migration is observed [8,95,130]. By acquiring time-series microscopic images of this collective migration and analyzing them by particle image velocimetry, the migration velocity of the cells and the strain (traction force) generated in the monolayer can be obtained to clarify the dynamic characteristics [105]. The wound healing assay, in which a monolayer of cells cultured on a dish is scratched and then the cells recovering the damaged area are observed, has frequently been used to investigate collective migration. In an experiment using a microfluidic device, a similar wound healing assay that chemically damages a monolayer of ECs by flowing trypsin solution under fluid control is proposed [139]. Moreover, collective migration of ECs is a necessary process in sprouting at the beginning of angiogenesis. By placing a hydrogel that mimics an ECM in a microfluidic device and forming a monolayer of ECs at the interface, the effects of interstitial flow in the ECM and shear stress on angiogenesis can be evaluated [140]. Furthermore, it is possible to quantitatively evaluate the permeability of a monolayer of ECs. The permeability can be measured by quantifying the diffusion of fluorescence-labeled dextran, which is added to the cell culture medium, goes through the monolayer formed on a hydrogel, and diffuses in the gel [114,133,138,141]. Alternatively, the same as the TEER method using a transwell, the electrical resistance between the upper and lower channels separated by a monolayer of ECs formed on a membrane with nano-sized pores can provide permeability [118,132,135].

3.4.4. Experiments with Microvascular Networks and Their Perspective

Research using microfluidic devices has also established a method to construct a three-dimensional microvascular network like capillaries [142]. By culturing ECs densely mixed in a hydrogel such as fibrin gel, vasculogenesis occurs and a microvascular network is formed [143]. To stabilize the microvascular network, cells such as pericytes and fibroblasts should be mixed with ECs at an appropriate density. Additionally, it has been proposed to mix astrocytes found in the central nervous system to reproduce the BBB of blood vessels in the human brain [144]. The permeability of the microvascular network formed in microfluidic devices has been evaluated, as well as the changes in response to cell composition, chemical stimuli, and shear stress. Additionally, microvascular networks are utilized to observe the intravascular invasion and extravasation of cancer cells in a cancer microenvironment [145,146]. Furthermore, nanoparticles are injected into microvascular networks for drug delivery purposes [147].

Cell experiments using microfluidic devices have evolved into organ-on-chips that reproduce the functions of organs and in vivo tissues by integrating various channel structures and cells. One notable example is lung-on-a-chip, which mimicked a microenvironment in an alveolus by coculture of endothelial and epithelial monolayers with sandwiching a porous membrane set in a microchannel [148]. The chip yielded observation of various cell dynamics in the presence of blood and air flows and stretching. Additionally, devices have been proposed to mimic the function of the entire human body by connecting the functions of multiple organs. The use of such microfluidic devices makes it possible to perform cellular experiments under conditions that reproduce the microenvironment in vivo, and it is expected to be utilized for drug screening for various diseases. The importance of microfluidic devices for cellular experiments will increase in the future as a research tool that contributes to the 3Rs (Replacement, Reduction, and Refinement) in animal experiments for medical research, including drug discovery.

4. Conclusions

This review revealed that the flow-inducing functions are connected to EC_S and SMCs in blood vessels as shown in Table 1. The studies using flow chambers are elucidating from the response of ECs themselves on surfaces in direct contact with flow to the response of cells in the wall to signals from ECs. These chamber studies have shown that the effects of flow are transmitted in a cascade from ECs to cells in the wall. The relationship with organs other than arteries is also spotted using the flow chambers.

Table 1. Chambers with flow character and EC responses.

	Signal and Response by Flow	Name of Chamber	Flow Character	ECs on ...
Monolayer	Inside ECs/With ECs	Parallel (Figure 2)	One direction	Rigid wall
		T-chamber (Figure 3)	WSSG	Rigid wall
		Step (Figure 4)	Vortex	Rigid wall
		Cone plate (Figure 5)	Couette flow	Rigid wall
		Stretch (Figure 6)	One direction	Deformed wall
Coculture with SMC	Cross-talk with SMC	Double-side flat (Figure 7A)	One direction	porous membrane with SMC in the oppposite side
		Double-side 3D (Figure 7B)	3D tubular	
		Single-side (Figure 7C)	One direction	Rigid wall with SMC
		Direct culture (Figure 7D)	One direction	Directly on SMC
		3D culture (Figure 7E)	One direction	Collagen type 1 gel with SMC
Microfluidic cell culture	Cross-talk with other cells via ECM	Another type (Figure 7F)	3D tubular	In hydrogel

Author Contributions: Conceptualization, M.O., N.S., K.F., H.A. and Z.W.; methodology, M.O., N.S., K.F., H.A., Y.K. and Z.W.; software, M.O., N.S., K.F., H.A., Y.K. and Z.W.; validation, M.O., N.S., K.F., H.A. and Y.K. and Z.W.; formal analysis, M.O., N.S., K.F., H.A., Y.K. and Z.W.; investigation, M.O., N.S., K.F., H.A., Y.K. and Z.W.; resources, M.O.; data curation, M.O., N.S., K.F., H.A., Y.K. and Z.W.; writing—original draft preparation, M.O., N.S., K.F., H.A., Y.K. and Z.W.; writing—review and editing, M.O., N.S., K.F., H.A., Y.K. and Z.W.; visualization, M.O., N.S., K.F., H.A., Y.K. and Z.W.; supervision, M.O., N.S., K.F. and H.A.; project administration, M.O.; funding acquisition, M.O., N.S., K.F. and H.A. All authors have read and agreed to the published version of the manuscript.

Funding: This research was funded by Grant-in-aid, Kakenhi [B] 20H04557, 19H04435 and [C] 22K12795. Tokyo Metropolitan Government Advanced Research Grant (R2-2).

Institutional Review Board Statement: Not applicable.

Informed Consent Statement: Not applicable.

Data Availability Statement: Not applicable.

Conflicts of Interest: The authors declare no conflict of interest.

Abbreviations

BBB, Blood-brain barrier; CFD, Computational fluid dynamics; EC, Endothelial cell; ECM, Extracellular matrix; SMC, Smooth muscle cell; TEER, Transendothelial electronic resistance; WSS, Wall shear stress.

References

1. Flaherty, J.T.; Pierce, J.E.; Ferrans, V.J.; Patel, D.J.; Tucker, W.K.; Fry, D.L. Endothelial nuclear patterns in the canine arterial tree with particular reference to hemodynamic events. *Circ. Res.* **1972**, *30*, 23–33. [CrossRef] [PubMed]
2. Smith, M.L.; Smith, M.J.; Lawrence, M.B.; Ley, K. Viscosity-independent velocity of neutrophils rolling on p-selectin in vitro or in vivo. *Microcirculation* **2002**, *9*, 523–536. [CrossRef] [PubMed]
3. Choi, H.W.; Ferrara, K.W.; Barakat, A.I. Modulation of ATP/ADP concentration at the endothelial surface by shear stress: Effect of flow recirculation. *Ann. Biomed. Eng.* **2007**, *35*, 505–516. [CrossRef] [PubMed]
4. Avari, H.; Savory, E.; Rogers, K.A. An In Vitro Hemodynamic Flow System to Study the Effects of Quantified Shear Stresses on Endothelial Cells. *Cardiovasc. Eng. Technol.* **2016**, *7*, 44–57. [CrossRef] [PubMed]
5. Viegas, K.D.; Dol, S.S.; Salek, M.M.; Shepherd, R.D.; Martinuzzi, R.M.; Rinker, K.D. Methicillin resistant Staphylococcus aureus adhesion to human umbilical vein endothelial cells demonstrates wall shear stress dependent behaviour. *Biomed. Eng. Online* **2011**, *10*, 20. [CrossRef]
6. Taba, Y.; Sasaguri, T.; Miyagi, M.; Abumiya, T.; Miwa, Y.; Ikeda, T.; Mitsumata, M. Fluid shear stress induces lipocalin-type prostaglandin D(2) synthase expression in vascular endothelial cells. *Circ. Res.* **2000**, *86*, 967–973. [CrossRef]

7. Frame, M.D.; Sarelius, I.H. Flow-induced cytoskeletal changes in endothelial cells growing on curved surfaces. *Microcirculation* **2000**, *7*, 419–427. [CrossRef]
8. Lafaurie-Janvore, J.; Antoine, E.E.; Perkins, S.J.; Babataheri, A.; Barakat, A.I. A simple microfluidic device to study cell-scale endothelial mechanotransduction. *Biomed. Microdevices* **2016**, *18*, 63. [CrossRef]
9. Wang, Y.X.; Xiang, C.; Liu, B.; Zhu, Y.; Luan, Y.; Liu, S.T.; Qin, K.R. A multi-component parallel-plate flow chamber system for studying the effect of exercise-induced wall shear stress on endothelial cells. *Biomed. Eng. Online* **2016**, *15*, 154. [CrossRef]
10. Levesque, M.J.; Nerem, R.M. The elongation and orientation of cultured endothelial cells in response to shear stress. *J. Biomech. Eng.* **1985**, *107*, 341–347. [CrossRef]
11. Prasad, A.R.; Logan, S.A.; Nerem, R.M.; Schwartz, C.J.; Sprague, E.A. Flow-related responses of intracellular inositol phosphate levels in cultured aortic endothelial cells. *Circ. Res.* **1993**, *72*, 827–836. [CrossRef] [PubMed]
12. Munn, L.L.; Melder, R.J.; Jain, R.K. Analysis of cell flux in the parallel plate flow chamber: Implications for cell capture studies. *Biophys. J.* **1994**, *67*, 889–895. [CrossRef]
13. Anzai, H.; Watanabe, T.; Han, X.; Putra, N.K.; Wang, Z.; Kobayashi, H.; Ohta, M. Endothelial cell distributions and migration under conditions of flow shear stress around a stent wire. *Technol. Health Care* **2020**, *28*, 345–354. [CrossRef] [PubMed]
14. Wang, Z.; Putra, N.K.; Anzai, H.; Ohta, M. Endothelial Cell Distribution after Flow Exposure with Two Stent Struts Placed in Different Angles. *Front. Physiol.* **2022**, *12*, 733547. [CrossRef] [PubMed]
15. Cadroy, Y.; Diquelou, A.; Dupouy, D.; Bossavy, J.P.; Sakariassen, K.S.; Sie, P.; Boneu, B. The thrombomodulin/protein C/protein S anticoagulant pathway modulates the thrombogenic properties of the normal resting and stimulated endothelium. *Arter. Thromb. Vasc. Biol.* **1997**, *17*, 520–527. [CrossRef]
16. Gopalan, P.K.; Smith, C.W.; Lu, H.; Berg, E.L.; McIntire, L.V.; Simon, S.I. Neutrophil CD18-dependent arrest on intercellular adhesion molecule 1 (ICAM-1) in shear flow can be activated through L-selectin. *J. Immunol.* **1997**, *158*, 367–375.
17. Gosgnach, W.; Challah, M.; Coulet, F.; Michel, J.B.; Battle, T. Shear stress induces angiotensin converting enzyme expression in cultured smooth muscle cells: Possible involvement of bFGF. *Cardiovasc. Res.* **2000**, *45*, 486–492. [CrossRef]
18. Han, Z.; Chen, Y.R.; Jones, C.I., 3rd; Meenakshisundaram, G.; Zweier, J.L.; Alevriadou, B.R. Shear-induced reactive nitrogen species inhibit mitochondrial respiratory complex activities in cultured vascular endothelial cells. *Am. J. Physiol. Cell Physiol.* **2007**, *292*, C1103–C1112. [CrossRef]
19. Popa, M.; Tahir, S.; Elrod, J.; Kim, S.H.; Leuschner, F.; Kessler, T.; Bugert, P.; Pohl, U.; Wagner, A.H.; Hecker, M. Role of CD40 and ADAMTS13 in von Willebrand factor-mediated endothelial cell-platelet-monocyte interaction. *Proc. Natl. Acad. Sci. USA* **2018**, *115*, E5556–E5565. [CrossRef]
20. Lawrence, M.B.; McIntire, L.V.; Eskin, S.G. Effect of flow on polymorphonuclear leukocyte/endothelial cell adhesion. *Blood* **1987**, *70*, 1284–1290. [CrossRef]
21. Ozdemir, T.; Zhang, P.; Fu, C.; Dong, C. Fibrin serves as a divalent ligand that regulates neutrophil-mediated melanoma cells adhesion to endothelium under shear conditions. *Am. J. Physiol. Cell Physiol.* **2012**, *302*, C1189–C1201. [CrossRef] [PubMed]
22. Kona, S.; Dong, J.F.; Liu, Y.; Tan, J.; Nguyen, K.T. Biodegradable nanoparticles mimicking platelet binding as a targeted and controlled drug delivery system. *Int. J. Pharm.* **2012**, *423*, 516–524. [CrossRef] [PubMed]
23. Rychak, J.J.; Lindner, J.R.; Ley, K.; Klibanov, A.L. Deformable gas-filled microbubbles targeted to P-selectin. *J. Control. Release* **2006**, *114*, 288–299. [CrossRef] [PubMed]
24. Takahashi, M.; Berk, B.C. Mitogen-activated protein kinase (ERK1/2) activation by shear stress and adhesion in endothelial cells. Essential role for a herbimycin-sensitive kinase. *J. Clin. Investig.* **1996**, *98*, 2623–2631. [CrossRef]
25. Li, S.; Kim, M.; Hu, Y.L.; Jalali, S.; Schlaepfer, D.D.; Hunter, T.; Chien, S.; Shyy, J.Y. Fluid shear stress activation of focal adhesion kinase. Linking to mitogen-activated protein kinases. *J. Biol. Chem.* **1997**, *272*, 30455–30462. [CrossRef]
26. Geiger, R.V.; Berk, B.C.; Alexander, R.W.; Nerem, R.M. Flow-induced calcium transients in single endothelial cells: Spatial and temporal analysis. *Am. J. Physiol.* **1992**, *262*, C1411–C1417. [CrossRef]
27. Chachisvilis, M.; Zhang, Y.L.; Frangos, J.A. G protein-coupled receptors sense fluid shear stress in endothelial cells. *Proc. Natl. Acad. Sci. USA* **2006**, *103*, 15463–15468. [CrossRef]
28. Plata, A.M.; Sherwin, S.J.; Krams, R. Endothelial nitric oxide production and transport in flow chambers: The importance of convection. *Ann. Biomed. Eng.* **2010**, *38*, 2805–2816. [CrossRef]
29. Surya, V.N.; Michalaki, E.; Fuller, G.G.; Dunn, A.R. Lymphatic endothelial cell calcium pulses are sensitive to spatial gradients in wall shear stress. *Mol. Biol. Cell* **2019**, *30*, 923–931. [CrossRef]
30. Michalaki, E.; Surya, V.N.; Fuller, G.G.; Dunn, A.R. Perpendicular alignment of lymphatic endothelial cells in response to spatial gradients in wall shear stress. *Commun. Biol.* **2020**, *3*, 57. [CrossRef]
31. Chiu, J.J.; Chen, C.N.; Lee, P.L.; Yang, C.T.; Chuang, H.S.; Chien, S.; Usami, S. Analysis of the effect of disturbed flow on monocytic adhesion to endothelial cells. *J. Biomech.* **2003**, *36*, 1883–1895. [CrossRef]
32. Bao, X.; Lu, C.; Frangos, J.A. Temporal gradient in shear but not steady shear stress induces PDGF-A and MCP-1 expression in endothelial cells: Role of NO, NF kappa B, and egr-1. *Arter. Thromb. Vasc. Biol.* **1999**, *19*, 996–1003. [CrossRef] [PubMed]
33. Bao, X.; Clark, C.B.; Frangos, J.A. Temporal gradient in shear-induced signaling pathway: Involvement of MAP kinase, c-fos, and connexin43. *Am. J. Physiol. Heart Circ. Physiol.* **2000**, *278*, H1598–H1605. [CrossRef] [PubMed]
34. Bao, X.; Lu, C.; Frangos, J.A. Mechanism of temporal gradients in shear-induced ERK1/2 activation and proliferation in endothelial cells. *Am. J. Physiol. Heart Circ. Physiol.* **2001**, *281*, H22–H29. [CrossRef] [PubMed]

35. Haidekker, M.A.; L'Heureux, N.; Frangos, J.A. Fluid shear stress increases membrane fluidity in endothelial cells: A study with DCVJ fluorescence. *Am. J. Physiol. Heart Circ. Physiol.* **2000**, *278*, H1401–H1406. [CrossRef]
36. Butler, P.J.; Norwich, G.; Weinbaum, S.; Chien, S. Shear stress induces a time- and position-dependent increase in endothelial cell membrane fluidity. *Am. J. Physiol. Cell Physiol.* **2001**, *280*, C962–C969. [CrossRef]
37. Butler, P.J.; Tsou, T.C.; Li, J.Y.; Usami, S.; Chien, S. Rate sensitivity of shear-induced changes in the lateral diffusion of endothelial cell membrane lipids: A role for membrane perturbation in shear-induced MAPK activation. *FASEB J.* **2002**, *16*, 216–218. [CrossRef]
38. Chien, S. Molecular and mechanical bases of focal lipid accumulation in arterial wall. *Prog. Biophys. Mol. Biol.* **2003**, *83*, 131–151. [CrossRef]
39. Dewey, C.F., Jr.; Bussolari, S.R.; Gimbrone, M.A., Jr.; Davies, P.F. The Dynamic Response of Vascular Endothelial Cells to Fluid Shear Stress. *J. Biomech. Eng.* **1981**, *103*, 177–185. [CrossRef]
40. Alevriadou, B.R.; Moake, J.L.; Turner, N.A.; Ruggeri, Z.M.; Folie, B.J.; Phillips, M.D.; Schreiber, A.B.; Hrinda, M.E.; McIntire, L.V. Real-time analysis of shear-dependent thrombus formation and its blockade by inhibitors of von Willebrand factor binding to platelets. *Blood* **1993**, *81*, 1263–1276. [CrossRef]
41. Galbusera, M.; Zoja, C.; Donadelli, R.; Paris, S.; Morigi, M.; Benigni, A.; Figliuzzi, M.; Remuzzi, G.; Remuzzi, A. Fluid shear stress modulates von Willebrand factor release from human vascular endothelium. *Blood* **1997**, *90*, 1558–1564. [CrossRef] [PubMed]
42. Fallgren, C.; Ljungh, A.; Shenkman, B.; Varon, D.; Savion, N. Venous shear stress enhances platelet mediated staphylococcal adhesion to artificial and damaged biological surfaces. *Biomaterials* **2002**, *23*, 4581–4589. [CrossRef]
43. Bongrazio, M.; Pries, A.R.; Zakrzewicz, A. The endothelium as physiological source of properdin: Role of wall shear stress. *Mol. Immunol.* **2003**, *39*, 669–675. [CrossRef]
44. Maroski, J.; Vorderwülbecke, B.J.; Fiedorowicz, K.; Da Silva-Azevedo, L.; Siegel, G.; Marki, A.; Pries, A.R.; Zakrzewicz, A. Shear stress increases endothelial hyaluronan synthase 2 and hyaluronan synthesis especially in regard to an atheroprotective flow profile. *Exp. Physiol.* **2011**, *96*, 977–986. [CrossRef]
45. Bretón-Romero, R.; González de Orduña, C.; Romero, N.; Sánchez-Gómez, F.J.; de Álvaro, C.; Porras, A.; Rodríguez-Pascual, F.; Laranjinha, J.; Radi, R.; Lamas, S. Critical role of hydrogen peroxide signaling in the sequential activation of p38 MAPK and eNOS in laminar shear stress. *Free Radic. Biol. Med.* **2012**, *52*, 1093–1100. [CrossRef] [PubMed]
46. Kim, B.; Lee, H.; Kawata, K.; Park, J.-Y. Exercise-Mediated Wall Shear Stress Increases Mitochondrial Biogenesis in Vascular Endothelium. *PLoS ONE* **2014**, *9*, e111409. [CrossRef] [PubMed]
47. Franzoni, M.; O'Connor, D.T.; Marcar, L.; Power, D.; Moloney, M.A.; Kavanagh, E.G.; Leask, R.L.; Nolan, J.; Kiely, P.A.; Walsh, M.T. The Presence of a High Peak Feature Within Low-Average Shear Stimuli Induces Quiescence in Venous Endothelial Cells. *Ann. Biomed. Eng.* **2020**, *48*, 582–594. [CrossRef]
48. O'Keeffe, L.M.; Muir, G.; Piterina, A.V.; McGloughlin, T. Vascular cell adhesion molecule-1 expression in endothelial cells exposed to physiological coronary wall shear stresses. *J. Biomech. Eng.* **2009**, *131*, 3148191. [CrossRef]
49. Parker, I.K.; Roberts, L.M.; Hansen, L.; Gleason, R.L., Jr.; Sutliff, R.L.; Platt, M.O. Pro-atherogenic shear stress and HIV proteins synergistically upregulate cathepsin K in endothelial cells. *Ann. Biomed. Eng.* **2014**, *42*, 1185–1194. [CrossRef]
50. Dai, G.; Kaazempur-Mofrad, M.; Kamm, R.; Zhang, Y.; Vaughn, S.; García-Cardeña, G.; A Gimbrone, M. Distinct endothelial phenotypes evoked by arterial waveforms derived from atherosclerosis-prone and atherosclerosis-protected regions of the human vasculature. *Cardiovasc. Pathol.* **2004**, *13*, 26. [CrossRef]
51. Franzoni, M.; Cattaneo, I.; Ene-Iordache, B.; Oldani, A.; Righettini, P.; Remuzzi, A. Design of a cone-and-plate device for controlled realistic shear stress stimulation on endothelial cell monolayers. *Cytotechnology* **2016**, *68*, 1885–1896. [CrossRef] [PubMed]
52. Feaver, R.E.; Gelfand, B.D.; Blackman, B.R. Human haemodynamic frequency harmonics regulate the inflammatory phenotype of vascular endothelial cells. *Nat. Commun.* **2013**, *4*, 1525. [CrossRef] [PubMed]
53. Shankaran, H.; Neelamegham, S. Nonlinear Flow Affects Hydrodynamic Forces and Neutrophil Adhesion Rates in Cone–Plate Viscometers. *Biophys. J.* **2001**, *80*, 2631–2648. [CrossRef]
54. Vedernikov, Y.P.; Aarhus, L.; Shepherd, J.T.; Vanhoutte, P.M. Postmortem changes in endothelium-dependent and independent responses of porcine coronary arteries. *Gen. Pharmacol. Vasc. Syst.* **1990**, *21*, 49–52. [CrossRef]
55. Rosales, O.R.; Isales, C.M.; Barrett, P.Q.; Brophy, C.; Sumpio, B.E. Exposure of endothelial cells to cyclic strain induces elevations of cytosolic Ca2+ concentration through mobilization of intracellular and extracellular pools. *Biochem. J.* **1997**, *326*, 385–392. [CrossRef]
56. Caille, N.; Tardy, Y.; Meister, J.J. Assessment of strain field in endothelial cells subjected to uniaxial deformation of their substrate. *Ann. Biomed. Eng.* **1998**, *26*, 409–416. [CrossRef]
57. Yamada, H.; Ando, H. Orientation of apical and basal actin stress fibers in isolated and subconfluent endothelial cells as an early response to cyclic stretching. *Mol. Cell Biomech.* **2007**, *4*, 1–12.
58. Takeda, H.; Komori, K.; Nishikimi, N.; Nimura, Y.; Sokabe, M.; Naruse, K. Bi-phasic activation of eNOS in response to uni-axial cyclic stretch is mediated by differential mechanisms in BAECs. *Life Sci.* **2006**, *79*, 233–239. [CrossRef]
59. Katanosaka, Y.; Bao, J.H.; Komatsu, T.; Suemori, T.; Yamada, A.; Mohri, S.; Naruse, K. Analysis of cyclic-stretching responses using cell-adhesion-patterned cells. *J. Biotechnol.* **2008**, *133*, 82–89. [CrossRef]
60. Hashimoto-Komatsu, A.; Hirase, T.; Asaka, M.; Node, K. Angiotensin II induces microtubule reorganization mediated by a deacetylase SIRT2 in endothelial cells. *Hypertens. Res.* **2011**, *34*, 949–956. [CrossRef]

1. Mendez-Barbero, N.; Gutierrez-Munoz, C.; Blanco-Colio, L.M. Cellular Crosstalk between Endothelial and Smooth Muscle Cells in Vascular Wall Remodeling. *Int. J. Mol. Sci.* **2021**, *22*, 7284. [CrossRef] [PubMed]
2. Chiu, J.J.; Chen, L.J.; Lee, P.L.; Lee, C.I.; Lo, L.W.; Usami, S.; Chien, S. Shear stress inhibits adhesion molecule expression in vascular endothelial cells induced by coculture with smooth muscle cells. *Blood* **2003**, *101*, 2667–2674. [CrossRef] [PubMed]
3. Hastings, N.E.; Simmers, M.B.; McDonald, O.G.; Wamhoff, B.R.; Blackman, B.R. Atherosclerosis-prone hemodynamics differentially regulates endothelial and smooth muscle cell phenotypes and promotes pro-inflammatory priming. *Am. J. Physiol. Cell Physiol.* **2007**, *293*, C1824–C1833. [CrossRef]
4. Ji, Q.; Wang, Y.L.; Xia, L.M.; Yang, Y.; Wang, C.S.; Mei, Y.Q. High shear stress suppresses proliferation and migration but promotes apoptosis of endothelial cells co-cultured with vascular smooth muscle cells via down-regulating MAPK pathway. *J. Cardiothorac. Surg.* **2019**, *14*, 216. [CrossRef] [PubMed]
5. Jia, L.; Wang, L.; Wei, F.; Li, C.; Wang, Z.; Yu, H.; Chen, H.; Wang, B.; Jiang, A. Effects of Caveolin-1-ERK1/2 pathway on endothelial cells and smooth muscle cells under shear stress. *Exp. Biol. Med. (Maywood)* **2020**, *245*, 21–33. [CrossRef] [PubMed]
6. Nackman, G.B.; Fillinger, M.F.; Shafritz, R.; Wei, T.; Graham, A.M. Flow modulates endothelial regulation of smooth muscle cell proliferation: A new model. *Surgery* **1998**, *124*, 353–361. [CrossRef]
7. Qi, Y.X.; Jiang, J.; Jiang, X.H.; Wang, X.D.; Ji, S.Y.; Han, Y.; Long, D.K.; Shen, B.R.; Yan, Z.Q.; Chien, S.; et al. PDGF-BB and TGF-{beta}1 on cross-talk between endothelial and smooth muscle cells in vascular remodeling induced by low shear stress. *Proc. Natl. Acad. Sci. USA* **2011**, *108*, 1908–1913. [CrossRef]
8. Rashdan, N.A.; Lloyd, P.G. Fluid shear stress upregulates placental growth factor in the vessel wall via NADPH oxidase 4. *Am. J. Physiol. Heart Circ. Physiol.* **2015**, *309*, H1655–H1666. [CrossRef]
9. Tsai, M.C.; Chen, L.; Zhou, J.; Tang, Z.; Hsu, T.F.; Wang, Y.; Shih, Y.T.; Peng, H.H.; Wang, N.; Guan, Y.; et al. Shear stress induces synthetic-to-contractile phenotypic modulation in smooth muscle cells via peroxisome proliferator-activated receptor alpha/delta activations by prostacyclin released by sheared endothelial cells. *Circ. Res.* **2009**, *105*, 471–480. [CrossRef]
10. Wang, Y.H.; Yan, Z.Q.; Qi, Y.X.; Cheng, B.B.; Wang, X.D.; Zhao, D.; Shen, B.R.; Jiang, Z.L. Normal shear stress and vascular smooth muscle cells modulate migration of endothelial cells through histone deacetylase 6 activation and tubulin acetylation. *Ann. Biomed. Eng.* **2010**, *38*, 729–737. [CrossRef]
11. Yao, Q.P.; Qi, Y.X.; Zhang, P.; Cheng, B.B.; Yan, Z.Q.; Jiang, Z.L. SIRT1 and Connexin40 Mediate the normal shear stress-induced inhibition of the proliferation of endothelial cells co-cultured with vascular smooth muscle cells. *Cell Physiol. Biochem.* **2013**, *31*, 389–399. [CrossRef] [PubMed]
12. Zhou, J.; Li, Y.S.; Nguyen, P.; Wang, K.C.; Weiss, A.; Kuo, Y.C.; Chiu, J.J.; Shyy, J.Y.; Chien, S. Regulation of vascular smooth muscle cell turnover by endothelial cell-secreted microRNA-126: Role of shear stress. *Circ. Res.* **2013**, *113*, 40–51. [CrossRef] [PubMed]
13. Redmond, E.M.; Cahill, P.A.; Sitzmann, J.V. Perfused transcapillary smooth muscle and endothelial cell co-culture–a novel in vitro model. *Vitr. Cell Dev. Biol. Anim.* **1995**, *31*, 601–609. [CrossRef] [PubMed]
14. Redmond, E.M.; Cahill, P.A.; Sitzmann, J.V. Flow-mediated regulation of endothelin receptors in cocultured vascular smooth muscle cells: An endothelium-dependent effect. *J. Vasc. Res.* **1997**, *34*, 425–435. [CrossRef] [PubMed]
15. Redmond, E.M.; Cullen, J.P.; Cahill, P.A.; Sitzmann, J.V.; Stefansson, S.; Lawrence, D.A.; Okada, S.S. Endothelial cells inhibit flow-induced smooth muscle cell migration: Role of plasminogen activator inhibitor-1. *Circulation* **2001**, *103*, 597–603. [CrossRef]
16. Cucullo, L.; Hossain, M.; Tierney, W.; Janigro, D. A new dynamic in vitro modular capillaries-venules modular system: Cerebrovascular physiology in a box. *BMC Neurosci.* **2013**, *14*, 18. [CrossRef]
17. Janke, D.; Jankowski, J.; Ruth, M.; Buschmann, I.; Lemke, H.D.; Jacobi, D.; Knaus, P.; Spindler, E.; Zidek, W.; Lehmann, K.; et al. The "artificial artery" as in vitro perfusion model. *PLoS ONE* **2013**, *8*, e57227. [CrossRef]
18. Niwa, K.; Kado, T.; Sakai, J.; Karino, T. The effects of a shear flow on the uptake of LDL and acetylated LDL by an EC monoculture and an EC-SMC coculture. *Ann. Biomed. Eng.* **2004**, *32*, 537–543. [CrossRef]
19. Yamamoto, M.; James, D.; Li, H.; Butler, J.; Rafii, S.; Rabbany, S. Generation of stable co-cultures of vascular cells in a honeycomb alginate scaffold. *Tissue Eng. Part A* **2010**, *16*, 299–308. [CrossRef]
20. Sakamoto, N.; Ohashi, T.; Sato, M. Effect of fluid shear stress on migration of vascular smooth muscle cells in cocultured model. *Ann. Biomed. Eng.* **2006**, *34*, 408–415. [CrossRef]
21. Wang, H.Q.; Huang, L.X.; Qu, M.J.; Yan, Z.Q.; Liu, B.; Shen, B.R.; Jiang, Z.L. Shear stress protects against endothelial regulation of vascular smooth muscle cell migration in a coculture system. *Endothelium* **2006**, *13*, 171–180. [CrossRef] [PubMed]
22. Chiu, J.J.; Chen, L.J.; Chang, S.F.; Lee, P.L.; Lee, C.I.; Tsai, M.C.; Lee, D.Y.; Hsieh, H.P.; Usami, S.; Chien, S. Shear stress inhibits smooth muscle cell-induced inflammatory gene expression in endothelial cells: Role of NF-kappaB. *Arter. Thromb. Vasc. Biol.* **2005**, *25*, 963–969. [CrossRef] [PubMed]
23. Sakamoto, N.; Ueki, Y.; Oi, M.; Kiuchi, T.; Sato, M. Fluid shear stress suppresses ICAM-1-mediated transendothelial migration of leukocytes in coculture model. *Biochem. Biophys. Res. Commun.* **2018**, *502*, 403–408. [CrossRef] [PubMed]
24. Lavender, M.D.; Pang, Z.; Wallace, C.S.; Niklason, L.E.; Truskey, G.A. A system for the direct co-culture of endothelium on smooth muscle cells. *Biomaterials* **2005**, *26*, 4642–4653. [CrossRef]
25. Truskey, G.A. Endothelial Cell Vascular Smooth Muscle Cell Co-Culture Assay For High Throughput Screening Assays For Discovery of Anti-Angiogenesis Agents and Other Therapeutic Molecules. *Int. J. High Throughput Screen* **2010**, *2010*, 171–181. [CrossRef]

86. Cao, L.; Wu, A.; Truskey, G.A. Biomechanical effects of flow and coculture on human aortic and cord blood-derived endothelial cells. *J. Biomech.* **2011**, *44*, 2150–2157. [CrossRef]
87. Sakamoto, N.; Kiuchi, T.; Sato, M. Development of an endothelial-smooth muscle cell coculture model using phenotype-controlled smooth muscle cells. *Ann. Biomed. Eng.* **2011**, *39*, 2750–2758. [CrossRef]
88. Meng, H.; Tutino, V.M.; Xiang, J.; Siddiqui, A. High WSS or low WSS? Complex interactions of hemodynamics with intracranial aneurysm initiation, growth, and rupture: Toward a unifying hypothesis. *AJNR Am. J. Neuroradiol.* **2014**, *35*, 1254–1262. [CrossRef]
89. Kimura, N.; Nakamura, M.; Komiya, K.; Nishi, S.; Yamaguchi, A.; Tanaka, O.; Misawa, Y.; Adachi, H.; Kawahito, K. Patient-specific assessment of hemodynamics by computational fluid dynamics in patients with bicuspid aortopathy. *J. Thorac. Cardiovasc. Surg.* **2017**, *153*, S52–S62.e3. [CrossRef]
90. Han, X.; Sakamoto, N.; Tomita, N.; Meng, H.; Sato, M.; Ohta, M. Influence of TGF-beta1 expression in endothelial cells on smooth muscle cell phenotypes and MMP production under shear stress in a co-culture model. *Cytotechnology* **2019**, *71*, 489–496. [CrossRef]
91. Siddique, A.; Pause, I.; Narayan, S.; Kruse, L.; Stark, R.W. Endothelialization of PDMS-based microfluidic devices under high shear stress conditions. *Colloids Surf. B Biointerfaces* **2021**, *197*, 111394. [CrossRef]
92. Meng, Q.; Wang, Y.; Li, Y.; Shen, C. Hydrogel microfluidic-based liver-on-a-chip: Mimicking the mass transfer and structural features of liver. *Biotechnol. Bioeng.* **2021**, *118*, 612–621. [CrossRef]
93. Shen, C.; Li, Y.; Wang, Y.; Meng, Q. Non-swelling hydrogel-based microfluidic chips. *Lab Chip* **2019**, *19*, 3962–3973. [CrossRef]
94. van der Meer, A.D.; Poot, A.A.; Feijen, J.; Vermes, I. Analyzing shear stress-induced alignment of actin filaments in endothelial cells with a microfluidic assay. *Biomicrofluidics* **2010**, *4*, 11103. [CrossRef]
95. Reinitz, A.; DeStefano, J.; Ye, M.; Wong, A.D.; Searson, P.C. Human brain microvascular endothelial cells resist elongation due to shear stress. *Microvasc. Res.* **2015**, *99*, 8–18. [CrossRef]
96. Satoh, T.; Narazaki, G.; Sugita, R.; Kobayashi, H.; Sugiura, S.; Kanamori, T. A pneumatic pressure-driven multi-throughput microfluidic circulation culture system. *Lab Chip* **2016**, *16*, 2339–2348. [CrossRef]
97. Park, D.Y.; Kim, T.H.; Lee, J.M.; Ahrberg, C.D.; Chung, B.G. Circular-shaped microfluidic device to study the effect of shear stress on cellular orientation. *Electrophoresis* **2018**, *39*, 1816–1820. [CrossRef]
98. Sonmez, U.M.; Cheng, Y.W.; Watkins, S.C.; Roman, B.L.; Davidson, L.A. Endothelial cell polarization and orientation to flow in a novel microfluidic multimodal shear stress generator. *Lab Chip* **2020**, *20*, 4373–4390. [CrossRef]
99. Baratchi, S.; Tovar-Lopez, F.J.; Khoshmanesh, K.; Grace, M.S.; Darby, W.; Almazi, J.; Mitchell, A.; McIntyre, P. Examination of the role of transient receptor potential vanilloid type 4 in endothelial responses to shear forces. *Biomicrofluidics* **2014**, *8*, 044117. [CrossRef]
100. Liu, X.F.; Yu, J.Q.; Dalan, R.; Liu, A.Q.; Luo, K.Q. Biological factors in plasma from diabetes mellitus patients enhance hyperglycaemia and pulsatile shear stress-induced endothelial cell apoptosis. *Integr. Biol.* **2014**, *6*, 511–522. [CrossRef]
101. Plouffe, B.D.; Njoka, D.N.; Harris, J.; Liao, J.; Horick, N.K.; Radisic, M.; Murthy, S.K. Peptide-mediated selective adhesion of smooth muscle and endothelial cells in microfluidic shear flow. *Langmuir* **2007**, *23*, 5050–5055. [CrossRef]
102. Rossi, M.; Lindken, R.; Hierck, B.P.; Westerweel, J. Tapered microfluidic chip for the study of biochemical and mechanical response at subcellular level of endothelial cells to shear flow. *Lab Chip* **2009**, *9*, 1403–1411. [CrossRef]
103. Tsou, J.K.; Gower, R.M.; Ting, H.J.; Schaff, U.Y.; Insana, M.F.; Passerini, A.G.; Simon, S.I. Spatial regulation of inflammation by human aortic endothelial cells in a linear gradient of shear stress. *Microcirculation* **2008**, *15*, 311–323. [CrossRef]
104. Feng, S.; Mao, S.; Zhang, Q.; Li, W.; Lin, J.M. Online Analysis of Drug Toxicity to Cells with Shear Stress on an Integrated Microfluidic Chip. *ACS Sens.* **2019**, *4*, 521–527. [CrossRef]
105. Galie, P.A.; van Oosten, A.; Chen, C.S.; Janmey, P.A. Application of multiple levels of fluid shear stress to endothelial cells plated on polyacrylamide gels. *Lab Chip* **2015**, *15*, 1205–1212. [CrossRef]
106. Perrault, C.M.; Brugues, A.; Bazellieres, E.; Ricco, P.; Lacroix, D.; Trepat, X. Traction Forces of Endothelial Cells under Slow Shear Flow. *Biophys. J.* **2015**, *109*, 1533–1536. [CrossRef]
107. Wang, L.; Zhang, Z.L.; Wdzieczak-Bakala, J.; Pang, D.W.; Liu, J.; Chen, Y. Patterning cells and shear flow conditions: Convenient observation of endothelial cell remoulding, enhanced production of angiogenesis factors and drug response. *Lab Chip* **2011**, *11*, 4235–4240. [CrossRef]
108. Akbari, E.; Spychalski, G.B.; Rangharajan, K.K.; Prakash, S.; Song, J.W. Flow dynamics control endothelial permeability in a microfluidic vessel bifurcation model. *Lab Chip* **2018**, *18*, 1084–1093. [CrossRef]
109. Inglebert, M.; Locatelli, L.; Tsvirkun, D.; Sinha, P.; Maier, J.A.; Misbah, C.; Bureau, L. The effect of shear stress reduction on endothelial cells: A microfluidic study of the actin cytoskeleton. *Biomicrofluidics* **2020**, *14*, 024115. [CrossRef]
110. Khan, O.F.; Sefton, M.V. Endothelial cell behaviour within a microfluidic mimic of the flow channels of a modular tissue engineered construct. *Biomed. Microdevices* **2011**, *13*, 69–87. [CrossRef]
111. Zhang, X.; Huk, D.J.; Wang, Q.; Lincoln, J.; Zhao, Y. A microfluidic shear device that accommodates parallel high and low stress zones within the same culturing chamber. *Biomicrofluidics* **2014**, *8*, 054106. [CrossRef]
112. Venugopal Menon, N.; Tay, H.M.; Pang, K.T.; Dalan, R.; Wong, S.C.; Wang, X.; Li, K.H.H.; Hou, H.W. A tunable microfluidic 3D stenosis model to study leukocyte-endothelial interactions in atherosclerosis. *APL Bioeng.* **2018**, *2*, 016103. [CrossRef]
113. Arora, S.; Lam, A.J.Y.; Cheung, C.; Yim, E.K.F.; Toh, Y.C. Determination of critical shear stress for maturation of human pluripotent stem cell-derived endothelial cells towards an arterial subtype. *Biotechnol. Bioeng.* **2019**, *116*, 1164–1175. [CrossRef] [PubMed]

114. Booth, R.; Noh, S.; Kim, H. A multiple-channel, multiple-assay platform for characterization of full-range shear stress effects on vascular endothelial cells. *Lab Chip* **2014**, *14*, 1880–1890. [CrossRef] [PubMed]
115. Hattori, K.; Munehira, Y.; Kobayashi, H.; Satoh, T.; Sugiura, S.; Kanamori, T. Microfluidic perfusion culture chip providing different strengths of shear stress for analysis of vascular endothelial function. *J. Biosci. Bioeng.* **2014**, *118*, 327–332. [CrossRef]
116. Liu, M.C.; Shih, H.C.; Wu, J.G.; Weng, T.W.; Wu, C.Y.; Lu, J.C.; Tung, Y.C. Electrofluidic pressure sensor embedded microfluidic device: A study of endothelial cells under hydrostatic pressure and shear stress combinations. *Lab Chip* **2013**, *13*, 1743–1753. [CrossRef]
117. Chen, Z.Z.; Yuan, W.M.; Xiang, C.; Zeng, D.P.; Liu, B.; Qin, K.R. A microfluidic device with spatiotemporal wall shear stress and ATP signals to investigate the intracellular calcium dynamics in vascular endothelial cells. *Biomech. Model. Mechanobiol.* **2019**, *18*, 189–202. [CrossRef]
118. Sei, Y.J.; Ahn, S.I.; Virtue, T.; Kim, T.; Kim, Y. Detection of frequency-dependent endothelial response to oscillatory shear stress using a microfluidic transcellular monitor. *Sci. Rep.* **2017**, *7*, 10019. [CrossRef]
119. Shao, J.; Wu, L.; Wu, J.; Zheng, Y.; Zhao, H.; Jin, Q.; Zhao, J. Integrated microfluidic chip for endothelial cells culture and analysis exposed to a pulsatile and oscillatory shear stress. *Lab Chip* **2009**, *9*, 3118–3125. [CrossRef]
120. Estrada, R.; Giridharan, G.A.; Nguyen, M.D.; Prabhu, S.D.; Sethu, P. Microfluidic endothelial cell culture model to replicate disturbed flow conditions seen in atherosclerosis susceptible regions. *Biomicrofluidics* **2011**, *5*, 32006–3200611. [CrossRef]
121. Lee, J.; Estlack, Z.; Somaweera, H.; Wang, X.; Lacerda, C.M.R.; Kim, J. A microfluidic cardiac flow profile generator for studying the effect of shear stress on valvular endothelial cells. *Lab Chip* **2018**, *18*, 2946–2954. [CrossRef]
122. Yang, Y.; Fathi, P.; Holland, G.; Pan, D.; Wang, N.S.; Esch, M.B. Pumpless microfluidic devices for generating healthy and diseased endothelia. *Lab Chip* **2019**, *19*, 3212–3219. [CrossRef] [PubMed]
123. Zhao, P.; Liu, X.; Zhang, X.; Wang, L.; Su, H.; He, N.; Zhang, D.; Li, Z.; Kang, H.; Sun, A.; et al. Flow shear stress controls the initiation of neovascularization via heparan sulfate proteoglycans within a biomimetic microfluidic model. *Lab Chip* **2021**, *21*, 421–434. [CrossRef] [PubMed]
124. Zukerman, H.; Khoury, M.; Shammay, Y.; Sznitman, J.; Lotan, N.; Korin, N. Targeting functionalized nanoparticles to activated endothelial cells under high wall shear stress. *Bioeng. Transl. Med.* **2020**, *5*, e10151. [CrossRef]
125. Lewis, D.M.; Abaci, H.E.; Xu, Y.; Gerecht, S. Endothelial progenitor cell recruitment in a microfluidic vascular model. *Biofabrication* **2015**, *7*, 045010. [CrossRef]
126. Lewis, D.M.; Mavrogiannis, N.; Gagnon, Z.; Gerecht, S. Microfluidic platform for the real time measurement and observation of endothelial barrier function under shear stress. *Biomicrofluidics* **2018**, *12*, 042202. [CrossRef]
127. Hsu, S.; Thakar, R.; Liepmann, D.; Li, S. Effects of shear stress on endothelial cell haptotaxis on micropatterned surfaces. *Biochem. Biophys. Res. Commun.* **2005**, *337*, 401–409. [CrossRef]
128. Didar, T.F.; Tabrizian, M. Generating multiplex gradients of biomolecules for controlling cellular adhesion in parallel microfluidic channels. *Lab Chip* **2012**, *12*, 4363–4371. [CrossRef]
129. Hirose, S.; Tabata, Y.; Sone, K.; Takahashi, N.; Yoshino, D.; Funamoto, K. P21-activated kinase regulates oxygen-dependent migration of vascular endothelial cells in monolayers. *Cell Adh. Migr.* **2021**, *15*, 272–284. [CrossRef]
130. Tabata, Y.; Yoshino, D.; Funamoto, K.; Koens, R.; Kamm, R.D. Migration of vascular endothelial cells in monolayers under hypoxic exposure. *Integr. Biol.* **2019**, *11*, 26–35. [CrossRef]
131. Gnecco, J.S.; Pensabene, V.; Li, D.J.; Ding, T.; Hui, E.E.; Bruner-Tran, K.L.; Osteen, K.G. Compartmentalized Culture of Perivascular Stroma and Endothelial Cells in a Microfluidic Model of the Human Endometrium. *Ann. Biomed. Eng.* **2017**, *45*, 1758–1769. [CrossRef]
132. Griep, L.M.; Wolbers, F.; de Wagenaar, B.; ter Braak, P.M.; Weksler, B.B.; Romero, I.A.; Couraud, P.O.; Vermes, I.; van der Meer, A.D.; van den Berg, A. BBB on chip: Microfluidic platform to mechanically and biochemically modulate blood-brain barrier function. *Biomed. Microdevices* **2013**, *15*, 145–150. [CrossRef]
133. Wang, Y.I.; Abaci, H.E.; Shuler, M.L. Microfluidic blood-brain barrier model provides in vivo-like barrier properties for drug permeability screening. *Biotechnol. Bioeng.* **2017**, *114*, 184–194. [CrossRef]
134. Park, T.E.; Mustafaoglu, N.; Herland, A.; Hasselkus, R.; Mannix, R.; FitzGerald, E.A.; Prantil-Baun, R.; Watters, A.; Henry, O.; Benz, M.; et al. Hypoxia-enhanced Blood-Brain Barrier Chip recapitulates human barrier function and shuttling of drugs and antibodies. *Nat. Commun.* **2019**, *10*, 2621. [CrossRef]
135. Terrell-Hall, T.B.; Ammer, A.G.; Griffith, J.I.; Lockman, P.R. Permeability across a novel microfluidic blood-tumor barrier model. *Fluids Barriers CNS* **2017**, *14*, 3. [CrossRef]
136. Huh, D.; Leslie, D.C.; Matthews, B.D.; Fraser, J.P.; Jurek, S.; Hamilton, G.A.; Thorneloe, K.S.; McAlexander, M.A.; Ingber, D.E. A human disease model of drug toxicity-induced pulmonary edema in a lung-on-a-chip microdevice. *Sci. Transl. Med.* **2012**, *4*, 159ra147. [CrossRef]
137. Lamberti, G.; Soroush, F.; Smith, A.; Kiani, M.F.; Prabhakarpandian, B.; Pant, K. Adhesion patterns in the microvasculature are dependent on bifurcation angle. *Microvasc. Res.* **2015**, *99*, 19–25. [CrossRef]
138. Franklin-Murray, A.L.; Mallya, S.; Jankeel, A.; Sureshchandra, S.; Messaoudi, I.; Lodoen, M.B. Toxoplasma gondii Dysregulates Barrier Function and Mechanotransduction Signaling in Human Endothelial Cells. *mSphere* **2020**, *5*, e00550-19. [CrossRef]
139. van der Meer, A.D.; Vermeul, K.; Poot, A.A.; Feijen, J.; Vermes, I. A microfluidic wound-healing assay for quantifying endothelial cell migration. *Am. J. Physiol. Heart Circ. Physiol.* **2010**, *298*, H719–H725. [CrossRef]

140. Galie, P.A.; Nguyen, D.H.; Choi, C.K.; Cohen, D.M.; Janmey, P.A.; Chen, C.S. Fluid shear stress threshold regulates angiogenic sprouting. *Proc. Natl. Acad. Sci. USA* **2014**, *111*, 7968–7973. [CrossRef]
141. Funamoto, K.; Yoshino, D.; Matsubara, K.; Zervantonakis, I.K.; Nakayama, M.; Masamune, J.; Kimura, Y.; Kamm, R.D. Endothelial monolayer permeability under controlled oxygen tension. *Integr. Biol.* **2017**, *9*, 529–538. [CrossRef]
142. Kim, S.; Lee, H.; Chung, M.; Jeon, N.L. Engineering of functional, perfusable 3D microvascular networks on a chip. *Lab Chip* **2013**, *13*, 1489–1500. [CrossRef]
143. Oh, S.; Ryu, H.; Tahk, D.; Ko, J.; Chung, Y.; Lee, H.K.; Lee, T.R.; Jeon, N.L. "Open-top" microfluidic device for zzzzi three-dimensional capillary beds. *Lab Chip* **2017**, *17*, 3405–3414. [CrossRef]
144. Campisi, M.; Shin, Y.; Osaki, T.; Hajal, C.; Chiono, V.; Kamm, R.D. 3D self-organized microvascular model of the human blood-brain barrier with endothelial cells, pericytes and astrocytes. *Biomaterials* **2018**, *180*, 117–129. [CrossRef]
145. Shirure, V.S.; Bi, Y.; Curtis, M.B.; Lezia, A.; Goedegebuure, M.M.; Goedegebuure, S.P.; Aft, R.; Fields, R.C.; George, S.C. Tumor-on-a-chip platform to investigate progression and drug sensitivity in cell lines and patient-derived organoids. *Lab Chip* **2018**, *18*, 3687–3702. [CrossRef]
146. Sobrino, A.; Phan, D.T.; Datta, R.; Wang, X.; Hachey, S.J.; Romero-López, M.; Gratton, E.; Lee, A.P.; George, S.C.; Hughes, C.C. 3D microtumors in vitro supported by perfused vascular networks. *Sci. Rep.* **2016**, *6*, 31589. [CrossRef]
147. Lee, S.W.L.; Campisi, M.; Osaki, T.; Possenti, L.; Mattu, C.; Adriani, G.; Kamm, R.D.; Chiono, V. Modeling Nanocarrier Transport across a 3D In Vitro Human Blood-Brain-Barrier Microvasculature. *Adv. Health Mater.* **2020**, *9*, e1901486. [CrossRef]
148. Huh, D.; Matthews, B.D.; Mammoto, A.; Montoya-Zavala, M.; Hsin, H.Y.; Ingber, D.E. Reconstituting organ-level lung functions on a chip. *Science* **2010**, *328*, 1662–1668. [CrossRef]

Article

Hemodynamic Analysis of the Geometric Features of Side Holes Based on GDK Catheter

Yang Yang [1], Yijing Li [1], Chen Liu [2], Jingyuan Zhou [3], Tao Li [1], Yan Xiong [1,*] and Ling Zhang [2,*]

1. College of Mechanical Engineering, Sichuan University, Chengdu 610065, China
2. Department of Nephrology, Kidney Research Institute, West China Hospital of Sichuan University, Chengdu 610041, China
3. Department of Applied Mechanics, Sichuan University, Chengdu 610065, China
* Correspondence: xy@scu.edu.cn (Y.X.); zhanglinglzy@163.com (L.Z.)

Abstract: Hemodialysis is an important means to maintain life in patients with end-stage renal disease (ESRD). Approximately 76.8% of patients who begin hemodialysis do so through catheters, which play vital roles in the delivery of hemodialysis to patients. During the past decade, the materials, structures, and surface-coating technologies of catheters have constantly been evolving to ameliorate catheter-related problems, such as recirculation, thrombosis, catheter-related infections, and malfunction. In this study, based on the commercial GDK catheter, six catheter models (GDK, GDK1, GDK2, GDK3, GDK4, and GDK5) with different lumen diameters and different geometric features of side holes were established, and computational flow dynamics (CFD) were used to measure flow rate, shear stress, residence time (RT), and platelet lysis index (PLI). These six catheters were then printed with polycarbonate PC using 3D printing technology to verify recirculation rates. The results indicated that: (1) the catheter with a 5.5 mm outer diameter had the smallest average shear stress in the arterial lumen and the smallest proportion of areas with shear stress > 10 pa. With increasing catheter diameter, the shear stress in the tip volume became lower, the average RT increased, and the PLI decreased due to larger changes in shear stress; (2) the catheters with oval-shaped side holes had smaller shear stress levels than those with circular-shaped holes, indicating that the oval design was more effective; (3) the catheter with parallel dual side holes had uniformly distributed flow around side holes and exhibited lower recirculation rates in both forward and reverse connections, while linear multi-side holes had higher shear stress levels due to the large differences in flow around side holes. The selection of the material and the optimization of the side holes of catheters have significant impacts on hemodynamic performances and reduce the probability of thrombosis, thus improving the efficiency of dialysis. This study would provide some guidance for optimizing catheter structures and help toward the commercialization of more efficient HD catheters.

Keywords: hemodynamic; geometric features; side holes; catheter; shear stress

1. Introduction

More than 1 million patients die each year worldwide due to ESRD (end-stage renal disease), and up to 1.7 million patients with acute kidney injury die due to lack of access to effective treatment, 85% of which deaths occur in developing countries [1]. Hemodialysis is one of the kidney replacement methods for the treatment of ESRD [2]. For dialysis treatment, building an ideal vascular pathway is the primary preparation before treatment, and it is also a necessary condition to successfully achieve the treatment effect [3]. In the United States, approximately 80% of patients began hemodialysis with a catheter in 2011 [4]. In China, more than 100,000 hemodialysis patients need to have a double-lumen catheter fitted at least once (temporary or long-term) for dialysis treatment [5].

Despite hemodialysis (HD) catheters being widely used and despite the low cost of their placement and replacement, they have certain defects, such as being associated with high rates of thrombosis, infection, and dysfunction [6]. Catheter materials have significant impacts on the prevention of catheter-related infection and the life of hemodialysis catheters. The ideal catheter material would be biocompatible, hemocompatible, biostable, chemically neutral, stable, and deformable in accordance with environmental forces [7]. Traditional biomaterials include polyurethane and silicone. Through advances in materials technology, there has been a transition to the use of polyurethane or carbon ethane (polyurethane/polycarbonate copolymer), which have better catheter strength and flexibility while maintaining larger internal diameters. The properties of the material used determine the mechanical properties of a catheter to some extent, while the surface coating on the catheter material is related to the biocompatibility and anticoagulation properties of the catheter. In addition, the design of catheters is also closely related to blood recirculation and thrombus formation [8].

In recent years, studies on catheter design have investigated the design of the lumen, the distal tip, and the side holes, along with hemodynamic performance by means of CFD, animal, or in vitro experiments [9]. For example, Ling et al. [10] compared the clinical and rheologic outcomes of HD catheters with step, split, or symmetrical tips in patients; Vesely et al. [6] found that symmetric-tip and step-tip designs had the advantage of more stable fluid flow patterns and lower recirculation rates compared to split-tip designs in a model simulating hemodialysis treatment; Ogawa et al. [11] observed and evaluated the recirculation and fluid characteristics of the catheter tips for three catheters with different tip shapes; Tal et al. [12] compared catheters with or without side holes by performing an analysis of flow rates, infection rates, and survival rates; Owen et al. [13] investigated the impact of different side-hole configurations on a symmetrical-tip catheter by evaluating the local hemodynamics and catheter performance through CFD. The above studies focused on hemodynamic analyses of the existing products and models, and only a few studies have put forward new ideas about catheter design to improve hemodynamic performance. For instance, Cho et al. [8] put forward three new HD catheter designs and compared the effects of the catheters' side holes and distal tips on hemodynamic factors with those of existing catheters, using CFD and in vitro methods, while Clark et al. [14] compared a new dialysis catheter (VectorFlow) with a Palindrome catheter in terms of shear stress, RT, PLI, and recirculation rate using the CFD method and a bench model of hemodialysis.

In the present study, the comparative analysis of new designs for catheter lumens and patterns of catheter side holes was the research focus. Through structural improvements based on the commercial GDK catheter (Gambro, Stockholm, Sweden), five newly designed catheter models are put forward, the CFD method and in vitro experiments having been used to study the effects of design parameters. The problems affecting catheter performance include platelet activation [15], flow stagnation regions [16], and the recirculation of dialyzed blood [17]. Platelet activation and aggregation are related to elevated regions of shear stress and prolonged blood transit time [18]. The higher the values of shear stress, PLI and RT lead to higher probabilities of thrombosis. Lower recirculation of dialyzed blood affects dialysis efficiency. Therefore, the variations in lumen diameter and the different geometric features of the side holes were assessed among these six types of catheters by comparing the values for flow rate, shear stress, RT, and PLI of inflowing blood at the tip of catheter along with the recirculation rates of these 3D-printed catheters to spot possible design issues.

This study could help achieve catheter designs with more optimized structures, longer life, and higher dialysis efficiencies. The results could ultimately lead to better-performing catheters to improve the life cycle of patients with kidney disease.

2. Methods

Six catheter models were simulated by the finite element method through three-dimensional modeling, meshing, and numerical simulation, and their hemodynamic parameters, such as flow rate, shear stress, RT, and PLI, were obtained. Then, these catheter models were 3D-printed with polycarbonate PC, and their recirculation rates were tested by dye tracing in in vitro experiments.

2.1. Geometry of Hemodialysis Catheters

The GDK catheter, as shown in Figure 1, has a symmetric design; the inner diameter is 2.6 mm and the outer is 3.6 mm, and the thickness of the board which separates the arterial and venous lumens is 0.5 mm. The three circular side holes with a vertical distribution and the distal tips with nozzle shapes are located on the venous lumen. Five circular side holes for the inflow are located on the venous lumen. All these circular side holes' diameters are 2 mm.

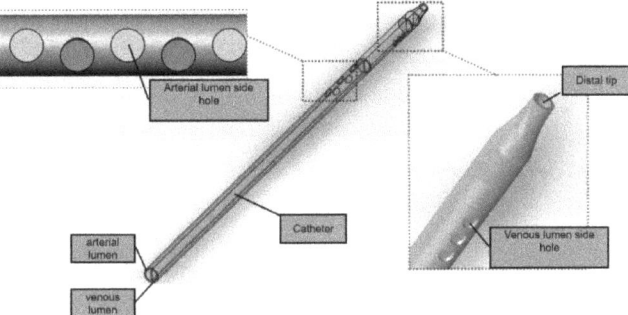

Figure 1. The GDK catheter.

Keeping other structural elements unchanged, the outer diameters of the GDK catheter were expanded from 3.6 mm to 4.3 mm (GDK1) and 5.5 mm (GDK2, 3, 4, and 5), as shown in Figure 2, so the inner diameter was changed to 3.3 mm (GDK1) and 4.5 mm (GDK2, 3, 4, and 5). For GDK1 and GDK2, the distributions of side holes were kept the same as those for the GDK catheter to investigate the effects of the internal cavity on the flow pattern. In GDK3, the shape of the side holes was changed to oval, but the hole area was kept the same to compare the effects of the shapes of the side holes. For GDK4, the area of the side holes was enlarged from 3.14 mm^2 to 6.28 mm^2 to observe the effect of the area of the side holes on the flow. The side holes of the venous lumen were named side-hole1, side-hole2, and side-hole3, according to the distance from the tip. For GDK5, the distribution of the side holes was changed from linear to a parallel dual-hole pattern.

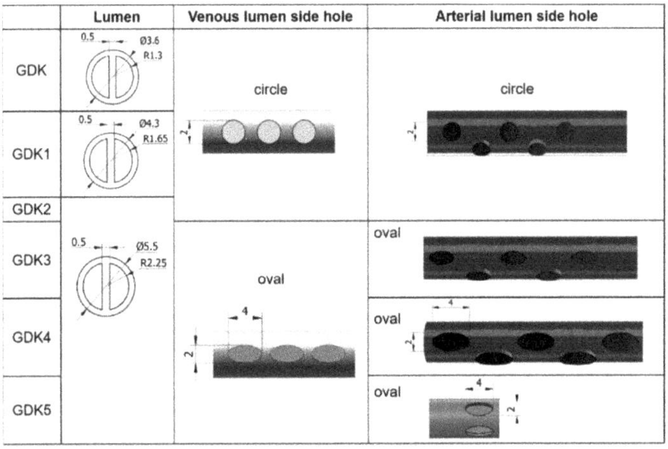

Figure 2. The side holes in the lumens of different configurations of GDK catheters.

In the model domain, the 3D model of the superior vena cava (SVC) and the catheter is shown in Figure 3. The SVC was considered to be a cylinder with a diameter of 20 mm. In order to eliminate the impact of unrelated variables on performance, the lengths of the SVC and each catheter were set to 340 mm and 200 mm, respectively, and the distance between the distal tip of each catheter and the SVC was fixed to 100 mm so that the blood inflows to the catheters were fully normalized and no outlet effects occurred.

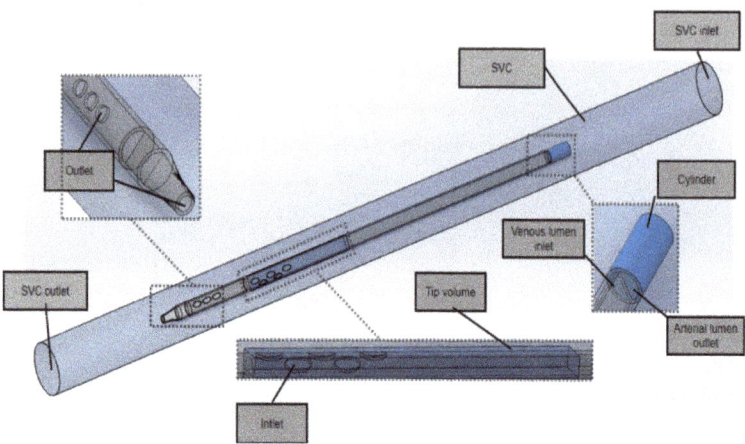

Figure 3. The 3D model of the SVC and the catheter.

In the SVC fluid domain, a cylindrical cavity with a length of 20 mm and the same diameter as the outer diameter of the catheter was dug out at the front end of the root catheter, thus imposing a boundary condition. In order to better compare inflow parameters, a cuboid 'tip volume' was defined at the tip of the venous lumen for each catheter. Its width and height were equal to the outer diameter and radius of each catheter, respectively, such that it included the entire inflow lumen. In order to focus on the flow characteristics of the different catheters, the tip volume extended from the most distal point of the catheter up to 50 mm, where the flow velocity would become stable and fully developed and there would be no further flow interference.

All geometries were created using Solidworks software (Dassault System SolidWorks, Concord, MA, USA) and exported into Fluent Meshing software (Fluent Inc., Lebanon, NH, USA).

2.2. Mesh of Hemodialysis Catheters

The grid consisted of a mixture of tetrahedral and hexahedral grids, the number of which depended on the geometry of the catheter, and an encrypted grid at the 'tip volume' and side holes. In order to obtain the best calculation time, the grid was verified independently. Table 1 shows the maximum velocity and the average shear stress at the tip volume of the GDK with different numbers of grids, resulting in a grid consisting of 320,187 cells, which were mainly tetrahedral, and the inner and outer lumens of the catheters were distributed to 6 boundary layers. A cross-section of the grid is shown in Figure 4.

Table 1. Various grid numbers and test results for independent tests.

Number of Grids	Maximum Velocity (m/s)	Average Shear Stress (1/s)
94,048	1.65	2.22
220,826	1.83	2.88
320,187	1.78	3.19
420,186	1.78	3.19

Figure 4. Cross section of grid.

2.3. Governing Equations and Boundary Conditions

The fluid relevant here is blood, which is identified as an incompressible, uniform, non-Newtonian fluid [19], the viscosity of which varies with the shear rate. In previous studies, incompressible Newtonian fluid models were established as the working fluids; this study considered the change in shear rate occurring during inflow and used an asymptotic shear-thinning Carreau model [20], which was defined using the shear rate of Equation (1) and defined by Equation (2).

$$\dot{\gamma} = \sqrt{2 \times D_{ij} \cdot D_{ij}} \quad (1)$$

where $\dot{\gamma}$ is the shear rate and D is the strain rate tensor, with $i, j = 1, 2, 3$ as the inner projects.

$$\mu = \mu_\infty + (\mu_\infty + \mu_0)\left[1 + (\lambda\dot{\gamma})^2\right]^{\frac{n-1}{2}}. \quad (2)$$

where μ is the viscosity of blood and $\mu_\infty = 0.0345$ Pa.s, $n = 0.25$, $\mu_0 = 0.025$ Pa.s, and $\lambda = 25$ s.

Boundary conditions: The SVC inlet was set to a 0.3 m/s velocity inlet, and the outlet was a pressure outlet, the gauge pressure of which was zero. The surfaces of the catheter and SVC were set as the wall without slip condition, and the inlet and outlet of the catheter were set as a mass flow inlet and mass flow outlet of 400 mL/min, respectively [21].

The platelet lysis index (PLI) (Equation (3)) was calculated to evaluate the possible damage occurring to the platelets. This index was first applied to the heart valve prostheses [22] and is widely used to evaluate the risk of platelet-activated aggregation and thrombosis in previous studies [14].

$$PLI = 3.66 \cdot 10^{-6} \cdot t_p^{0.77} \cdot \tau_p^{3.075}. \quad (3)$$

where t_p is the residence time of the platelet and τ_p is the shear stress acting on the platelet. For the GDK catheter, side holes away from the tip were taken as the calculation path.

Each path flowed into the Poiseuillean flow zone from the side hole of the catheter and through the disturbance zone. The shear stress, velocity, and retention time were outputted at each point the path passed through, and the PLI was calculated at each step (0.5 μm).

Solution settings: The ANSYS Fluent COUPLED solver was used to solve the fluid numerically. The flow in the catheter was similar to that in other catheter studies [23,24], assuming laminar flow. The mass continuity residual magnitude was less than 10^{-6}, and the combined flow rate of the tip and the side hole was equal to 400 ± 1 mL/min, these values being used to assess convergence.

2.4. In Vitro Experiment for Recirculation

The recirculation test bench for the dye tracing experiment is shown in Figure 5. The device simulates the superior vena cava mainly by a plexiglass tube (diameter 20 mm, length 500 mm). A steady flow is maintained in the plexiglass tube by connecting it to a peristaltic pump at the upper and lower adapters. The catheter gland can be adjusted to ensure that catheters with different external diameters are inserted into the plexiglass tube. The catheter gland is connected vertically to the catheter with a silicone rubber seal to achieve a complete seal inside the vessel.

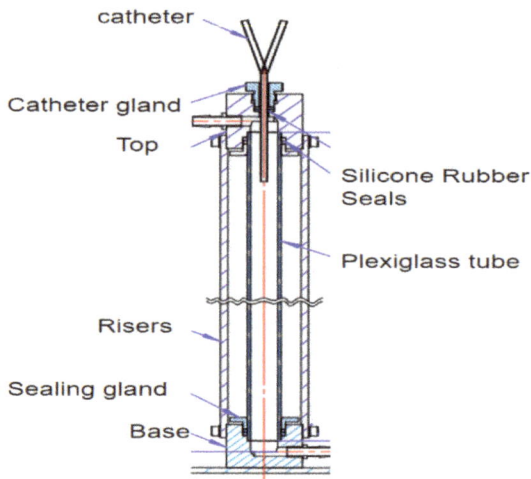

Figure 5. Recirculation test bench for the dye tracing experiment.

According to decreasing order of thrombus formation, the catheter materials used were polyvinyl chloride, polyethylene, polyurethane, and silica gel. Silicone or polyethylene carbamate catheters are to be preferred because of their high smoothness, strong adhesion to resist fibers and pathogens, good histocompatibility, small vascular stimulation, lower chance of thrombosis, and reduced chances of infection and intravascular injury [25]. The six catheters were printed with polycarbonate PC material using 3D printing technology to verify the recirculation rates. A 3D print of the GDK catheter is shown in Figure 6. Polycarbonate PC is widely used in artificial kidney hemodialysis equipment and other medical equipment that needs to be operated under transparent, intuitive conditions and requires repeated sterilization.

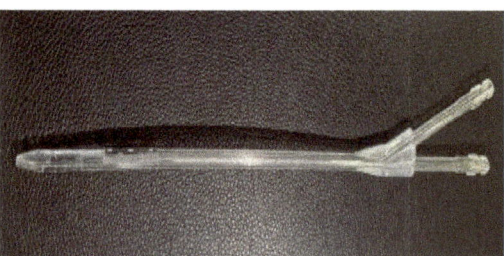

Figure 6. A 3D print of the GDK catheter.

The overall flow rate of the device is controlled by three dual-channel peristaltic pumps, as shown in Figure 7. The Rombauer BT100 peristaltic pump is connected to the arteriovenous lumen with a maximum flow rate of 570 mL/min, pumping fluid from Reservoir 3 into the glass column at a rate of 400 mL/min through the venous lumen of the hemodialysis catheter. Fluid is drawn from the glass column through the arterial lumen at 400 mL/min and then outputted to Reservoir 4. The BT-CA JIHPUMP BT-600CA peristaltic pump was used to maintain fluid flow through the glass column at 2400 mL/min to simulate the flow of blood from the superior vena cava. In practice, surgeons would resolve a catheter malfunction by reversing the direction of blood flow in the arterial and venous lumens [26]. In the experiment, the peristaltic direction of BT-100 was reversed, and the fluid from Reservoir 4 passed through the arterial lumen into the plexiglass tube at a flow rate of 400 mL/min, while the fluid from Reservoir 3 was withdrawn from the venous lumen at a flow rate of 400 mL/min to realize the reverse connection of the dialysis catheter.

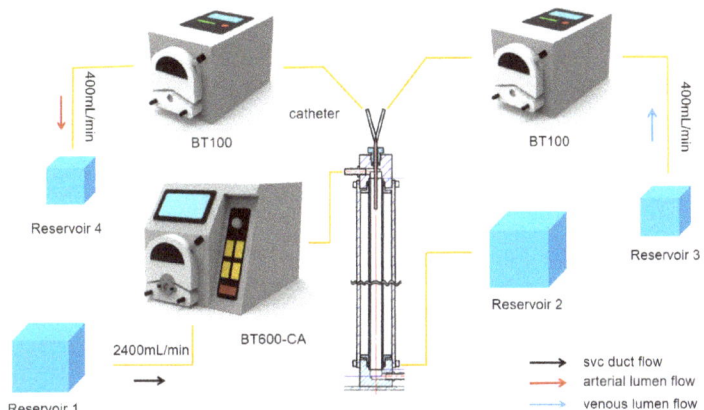

Figure 7. Schematic diagram of the dye tracing device (forward).

Recirculation in the catheter means that when the dialyzed blood passes through the venous lumen and returns to the body's superior vena cava to be withdrawn again by the arterial lumen the blood gets dialyzed again, which reduces the efficiency of hemodialysis [17].

Taking the forward connection as an example, we first ran the BT600-CA peristaltic pump to maintain a flow rate of 2400 mL/min in the Plexiglass tube, then simultaneously started the BT100 peristaltic pump to inject the dye liquid at a concentration of 1% into Reservoir 3, which flowed into the venous lumen. After a one-minute test, 50 mL samples were extracted from Reservoir 4. When the absorbance coefficient and the optical path of the dye solution are unchanged, the absorbance of the dye solution is proportional to the concentration of the dye solution [27], and the concentration of red dye in the fluid in the arterial lumen was measured using a UV–Visible spectrophotometer. After the completion of each experiment, the pump channel was connected to a peristaltic pump and pure water was used to discharge the dye reagents in the pump channel so as to avoid the impact of residual dye reagents and errors in subsequent experiments. Each experiment was repeated five times, and the average values and standard errors of the means (SEMs) were calculated to improve the accuracy of the experimental data. The recirculation rate (RR) values were calculated as shown in Equation (4).

$$RR(\%) = \frac{Q_a \times C_a}{Q_v \times C_v} \times 100 \tag{4}$$

where Q_a and Q_v are the flow rates of the arterial lumen and the venous lumen in the catheter, respectively, and C_v and C_a are the concentrations of the dye from the venous lumen and arterial lumen, respectively. In order to determine statistical significance, the non-parametric Kruskal–Wallis test was used, with $P < 0.05$, to compare the recirculation of the different catheters.

3. Results

3.1. Analysis of Flow Rate

Blood, after dialysis, flows out from the distal tip and the side holes in the venous lumen. The function of the side holes is to reduce flow velocity at the distal tip of the catheter (Q_{tip}), leading to a lower recirculation rate.

The flow rates of the side holes and the distal tip in the venous lumen of the GDK, GDK1, GDK2, and GDK3 catheters are shown in Table 2, which also presents the percentage of flow rates through side holes/distal tips based on total flow for these catheters.

Table 2. The flow rate through the distal tip and side holes in the venous lumen.

	$Q_{side-hole1}$ (mL/min)	$Q_{side-hole2}$ (mL/min)	$Q_{side-hole3}$ (mL/min)	Q_{side} (mL/min)/ Percentage of Flow Rate through Side Holes	Q_{tip} (mL/min)/ Percentage of Flow Rate through Distal Tip
GDK	89.71	38.74	14.57	143.02 (35.76%)	257.14 (64.28%)
GDK1	89.83	77.41	58.54	225.78 (56.44%)	173.98 (43.49%)
GDK2	74.14	72.34	62.29	208.77 (52.19%)	191.2 (47.8%)
GDK3	135.66	80.57	44.57	260.8 (65.2%)	139.08 (34.77%)

The flow rate of the side holes decreased with the distance of the side holes from the tip. In the case of the GDK catheter, the flow rate of side-hole1 near the distal tip was 89.71 mL/min, while those of side-hole2 and 3, away from the distal tip, were 38.74 and 14.57 mL/min, respectively. Similar trends were also seen for other catheters (GDK1, GDK2, and GDK3). When the outer diameter of the catheter was increased from 3.6 mm (GDK) to 4.3 mm (GDK1), the flow rate of the distal tip decreased from 257.14 mL/min to 173.98 mL/min, and the total flow of these side holes (Q_{side}) increased from 143.02 mL/min to 225.78 mL/min. The Q_{tip}'s proportion of total flow decreased from 64.28% to 43.49%, thereby increasing the cross-sectional areas of the inner cavity, which proved to be effective for improving the flow distribution in the catheter.

When the outer diameter was 5.5 mm, the Q_{tip} in GDK2 increased to 47.8%. When the side holes were changed from being circular-shaped with an area of 3.14 mm^2 (GDK2) to oval-shaped with an area of 6.28 mm^2 (GDK3), the flow rate of the distal tip decreased from 191.2 mL/min to 139.1 mL/min, and the Q_{tip}'s proportion of total flow decreased from 47.8% to 34.77%. The flow rate of these side holes (Q_{side}) increased from 208.8 mL/min to 260.9 mL/min. This indicates that the flow from the distal tip in the venous lumen can be varied by modifying the shape and area of the side holes, changing the flow through it. This suggests that larger outside diameters and oval side holes with larger areas on the venous lumen are all effective ways to improve the flow configuration in the catheter.

Although the arrangements are similar for GDK2 and GDK3, there are large differences in the flow distributions of each side hole. The flow rates for GDK2 with circular-shaped side holes along the distance to the tip were 188.1 mL/min, 109.71 mL/min, 61.7 mL/min, 27.65 mL/min, and 12.11 mL/min, respectively, from the far to near ends; the first two side holes took 74.4% of dialysis blood flow inlet in the arterial lumen. For GDK3, the values of flow rate for each side hole were changed to 194.5 mL/min, 112.1 mL/min, 59.66 mL/min, 24.85 mL/min, and 8.8 mL/min, respectively, because of the changes in the shape of the side holes. The hole flow distribution of the GDK4 catheter was changed further by the increased lateral hole area; the flow rate of two oval-shaped side holes away from the tip reached 95.4% of the total inlet flow. When the arrangement of the side holes was changed to parallel dual holes, both side holes would split inlet flow, as shown in Figure 8d; the flow rates for each side were about 200 mL/min in the arterial lumen.

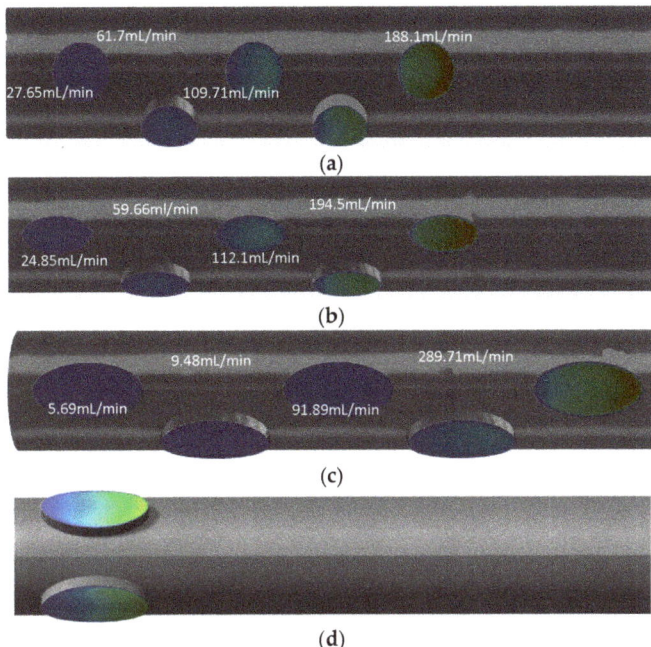

Figure 8. Flow distribution of side holes in the arterial lumen for GDK2 (**a**), GDK3 (**b**), GDK4 (**c**), and GDK5 (**d**).

The catheters with parallel side holes had lower flow rates than that with linear multi-side holes because of the flow being split evenly across each hole. It can be concluded that the structure of the side holes has an important effect on the flow distribution of the catheter.

3.2. Analysis of Shear Stress

The flow of blood through the catheter is considered laminar [28]. It is believed that shear stresses over 10 pa would cause damage to platelets and lead to thrombosis [29]. So, the maximum flow rate, the average shear stress, and the proportion of the area of shear stress over 10 Pa were determined from the tip volume of each of the six catheters (as shown in Table 3) to quantitatively assess hemodynamic performance.

Table 3. The maximum flow velocity and average shear stress values at the tip volume for each of the six catheters.

	Max Velocity (m/s)	Average Shear Stress (Pa)	Percentage of Shear Stress > 10 Pa
GDK	5.14	13.6	22.6%
GDK1	3.18	11.0	27.4%
GDK2	1.71	3.43	12.8%
GDK3	1.81	3.15	12.2%
GDK4	1.78	3.19	12.6%
GDK5	1.72	3.16	12.5%

It is shown that the percentage of shear stress regions > 10 Pa was 22.6% and that the maximum flow velocity was 5.14 m/s while the average shear stress was 13.6 pa at the tip volume of the commercial GDK catheter. When the outer diameter of the GDK catheter was enlarged from 3.6 mm (GDK) to 4.3 mm (GDK1), the percentage of regions where the shear stress at the tip volume exceeded 10 Pa in the arterial lumen increased to 27.4%, and the average shear stress decreased to 11 pa, with little change in the shear stress level.

When the outer diameter was 5.5 mm (GDK2), the shear stress at the tip volume in the arterial lumen changed significantly, with average shear stress decreasing from 11 pa to 3.43 pa, and the shear stress region > 10 pa also decreased to 12.8%. It was presumed that increasing the outer diameter would reduce the shear stress in the tip volume of the arterial lumen. The catheters (GDK2, GDK3, GDK4, and GDK5) with 5.5 mm outer diameters were all associated with similar levels of mean shear stress and percentage shear stress > 10 Pa.

When the area of the side holes in the arterial lumen was kept the same and the shape was changed from circular to oval, the shear stress level in the tip volume was changed, and the average shear stress in GDK3 with oval-shaped side holes was a little smaller than that in GDK2 with circular-shaped holes (3.15 pa < 3.43 pa), while, when the shape of the oval-shaped side holes was kept unchanged and the area was enlarged from 3.14 mm^2 (GDK3) to 6.28 mm^2 (GDK4), the average shear stress became slightly larger (3.15 pa < 3.19 pa) and the percentage of shear stress > 10 Pa at the tip volume for these catheters maintained similar levels (12.8%, 12.2%, and 12.6%). This indicates that when considering the variation in the shape and area of side holes acting on the level of shear stress at the tip volume of the catheter arterial lumen, oval-shaped side holes with smaller areas have a positive impact on reducing platelet activation as well as the risk of blood damage.

The GDK5 catheter at the tip volume was associated with lower average shear stress (3.16 Pa) compared with that of the GDK4 catheter, at 3.19 Pa. Since the flow rate of each side hole was so different, the average shear stress of the GDK4 catheter with linear multi-side holes was a little higher than that with the parallel double side-hole structure, which also reduced the maximum velocity at the tip volume from 1.78 m/s to 1.72 m/s.

3.3. Analysis of PLI

PLI, as a weighting of platelet models experiencing high shear stress and residence time, can reflect the activation state of platelets to assess the merit of catheter tip design. The values for the average residence time (RT), PLI, average shear stress, and percentage of shear stress >10 Pa at the tip volume of the six catheters are presented in Table 4.

Table 4. The values for average RT, PLI, average shear stress, and percentage shear stress > 10 Pa at the tip volume of the six catheters.

	Average RT (s)	PLI	Average Shear Stress (Pa)	Percentage Shear Stress > 10 Pa
GDK	0.008	0.9225	23.107	66.05
GDK1	0.01317	0.173	13.684	45.71
GDK2	0.0304	0.0326	5.867	14.60
GDK3	0.0179	0.0253	5.81	14.3
GDK4	0.0308	0.0174	5.388	13.00
GDK5	0.0409	0.0124	5.386	12.98

Shear stress and retention time need to be comprehensively considered to analyze the effects of lumen diameter on catheter function. The PLI of the GDK2 catheter with an outer diameter of 5.5 mm is 3.5% of that of the catheter with an outer diameter of 3.6 mm (GDK); it can be observed that the outer diameter is positively correlated with the average residence time of the platelet model, while it is negatively correlated with the mean shear stress and PLI values.

In contrast to the circular-shaped side holes, the PLI value for the oval-shaped side holes was reduced from 0.0326 (GDK2) to 0.0253 (GDK3) and the average shear stress was also reduced from 5.867 Pa to 5.81 Pa. The changes in these values indicated that the change of side-hole shape would slightly reduce the average shear stress acting on the platelets and PLI, but reduce average residence time in the tip volume in these two catheters.

As shown in Table 4, low PLI was seen with the GDK4 (PLI = 0.0174) and GDK5 (PLI = 0.0124) catheters. GDK5 with parallel dual-side holes experienced slightly longer residence times (0.0409) than the design with linear multi-side holes but had a slightly lower PLI because of a small reduction in high shear stress regions > 10 Pa. It is also clear that the arterial lumen with parallel dual oval-shaped side holes is the most reasonable design and has the lowest PLI and average shear stress values.

Although a large amount of blood is delivered to the downstream region of the lumen, there is still blood circulating in the lumen due to an insufficient pressure gradient, which contributes to the maximum retention time, and the blood circulating in the lumen is also a trigger for thrombosis. The platelet inflows for the GDK2, GDK3, GDK4, and GDK5 catheters are visualized in Figure 6.

It can be seen from Figure 9a,b, with the same arrangement of side holes and side-hole areas, that the flow characteristics of the GDK3 catheter are better than those of GDK2 and that the average RT of the catheter with oval-shaped side holes is lower than that with circular-shaped holes (0.0253 m/s < 0.0326 m/s). This is due to the fact that, with the linear arrangement of side holes, the inflow from the second side hole is obstructed as it flows down the lumen by the inflow from the first side hole. When the shape of the side holes was changed from circular to oval, the blocking effect was reduced, and the flow from the second side hole entered the lumen downstream more quickly, so that, regarding the shape of the side holes in the arterial lumen, oval-shaped side holes have the effect of reducing the RT of blood compared to circular-shaped ones and reduce the risk of thrombosis. As shown in Figure 8b,c, because the areas of the side holes in the GDK4 catheter are larger than those in the GDK3, the pressure gradient in the GDK4 is not sufficient to allow all the blood to enter the downstream region of the lumen, resulting in the circulation of blood at the entrance to the catheter upstream of the lumen, thus making the average RT of the GDK4 catheter larger than that of the GDK3 catheter (0.0308 m/s > 0.0179 m/s). The GDK5 catheter with a parallel dual side-hole arrangement had theoretically lower average RT values than the linear side-hole design, as the inflow was split equally, allowing undisturbed inflow from both holes. However, because of the obvious area of blood self-circulation upstream of the arterial lumen, as can be seen in Figure 9d, which makes the blood turbulent and stagnant in the lumen, the average RT of the GDK5 catheter was larger than that of the GDK4 (0.0409 m/s > 0.0308 m/s).

The recirculation rates for the GDK catheter after structural changes were obtained by a dye tracing experiment. The catheter was connected forward and backward to evaluate the change in catheter dialysis efficiency with different connection methods. The reverse connection is the reversal of the direction of blood flow in the arteriovenous lumen in order to eliminate a thrombus. Each experiment was repeated five times, and the averages and standard errors of the mean values (SEM) were calculated to improve the accuracy of the experimental data. The results of the dye tracing experiment, as shown in Figure 10, indicated that the RRs of GDK, GDK1, and GDK2, which worked with forward and reverse usage, were $0.68 \pm 0.28\%$, $4.69 \pm 1.41\%$, $0.72 \pm 0.24\%$, $5.09 \pm 1.23\%$, $0.46 \pm 0.26\%$, and $4.2 \pm 1.14\%$, respectively; those of the dye tracing experiments showed that the values of RR for GDK3, GDK4, and GDK5, which worked with forward and reverse usage, were $0.26 \pm 0.14\%$, $3.23 \pm 0.96\%$, $0.86 \pm 0.27\%$, $6.64 \pm 1.27\%$, $0.33 \pm 0.12\%$, and $2.86 \pm 0.50\%$, respectively. The recirculation rates of the catheters with the reverse connection between the forward connection were statistically significant ($p < 0.05$, Kruskal–Wallis test). The recirculation rates of GDK5 with the reverse connection were statistically significant with respect to all catheters except GDK3 with the reverse connection. The method of connection and the distribution of the side holes play important roles in recirculation in catheters.

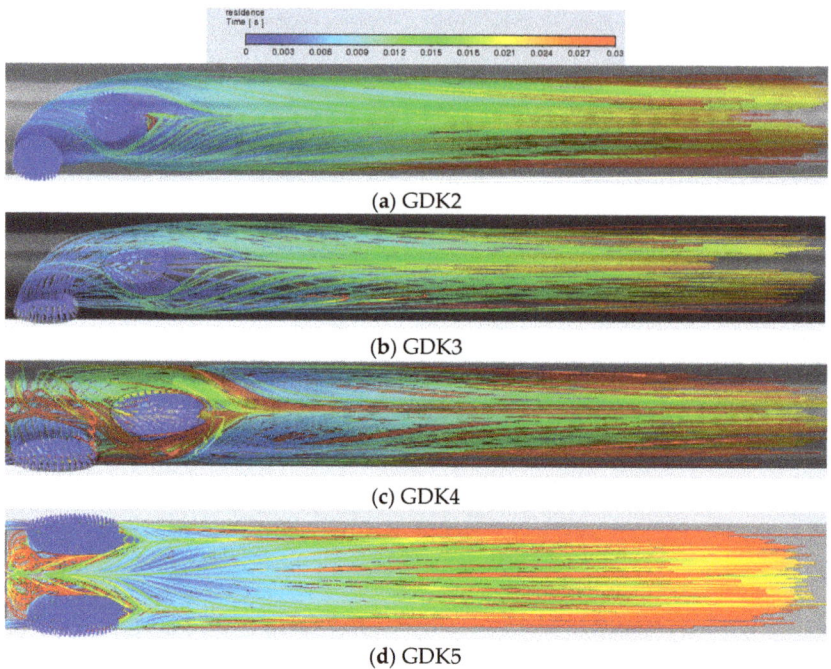

Figure 9. Residence time of platelets in the arterial lumen for GDK2 (**a**), GDK3 (**b**), GDK4 (**c**), and GDK5 (**d**).

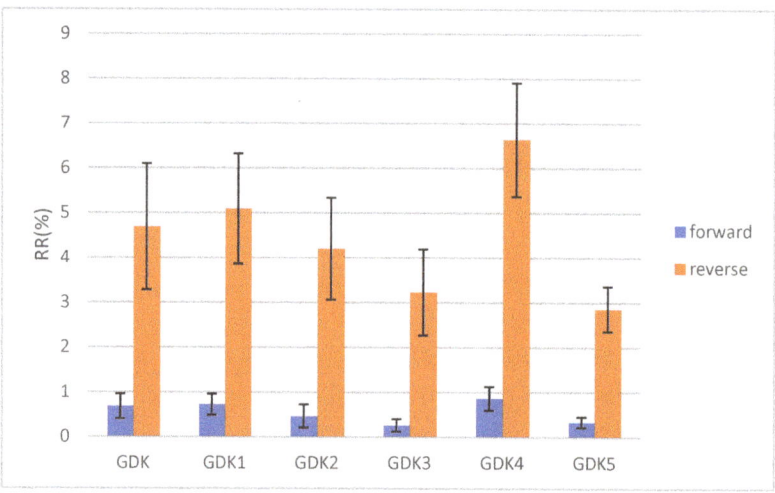

Figure 10. Forward and reverse recirculation rates for the six catheters. The blue columns show RRs (%) for the forward connection and the orange columns show RRs (%) for the reverse connection. The error bars are for RR values (%).

As can be seen from the experiment, all six catheters showed that the recirculation rate of the reverse connection was higher than that of the forward connection. The GDK3 catheter had the lowest recirculation rate for the forward connection, the GDK5 catheter had the lowest recirculation rate for the reverse connection, the GDK5 arterial lumen with double side holes had a lower recirculation rate compared to the other ones, and the GDK4 catheter exhibited the highest recirculation rate for the forward and reverse connections.

4. Discussion

With the continuous development of science and technology, new technologies, new materials, and new structures have been introduced, and new developments in materials, surface coatings, and structures have been introduced in clinical treatment. Catheter material is an important determinant in the prevention of catheter-related infection [7] and usually consists of high polymers (usually polyurethane or silicone). Catheter structural considerations mainly relate to the structure of the lumen, side hole, and tip.

In this study, we took the GDK catheter as a prototype, which has a symmetric tip with the same length as the arterial and venous lumens [9]. The flow characteristics of five types of new HDs, designated based on the GDK catheter were evaluated with CFD and in vitro experiments to investigate the effects of the outer diameter of the catheter, the geometry, and the arrangement of side holes on catheter hemodynamics. The results showed that the catheter with an outside diameter of 5.5 mm and parallel double oval-shaped side holes achieved relatively low shear stress levels and recirculation rates. These results would help catheter structural optimization design in the future, which aims to decrease the risk of dialysis treatment interruption caused by catheter-related infections and insufficient flow and reduce the economic burden and increase the life cycle of patients with kidney disease.

For the commercial GDK catheter, dialysis blood mainly flows out from the distal tip, up to 64.8% of total flow, and the flow rate of the side holes could not be kept high enough to prevent recirculation. Increasing the flow of side holes and decreasing the flow of distal tips helps to reduce the recirculation rate of the catheter and improve dialysis efficiency in symmetric-tip catheters [8]. In the study, two methods were put forward to optimize catheter structure: one is to enlarge the lumen of the catheter, and the other is to change the geometric features of the side holes. The results showed that when the outside diameter of the GDK catheter was enlarged to 5.5 mm and the side-hole shape was oval with a larger area a greater outlet flow through the side holes and a high flow-rate region located in the tip volume would achieve a lower rate of catheter recirculation. The experiments also verified that the GDK3 catheter connection had the lowest recirculation rate. These findings are consistent with those of Cho et al. (2021), who found that larger side holes and a nozzle-shaped distal tip could reduce the flow rate and high shear stress region and improve catheter effectiveness, while Mareels et al. [15] found that the side holes helped reduce shear stress and RT. Shear stress as a risk factor for platelet activation leads to thrombosis. A threshold of 10 Pa of shear stress for platelet activation has been identified, and areas of flow stagnation or recirculation can induce platelet aggregation. It is necessary for HD catheters to reduce blood recirculation and shear stress to ensure performance [6]. For the arterial lumen, it was found that as the outer diameter of the catheter increases, the flow rate of dialysis blood within the catheter decreases, the blood RT increases, the shear stress level in the tip volume decreases, and the PLI decreases further. Therefore, for the design of a catheter, although a larger outer diameter may cause a certain degree of increase in blood RT, due to the reduced level of shear stress in the arterial lumen, increasing the outer diameter would help to reduce the probability of thrombosis and thus increase catheter life.

Side hole design has been a controversial issue. The number and size of side holes play an important role in blood recirculation [24]. Multiple side holes are used in the arterial lumen design for the GDK catheter so that when the side holes of the arterial lumen are blocked by a generated thrombus during the dialysis process, the remaining side holes can be used as a backup entrance to continue the dialysis process. However, in the study, the flow rate of the side holes increased, as the side holes were positioned away from the tip, and the two side holes far from the tip took up as much as 74.4% of the flow rate. As shown in Figure 8, the side holes far from the distal tip were more associated with increased turbulence and subsequent intra-luminal thrombosis [30], so these side holes cannot be used as backup inlets to ensure the normal operation of dialysis after thrombus formation. Therefore, this structure has to be optimized to parallel dual side holes (GDK5) to keep each side hole in the arterial lumen serving as a means of blood suction. It was found that both side holes were assumed to be taken as the blood inlet end, and the flow was evenly divided, which indicated that the parallel dual side holes served to equalize the flow; when one side hole was clogged by a thrombus, the other side hole could continue to provide suction to reduce catheter malfunction. In this study, changing the side holes in the arterial lumen from circular to oval further improved the hemodynamics of the catheter. The large size of the side holes was advantageous, such that GDK4, with oval-shaped side holes, had a better performance, and the oval-shaped side holes as means of inflow allowed for a slow flow of blood into the catheter, indicating that these oval-shaped side holes are more suitable for long-term use compared to circular-shaped ones.

Since the side holes are mostly boreholes with rough walls, the more side holes there are, the more thrombi are generated at the surface of the side holes, which block the catheter in the inner lumen, leading to lower blood performance and shorter dialysis lifespan [31]. The GDK catheter is based on the idea that when blood flow is blocked by thrombus generation in the side holes, more side holes represent more alternate entrances to reduce catheter power loss. In contrast, the design of parallel side holes in the arterial lumen is superior to the original GDK catheter design because blood is evenly distributed and flows do not affect each other, such that dialysis function is not affected when one side hole is blocked due to thrombosis.

Blood recirculation after dialysis is prone to occur in the presence of reverse connections, thus affecting catheter dialysis efficiency. In this paper, the recirculation rates of different catheters were measured by dye tracing experiments. The variation of recirculation depends on the area, geometry, and arrangement of side holes. Most recirculation in dialysis catheters occurs at the distal tip, so it is necessary to reduce the flow rate at the distal tip and increase the proportion of flow through the side holes to reduce recirculation. Through dye tracing experiments, we found that the area of side holes has a certain effect on the recirculation rate and that the GDK4 catheter has a higher forward and reverse recirculation rate than the other catheters, which is due to the fact that the blood after dialysis does not enter into the main circulation but flows from the venous lumen and enters again from the arterial lumen and the purified blood is dialyzed twice, thus reducing the efficiency of hemodialysis.

The arrangement of the side holes has an effect on the recirculation rate of the catheter, and in the experiment catheters with parallel dual side-hole arrangements exhibited lower recirculation rates in both forward and reverse connections. Therefore, in the future commercialization of dialysis catheters, the parallel dual side-hole structure can be a new direction for structural optimization. These improvements would improve hemodialysis efficiency and thus reduce the economic and physical burden of patients with chronic renal failure.

There are several limitations to the study. The first is the that the study only focused on a single manufacturer's catheter; the design and analysis of the other five new catheter models were based on commercial GDK catheters. Another limitation of this study is that in vitro experiments do not allow for the complete evaluation of catheter function, such that experiments with animals should be required in future studies.

5. Conclusions

Optimal catheter design is an urgent need for patients with nephropathy and its ultimate goal is to reduce the failure of dialysis treatment due to unreasonable catheter structure. In this study, we analyzed the design of HD catheters, including changes to diameters and side holes based on the GDK catheter, in order to evaluate their effects on hemodynamic factors, such as flow rate, shear stress, PLI, RT, and RR, through numerical simulation and in vitro experiments. The results indicated that larger outer diameters and oval-shaped side holes can reduce average shear stress in the arterial lumen and that oval-shaped side holes are effective for reducing RRs. The parallel dual side-hole structures have uniformly distributed side hole flow and the combination of a 5.5 mm diameter and parallel dual oval shapes for side holes exhibited better hemodynamic properties, thus providing better performance than the existing models.

Author Contributions: Conceptualization, Y.Y., Y.X. and L.Z.; methodology, Y.Y. and Y.X.; software, Y.Y., Y.L. and J.Z.; validation, Y.Y., C.L. and T.L.; formal analysis, Y.Y. and Y.L.; data curation, Y.Y. and C.L.; writing—original draft preparation, Y.X.; writing—review and editing, Y.Y. and Y.X.; visualization, Y.Y. and L.Z.; supervision, Y.X. and L.Z.; project administration, Y.X. and L.Z.; funding acquisition, L.Z. All authors have read and agreed to the published version of the manuscript.

Funding: This research was funded by Sichuan Provincial Science and Technology Department grant number [2020YFG0105] And The APC was funded by Zhang, L.

Institutional Review Board Statement: Not applicable.

Informed Consent Statement: Not applicable.

Data Availability Statement: Not applicable.

Conflicts of Interest: The authors declare no conflict of interest.

References

1. Mehta, R.L.; Cerdá, J.; Burdmann, E.A.; Tonelli, M.; García-García, G.; Jha, V.; Susantitaphong, P.; Rocco, M.; Vanholder, R.; Sever, M.S.; et al. International Society of Nephrology's 0by25 initiative for acute kidney injury (zero preventable deaths by 2025): A human rights case for nephrology. *Lancet* **2015**, *385*, 2616–2643. [CrossRef]
2. Xu, X.D.; Han, X.; Yang, Y.; Li, X. Comparative study on the efficacy of peritoneal dialysis and hemodialysis in patients with end-stage diabetic nephropathy. *Pak. J. Med. Sci.* **2020**, *36*, 1484. [CrossRef]
3. Hirakata, T. Japanese Society for Dialysis Therapy Guidelines for Management of Cardiovascular Diseases in Patients on Chronic Hemodialysis. *Ther. Apher. Dial.* **2012**, *16*, 387–435. [CrossRef]
4. U.S. Renal Data System. USRDS 2013 Annual Data Report: Atlas of Chronic Kidney Disease and End-Stage Renal Disease in the United States (Table D7). *Natl. Inst. Health Natl. Inst. Diabetes Dig. Kidney Dis.* **2013**, *2014*, 83–95.
5. Wang, Y.L.; Ye, Z.Y.; Jin, Q.Z. *Expert Consensus on Vascular Access for Hemodialysis in China*, 1st ed.; China Blood Purification: Bei Jing, China, 2014; Volume 13, pp. 549–558. (In Chinese)
6. Depner, T.A. Catheter performance. In *Seminars in Dialysis*; Blackwell Science Inc: New York, NY, USA, 2001; Volume 14, pp. 425–431.
7. Frasca, D.; Dahyot-Fizelier, C.; Mimoz, O. Prevention of central venous catheter-related infection in the intensive care unit. *Crit. Care* **2010**, *14*, 212. [CrossRef] [PubMed]
8. Cho, S.; Song, R.; Park, S.C.; Park, H.S.; Abbasi, M.S.; Lee, J. Development of New Hemodialysis Catheter Using Numerical Analysis and Experiments. *ASAIO J.* **2021**, *67*, 817–824. [CrossRef] [PubMed]
9. Tal, M.G. Comparison of Recirculation Percentage of the Palindrome Catheter and Standard Hemodialysis Catheters in a Swine Model. *J. Vasc. Interv. Radiol.* **2005**, *16*, 1237–1240. [CrossRef]
10. Ling, X.C.; Lu, H.P.; Loh, E.W.; Lin, Y.K.; Li, Y.S.; Lin, C.H.; Ko, Y.C.; Wu, M.Y.; Lin, Y.F.; Tam, K.W. A systematic review and meta-analysis of the comparison of performance among step-tip, split-tip, and symmetrical-tip hemodialysis catheters. *J. Vasc. Surg.* **2019**, *69*, 1282–1292. [CrossRef]
11. Ogawa, T.; Sasaki, Y.; Kanayama, Y.; Yasuda, K.; Okada, Y.; Kogure, Y.; Sano, T.; Hatano, M.; Hara, H.; Kanozawa, K.; et al. Evaluation of the functions of the temporary catheter with various tip types. *Hemodial. Int.* **2017**, *21*, S10–S15. [CrossRef]
12. al, M.G.; Peixoto, A.J.; Crowley, S.T.; Denbow, N.; Eliseo, D.; Pollak, J. Comparison of side hole versus non side hole high flow hemodialysis catheters. *Hemodial. Int.* **2006**, *10*, 63–67.
13. Owen, D.G.; de Oliveira, D.C.; Qian, S.; Green, N.C.; Shepherd, D.E.; Espino, D.M. Impact of side-hole geometry on the performance of hemodialysis catheter tips: A computational fluid dynamics assessment. *PLoS ONE* **2020**, *15*, e0236946. [CrossRef]
14. Clark, T.W.I.; Van Canneyt, K.; Verdonck, P. Computational flow dynamics and preclinical assessment of a novel hemodialysis catheter. In *Seminars in Dialysis*; Blackwell Publishing Ltd.: Oxford, UK, 2012; Volume 25, pp. 574–581.
15. Nobili, M.; Sheriff, J.; Morbiducci, U.; Redaelli, A.; Bluestein, D. Platelet activation due to hemodynamic shear stresses: Damage accumulation model and comparison to in vitro measurements. *ASAIO J.* **2008**, *54*, 64. [CrossRef]
16. Tan, J.; Mohan, S.; Herbert, L.; Anderson, H.; Cheng, J.T. Identifying hemodialysis catheter recirculation using effective ionic dialysance. *ASAIO J.* **2012**, *58*, 522–525. [CrossRef] [PubMed]
17. Vesely, T.M.; Ravenscroft, A. Hemodialysis catheter tip design: Observations on fluid flow and recirculation. *J. Vasc. Access* **2016**, *17*, 29. [CrossRef] [PubMed]
18. Strony, J.; Beaudoin, A.; Brands, D.; Adelman, B. Analysis of shear stress and hemodynamic factors in a model of coronary artery stenosis and thrombosis. *Am. J. Physiol.* **1993**, *265*, H1787–H1796. [CrossRef] [PubMed]

19. Carty, G.; Chatpun, S.; Espino, D.M. Modeling Blood Flow Through Intracranial Aneurysms: A Comparison of Newtonian and Non-Newtonian Viscosity. *J. Med. Biol. Eng.* **2016**, *36*, 396–409. [CrossRef]
20. Lee, S.W.; Steinman, D.A. On the relative importance of rheology for image-based CFD models of the carotid bifurcation. *J. Biomech. Eng.* **2007**, *129*, 273–278. [CrossRef]
21. Trerotola, S.O. Hemodialysis catheter placement and management. *Radiology* **2000**, *215*, 651–658. [CrossRef]
22. Giersiepen, M.; Wurzinger, L.; Opitz, R.; Reul, H. Estimation of shear stress-related blood damage in heart valve prostheses-in vitro comparison of 25 aortic valves. *Int. J. Artif. Organs* **1990**, *13*, 300–306. [CrossRef]
23. Mareels, G.; Kaminsky, R.; Eloot, S.; Verdonck, P.R. Particle image velocimetry–validated, computational fluid dynamics–based design to reduce shear stress and residence time in central venous hemodialysis catheters. *ASAIO J.* **2007**, *53*, 438–446. [CrossRef]
24. Clark, T.W.; Isu, G.; Gallo, D.; Verdonck, P.; Morbiducci, U. Comparison of symmetric hemodialysis catheters using computational fluid dynamics. *J. Vasc. Interv. Radiol.* **2015**, *26*, 252–259. [CrossRef] [PubMed]
25. Yi, L.P.; Zhang, Y.F. Hemodialysis catheter choice and infection control. *Zhongguo Zuzhi Gongcheng Yanjiu Yu Linchuang Kangfu* **2011**, *15*, 5421. (In Chinese)
26. Ash, S.R. Fluid mechanics and clinical success of central venous catheters for dialysis—Answers to simple but persisting problems. In *Seminars in Dialysis*; Blackwell Publishing Ltd.: Oxford, UK, 2007; Volume 20, pp. 237–256.
27. Luo, J.; Wang, Y.; Zhang, Y.; Jiang, J. Concentration detection of dye components with dual-wavelength spectrophotometry. *J. Silk* **2016**, *53*, 16–20.
28. Mareels, G.; De Wachter, D.S.; Verdonck, P.R. Computational fluid dynamics-analysis of the Niagara hemodialysis catheter in a right heart model. *Artif. Organs* **2004**, *28*, 639–648. [CrossRef] [PubMed]
29. Hellums, J.D. Response of platelets to shear stress: A review. *Rheol. Blood Vessel. Assoc. Tissues* **1981**, *214*, 104.
30. El Khudari, H.; Ozen, M.; Kowalczyk, B.; Bassuner, J.; Almehmi, A. Hemodialysis Catheters: Update on Types, Outcomes, Designs and Complications. In *Seminars in Interventional Radiology*; Thieme Medical Publishers, Inc: New York, NY, USA, 2022; Volume 39, pp. 90–102.
31. Twardowski, Z.J.; Moore, H.L. Side holes at the tip of chronic hemodialysis catheters are harmful. *J. Vasc. Access* **2001**, *2*, 8–16. [CrossRef]

Review

Image-Based Finite Element Modeling Approach for Characterizing In Vivo Mechanical Properties of Human Arteries

Liang Wang [1], Akiko Maehara [2], Rui Lv [1], Xiaoya Guo [3], Jie Zheng [4], Kisten L. Billiar [5], Gary S. Mintz [2] and Dalin Tang [1,6,*]

1. School of Biological Science and Medical Engineering, Southeast University, Nanjing 210096, China
2. The Cardiovascular Research Foundation, Columbia University, New York, NY 10019, USA
3. School of Science, Nanjing University of Posts and Telecommunications, Nanjing 210023, China
4. Mallinckrodt Institute of Radiology, Washington University, St. Louis, MO 63110, USA
5. Department of Biomedical Engineering, Worcester Polytechnic Institute, Worcester, MA 01609, USA
6. Mathematical Sciences Department, Worcester Polytechnic Institute, Worcester, MA 01609, USA
* Correspondence: dtang@wpi.edu; Tel.: +1-508-831-5332

Abstract: Mechanical properties of the arterial walls could provide meaningful information for the diagnosis, management and treatment of cardiovascular diseases. Classically, various experimental approaches were conducted on dissected arterial tissues to obtain their stress–stretch relationship, which has limited value clinically. Therefore, there is a pressing need to obtain biomechanical behaviors of these vascular tissues in vivo for personalized treatment. This paper reviews the methods to quantify arterial mechanical properties in vivo. Among these methods, we emphasize a novel approach using image-based finite element models to iteratively determine the material properties of the arterial tissues. This approach has been successfully applied to arterial walls in various vascular beds. The mechanical properties obtained from the in vivo approach were compared to those from ex vivo experimental studies to investigate whether any discrepancy in material properties exists for both approaches. Arterial tissue stiffness values from in vivo studies generally were in the same magnitude as those from ex vivo studies, but with lower average values. Some methodological issues, including solution uniqueness and robustness; method validation; and model assumptions and limitations were discussed. Clinical applications of this approach were also addressed to highlight their potential in translation from research tools to cardiovascular disease management.

Keywords: finite element updating approach; arterial material properties; in vivo; material parameters estimation

1. Introduction

Mechanical forces play a fundamental role in the initiation, development and final critical clinical events of cardiovascular diseases (CVD) such as stroke and heart attack [1,2]. As diseases progress, biochemical compositions of cardiovascular tissues alter as well as their mechanical properties [3,4]. Clinical observations have shown that elevated tissue stiffness associated with pathology often represents an early warning sign of diseases, as in atherosclerosis [4], heart failure [5] and even cancer diseases [3]. Therefore, accurate determination of the mechanical properties of the arterial wall could provide meaningful information for cardiovascular research in multifaceted ways: (a) estimating the material stiffness of the vascular tissues to assess the severity of cardiovascular diseases such as atherosclerosis [6]; (b) being an essential element of the computational models to simulate biomechanical conditions for better understanding cardiovascular physiology and pathophysiology such as stress-based aortic aneurysm rupture risk assessment [7]; (c) searching for plausible substitutes with proper mechanical properties to replace diseased arterial segments, such as tissue engineering vascular grafts [8]; (d) predicting the mechanical interactions between arteries and implanted devices for better treatment prognosis [9]; and many other applications.

Great effort has been exerted to perform ex vivo experiments on human arterial walls in health and disease, and important conclusions on their mechanical behaviors were drawn [10–12]. Ex vivo experiments includes inflation-extension testing, indentation testing, uniaxial extension and planar biaxial testing [13–15]. These experiments record the deformation responses (stretch, strain, displacement or elongation) of the arterial tissues corresponding to given loading conditions (stress, force or pressure). Several constitutive models were proposed to fit these experimental stress–strain or stress–stretch ratio data [11]. These theoretical and experimental analyses have deepened our understanding on the elastic mechanical properties of vascular tissues. Early experimental studies on healthy arteries showed that the stress–stretch ratio curve of the cardiovascular tissue was typically in exponential form. Therefore, a Fung-type-material model was proposed to describe the material properties for these tissues [10]. To study the mechanical properties of diseased tissues, Holzapfel et al., examined atherosclerotic plaque tissues in the iliac artery ex vivo using uniaxial testing [16]. Experimental data indicated that tissue properties were highly nonlinear and anisotropic. An anisotropic Mooney–Rivlin material model was introduced to describe the mechanical properties of atherosclerotic plaques [11]. Furthermore, layer-specific and component-specific material properties of carotid plaque were also documented, and a large inter-specimen variation was reported [17]. Location-specific material properties along the aortic segments were also systematically investigated, and Peña et al., reported that the healthy aortic tissue became more anisotropic and stiffer as the distance to the heart increased [18]. The experimental data collected from these ex vivo studies are fundamental to formulating and testing the constitutive models. Excellent reviews on these experimental methods could be found in the literature [13–15].

Even though considerable ex vivo experimental data have accumulated over the years, they only provided biomechanical information of arterial diseases at one time-point, which is very limited for clinical applications for two reasons: (a) ex vivo arterial tissues isolated from the vascular tree might not represent the exact mechanical properties as in the living subjects [19]; (b) characterization of patient-specific tissue material properties in vivo is more suitable for monitoring arterial stiffness over the long term for disease management, and predicting mechanical responses of the arteries when medical devices are implanted [6]. Therefore, there is a pressing need to characterize patient-specific tissue material properties in vivo.

This review paper aims to review the studies to characterize subject-specific arterial wall properties in vivo using a finite-element-model-based updating approach (FEMBUA). Material properties of the aorta, carotid and coronary arteries determined from the novel FEMBUA and classical ex vivo experimental methods were compared to investigate whether any discrepancy in material properties exists for both approaches. Section 2 will present other simple methods to characterize arterial tissue material properties in vivo other than the FEMBUA. The comprehensive framework of the FEMBUA and its elaborate procedure will be detailed in Section 3. Section 4 will report the mechanical properties of different arterial walls, including aorta, carotid and coronary, published in in vivo studies, and their results will be further compared to those from ex vivo experimental studies. Section 5 will discuss some methodological issues with respect to the FEMBUA, followed by the conclusion remarks and future directions in Section 6.

2. In Vivo Methods to Quantify Material Properties of Arterial Walls

Several in vivo methods were proposed to characterize subject-specific material properties of arterial walls in vivo or even in situ. The principle of these methods is to acquire the arterial wall deformation and corresponding loading conditions from clinical data, and link them to obtain the arterial material properties. Thanks to the considerable advances in medical imaging technologies, dynamic vessel motion can be recorded in vivo via time-resolved imaging modalities, such as ultrasound (US), cine magnetic resonance imaging (MRI), cine intravascular ultrasound (IVUS) and electrocardiogram (ECG)-gated computed tomography (CT). Blood pressure conditions are also measured as the loading conditions to drive the vessel to pulsating deformations. Simple and sophisticated methods were proposed in the existing literature based on different assumptions.

Early in vivo studies of vessel material properties assumed that the arterial wall was a straight thin-walled circular cylinder with linear elastic material properties, and its stiffness could be estimated by Peterson's modulus (denoted as Ep) using the formula [19,20]:

$$Ep = \Delta P \times d / \Delta d \tag{1}$$

where d is the diameter, and ΔP and Δd are the differences in pressure and diameter at diastole and systole, respectively. This method is very intuitive and provides a simple way to measure arterial stiffness using pressure and strain data. However, a more rigorous index for material stiffness is called Young's modulus (E), which is the ratio of stress over strain. Formula (1), together with Laplace's law, is employed to calculate E as [21]:

$$E = \Delta P \times d^2 / 2h\Delta d \ (= d/2h \times Ep) \tag{2}$$

where h is the wall thickness.

Another simple and commonly used formula to estimate the Young's modulus of the arterial wall is the Moens–Kortoweg equation, which relates the Young's modulus of the blood vessel with pulse wave velocity (PWV) from the heart to the peripheral vascular [20]:

$$PWV = \sqrt{Eh/\rho d} \tag{3}$$

where ρ is the density of the blood. More details on the derivation of the abovementioned formulas can be found in [20]. These simple in vivo methods have already been applied into clinical settings to assess the risk of cardiovascular diseases [22,23]. However, they rely on strong assumptions without considering arterial thickness and geometry, non-linear anisotropic elastic properties, tissue heterogeneity and location-specific material property variations.

To obtain more sophisticated nonlinear anisotropic mechanical behavior of arterial walls, an analytical approach was employed with constitutive models incorporated to characterize their nonlinear stress–strain relationship by identifying the material parameters in the constitutive models [24]. In this approach, material parameters of the constitutive models could be determined as follows: (1) pressure and vessel deformation were measured simultaneously from the individuals in clinical practice; (2) classical solid mechanics theory was used to establish the relationship between vessel stress and pressure analytically to obtain stress conditions using measured pressure conditions [24–26]; (3) parameter values in selected constitutive models would be chosen to fit vessel stress and deformation data. However, this approach treated the blood vessel as ideal circular geometry, and was based on classical solid mechanics theory (like Laplace's law) which excluded the discrepancy of deformation in the arterial tissue. This simplification would lead to inaccurate calculations of pressure estimation and stress/strain distributions in the arteries, and wipe out the local stress concentration, which are closely related to atherosclerotic plaque rupture and aneurysm rupture. To overcome this limitation, a finite-element-model-based updating approach (FEMBUA) was introduced by several groups to quantify the complex mechanical properties of patient-specific arterial tissues. The following section will introduce the sophisticated framework of this approach.

3. Framework of Finite-Element-Model-Based Updating Approach

The in vivo quantification of mechanical properties of arterial tissues based on in vivo medical images is intrinsically an inverse problem [27,28]. The steps of the FEMBUA to solve the inverse problem are outlined below (Figure 1):

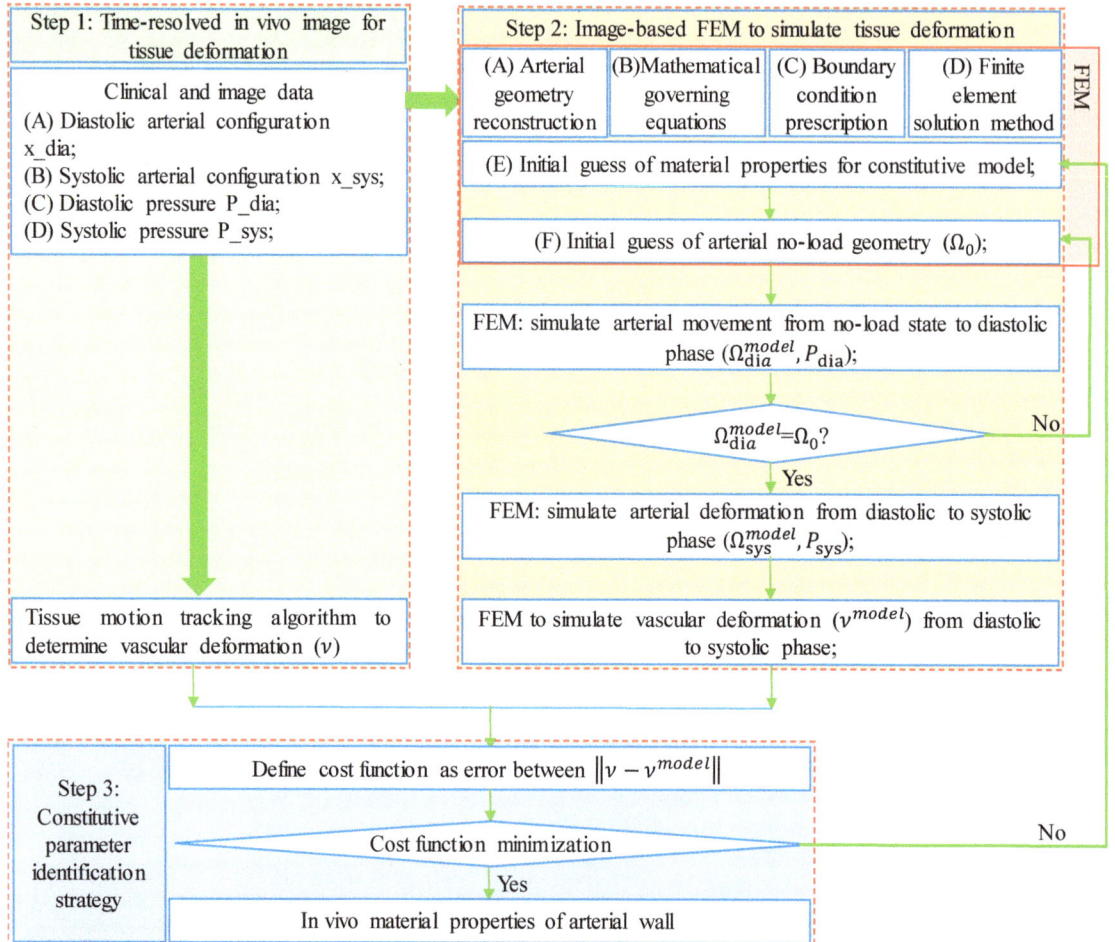

Figure 1. Flowchart of finite-element-model-based updating method to identify the in vivo material properties based on clinical data at diastolic and systolic phases using deformation as criterion.

Step 1, time-resolved in vivo medical image acquisition with tissue deformation under dynamic loading conditions;

Step 2, image-based FEM to simulate the tissue deformation corresponding to in vivo loading conditions with tissue material properties to be determined;

Step 3, constitutive parameter identification strategy to find the correct tissue material properties so that tissue deformation in FEM would recover that on in vivo images. Full descriptions of these steps are given in the following subsections.

3.1. Data Acquisition and Vessel/Tissue Motion Tracking

In Vivo Clinical Data Acquisition: To visualize the vascular deformation in vivo, time-resolved imaging modality was employed to obtain a series of medical images to track arterial deformation during one cardiac cycle. Table 1 summarizes some commonly used imaging technologies to detect vessel motion for various arterial walls in the clinical setting. More specifically, time-resolved 3D ultrasound (t+3D US) [29] and ECG-gated CT [30] were used to track aortic tissue motion with acceptable image resolution. Cine-MRI [31] and Cine-IVUS [32] were used to obtain carotid and coronary motion data, respectively. Besides the image data, simultaneous loading conditions, such as on-site pulsating blood pressure waveform, were also acquired. Noninvasively measured arm cuff blood pressures were obtained in some studies in lieu of on-site blood pressure to avoid invasive procedures [32]. Other loading information such as external compression force and active stress from smooth muscle cells cannot be estimated in vivo. These forces were normally not considered in FEM [26].

Table 1. Selected time-resolved image modalities to visualize human vascular motion and deformation in clinical setting. Abbreviations: Time-resolved 3D ultrasound (t+3D US); electrocardiogram (ECG)-gated computed tomography (ECG-gated CT); cine magnetic resonance imaging (MRI); cine intravascular ultrasound (IVUS).

Image Modality	Temporal Resolution	Spatial Resolution	Artery	Strength and Weakness in Arterial Wall Detection	Reference
t + 3D (4D) US	~10 frames/s	~0.5 mm	Aorta	Cheap, fast and easy way to detect arterial boundaries and tissue compositions, but inter- and intra-observer variability in image interpretation;	[29,33]
ECG-gated CT	~10 frames/cardiac cycle	~0.5 mm	Aorta	Superb calcified tissue detection and lumen detection; limited in detecting other plaque compositions, such as lipid and vessel wall;	[30,34,35]
Cine MRI	~50 frames/cardiac cycle	~0.6 mm	Carotid	Detection of the whole vascular cross-section with superior soft-tissue contrast, but long scanning time;	[31,36,37]
Cine IVUS	~30 frames/s	100 μm	Coronary	High resolution and large penetration depth for arterial tissue detection, also can detect arterial tissue compositions;	[32,38]

Vessel/Tissue Motion Tracking: Arteries deform under time-varying loading conditions. One simple way to approximate the deformation is to calculate the changes in lumen circumference, lumen area or lumen volume between two cardiac phases and consider it as vessel deformation under two different pressure conditions. Typically, diastolic and systolic phases were selected to quantify the deformation of the arterial wall, with the diastolic phase often treated as the "reference" phase and systolic phase as the deformed phase (see Figure 1). This way, we could quantify the average deformation of the arterial wall by treating it as a homogeneous material. However, more sophisticated methods, such as the speckle tracking algorithm [39] and digit image correlation [40], were introduced to track regional tissue displacement by examining the cross-correlation of the speckle patterns in the medical images from diastolic to systolic phases. The accuracy of these algorithms has been validated in vivo and in vitro, with good agreements found [29]. More details on the algorithms could be found in these excellent reviews [40,41]. Franquet et al., also investigated vessel material properties with more cardiac phases taken into consideration [37].

3.2. Image-Based Finite Element Models

In the FEMBUA, image-based finite element models (FEMs) are constructed to calculate plaque stress/strain conditions, while the parameter values in those models are determined iteratively so that model solutions can satisfy measured vessel/tissue deformation conditions. The FEM models constructed to identify the material properties of the arterial wall in the literature were mainly structure-only models to expedite the updating approach detailed in Section 3.3. The essential elements to construct such a model contain the following procedures:

Arterial Wall Geometry Reconstruction: 3D US, CT, MRI and IVUS images can show the cross-section of the arterial wall, and the images corresponding to the diastolic phase were used to reconstruct the referenced geometry of the arterial wall [34,42,43]. For some imaging modalities, some heterogeneous components (e.g., atherosclerotic plaque compositions, intraluminal thrombus) can also be detected and reconstructed for more accurate representation of vessel wall structure [43].

Pre-Stressing Geometry Estimation: Since medical images were acquired in vivo, the referenced geometry reconstructed were loaded with the physiological pressure on the luminal surface, axially stretched, and other loading conditions (such as circumferential residual stress). Thus, a pre-stressing algorithm should be performed to obtain no-load geometry (corresponding to zero-pressure condition) based on the referenced geometry as the initial geometry to start the computational simulation. To this end, Guo et al., employed a pre-shrink stretch procedure by shrinking the referenced coronary artery geometry circumferentially and axially to obtain the no-load geometry [32]. Speelman et al., proposed a backward incremental method to estimate the arterial geometry under the no-load state [44]. Other patient-specific algorithms were also developed and can be found in the relevant reference [45]. It should be noted that the no-load geometry should be estimated with the prerequisite of known material properties of the arterial wall. However, since the arterial material properties are unknown at this stage, no-load geometry would be determined along with in vivo arterial material properties following the constitutive parameter identification strategy specified in Section 3.3.

Constitutive Models for Arterial Wall: Arterial walls are generally treated as elastic, either anisotropic or isotropic, nearly-incompressible, homogeneous material. They could also be considered as heterogeneous material if different tissue components were included [46]. Several constitutive models were proposed to describe their mechanical properties ranging from the simple Hookean model to the more sophisticated nonlinear anisotropic ones [11,47]. A list of commonly used material models including the Hookean model [31], NeoHookean model [48], Yeoh model [49], Demiray model [30], Mooney–Rivlin (MR) model [11], Gasser–Ogden–Holzapfel (GOH) model [50], Holzapfel2005 model [51], Fung-type material [10,52] and their strain energy density functions are provided in the supplement (see Supplementary File). The constitutive parameters in the material models were to be determined following the strategy in Section 3.3.

Mathematical Equations Governing Arterial Wall Motion: The mathematical equations governing arterial wall motion consist of equations of motion, strain-displacement relations and the stress–strain relations that could be derived from strain energy density functions for hyperelastic materials [53]. With a proper prescription of boundary conditions, this equation system could be solved to obtain vessel biomechanical conditions, such as arterial wall deformation and stress/strain conditions, which would be used to compare with corresponding clinical measurements.

Boundary Conditions: Proper boundary conditions were applied to FEM to mimic the loading conditions on the arterial wall in vivo. The most important loading condition that triggers vascular deformation is pulsating pressure prescribed on the luminal surface on the arterial wall [42]. It corresponds to the differential blood pressure measured from each individual between the referenced state and the deformed state. In addition to the pressure conditions, some other loading conditions, such as external pressure conditions and axial stretch, were also considered in some studies for more accurate simulations [24,26]. Proper fixity boundary conditions should be applied to avoid unexpected rigid body movement of the arterial wall [54].

Solution Method for the Finite Element Model: Finite element mesh could be generated using commercial finite element software such as ANSYS, ADINA, Abaqus or self-developed in-house software. FEM models were further solved using sophisticated numerical schemes built in these software. Mesh analysis should be performed to guarantee the accuracy of the model solution, which would affect the accuracy of material parameter estimation [42].

3.3. Constitutive Parameter Identification Strategy

To obtain vessel material properties correctly, vessel deformation from FEM should match those from in vivo medical images [55]. A cost function was introduced to measure the discrepancy between arterial wall deformation from FEM and from in vivo images. Then, an optimization algorithm was utilized to find the optimal material parameters along with the no-load vessel geometry by minimizing the cost function.

Cost Function Definition: Most studies constructed the cost function as the sum of the squares of the nodal deformation differences between FEM vessel morphology and vessel geometry reconstructed from images [34,36]. Some studies also simply defined the cost function as the square difference in lumen circumference or lumen area or lumen volume [32]. These measurements are more available, and easier to calculate. However, the drawback is that these measurements only provide one quantity from one FEM, so arterial tissues must be assumed to be uniform homogeneous materials. Variations of the cost functions also exist by comparing the difference in pressure or stress from FEM and from clinical measurements in prior studies [26,36].

Optimization Algorithm to Search Correct Material Properties: The value of the cost function abovementioned is dependent on the material parameters in FEM models. It is a nonlinear, multivariate optimization problem to find the correct constitutive parameters by minimizing the cost function [28,29], especially in the case considering the arterial wall as heterogeneous material [56]. The number of material parameters increases linearly as the arterial wall is divided into several subdomains with different material parameters.

There are two essential difficulties [24] inherent in this type of nonlinear and nonconvex optimization problem: (1) the cost function has multiple local minima, and simple gradient-based algorithms may not be able to find a global minimum; (2) a second more fundamental difficulty is over-parameterization, and solutions of material parameters may not be unique. To address these difficulties, a combined stochastic/deterministic approach was recommended in some studies with a two-step approach: Step 1, hundreds of sets of material parameters were chosen randomly by the Monte Carlo algorithm to evaluate the cost function; Step 2, a deterministic nonlinear algorithm, such as the Nelder–Mead simplex algorithm [57], was applied to obtain the final material parameters, using the parameter set with the minimal cost function value determined from Step 1 as initial parameters.

Once the constitutive parameters were found, a stress–strain (or stress–stretch ratio) relationship could be derived from the strain energy density functions. More details on the derivation can be found in the existing literature and are omitted here [11,18,52]. For comparison purpose, the effective Young's modulus was defined as the slope of the proportional function to fit nonlinear stress–stretch ratio material curves on the stretch interval [1.0 1.1] to measure the tissue stiffness [58]. For anisotropic material models, effective Young's moduli were calculated for the material curves along both the circumferential and longitudinal directions (by fixing the stretch ratio to 1.0 in the other direction) and denoted as E_c and E_a, respectively.

4. In Vivo Mechanical Properties of Individual-Specific Arterial Wall Tissue

The framework of the FEMBUA was employed to determine individual-specific material properties of arterial tissues for mostly middle- or large-size arteries in vivo [59]. To examine the difference in material properties obtained from the FEMBUA and classical experimental approaches, some representative ex vivo experimental studies were selected, and their results were compared with those from in vivo studies using the FEMBUA. Table 2 lists some prior studies on human aorta, carotid and coronary arteries using this in vivo identification approach, as well as some ex vivo experimental studies for comparison purpose. Subject information of these in vivo and ex vivo studies is also provided. Furthermore, one representative set of material parameters (average values or median values of the material parameters for all samples from each study) were chosen to plot the material curves and calculate the effective Young's modulus. More details on each vascular bed are given in the following subsections.

Table 2. Subject information, study details and mechanical properties results from in vivo and some representative ex vivo studies. Abbreviations: AA, abdominal aorta; AAA, abdominal aortic aneurysm; AsA, ascending thoracic aorta; AsAA, ascending thoracic aortic aneurysm; DsA: descending thoracic aorta; Ec, effective Young's modulus in circumferential direction; Ea, effective Young's modulus in longitudinal direction.

Reference	Tissue Sample Information	Material Model	Imaging/ Experiment Techniques	Effective Young's Modulus
		In Vivo Aorta		
[29]	5 AA samples from 5 healthy subjects	GOH model	t + 3D US	Ec = 969.5 kPa Ea = 843.7 kPa
[30]	5 AsAA samples from 5 patients	Demiray model	ECG-gated CT	Ec = Ea = 180.3 kPa
[33]	1 AA sample from 1 healthy subject	GOH model	t + 3D US	Ec = 605.7 kPa Ea = 605.4 kPa
	1 AAA sample from 1 patient			Ec = 5576.7 kPa Ea = 1770.2 kPa
[34]	4 AsAA samples from 4 patients	GOH model	ECG-gated CT	Ec = 270.2 kPa Ea = 276.5 kPa
[35]	4 AsAA samples from 4 patients	GOH model	ECG-gated CT	Ec = 363.1 kPa Ea = 355.7 kPa
[55]	9 AsAA samples from 9 patients	Yeoh model	ECG-gated CT	Ec = Ea = 573.9 kPa
		Ex Vivo Aorta		
[49]	69 AAA specimens	Yeoh model	Uniaxial testing	Ec = 2382.4 kPa Ea = 1856.3 kPa
[60]	6 AsA specimens from donors with age 0 to 30	GOH model	Biaxial testing	Ec = 1268.4 kPa Ea = 1182.1 kPa
	6 AsA specimens from donors with age 31 to 60			Ec = 1025.5 kPa Ea = 905.9 kPa
	17 AsA specimens from donors with age above 61			Ec = 2365.8 kPa Ea = 1698.6 kPa
[61]	5 DsA specimens from 5 young donors with age 20 to 36	MR model	Uniaxial testing	Ec = 181.5 kPa Ea = 176.0 kPa
	5 DsA specimens from 5 old donors with age 45 to 60			Ec = 232.0 kPa Ea = 186.5 kPa
		In Vivo Carotid		
[36]	12 atherosclerotic carotid samples from 12 patients	MR model	Cine MRI	Ec = Ea = 422.6 kPa
[31]	2 carotid samples from 2 healthy subjects	Hookean model	Cine MRI	Ec = Ea = 781.8 kPa
[37]	4 carotid samples from 4 young healthy subjects with age 24 to 26	Hookean model	Cine MRI	Ec = Ea = 833.7 kPa
	5 carotid samples from 5 middle-age healthy subjects with age 51 to 63			Ec = Ea = 1815.3 kPa
	4 atherosclerotic carotid samples from 4 old patients with age 68 to 76			Ec = Ea = 6926.2 kPa
[58]	81 atherosclerotic carotid samples from 8 patients	MR model	Cine MRI	Ec = Ea = 555.1 kPa
		Ex Vivo Carotid		
[62]	14 atherosclerotic carotid specimens from 14 patients	Yeoh model	Uniaxial testing	Ec = Ea = 606.2 kPa
[63]	11 common carotid specimens from 11 relatively healthy subjects	Hozapfel2005 model	Extension-inflation tests	Ec = 1235.7 kPa Ea = 176.7 kPa
[17]	59 atherosclerotic carotid specimens of fibrous cap	MR model	Uniaxial testing	Ec = Ea = 1245.4 kPa
		In Vivo Coronary		
[32]	2 atherosclerotic coronary samples from 1 patient	MR model	Cine IVUS	Ec = 484.6 kPa Ea = 279.8 kPa
[38]	20 atherosclerotic coronary samples from 13 patients	MR model	Cine IVUS	Ec = 1022.5 kPa Ea = 590.6 kPa

Table 2. Cont.

Reference	Tissue Sample Information	Material Model	Imaging/Experiment Techniques	Effective Young's Modulus
		Ex Vivo Coronary		
[51]	13 coronary intima specimens from 13 relatively healthy subjects	Hozapfel2005 model	Uniaxial testing	Ec = 497.5 kPa Ea = 862.6 kPa
[64]	4 coronary specimens from 2 relatively healthy subjects	MR model	Biaxial testing	Ec = 1602.5 kPa Ea = 925.3 kPa
[65]	14 healthy coronary specimens 8 atherosclerotic coronary specimens	Hookean model	Uniaxial testing	Ec = Ea = 1909.5 kPa Ec = Ea = 4864.1 kPa

4.1. Aortic Tissue

The human aorta contains a wide range of vascular course, including ascending thoracic aorta (AsA), aortic arch, descending thoracic aorta (DsA) and abdominal aorta (AA). Aortic aneurysm is a common pathological condition influencing the health state of the aorta. The stress–stretch ratio curves of healthy and aneurysmal aortic tissues listed in Table 2 using the in vivo FEMBUA method and ex vivo experimental methods are plotted in Figure 2. Material curves in the circumferential and longitudinal directions were given for anisotropic material models.

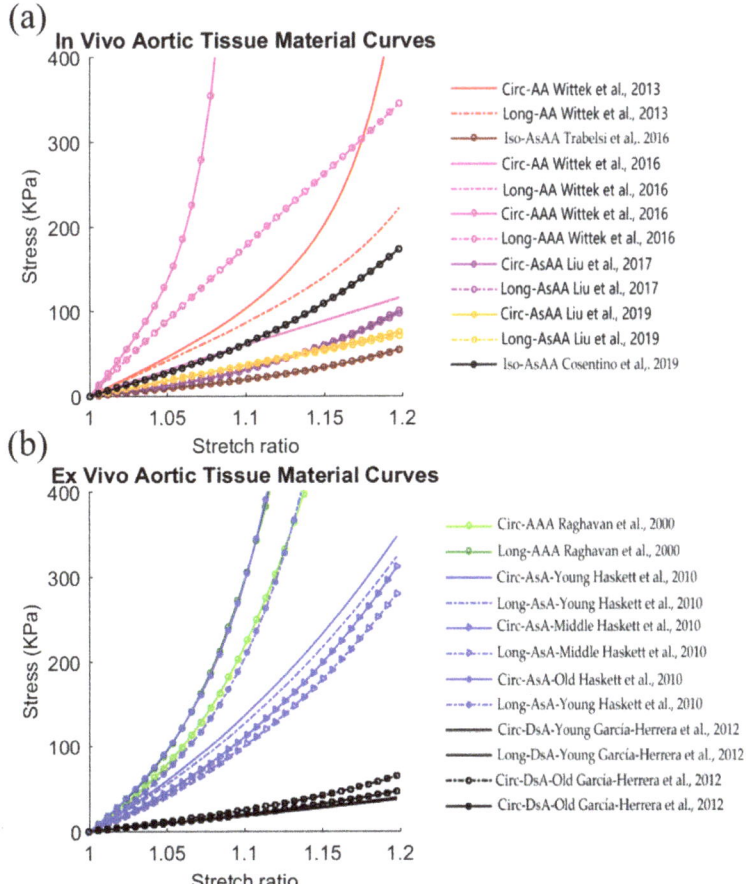

Figure 2. Stress–stretch ratio curves of healthy and diseased aortic tissues from (**a**) in vivo studies and (**b**) ex vivo studies. Abbreviations: AA, abdominal aorta; AAA, abdominal aortic aneurysm;

AsA, ascending thoracic aorta; AsAA, ascending thoracic aortic aneurysm; DsA: descending thoracic aorta; Ec, effective Young's modulus in circumferential direction; Ea, effective Young's modulus in longitudinal direction; Iso, isotropic material; Circ, material curves in circumferential direction; Long, material curves in longitudinal direction [29,30,33–35,49,55,60,61].

Based on the material curves in Figure 2a,b, large variations in aortic tissue stiffness could be observed for both in vivo studies and ex vivo studies. For in vivo studies, the Ec ranged from 180.3 kPa to 5576.7 kPa (Ea from 180.3kPa to 1770.2 kPa) whereas in ex vivo studies, the Ec ranged from 181.5 kPa to 2382.4 kPa (Ea from 176.0 kPa to 1856.3 kPa). This demonstrated that the aortic tissue stiffness values from in vivo and ex vivo studies were in the comparable ranges. However, compared to in vivo studies, all aortic tissues (including healthy and diseased tissues) from ex vivo studies yielded higher average tissue stiffness weighted by number of samples, with 201.5% higher stiffness for the Ec (2040.4 kPa vs. 676.9 kPa), and 267.1% higher for the Ea (1922.8 kPa vs. 523.8 kPa), respectively). It could also be observed that most listed studies had higher tissue stiffness in the circumferential direction than in the longitudinal direction [49]. Moreover, tissue anisotropy was clear but less significant in ex vivo studies, according to the difference in circumferential and longitudinal material curves from the same study [29,34,60].

Among the listed in vivo studies, one study determined the material properties of both healthy and aneurysmal tissues to study the impact of pathological conditions on the material properties following the same method. This study showed that aneurysmal aortic tissues tend to have higher stiffness than non-aneurysmal aortic tissues [33]. This conclusion was consistent with other ex vivo studies [66]. Furthermore, it is fortunate that García-Herrera et al. [61] and Haskett et al. [60] harvested enough specimens from healthy donors to investigate the aging effect on tissue material properties. García-Herrera et al. [61] reported that aortic stiffness increased as the age increased. For the study from Haskett et al. [60], they classified the donors into young, middle age and old groups. They found that the oldest people had the highest aortic stiffness, whereas middle-age donors had close but slightly lower aortic stiffness than the young group.

4.2. Carotid and Coronary Arterial Tissues

Fewer studies were performed to identify the material properties of carotid and coronary arteries in vivo. Figure 3 gives the plots of the material curves of healthy and diseased carotid and coronary tissues listed in Table 2. These studies have shown similar results to aortic tissue. Compared to aortic studies, tissue anisotropy was more obvious in carotid and coronary specimens, with the Ec higher than the Ea for the listed studies. All carotid studies in Table 2 showed that the Ec and Ea both ranged from 422.6 kPa to 6926.2 kPa for in vivo studies whereas the Ec ranged from 606.2 kPa to 1245.4 kPa (Ea from 176.7 kPa to 1245.4 kPa) for ex vivo studies. Therefore, tissue stiffness values from in vivo studies were generally in the same magnitude as that from ex vivo studies. Coronary studies yielded a similar conclusion. Similarly, compared to ex vivo studies, carotid/coronary tissues from ex vivo studies yielded a higher average tissue stiffness weighted by number of samples. For carotid tissues, the average tissue stiffness values from ex vivo and in vivo studies were 1137.6 kPa vs. 849.2 kPa (34% higher) for the Ec, and 999.0 kPa vs. 849.2 kPa (17.6% higher) for the Ea, respectively. For coronary tissues, the numbers were 1975.2 kPa vs. 973.6 kPa (102.9% higher) for the Ec, and 1959.6 kPa vs. 562.3 kPa (248.5% higher) for the Ea, respectively. Atherosclerosis is the most common disease occurring in both carotid and coronary arteries. This disease elevates the tissue stiffness in carotid and coronary, as demonstrated by Franquet et al. [37] and Kaimi et al. [65]. This phenomenon was also observed by ex vivo studies [52]. An aging effect was also investigated by Franquet et al. [37], and their results demonstrated an increase in the elastic modulus of the common carotid artery as age increases.

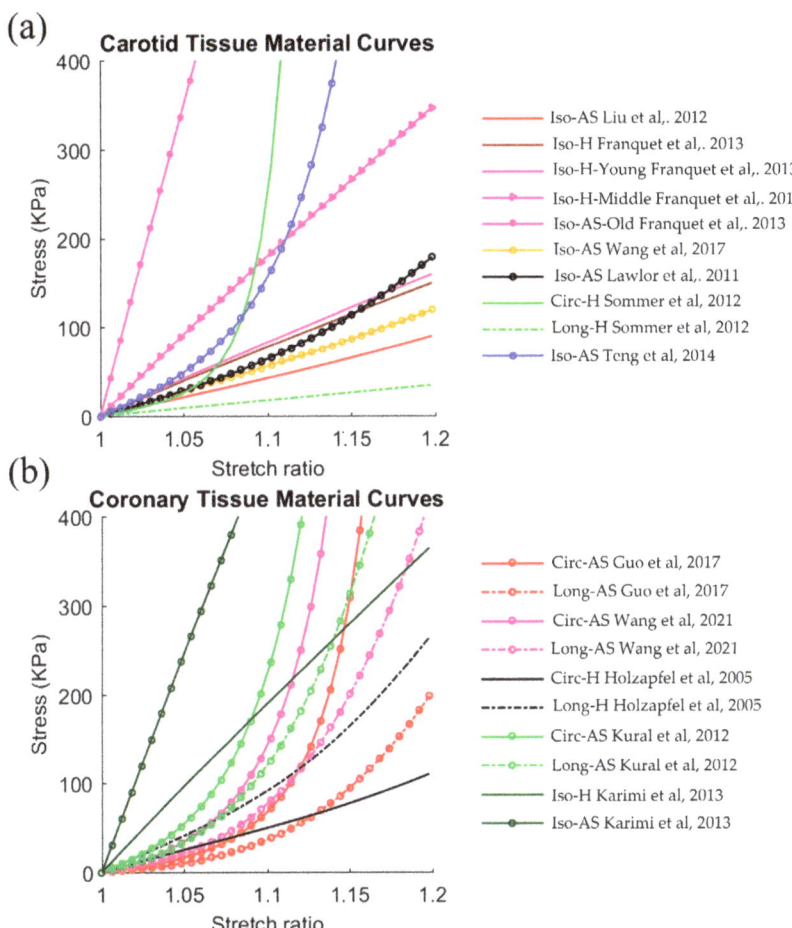

Figure 3. Stress–stretch ratio curves of healthy and diseased aortic tissues for all (**a**) carotid studies and (**b**) coronary studies. Abbreviations: Iso, isotropic material; Circ, material curves in circumferential direction; Long, material curves in longitudinal direction; H, healthy; AS, atherosclerosis [17,31,32,36–38,51,58,62–65].

5. Some Methodological Issues in Finite-Element-Model-Based Updating Approach

5.1. Significance of In Vivo Identification Framework

Classically, mechanical experiments are conducted to quantify the mechanical properties using arterial tissues ex vivo [14]. However, the FEMBUA provides another way to determine tissue properties in vivo. This in vivo method could be easily modified to successfully apply to other biological tissues, and even non-biomaterials such as cerebral aneurysmal tissue [67]; cardiac tissue [68]; thigh muscle tissue [69]; human skin [70]; silicone gel soft tissue [71]; and metal [72].

Since tissue samples are often not available for in vivo studies and the material properties of arterial tissues alter when taken out of living subjects, the image-based finite element modeling approach is more suitable for studies under in vivo conditions with potential clinical implementations [22,23]. Prior studies have demonstrated that patient-specific in vivo tissue material properties had a significant influence on cardiovascular biomechanics, especially in strain calculation compared to ex vivo material properties [38]. Therefore, patient-specific in vivo material properties are desirable for personalized treatment.

Even though in vivo and ex vivo methods follow different approaches in determining tissue properties, they are not exclusive to each other. Rather, they are complementary to each other. A hybrid approach that uses a finite element updating approach to match stress–strain data from biaxial/uniaxial experiments has been proposed to quantify the aortic aneurysmal tissue [73].

5.2. Comparison in Tissue Stiffness from FEMBUA and Ex Vivo Experimental Approaches

To assess its accuracy and efficacy, comparison analysis between the novel FEMBUA and classical ex vivo experimental approaches were performed. Some representative ex vivo studies were selected, and the tissue stiffness values from these studies were compared to those from in vivo studies using the FEMBUA. For all types of arteries (aorta, carotid or coronary), the ranges of tissue stiffness were generally in the same magnitude for both methods. However, the dissected tissues from ex vivo studies were stiffer than tissue samples from in vivo studies by average. It should be noted that the limited number of ex vivo studies could influence the conclusions abovementioned, because the average tissue stiffness would change if a different set of ex vivo representative studies were chosen. Thus, the comparison was evaluated numerically, not statistically for all tissues, not for just healthy or diseased ones, in which case, the selected studies would have more influence. Nevertheless, these conclusions suggested that the FEMBUA is an accurate and effective approach for quantifying the material properties of arterial walls. However, more rigorous comparison analysis should be performed by conducting both methods on the same tissue samples as detailed in the following section.

5.3. Validation of In Vivo Identification Approach

Several research groups investigated vessel material properties using both the in vivo FEMBUA method and ex vivo experiments on the same arterial tissue for validation purposes [30,35,55]. Based on five aortic aneurysmal specimens from two patients, Liu M et al., compared the mechanical properties of aortic tissue from in vivo and ex vivo biaxial testing methods. The authors found that material curves from both methods were close to the average value of the mean absolute percentage error less than 5% [35]. Based on a larger sample size (n = 10), Cosentino et al., reported similar observations. They stated that at strain of 0.14, the relative difference in stress response from the stress–strain curves of both methods was less than 24% [55]. Additionally, Trabelsi et al., performed FEMs with the material parameters determined from both in vivo and ex vivo methods for the same tissue. Their simulation showed that the difference in peak wall stress between the two methods was less than 20% [30]. These studies supported that, using ex vivo experiment testing as the gold standard, the FEMBUA yields mild difference in biomechanical results for the same specimen.

5.4. Method Reproducibility and Noise Sensitivity Analysis

Reproducible, accurate determination of arterial tissue material properties is an important prerequisite for clinical applications. In a methodological study, Narayanan et al., demonstrated the reproducibility of the FEMBUA by repeating the approach for three patients to obtain their material properties. The results of material parameters from different runs were recovered with errors of $3.0 \pm 4.7\%$ [28].

The robustness of the solution to the inverse problem remains to be an important issue. There are many uncertainties that would impact the results of mechanical property determination which include medical image resolution, on-site pressure measurement, etc. Narayanan et al., performed sensitivity analysis by applying random Gaussian noise to the original medical images. Their results showed that the errors for the material parameters were $1.3 \pm 1.6\%$ for the 5% noise addition. In addition to the noise inherent to the image-based geometry, Narayanan et al., also investigated the impact of such pressure perturbations on material property recovery. They claimed that such perturbation would result in controllable error in tissue stiffness estimation. More specifically, the relative error was equal to the relative error of the applied perturbation [28].

5.5. Modeling Assumptions and Limitations

There are some assumptions involved in FEM for in vivo indentation of the material properties of arterial walls, which would impact the results from the FEMBUA: (a) Axial stretch has considerable impact on the result of mechanical parameter values [38,48]. Therefore, patient- and vessel-specific axial stretch data are needed to determine more accurate material properties. Currently, due to a lack of in vivo axial stretch data, prior studies just set it to a given stretch ratio in the computational models when determining in vivo material properties. Wang et al. [38] investigated the impact of axial stretch on material properties. Their results indicated that smaller axial stretch led to greater slice shrinkage and softer material stiffness estimation. (b) Neglecting perivascular pressure conditions could lead to over- or under-estimation of material properties, depending on vascular bed. That is because it is the transmural pressure, not blood pressure alone, that drives the arteries to expand and contract. Due to the perivascular tissue and environment that the aorta, carotid and coronary arteries are situated in, perivascular pressure conditions are different, and physiologically not equal to zero [26]. Not considering a positive perivascular pressure in the FEM would lead to an overestimation of the material stiffness, whereas ignoring a negative one would lead to underestimation. (c) Active stress from smooth muscle cells has an impact. The function of the smooth muscle in the arterial wall is to produce active tension, relatively independent of stretch. However, its impact on arterial stiffness remains controversial. Early evidence supported that the contribution of smooth muscle cell to the elastic properties of living blood vessel was very small [74]. More recently, Tremblay et al., considered the active stress in their method to estimate the material properties, and claimed that the effect of smooth muscle cell activation was non-negligible and could increase both the circumferential and axial stiffness of the tissue [75]. More attempts have to be made to set down the role of active stress played in tissue mechanical properties. (d) Residual stress was not included as no patient-specific opening angle data were available [76,77]. Currently, there are no studies that includes residual stress in their computational finite element models to characterize arterial wall properties. More efforts are needed to understand the impact of residual stress on results of mechanical properties. (e) Structure-only models instead of fluid-structure interaction models were typically utilized in the FEMBUA, because it is more computationally efficient, especially when multiple iterations were needed to obtain the optimal material properties in the updating approach. Furthermore, prior studies have demonstrated that structure-only models could close biomechanical conditions to the fluid–structure interaction models [53].

6. Conclusions Remarks and Future Directions

The image-based FEMBUA has been proven to accurately and effectively determine the material properties of arterial walls in vivo. In addition, the tissue stiffness from this method was consistent with that from ex vivo experimental approaches. However, current studies mainly use image and pressure data at two cardiac phases (typically systolic and diastolic phases). Imaging technologies with higher temporal resolution are desirable to obtain clinical images at more cardiac phases, so that more data points will be available to fit the complex nonlinear material models (such as the anisotropic Mooney–Rivlin model or Fung-type model with several material constants) in a least square sense. Thus, the image-based FEMBUA would be more robust and not sensitive to image noise or pressure measurement error.

Biomechanical properties provide vast information regarding cardiovascular tissues in living individuals, which could guide us to a better understanding of arterial mechanics and physiology, as well as for the analysis of the mechanisms of vascular diseases.

In vivo arterial tissue stiffness has already been employed in clinical settings as a risk factor for cardiovascular diseases, and has been proven to have clinical significance [22,23].

Since the FEMBUA could provide more detailed information on mechanical properties under in vivo conditions, it holds great potential in clinical applications for personalized treatment and precision medicine: (1) sophisticated material properties from this method, rather than simple arterial stiffness, are essential to accurately estimate stress/strain conditions for possible clinical applications, such as a stress-based diagnosis strategy to refine current diameter-based diagnosis criteria in aortic aneurysm assessment [7,78]; and (2) in vivo identification of the nonlinear anisotropic material properties of the cardiovascular tissue is also a prerequisite for predicting its interaction with implanted devices, such as coronary stents. Now, patient-specific computational modeling of coronary stents has been performed for individualized pre-procedural planning and predicting stenting prognosis [79]. Large-scale clinical studies are needed to verify the efficiency of in vivo material properties for clinical decision-making in these applications. Lastly, successful applications of computational modeling incorporating in vivo mechanical properties are based on the solid ground of an automated implementation of this in vivo approach with computational efficiency. With further validations, the FEMBUA could be further developed and automated to provide vessel material properties, which are an essential part for arterial models.

Supplementary Materials: The following supporting information can be downloaded at: https://www.mdpi.com/article/10.3390/jfb13030147/s1.

Author Contributions: Conceptualization, D.T. and L.W.; methodology, D.T., L.W., J.Z. and K.L.B.; formal analysis, L.W., R.L. and X.G.; writing—original draft preparation, D.T. and L.W.; writing—review and editing, D.T., L.W., G.S.M., A.M., J.Z., K.L.B., R.L. and X.G.; project administration, D.T., G.S.M. and A.M.; funding acquisition, L.W. All authors have read and agreed to the published version of the manuscript.

Funding: This research was supported in part by: the National Natural Science Foundation of China grants 11972117, 11802060; the Natural Science Foundation of Jiangsu Province under grant number BK20180352; a Jiangsu Province Science and Technology Agency under grant number BE2016785; and the Fundamental Research Funds for the Central Universities and the Zhishan Young Scholars Fund administrated by Southeast University (grant number 2242021R41123).

Institutional Review Board Statement: Not applicable.

Informed Consent Statement: Not applicable.

Data Availability Statement: Not applicable.

Conflicts of Interest: The authors declare no conflict of interest.

References

1. Malek, A.M.; Alper, S.L.; Izumo, S. Hemodynamic shear stress and its role in atherosclerosis. *JAMA* **1999**, *282*, 2035–2042. [CrossRef] [PubMed]
2. Kwak, B.R.; Bäck, M.; Bochaton-Piallat, M.; Caligiuri, G.; Daemen, M.J.A.P.; Davies, P.F.; Hoefer, I.E.; Holvoet, P.; Jo, H.; Krams, R.; et al. Biomechanical factors in atherosclerosis: Mechanisms and clinical implications. *Eur. Heart J.* **2014**, *35*, 3013–3020. [CrossRef] [PubMed]
3. James, F.G.; Mostafa, F.; Michael, I. Selected methods for imaging elastic properties of biological tissues. *Annu. Rev. Biomed. Eng.* **2003**, *5*, 57–78. [CrossRef]
4. Susan, J.Z.; Vojtech, M.; David, A.K. Mechanisms, pathophysiology, and therapy of arterial stiffness. *Arter. Thromb. Vasc. Biol.* **2005**, *25*, 932–943. [CrossRef]
5. Dirk, W.; Mario, K.; Paul, S.; Frank, S.; Alexander, R.; Kerstin, W.; Wolfgang, H.; Wolfgang, P.; Matthias, P.; Heinz-Peter, S.; et al. Role of left ventricular stiffness in heart failure with normal ejection fraction. *Circulation* **2008**, *117*, 2051–2060. [CrossRef]
6. Stephane, L.; John, C.; Luc, V.B.; Pierre, B.; Cristina, G.; Daniel, H.; Bruno, P.; Charalambos, V.; Ian, W.; Harry, S.-B.; et al. Expert consensus document on arterial stiffness: Methodological issues and clinical applications. *Eur. Heart J.* **2006**, *27*, 2588–2605. [CrossRef]
7. Vorp, D.A.; Vande Geest, J.P. Biomechanical determinants of abdominal aortic aneurysm rupture. *Arter. Thromb. Vasc. Biol.* **2005**, *25*, 1558–1566. [CrossRef]
8. Camasão, D.B.; Mantovani, D. The mechanical characterization of blood vessels and their substitutes in the continuous quest for physiological-relevant performances. A critical review. *Mater. Today Bio* **2021**, *10*, 100106. [CrossRef]

9. Moore, J., Jr.; Berry, J.L. Fluid and Solid Mechanical Implications of Vascular Stenting. *Ann. Biomed. Eng.* **2002**, *30*, 498–508. [CrossRef]
10. Fung, Y.C. *Biomechanics: Mechanical Properties of Living Tissues*; Springer: New York, NY, USA, 1993.
11. Holzapfel, G.A.; Gasser, T.C.; Ogden, R.W. A New Constitutive Framework for Arterial Wall Mechanics and a Comparative Study of Material Models. *J. Elast.* **2000**, *61*, 1–48. [CrossRef]
12. Walsh, M.T.; Cunnane, E.M.; Mulvihill, J.J.; Akyildiz, A.C.; Gijsen, F.J.H.; Holzapfel, G.A. Uniaxial tensile testing approaches for characterization of atherosclerotic plaques. *J. Biomech.* **2014**, *47*, 793–804. [CrossRef] [PubMed]
13. Macrae, R.A.; Miller, K.; Doyle, B.J. Methods in Mechanical Testing of Arterial Tissue: A Review. *Strain* **2016**, *52*, 380–399. [CrossRef]
14. Hayashi, K. Experimental approaches on measuring the mechanical properties and constitutive laws of arterial walls. *J. Biomech. Eng.* **1993**, *115*, 481–488. [CrossRef] [PubMed]
15. Sacks, M.S.; Sun, W. Multiaxial mechanical behavior of biological materials. *Annu. Rev. Biomed. Eng.* **2003**, *5*, 251–284. [CrossRef]
16. Holzapfel, G.A.; Sommer, G.; Regitnig, P. Anisotropic Mechanical Properties of Tissue Components in Human Atherosclerotic Plaques. *J. Biomech. Eng.* **2004**, *26*, 657–665. [CrossRef]
17. Teng, Z.; Zhang, Y.; Huang, Y.; Feng, J.; Yuan, J.; Lu, Q.; Sutcliffe, M.P.F.; Brown, A.J.; Jing, Z.; Gillard, J.H. Material properties of components in human carotid atherosclerotic plaques: A uniaxial extension study. *Acta. Biomater.* **2014**, *10*, 5055–5063. [CrossRef]
18. Peña, J.A.; Corral, V.; Martínez, M.A.; Peña, E. Over length quantification of the multiaxial mechanical properties of the ascending, descending and abdominal aorta using Digital Image Correlation. *J. Mech. Behav. Biomed. Mater.* **2018**, *77*, 434–445. [CrossRef]
19. Peterson, L.H.; Jensen, R.E.; Parnell, J. Mechanical Properties of Arteries in Vivo. *Circ. Res.* **1960**, *8*, 622–639. [CrossRef]
20. Gosling, R.G.; Budge, M.M. Terminology for Describing the Elastic Behavior of Arteries. *Hypertension* **2003**, *41*, 1180–1182. [CrossRef]
21. Claridge, M.W.; Bate, G.R.; Hoskins, P.R.; Adam, D.J.; Bradbury, A.W.; Wilmink, A.B. Measurement of arterial stiffness in subjects with vascular disease: Are vessel wall changes more sensitive than increase in intima–media thickness? *Atherosclerosis* **2009**, *205*, 477–480. [CrossRef]
22. Sutton-Tyrrell, K.; Najjar, S.S.; Boudreau, R.M.; Venkitachalam, L.; Kupelian, V.; Simonsick, E.M.; Havlik, R.; Lakatta, E.G.; Spurgeon, H.; Kritchevsky, S.; et al. Elevated Aortic Pulse Wave Velocity, a Marker of Arterial Stiffness, Predicts Cardiovascular Events in Well-Functioning Older Adults. *Circulation* **2005**, *111*, 3384–3390. [CrossRef] [PubMed]
23. Willum-Hansen, T.; Staessen, J.A.; Torp-Pedersen, C.; Rasmussen, S.; Thijs, L.; Ibsen, H.; Jeppesen, J. Prognostic Value of Aortic Pulse Wave Velocity as Index of Arterial Stiffness in the General Population. *Circulation* **2006**, *113*, 664–670. [CrossRef] [PubMed]
24. Stålhand, J.; Klarbring, A.; Karlsson, M. Towards in vivo aorta material identification and stress estimation. *Biomech. Model Mechan.* **2004**, *2*, 169–186. [CrossRef] [PubMed]
25. Schulze-Bauer, C.A.J.; Holzapfel, G.A. Determination of constitutive equations for human arteries from clinical data. *J. Biomech.* **2003**, *36*, 165–169. [CrossRef]
26. Masson, I.; Boutouyrie, P.; Laurent, S.; Humphrey, J.D.; Zidi, M. Characterization of arterial wall mechanical behavior and stresses from human clinical data. *J. Biomech.* **2008**, *41*, 2618–2627. [CrossRef]
27. Avril, S.; Evans, S.; Miller, K. Inverse problems and material identification in tissue biomechanics. *J. Mech. Behav. Biomed. Mater.* **2013**, *27*, 129–131. [CrossRef]
28. Narayanan, B.; Olender, M.L.; Marlevi, D.; Edelman, E.R.; Nezami, F.R. An inverse method for mechanical characterization of heterogeneous diseased arteries using intravascular imaging. *Sci. Rep.* **2021**, *11*, 22540. [CrossRef]
29. Wittek, A.; Karatolios, K.; Bihari, P.; Schmitz-Rixen, T.; Moosdorf, R.; Vogt, S.; Blasé, C. In vivo determination of elastic properties of the human aorta based on 4D ultrasound data. *J. Mech. Behav. Biomed. Mater.* **2013**, *27*, 167–183. [CrossRef]
30. Trabelsi, O.; Duprey, A.; Favre, J.-P.; Avril, S. Predictive Models with Patient Specific Material Properties for the Biomechanical Behavior of Ascending Thoracic Aneurysms. *Ann. Biomed. Eng.* **2016**, *44*, 84–98. [CrossRef]
31. Franquet, A.; Avril, S.; Le Riche, R.; Badel, P.; Schneider, F.C.; Li, Z.; Boissier, C.; Pierre Favre, J. A New Method for the In Vivo Identification of Mechanical Properties in Arteries From Cine MRI Images: Theoretical Framework and Validation. *IEEE Trans. Med. Imaging* **2013**, *32*, 1448–1461. [CrossRef]
32. Guo, X.; Zhu, J.; Maehara, A.; Monoly, D.; Samady, H.; Wang, L.; Billiar, K.L.; Zheng, J.; Yang, C.; Mintz, G.S.; et al. Quantify patient-specific coronary material property and its impact on stress/strain calculations using in vivo IVUS data and 3D FSI models: A pilot study. *Biomech. Model Mechan.* **2017**, *16*, 333–344. [CrossRef] [PubMed]
33. Wittek, A.; Derwich, W.; Karatolios, K.; Fritzen, C.P.; Vogt, S.; Schmitz-Rixen, T.; Blasé, C. A finite element updating approach for identification of the anisotropic hyperelastic properties of normal and diseased aortic walls from 4D ultrasound strain imaging. *J. Mech. Behav. Biomed. Mater.* **2016**, *58*, 122–138. [CrossRef]
34. Liu, M.; Liang, L.; Sun, W. A new inverse method for estimation of in vivo mechanical properties of the aortic wall. *J. Mech. Behav. Biomed. Mater.* **2017**, *72*, 148–158. [CrossRef] [PubMed]
35. Liu, M.; Liang, L.; Sulejmani, F.; Lou, X.; Iannucci, G.; Chen, E.; Leshnower, B.; Sun, W. Identification of in vivo nonlinear anisotropic mechanical properties of ascending thoracic aortic aneurysm from patient-specific CT scans. *Sci. Rep.* **2019**, *9*, 12983. [CrossRef] [PubMed]

36. Liu, H.; Canton, G.; Yuan, C.; Yang, C.; Billiar, K.; Teng, Z.; Hoffman, A.H.; Tang, D. Using In Vivo Cine and 3D Multi-Contrast MRI to Determine Human Atherosclerotic Carotid Artery Material Properties and Circumferential Shrinkage Rate and Their Impact on Stress/Strain Predictions. *J. Biomech. Eng.* **2012**, *134*, 011008. [CrossRef]
37. Franquet, A.; Avril, S.; Le Riche, R.; Badel, P.; Schneider, F.C.; Boissier, C.; Favre, J.-P. Identification of the in vivo elastic properties of common carotid arteries from MRI: A study on subjects with and without atherosclerosis. *J. Mech. Behav. Biomed. Mater.* **2013**, *27*, 184–203. [CrossRef]
38. Wang, L.; Zhu, J.; Maehara, A.; Lv, R.; Qu, Y.; Zhang, X.; Guo, X.; Billiar, K.L.; Chen, L.; Ma, G.; et al. Quantifying Patient-Specific in vivo Coronary Plaque Material Properties for Accurate Stress/Strain Calculations: An IVUS-Based Multi-Patient Study. *Front. Physiol.* **2021**, *12*, 721195. [CrossRef]
39. Seo, Y.; Ishizu, T.; Enomoto, Y.; Sugimori, H.; Yamamoto, M.; Machino, T.; Kawamura, R.; Aonuma, K. Validation of 3-dimensional speckle tracking imaging to quantify regional myocardial deformation. *Circ. Cardiovasc. Imaging* **2009**, *2*, 451–459. [CrossRef]
40. Pan, B.; Qian, K.; Xie, H.; Asundi, A. Two-dimensional digital image correlation for in-plane displacement and strain measurement: A review. *Meas. Sci. Technol.* **2009**, *20*, 6. [CrossRef]
41. Mondillo, S.; Galderisi, M.; Mele, D.; Cameli, M.; Lomoriello, V.S.; Zacà, V.; Ballo, P.; D'Andrea, A.; Muraru, D.; Losi, M.; et al. Echocardiography Study Group Of The Italian Society Of Cardiology (Rome, Italy). Speckle-tracking echocardiography: A new technique for assessing myocardial function. *J. Ultrasound. Med.* **2011**, *30*, 71–83. [CrossRef]
42. Yang, C.; Bach, R.G.; Zheng, J.; Naqa, E.I.; Woodard, P.K.; Teng, Z.; Billiar, K.; Tang, D. In vivo IVUS-based 3-D fluid-structure interaction models with cyclic bending and anisotropic vessel properties for human atherosclerotic coronary plaque mechanical analysis. *IEEE Trans. Bio-Med. Eng.* **2009**, *56*, 2420–2428. [CrossRef] [PubMed]
43. Carpenter, H.J.; Gholipour, A.; Ghayesh, M.H.; Zander, A.C.; Psaltis, P.J. A review on the biomechanics of coronary arteries. *Int. J. Eng. Sci.* **2020**, *147*, 103201. [CrossRef]
44. Speelman, L.; Bosboom, E.M.; Schurink, G.W.; Buth, J.; Breeuwer, M.; Jacobs, M.J.; van de Vosse, F.N. Initial stress and nonlinear material behavior in patient-specific AAA wall stress analysis. *J. Biomech.* **2009**, *7*, 1713–1719. [CrossRef] [PubMed]
45. Gee, M.W.; Förster, C.H.; Wall, W.A. A Computational Strategy for Prestressing Patient-Specific Biomechanical Problems Under Finite Deformation. *Int. J. Numer. Methods Biomed. Eng.* **2010**, *26*, 52–72. [CrossRef]
46. Tang, D.; Yang, C.; Zheng, J.; Woodard, P.K.; Sicard, G.A.; Saffitz, J.E.; Yuan, C. 3D MRI-Based Multicomponent FSI Models for Atherosclerotic Plaques. *Ann. Biomed. Eng.* **2004**, *32*, 947–960. [CrossRef]
47. Avril, S.; Badel, P.; Duprey, A. Anisotropic and hyperelastic identification of in vitro human arteries from full-field optical measurements. *J. Biomech.* **2010**, *43*, 2978–2985. [CrossRef]
48. Maso Talou, G.D.; Blanco, P.J.; Ares, G.D.; Guedes Bezerra, C.; Lemos, P.A.; Feijóo, R.A. Mechanical Characterization of the Vessel Wall by Data Assimilation of Intravascular Ultrasound Studies. *Front. Physiol.* **2018**, *28*, 292. [CrossRef]
49. Raghavan, M.L.; Vorp, D.A. Toward a biomechanical tool to evaluate rupture potential of abdominal aortic aneurysm: Identification of a finite strain constitutive model and evaluation of its applicability. *J. Biomech.* **2000**, *33*, 475–482. [CrossRef]
50. Gasser, T.C.; Ogden, R.W.; Holzapfel, G.A. Hyperelastic modelling of arterial layers with distributed collagen fibre orientations. *J. R. Soc. Interface* **2006**, *3*, 15–35. [CrossRef]
51. Holzapfel, G.A.; Sommer, G.; Gasser, C.T.; Regitnig, P. Determination of layer-specific mechanical properties of human coronary arteries with nonatherosclerotic intimal thickening and related constitutive modeling. *Am. J. Physiol. Heart Circ. Physiol.* **2005**, *289*, 2048–2058. [CrossRef]
52. Jankowska, M.A.; Bartkowiak-Jowsa, M.; Bedzinski, R. Experimental and constitutive modeling approaches for a study of biomechanical properties of human coronary arteries. *J. Mech. Behav. Biomed. Mater.* **2015**, *50*, 1–12. [CrossRef] [PubMed]
53. Huang, X.; Yang, C.; Zheng, J.; Bach, R.; Muccigrosso, D.; Woodard, P.K.; Tang, D. 3D MRI-based multicomponent thin layer structure only plaque models for atherosclerotic plaques. *J. Biomech.* **2016**, *49*, 2726–2733. [CrossRef]
54. Baldewsing, R.A.; Danilouchkine, M.G.; Mastik, F.; Schaar, J.A.; Serruys, P.W.; van der Steen, F.W.A. An Inverse Method for Imaging the Local Elasticity of Atherosclerotic Coronary Plaques. *IEEE Trans. Inf. Technol. Biomed.* **2008**, *12*, 277–289. [CrossRef] [PubMed]
55. Cosentino, F.; Agnese, V.; Raffa, G.M.; Gentile, G.; Bellavia, D.; Zingales, M.; Pilato, M.; Pasta, S. On the role of material properties in ascending thoracic aortic aneurysms. *Comput. Biol. Med.* **2019**, *109*, 70–78. [CrossRef] [PubMed]
56. Chandran, K.B.; Mun, J.H.; Choi, K.K.; Chen, J.S.; Hamilton, A.; Nagaraj, A.; McPherson, D.D. A method for in-vivo analysis for regional arterial wall material property alterations with atherosclerosis: Preliminary results. *Med. Eng. Phys.* **2003**, *25*, 289–298. [CrossRef]
57. Nelder, J.A.; Mead, R. A Simplex Method for Function Minimization. *Comput. J.* **1965**, *7*, 308–313. [CrossRef]
58. Wang, Q.; Canton, G.; Guo, J.; Guo, X.; Hatsukami, T.S.; Billiar, K.L.; Yuan, C.; Wu, Z.; Tang, D. MRI-based patient-specific human carotid atherosclerotic vessel material property variations in patients, vessel location and long-term follow up. *PLoS ONE* **2017**, *12*, e0180829. [CrossRef]
59. Ghassan, S.K. Biomechanics of the cardiovascular system: The aorta as an illustratory example. *J. R. Soc. Interface* **2006**, *3*, 719–740. [CrossRef]
60. Haskett, D.; Johnson, G.; Zhou, A.; Utzinger, U.; Vande Geest, J. Microstructural and biomechanical alterations of the human aorta as a function of age and location. *Biomech. Model Mechanobiol.* **2010**, *9*, 725–736. [CrossRef]

1. García-Herrera, C.M.; Celentano, D.J.; Cruchaga, M.A.; Rojo, F.J.; Atienza, J.M.; Guinea, G.V.; Goicolea, J.M. Mechanical characterisation of the human thoracic descending aorta: Experiments and modelling. *Comput. Methods Biomech. Biomed. Engin.* **2012**, *15*, 185–193. [CrossRef]
2. Lawlor, M.G.; O'Donnell, M.R.; O'Connell, B.M.; Walsh, M.T. Experimental determination of circumferential properties of fresh carotid artery plaques. *J. Biomech.* **2011**, *44*, 1709–1715. [CrossRef] [PubMed]
3. Sommer, G.; Holzapfel, G.A. 3D constitutive modeling of the biaxial mechanical response of intact and layer-dissected human carotid arteries. *J. Mech. Behav. Biomed. Mater.* **2012**, *5*, 116–128. [CrossRef] [PubMed]
4. Kural, M.H.; Cai, M.; Tang, D.; Gwyther, T.; Zheng, J.; Billiar, K.L. Planar biaxial characterization of diseased human coronary and carotid arteries for computational modeling. *J. Biomech.* **2012**, *45*, 790–798. [CrossRef] [PubMed]
5. Karimi, A.; Navidbakhsh, M.; Shojaei, A.; Faghihi, S. Measurement of the uniaxial mechanical properties of healthy and atherosclerotic human coronary arteries. *Mater. Sci. Eng. C. Mater Biol. Appl.* **2013**, *33*, 2550–2554. [CrossRef] [PubMed]
6. Niestrawska, J.A.; Viertler, C.; Regitnig, P.; Cohnert, T.U.; Sommer, G.; Holzapfel, G.A. Microstructure and mechanics of healthy and aneurysmatic abdominal aortas: Experimental analysis and modelling. *J. R. Soc. Interface* **2016**, *13*, 20160620. [CrossRef]
7. Zhao, X.; Raghavan, M.L.; Lu, J. Identifying heterogeneous anisotropic properties in cerebral aneurysms: A pointwise approach. *Biomech. Model. Mechanobiol.* **2011**, *10*, 177–189. [CrossRef]
8. Yu, H.; Del Nido, P.J.; Geva, T.; Yang, C.; Tang, A.; Wu, Z.; Rathod, R.H.; Huang, X.; Billiar, K.L.; Tang, D. Patient-specific in vivo right ventricle material parameter estimation for patients with tetralogy of Fallot using MRI-based models with different zero-load diastole and systole morphologies. *Int. J. Cardiol.* **2019**, *276*, 93–99. [CrossRef]
9. Affagard, J.S.; Feissel, P.; Bensamoun, S.F. Identification of hyperelastic properties of passive thigh muscle under compression with an inverse method from a displacement field measurement. *J. Biomech.* **2015**, *48*, 4081–4086. [CrossRef]
10. Tran, H.V.; Charleux, F.; Rachik, M.; Ehrlacher, A.; Ho Ba Tho, M.C. In vivo characterization of the mechanical properties of human skin derived from MRI and indentation techniques. *Comput. Methods Biomech. Biomed. Eng.* **2007**, *10*, 401–407. [CrossRef]
11. Moerman, K.M.; Holt, C.A.; Evans, S.L.; Simms, C.K. Digital image correlation and finite element modelling as a method to determine mechanical properties of human soft tissue in vivo. *J. Biomech.* **2009**, *42*, 1150–1153. [CrossRef]
12. Meuwissen, M.H.H.; Oomens, C.W.J.; Baaijens, F.P.T.; Petterson, R.; Janssen, J.D. Determination of the elasto-plastic properties of aluminium using a mixed numerical–experimental method. *J. Mater Process. Technol.* **1998**, *75*, 204–211. [CrossRef]
13. Davis, F.M.; Luo, Y.; Avril, S.; Duprey, A.; Lu, J. Local mechanical properties of human ascending thoracic aneurysms. *J. Mech. Behav. Biomed. Mater.* **2016**, *61*, 235–249. [CrossRef] [PubMed]
14. Roach, M.R.; Burton, A.C. The reason for the shape of the distensibility curves of arteries. *Can. J. Biochem. Physiol.* **1957**, *35*, 681–690. [CrossRef] [PubMed]
15. Tremblay, D.; Cartier, R.; Mongrain, R.; Leask, R.L. Regional dependency of the vascular smooth muscle cell contribution to the mechanical properties of the pig ascending aortic tissue. *J. Biomech.* **2010**, *43*, 2448–2451. [CrossRef]
16. Fung, Y.C.; Liu, S.Q. Strain distribution in small blood vessel with zero-stress state taken into consideration. *Am. J. Physiol.* **1992**, *262*, 544–552. [CrossRef] [PubMed]
17. Ohayon, J.; Dubreuil, O.; Tracqui, P.; Le Floc'h, S.; Rioufol, G.; Chalabreysse, L.; Thivolet, F.; Pettigrew, R.I.; Finet, G. Influence of residual stress/strain on the biomechanical stability of vulnerable coronary plaques: Potential impact for evaluating the risk of plaque rupture. *Am. J. Physiol. Heart Circ. Physiol.* **2007**, *293*, 1987–1996. [CrossRef]
18. Singh, T.P.; Moxon, J.V.; Gasser, T.C.; Golledge, J. Systematic Review and Meta-Analysis of Peak Wall Stress and Peak Wall Rupture Index in Ruptured and Asymptomatic Intact Abdominal Aortic Aneurysms. *J. Am. Heart Assoc.* **2021**, *10*, e019772. [CrossRef] [PubMed]
19. Zhao, S.; Wu, W.; Samant, S.; Khan, B.; Kassab, G.S.; Watanabe, Y.; Murasato, Y.; Sharzehee, M.; Makadia, J.; Zolty, D.; et al. Patient-specific computational simulation of coronary artery bifurcation stenting. *Sci. Rep.* **2021**, *11*, 16486. [CrossRef]

Article

Combining IVUS + OCT Data, Biomechanical Models and Machine Learning Method for Accurate Coronary Plaque Morphology Quantification and Cap Thickness and Stress/Strain Index Predictions

Rui Lv [1], Liang Wang [1,*], Akiko Maehara [2], Mitsuaki Matsumura [2], Xiaoya Guo [3], Habib Samady [4], Don P. Giddens [4,5], Jie Zheng [6], Gary S. Mintz [2] and Dalin Tang [1,7,*]

1. School of Biological Science and Medical Engineering, Southeast University, Nanjing 210096, China
2. The Cardiovascular Research Foundation, Columbia University, New York, NY 10019, USA
3. School of Science, Nanjing University of Posts and Telecommunications, Nanjing 210023, China
4. Department of Medicine, Emory University School of Medicine, Atlanta, GA 30322, USA
5. The Wallace H. Coulter Department of Biomedical Engineering, Georgia Institute of Technology, Atlanta, GA 30332, USA
6. Mallinckrodt Institute of Radiology, Washington University, St. Louis, MO 63110, USA
7. Mathematical Sciences Department, Worcester Polytechnic Institute, Worcester, MA 01609, USA
* Correspondence: liangwang@seu.edu.cn (L.W.); dtang@wpi.edu (D.T.); Tel.: +1-508-831-5332 (D.T.)

Citation: Lv, R.; Wang, L.; Maehara, A.; Matsumura, M.; Guo, X.; Samady, H.; Giddens, D.P.; Zheng, J.; Mintz, G.S.; Tang, D. Combining IVUS + OCT Data, Biomechanical Models and Machine Learning Method for Accurate Coronary Plaque Morphology Quantification and Cap Thickness and Stress/Strain Index Predictions. *J. Funct. Biomater.* **2023**, *14*, 41. https://doi.org/10.3390/jfb14010041

Academic Editor: Mahdi Bodaghi

Received: 27 November 2022
Revised: 25 December 2022
Accepted: 9 January 2023
Published: 11 January 2023

Copyright: © 2023 by the authors. Licensee MDPI, Basel, Switzerland. This article is an open access article distributed under the terms and conditions of the Creative Commons Attribution (CC BY) license (https:// creativecommons.org/licenses/by/ 4.0/).

Abstract: Assessment and prediction of vulnerable plaque progression and rupture risk are of utmost importance for diagnosis, management and treatment of cardiovascular diseases and possible prevention of acute cardiovascular events such as heart attack and stroke. However, accurate assessment of plaque vulnerability assessment and prediction of its future changes require accurate plaque cap thickness, tissue component and structure quantifications and mechanical stress/strain calculations. Multi-modality intravascular ultrasound (IVUS), optical coherence tomography (OCT) and angiography image data with follow-up were acquired from ten patients to obtain accurate and reliable plaque morphology for model construction. Three-dimensional thin-slice finite element models were constructed for 228 matched IVUS + OCT slices to obtain plaque stress/strain data for analysis. Quantitative plaque cap thickness and stress/strain indices were introduced as substitute quantitative plaque vulnerability indices (PVIs) and a machine learning method (random forest) was employed to predict PVI changes with actual patient IVUS + OCT follow-up data as the gold standard. Our prediction results showed that optimal prediction accuracies for changes in cap-PVI (C-PVI), mean cap stress PVI (meanS-PVI) and mean cap strain PVI (meanSn-PVI) were 90.3% (AUC = 0.877), 85.6% (AUC = 0.867) and 83.3% (AUC = 0.809), respectively. The improvements in prediction accuracy by the best combination predictor over the best single predictor were 6.6% for C-PVI, 10.0% for mean S-PVI and 8.0% for mean Sn-PVI. Our results demonstrated the potential using multi-modality IVUS + OCT image to accurately and efficiently predict plaque cap thickness and stress/strain index changes. Combining mechanical and morphological predictors may lead to better prediction accuracies.

Keywords: coronary vulnerable plaque; plaque models; fibrous cap thickness; vulnerable plaque model; plaque vulnerability prediction

1. Introduction

Vulnerable plaque progression and rupture are closely related to cardiovascular disease which is the leading cause of death worldwide [1]. Accurate assessment of plaque cap thickness and prediction require accurate plaque tissue component and structure quantifications and mechanical stress/strain calculations. Plaque vulnerability is commonly understood as the likelihood of a plaque rupture causing drastic clinical events such as heart attack or stroke. While plaque vulnerability is a well-known concept, its quantitative measure is nearly impossible due to lack of plaque rupture and clinical data, which hinders its application in clinical scenarios. American Heart Association (AHA) classified plaques into Types I-VI based on histological data [2,3]. The AHA plaque classifications have been considered as the gold standard in the research community. However, it is of qualitative nature and not convenient for quantitative vulnerability tracking and predictions. Some morphological and biomechanical plaque vulnerability indices (PVIs) have been introduced based on imaging data and biomechanical factors to overcome this limitation [4,5]. Tang et al. introduced a stress-based PVI (SPVI) using 34 in vivo magnetic resonance imaging (MRI) slices from 14 human coronary plaque samples. Their SPVI plaque vulnerability assessment had an 85% agreement rate with assessment performed by histopathological analysis [4]. Goncalves et al. introduced a vulnerability index (VI) and calculated VI values for 194 patients based on histological analysis. Their follow-up data (60 months, 45 postoperative cardiovascular events registered) showed that patients with a plaque VI in the fourth quartile compared with the first to third quartiles had significantly higher risk to suffer from a future cardiovascular event ($p = 0.0002$) [5]. Wang L et al. used intravascular ultrasound (IVUS)-based morphology PVI to assess and predict plaque vulnerability [6]. Due to MRI and IVUS image resolution limitations and difficulty in recruiting large number of patients with follow-up data, accurate and reliable plaque cap thickness measurements are still difficult to obtain in vivo and PVIs still require more effort to obtain acceptance in research community and clinical practice.

Accurate and reliable image data have been employed to visualize plaque morphology and serve as the basis for plaque cap thickness predictions. Atherosclerotic plaque progression is a long and slow process lasting some 30 to 50+ years. Rapid plaque progression could also be caused by plaque destabilization followed by thrombus formation and subsequent healing. From available patient follow-up data, plaque vessel wall thickness changes were mostly under 100 μm in a year [7]. With a 150–200 μm resolution from IVUS and 200–300 μm resolution from MRI, plaque progression and morphology changes cannot be quantified for prediction with admissible reliability and accuracy. With its superior resolution of approximately 10 μm, optical coherence tomography (OCT) is able to detect thin fibrous cap (the well-known 65 μm threshold cap thickness) of vulnerable plaques [3,8,9]. Liu et al. further demonstrated that plaques with thinner fibrous cap had higher probability to have plaque rupture and thrombosis events [10]. Reith et al. determined that compared to patients with stable angina pectoris, patients with acute coronary syndrome tended to have a smaller minimal fibrous cap thickness within lipid-rich lesions [11]. Among the morphological factors characterizing vulnerable plaques such as positive remodeling, large lipid size, and macrophages infiltration, plaque cap thickness is one of the most-watched measurable characteristics for plaque prone to rupture. Efforts combining IVUS and OCT to study vulnerable plaques have been reported, and impressive results suggest that integrating two imaging modalities could be used for more accurate cap stress/strain calculations and to better evaluate plaque progression and regression [12,13]. Therefore, it can be considered that the IVUS + OCT merged data could provide detailed plaque morphological information (especially cap thickness) which forms a reliable basis for further biomechanical analysis and cap thickness and stress/strain index change predictions.

It has been hypothesized that mechanical forces play an important role in plaque progression and rupture [14–16]. From a mechanical point of view, rupture occurs when plaque stress and strain at the fibrous cap exceed its tensile strength. Therefore, precise plaque stress and strain conditions may be helpful in predicting plaque rupture and critical clinical events [17,18]. Schaar et al. defined a vulnerable plaque as a plaque with a high strain region at the surface with adjacent low strain regions [19]. In addition, Zhang et al. calculated strain from in vivo image for the assessment of vulnerable plaques [20]. With the help of different evaluation methods, stress/strain variables under different definitions were calculated to detect vulnerable plaques and assess their vulnerability [4,21,22]. Those studies suggested that plaque stress and strain are closely related to plaque behaviors and could be utilized in vulnerability predictions.

In the field of predicting plaque behaviors, most of the references adopt mixed-effect logistic regression models or Cox regression models (proportional risk regression model) [23,24]. Recently, machine learning approaches were employed in plaque progression prediction studies due to their strong predictive power and time efficiency [25]. A risk stratification model based on machine learning was used to predict all-cause death, recurrent acute myocardial infarction, and massive hemorrhage after acute coronary syndrome [26]. Lin et al. developed and validated a deep learning algorithm based on face photos to evaluate the relationship between face features and CAD [27]. By classifying the behavior of plaques, Lv et al. successfully performed a binary prediction of plaque progression based on the generalized linear mixed model and the least squares support vector machine [28].

Plaque vulnerability quantification and predictions have several challenges: (a) lack of rupture and clinical data to establish the gold standard for assessment and prediction; (b) selection of proper predictors, vulnerability measurements and indices to perform predictions; (c) proper biomechanical models with acceptable labor cost for potential clinical implementations. With considerable effort, we have obtained multi-modality images from ten patients with follow-up scan. The data set for each patient include IVUS + OCT + Angiography data at both baseline and follow-up. IVUS + OCT (IO) data provide us with reliable and accurate plaque morphologies which is the basis for modeling and prediction effort. Without plaque rupture and clinical events to serve as the gold standard, a cap thickness-based plaque vulnerability index (C-PVI) was introduced using IO data to serve as an alternative gold standard in this paper. While this is not the best "gold standard", it is measurable in vivo with OCT accuracy and has the potential to be implemented in clinical practice. Several stress- and strain-based PVIs were introduced and their prediction results were compared. Time-saving 3D thin-slice models were constructed to obtain plaque stress/strain values. Values of nine morphological and biomechanical risk factors (list to be provided later) were extracted from IO images and computational models and used for prediction analysis. The random forest (RF) was adopted in this paper to predict the binary outcomes of PVI changes from baseline to follow-up. Results were compared and analyzed to identify the best single and combination predictors and the best performing PVI(s).

2. Materials and Methods

2.1. Data Acquisition, Segmentation, Slice Co-Registration and Merging

Existing de-identified IVUS, OCT and angiography data with follow-up from 10 patients (4F; mean age 70.4) were obtained from Cardiovascular Research Foundation (CRF), Columbia University, New York, NY, USA. Data were collected between April 2017 and November 2018 (mean follow-up time: 251 days) using protocol approved by local institutional review board following the rules of the Declaration of Helsinki of 1975, with informed consent obtained. Ten patients with stable angina pectoris were selected for analysis. Patients with acute coronary syndrome, severe calcified lesion, chronic total occlusion, or chronic kidney disease (Cr > 1.5 mg/dL) were excluded. Patient demographic data are shown in Table 1. Patient's arm blood pressure was collected and used as pressure conditions in computational models. OCT images of coronary arteries were acquired using commercially available ILUMIEN OPTIS System (St. Jude Medical, Westford, MA, USA). IVUS imaging was performed using OptiCross System (OptiCross, Boston Scientific Corporation, Natick, MA, USA) with a 40 MHz IVUS catheter motorized pullback at 0.5 mm/s. Coronary angiography data were obtained for both baseline and follow-up.

Co-registration of IVUS and OCT slices was performed by using fiduciary points such as side branches, bifurcations, and calcifications with the assistance of coronary angiography images (matched IVUS and OCT, matched baseline and follow-up) [29] (see Figure 1). Then, by mapping IVUS and OCT frames to the same coronary vessel segment, IVUS and OCT segmented slices were merged together to form IO slices to extract geometric contours for model construction and analysis (See Figure 2). OCT provided accurate information for plaque cap, lumen and calcification region. Plaque components included: (1) fibrous tissue (homogeneous, high backscattering region); (2) lipid-rich core (low-signal region with diffuse border) and (3) calcification (low backscattering region with sharp border). For large lipid components with a thin fibrous cap, OCT can "see" the cap clearly, but may not detect the lipid border far away from lumen due to its limited penetration. In that case, IVUS was used to obtain the lipid out-border and vessel outer-boundary. Segmentation was performed by ImageJ 1.52v software. One hundred and fourteen matched IO slices at baseline and follow-up (228 IO slices in total) were obtained from the ten patients to quantify plaque morphology and track cap thickness changes. More details about extracting plaque morphological and mechanical stress and strain data are given in Section 2.2. Sample slices with segmented contours for IVUS and OCT images were provided as supplemental material.

Table 1. Patient demographic data. F: female; M: male; BP: blood pressure; HT: hypertension; DM: diabetes mellitus; HL: hyperlipidemia; FU: follow-up.

Patient ID	Age	Sex	BP (mmHg)	Diagnosis History	FU Days
P1	80	F	71–138	HT DM	304
P2	70	M	84–155	HT	273
P3	65	F	63–149	DM	220
P4	66	M	89–150	DM	290
P5	81	M	69–112	HT	182
P6	73	M	55–150	HT HL	248
P7	74	F	62–151	HT DM HL	244
P8	62	F	79–117	HL	195
P9	61	M	78–128	HT DM HL	283
P10	72	M	80–143	HT DM HL	272

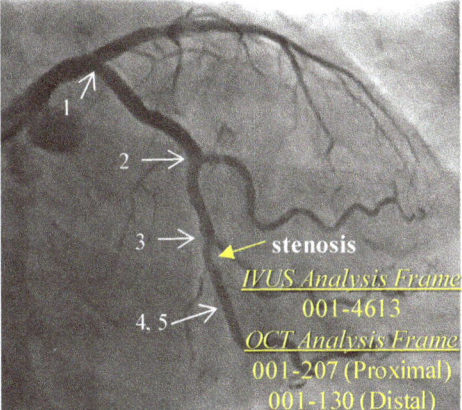
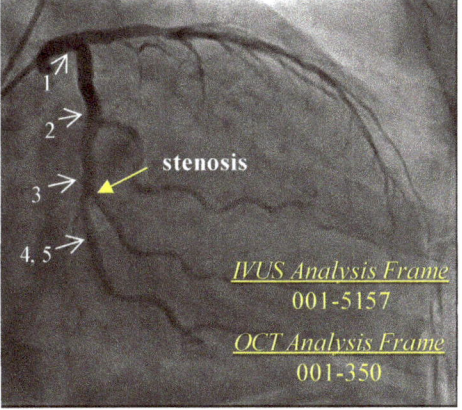

Figure 1. Registration of baseline and follow-up vessel segment using landmarks and vessel features (bifurcation, stenosis, and plaque components from IVUS/OCT). Only angiography is shown. (**a**) Baseline angiography; (**b**) Follow-up angiography. 1: left circumflex artery ostium, 2: 1st obtuse marginal branch, 3: 2nd obtuse marginal branch, 4: IVUS analysis start point, 5: OCT analysis start point.

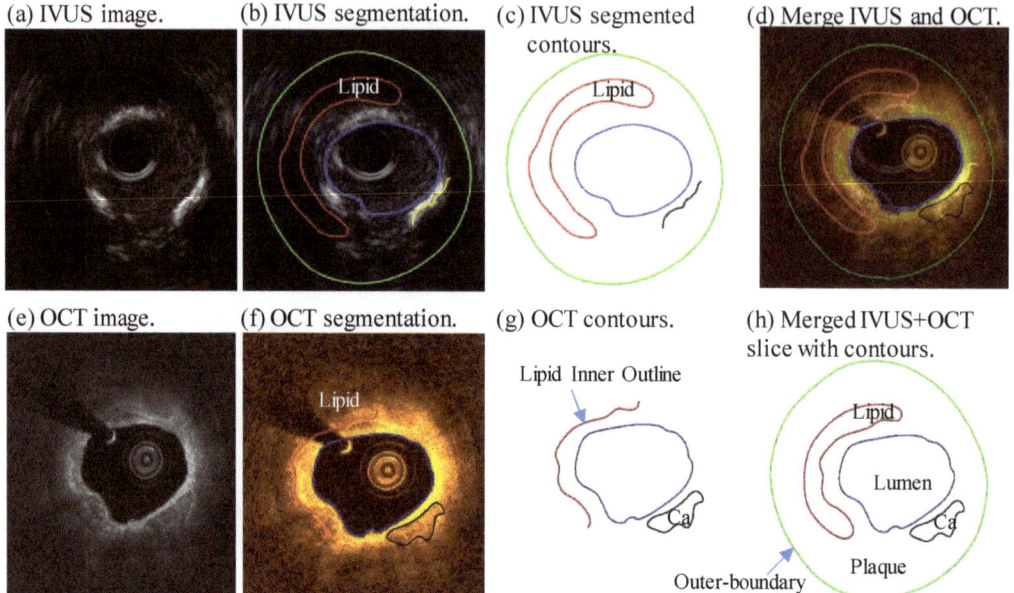

Figure 2. Merging OCT and IVUS contours to generate combined IVUS + OCT slice with contours. (**a**) IVUS images; (**b**) IVUS segmentation; (**c**) IVUS segmented contours; (**d**) Merge IVUS and OCT; (**e**) OCT image; (**f**) OCT segmentation; (**g**) OCT contours; (**h**) Merged IVUS + OCT slice with contours. Red: lipid; Green: outer-boundary; Blue: lumen; Black: calcification (Ca).

2.2. Thin-Slice Models, Morphological and Biomechanical Predictors, Data Extraction for Analysis

Three-dimensional (3D) thin-slice models were used to obtain plaque stress and strain data which will be used in plaque vulnerability predictions. A thickness of 0.5 mm was added to each IO slice to reconstruct the plaque geometry of 3D thin-slice model. A total of 228 models were constructed at baseline and follow-up. Under in vivo condition, arteries were subjected to blood pressure and axially stretched. Computational 3D thin-slice models need to start from zero-load geometries with zero pressure and stress/strain conditions. Therefore, axial and circumferential shrinking was applied to in vivo IO slices to obtain their zero-load state. Axial shrinkage was assumed to be 5% for all plaques while circumferential shrinkage rate was determined for each slice to match its in vivo morphology. Details of the pre-shrink–stretch process were described in our previous studies [30–32]. Vessel tissue was assumed to be hyperelastic, anisotropic, nearly-incompressible, and homogeneous. Plaque components (lipid and calcification) were assumed to be hyperelastic, isotropic, and nearly-incompressible [13]. The strain energy density functions for the isotropic and anisotropic modified Mooney–Rivlin material models are given below:

$$W_{iso} = c_1(I_1 - 3) + c_2(I_2 - 3) + D_1[exp(D_2(I_1 - 3)) - 1], \qquad (1)$$

$$W_{aniso} = W_{iso} + \frac{K_1}{K_2}\{exp[K_2(I_4 - 1)^2] - 1\}, \qquad (2)$$

where $I_1 = \Sigma(C_{ii})$, $I_2 = \frac{1}{2}[I_1^2 - C_{ij}C_{ij}]$, I_1 and I_2 are the first and second invariants of right Cauchy–Green deformation tensor $\mathbf{C} = [C_{ij}] = \mathbf{X}^T\mathbf{X}$, $\mathbf{X} = [X_{ij}] = [\partial x_i/\partial a_j]$, (x_i) is current position, (a_j) is original position, $I_4 = C_{ij}(n_c)_i(n_c)_j$, n_c is the unit vector in the circumferential direction of the vessel, c_1, c_2, D_1, D_2, K_1 and K_2 are material parameters. Material constants of isotropic Mooney–Rivlin model from the existing literature were used: Lipid: $c_1 = 0.5$ kPa, $c_2 = 0$ kPa, $D_1 = 0.5$ kPa, $D_2 = 1.5$; Calcification: $c_1 = 92$ kPa, $c_2 = 0$ kPa, $D_1 = 36$ kPa and $D_2 = 2.0$; Vessel/Fibrous tissue: $c_1 = -262.6$ kPa,

c_2 = 22.9 kPa, D_1 = 125.9 kPa, D_2 = 2.0, K_1 = 7.19 kPa, K_2 = 23.5 [13]. The 3D thin-slice models were solved by a finite element software ADINA 9.0 (Adina R&D, Watertown, MA, USA) to obtain plaque stress/strain distributions following our established procedures [31]. Nonlinear incremental iterative procedures were used to solve the models. Mesh analysis was performed by refining mesh density by 10% until changes in solutions became less than 2%.

Nine morphological and mechanical risk factors were selected as predictors to predict plaque vulnerability changes from baseline to follow-up: lumen area (LA), plaque area (PA), plaque burden (PB), minimum cap thickness (MinCapT), mean cap thickness (MeanCapT), maximum cap stress (MaxCapS), mean cap stress (MeanCapS), maximum cap strain (MaxCapSn), and mean cap strain (MeanCapSn). PB was defined as

$$PB = \frac{PA}{PA + LA'} \quad (3)$$

where PA is the area between the outer boundary contour and lumen contour and LA is the area inside lumen contour (see Figure 3). Values of the risk factors were extracted for each slice using a Four-Quarter Even-Spacing method [31,32]. For each matched slice, 100 evenly spaced points from the lumen were selected and morphological and biomechanical factors from the IO slice and 3D thin-slice model at each point were obtained for analysis. Figure 3b shows a simple illustration of the method. Data extraction of the nine risk factors and following statistical analysis were implemented by MATLAB (MATLAB R2018a, MathWorks, Natick, MA, USA).

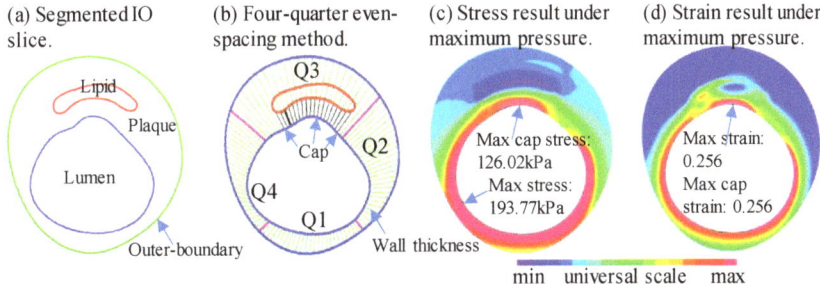

Figure 3. Illustration of slice contours, sample slices with Four-Quarter Even-Spacing method showing the definition and extraction of morphological and mechanical predictor data. (**a**) Segmented IO slice; (**b**) Four-Quarter Even-Spacing method. Bold black line shows the minimum cap thickness; (**c**) Stress result under maximum pressure; (**d**) Strain result under maximum pressure.

2.3. Plaque Vulnerability Indices

2.3.1. Cap Thickness Plaque Vulnerability Index (C-PVI)

Wang L et al. introduced 3 morphology-based indices (cap index, lipid index, and morphological index) and predicted their changes using patient IVUS follow-up data [6]. IO data can provide accurate cap thickness data, but there is no reliable information on lipid size. Bearing those in mind, and with the assumption that plaque rupture may be linked most closely to minimum cap thickness, C-PVI was introduced using MinCapT as a quantitative measure for plaque vulnerability with values 1, 2, 3 and 4 (see Table 2). Category 4 (C-PVI = 4) has the thinnest fibrous cap thickness, while Category 1 (C-PVI = 1) has the thickest fibrous cap thickness. Slice distributions for the 4 C-PVI values are given in Table 2. The cap thickness interval for each C-PVI value was chosen so that each category had some samples [33]. Some explanation is given in Section 4.1.

Table 2. Cap- and Stress/Strain-based PVI definitions and slice divisions by PVI values.

PVI Index Values	1	2	3	4
C-PVI Min-CapT Range (mm)	(0.36, 2)	(0.26, 0.36]	(0.20, 0.26]	(0.0, 0.20]
Slice Distributions	79	21	7	7
MaxS-PVI Max Stress Range (kPa)	(20, 80]	(80, 101]	(101, 110]	(110, ∞)
Slice Distributions	73	22	8	11
MeanS-PVI Mean Stress Range (kPa)	(20, 70]	(70, 88]	(88, 93]	(93, ∞)
Slice Distributions	82	23	8	1
MaxSn-PVI Max Strain Range	(0.05, 0.17]	(0.17, 0.18]	(0.18, 0.2]	(0.2, ∞)
Slice Distributions	80	15	13	6
MeanSn-PVI Mean Strain Range	(0.05, 0.18]	(0.18, 0.2]	(0.2, 0.21]	(0.21, ∞)
Slice Distributions	95	13	3	3

2.3.2. Stress Plaque Vulnerability Index (S-PVI)

It is believed that plaque cap stress is closely related to plaque progression rupture and could be used as another measurement for plaque vulnerability. Two stress-based plaque vulnerability indices (S-PVI) were introduced in this paper for our quantitative vulnerability analysis: one is based on MaxCapS denoted by MaxS-PVI, one is based on MeanCapS denoted by MeanS-PVI. Stress intervals for each index values are given in Table 2. Stress interval divisions were determined so that these index values had the best match rate with C-PVI. MaxS-PVI was introduced since plaque rupture is closely linked to MaxCapS. MeanS-PVI is also considered since MeanCapS is an averaged stress value and may provide plaque stress information in a more collective way. Both stress-based indices were used for plaque vulnerability investigation in this paper to compare which one will provide better prediction results.

2.3.3. Strain Plaque Vulnerability Index (Sn-PVI)

While most researchers concentrated on plaque stress (more focused on cap stress) for vulnerable plaque investigations, plaque cap strain measures whether plaque cap is stretched hard and may be a better indicator for cap mechanical conditions and plaque vulnerability. Similar to stress indices, two strain-based plaque vulnerability indices (Sn-PVI) were introduced in this paper for analysis: one is based on MaxCapSn denoted by MaxSn-PVI, one is based on MeanCapSn denoted by MeanSn-PVI. Strain intervals for each index values are given in Table 2. All arrangements for strain indices were similar to those for stress indices and are omitted for simplicity.

2.3.4. Prediction Methods and Plaque Vulnerability Predictions

All possible combinations (511 combinations) of the nine risk factors (see Section 2.2) with their values at baseline were used to predict 5 PVI changes. PVI changes were measured by changes in PVI between baseline and follow-up. Using C-PVI as an example for illustration, C-PVI change (ΔC-PVI) between baseline and follow-up for a given IO slice was defined as

$$\Delta\text{C-PVI}(\text{Slice \#}) = (\text{C-PVI at follow-up}) - (\text{C-PVI at baseline}). \tag{4}$$

For simplicity, its binary outcomes BΔC-PVI (defined below) served as the target variable in our prediction models:

$$\text{B}\Delta\text{C-PVI}(\text{Slice \#}) = \begin{cases} 1, & \text{if } \Delta\text{C-PVI} > 0; \\ -1, & \text{if } \Delta\text{C-PVI} \leq 0. \end{cases} \tag{5}$$

The same definition was used for other 4 PVIs. For each PVI, the values of the 9 risk factors at baseline and the PVI change binary outcomes of 114 slices were stored in a 10 × 114 matrix which was used as input file for the prediction methods. The RF was adopted in this paper to perform prediction. Figure 4 shows the schematic diagram of RF model. In each test, 9 risk factors and the PVI change binary outcomes from 114 slices were fed to the RF method. A standard five-fold cross-validation procedure was performed for model fitting and testing [31]. To be more specific, the data set (114 slices with baseline and follow-up scans) was randomly divided into five equal parts, with four parts used as the training set and the remaining one part used as the validation set. Then, the RF method was run five times so that each of the five parts had a chance to serve as the validation set. This five-fold cross-validation procedure was repeated 100 times (each time with new randomly divided 5 parts), and the results were averaged to obtain stable and accurate prediction results. The RF method was implemented by calling TreeBagger function in MATLAB (v2018 The MathWorks Inc., Natick, MA, USA). The number of trees in random forest was set to 50 since the prediction results (sensitivity and specificity) become stable, and further increase in the number (doubling) showed little difference in results. The procedure was repeated 100 times and the results were averaged to stabilize the prediction results. The output of the prediction was a True or False value (defined as True = ΔPVI > 0 and False = ΔPVI ≤ 0) corresponding to the optimal cutoff threshold probability) for each slice of the test set. The prediction results were compared with actual measurements of PVI changes based on IO image data (gold standard) to calculate prediction accuracy (Acc), sensitivity (Sen), and specificity (Spe) defined as follows:

$$Acc = \frac{TP + TN}{TP + FP + TN + FN}, \qquad (6)$$

$$Sen = \frac{TP}{TP + FN}, \qquad (7)$$

$$Spe = \frac{TN}{TN + FP}, \qquad (8)$$

where TP is the number of true positive outcomes (ΔPVI > 0 predicted as such), FP is the number of false positive outcomes (ΔPVI ≤ 0 predicted as ΔPVI > 0), TN is the number of true negative outcomes (ΔPVI ≤ 0 predicted as such), and FN is the number of false negative outcomes (ΔPVI > 0 predicted as ΔPVI ≤ 0). The abscissa of the receiver operating characteristic (ROC) curve is "1-Specificity" and the ordinate is Sensitivity. The area under ROC curve is the value of AUC. Five PVIs (C-PVI, 2 S-PVIs, and 2 Sn-PVIs) were used to identify which one would have better prediction accuracies. Details of the prediction methods and procedures were published before and are omitted here [34].

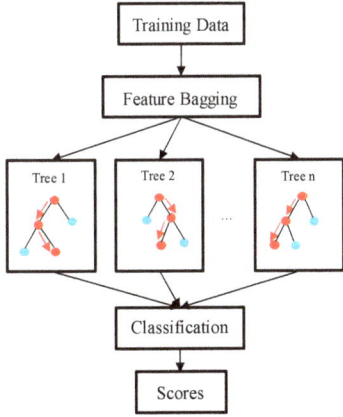

Figure 4. The schematic diagram of random forest model.

3. Results

3.1. Prediction Results for the 5 PVIs Using Combination Predictors

Table 3 lists the prediction accuracy, sensitivity, specificity, and AUC of the respective optimal combination predictors for the five PVIs. Among the five PVIs, C-PVI had the best prediction accuracy (90.3%) with the optimal predictor as a combination of PA + PB + MinCapT + MeanCapT + MeanCapSn. It also had the best specificity (95.8%). However, its sensitivity was only 56.7%. MaxSn-PVI had the best AUC (0.935) and the best Sensitivity + Specificity (1.745), with the optimal predictor as a combination of LA + PA + MaxCapSn. Its ROC curve is shown by Figure 5.

Table 3. Combination predictor prediction accuracy, sensitivity, specificity and AUC values for the 5 PVIs considered.

PVI Index	Best Predictor	Acc	Sen	Spe	Sen + Spe	AUC
C-PVI	PA + PB + MinCapT + MeanCapT + MeanCapSn	0.903	0.567	0.958	1.525	0.877
MaxS-PVI	MinCapT + MeanCapT + MaxCapS	0.779	0.617	0.844	1.461	0.776
MeanS-PVI	PA + MeanCapS	0.856	0.730	0.888	1.617	0.867
MaxSn-PVI	LA + PA + MaxCapSn	0.871	0.876	0.869	1.745	0.935
MeanSn-PVI	PA + PB + MaxCapSn + MeanCapSn	0.833	0.568	0.876	1.444	0.809

Note: Total number of slices with ΔC-PVI > 0 = 16; Total number of slices with ΔC-PVI $\leq$ 0 = 98.

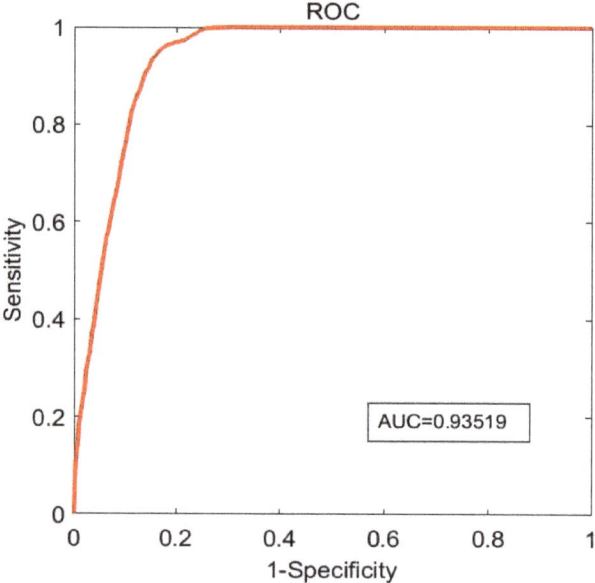

Figure 5. ROC curve with AUC = 0.935 for prediction of ΔMaxSn-PVI.

3.2. Prediction Results for the 5 PVIs Using Single Predictors

Prediction results of the nine single predictors for the five PVIs are shown in Table 4. Only one single predictor was used in each prediction here. Among the five PVIs, MaxSn-PVI had the best performance with MaxCapSn delivering best prediction AUC (0.909) and accuracy (79.8%). Among all nine predictors for C-PVI, PB had the best pre-diction accuracy (83.7%) and AUC (0.827). The two stress-based PVIs had lower pre-diction accuracies and AUC values. The best single predictor for MaxS-PVI was MeanCapT with accuracy of 65.4% and AUC 0.675. The best single predictor for MeanS-PVI was PA with accuracy of 75.6% and AUC 0.781.

Table 4. Single predictor prediction accuracy, sensitivity, specificity and AUC values for the 5 PVIs considered.

	C-PVI		MaxS-PVI
Predictor	(Acc, Sen, Spe, AUC)	Predictor	(Acc, Sen, Spe, AUC)
LA	(0.702, 0.136, 0.794, 0.416)	LA	(0.566, 0.288, 0.679, 0.483)
PA	(0.738, 0.068, 0.847, 0.434)	PA	(0.568, 0.309, 0.674, 0.465)
PB	(0.837, 0.702, 0.859, 0.827)	PB	(0.634, 0.445, 0.711, 0.632)
MinCapT	(0.756, 0.199, 0.847, 0.674)	MinCapT	(0.506, 0.351, 0.569, 0.463)
MeanCapT	(0.696, 0.220, 0.774, 0.571)	MeanCapT	(0.654, 0.388, 0.762, 0.675)
MaxCapS	(0.715, 0.235, 0.793, 0.524)	MaxCapS	(0.600, 0.323, 0.713, 0.610)
MeanCapS	(0.789, 0.279, 0.872, 0.687)	MeanCapS	(0.451, 0.184, 0.559, 0.356)
MaxCapSn	(0.778, 0.293, 0.858, 0.672)	MaxCapSn	(0.559, 0.349, 0.644, 0.486)
MeanCapSn	(0.785, 0.177, 0.885, 0.603)	MeanCapSn	(0.587, 0.386, 0.669, 0.540)
	MeanS-PVI		MaxSn-PVI
Predictor	(Acc, Sen, Spe, AUC)	Predictor	(Acc, Sen, Spe, AUC)
LA	(0.735, 0.316, 0.841, 0.664)	LA	(0.576, 0.551, 0.587, 0.627)
PA	(0.756, 0.513, 0.818, 0.781)	PA	(0.704, 0.535, 0.776, 0.699)
PB	(0.676, 0.355, 0.757, 0.640)	PB	(0.615, 0.302, 0.749, 0.523)
MinCapT	(0.704, 0.171, 0.839, 0.509)	MinCapT	(0.498, 0.353, 0.560, 0.474)
MeanCapT	(0.564, 0.238, 0.646, 0.436)	MeanCapT	(0.521, 0.331, 0.602, 0.504)
MaxCapS	(0.646, 0.179, 0.764, 0.492)	MaxCapS	(0.641, 0.506, 0.698, 0.641)
MeanCapS	(0.580, 0.294, 0.653, 0.513)	MeanCapS	(0.606, 0.422, 0.684, 0.543)
MaxCapSn	(0.717, 0.277, 0.828, 0.660)	MaxCapSn	(0.798, 0.593, 0.885, 0.909)
MeanCapSn	(0.716, 0.403, 0.795, 0.628)	MeanCapSn	(0.623, 0.306, 0.757, 0.585)
	MeanSn-PVI		
Predictor	(Acc, Sen, Spe, AUC)	Predictor	(Acc, Sen, Spe, AUC)
LA	(0.743, 0.047, 0.857, 0.508)	PA	(0.734, 0.200, 0.821, 0.451)
PB	(0.748, 0.361, 0.811, 0.579)	MinCapT	(0.710, 0.201, 0.793, 0.560)
MeanCapT	(0.714, 0.306, 0.781, 0.587)	MaxCapS	(0.701, 0.041, 0.808, 0.326)
MeanCapS	(0.715, 0.168, 0.805, 0.477)	MaxCapSn	(0.753, 0.266, 0.833, 0.693)
MeanCapSn	(0.659, 0.518, 0.682, 0.650)		

3.3. Combination Predictors Had Better Prediction Accuracies Than Those from Single Predictors

Figure 6 presents prediction accuracies of the best combination predictors and the best single predictors for the five PVIs. For C-PVI, the best combination predictor increased accuracy by 6.6% compared to the best single predictor (90.3% vs. 83.7%). For MaxS-PVI, the best combination predictor had a prediction accuracy which was 12.5% over that of the best single predictor (77.9% vs. 65.4%). For MeanS-PVI, the improvement of accuracy by the best combination predictor over the best single predictor was 10.0% (85.6% vs. 75.6%). Considering MaxSn-PVI and MeanSn-PVI, the best combination predictors improved predictor accuracies by 7.3% and 8.0% over those from the best single predictors, respectively (87.1% vs. 79.8%, and 83.3% vs. 75.3%). AUC values by the best combination predictors also improved over the best single predictors by 0.050, 0.101, 0.086, 0.026, and 0.116 for the five PVIs, respectively (see Figure 7). Overall, it was observed that prediction accuracies of the best combination predictors were higher than those from the best single predictors.

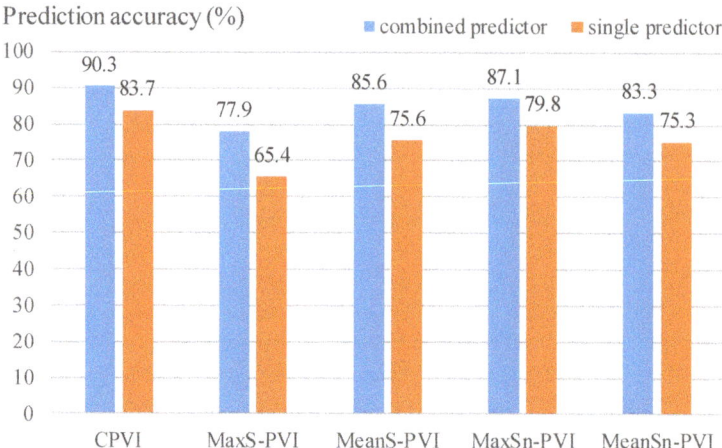

Figure 6. Comparison of prediction accuracies by best combination and single predictors for the five PVIs. Best combination predictors are provided in Table 3, and best single predictors are PB, MeanCapT, PA, MaxCapSn and MaxCapSn, respectively.

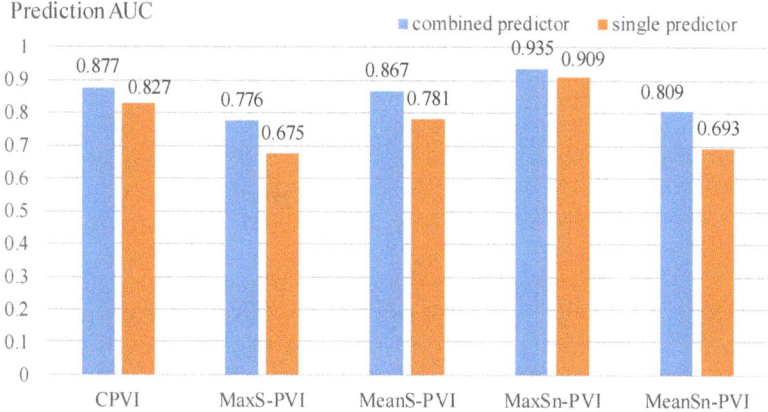

Figure 7. Comparison of AUC values by best combination and single predictors for the five PVIs. Best combination predictors are provided in Table 3, and best single predictors are PB, MeanCapT, PA, MaxCapSn and MaxCapSn, respectively.

4. Discussion

As important as vulnerable plaque research is to the health of the general public, progress has been limited by several key factors: (a) lack of quantitative measure of plaque vulnerability; without quantitative measure, it is hard to say whether plaque vulnerability is improving or not and it is difficult to perform prediction analysis; (b) lack of accurate medical images with acceptable resolution to provide exact plaque morphology for assessment and mechanical model construction; (c) lack of "gold standard" for plaque rupture or clinical events to validate plaque progression and vulnerability predictions. In the following, we will attempt to discuss the ways in which we tried to address those limitations in this paper.

4.1. Introducing Quantitative Plaque Vulnerability Indices for Vulnerability Predictions

Our research effort has been focused on introducing morphological and mechanical indices for plaque classification, comparison, and prediction [4,6]. Five PVIs were introduced in this paper as measures of plaque vulnerability. With the high resolution from OCT,

C-PVI was considered the "gold standard". Our criterion for an index was that it should be based on reliable data and it should be measurable. While C-PVI may be missing some other important factors such as cell activities on the lumen surface (inflammation, erosion), it focuses on cap thickness which may be one of the most watched items for vulnerable plaques. Stress and strain indices were included because we do believe that mechanical forces play an important role in plaque progression and rupture and monitoring them may provide useful information which is not included in plaque morphology alone. Our results actually demonstrated that strain-based PVIs had better prediction accuracy compared to stress-based PVIs (see Table 3).

Compared with the current literature, Mortensen et al. noted that PB may be a major predictor of cardiovascular event and mortality risk compared to coronary stenosis [35]. The prediction results using PB for C-PVI changes (Accuracy = 83.7%, see Table 4) are in good agreement with the statement, and the best combined predictive accuracy achieved 90.3%. It should be noted that due to our data limitation, the cap thickness threshold for category 4 plaque was set to 0.2 mm in Table 2 instead of the generally accepted threshold (65 µm) of vulnerable plaques. This value was chosen for two reasons: (1) It is larger than 65 µm used in other studies based on ex vivo histology data. This is reasonable since the fibrous cap thickness in in vivo data is higher than that in histological sections [36]; (2) We could have some number of slices in Category 4 when 0.2 mm was selected. If 65 µm threshold was adopted, the number of Category 4 plaques in our data set would be zero and prediction analysis would not be possible. Hence, the well-accepted cap thickness of 65 µm for highly vulnerable plaques was not adopted due to realistic in vivo data limitations compared to results based on histological data [3]. Another point to note is that if a slice does not have a lipid core, then the slice does not have a fibrous cap and is not included in the data set in this work.

In mechanics, various evidences indicated that high plaque stresses are indeed linked to plaque rupture which is more likely to occur near thin fibrous cap, so the cap position should attract more attention [4,17]. By setting fibrous cap and its shoulder as critical region, Wang L et al. clearly explained the use of morphological factors and S-PVI to predict plaque composition changes [37]. For MeanS-PVI, the accuracy of the single predictor (PA) was 75.6%, but the optimal combination predictor was 85.6%, showing a significant improvement. Numerous studies have attempted to establish a solid link between plaque strain values and their vulnerability by solving circumferential strain directly from in vivo images [21]. Since in vivo images do not have zero-load state, strain values calculated using in vivo images used a difference reference frame and would be smaller than true strain values using zero-load reference frames [21,22]. Our models included a pre-shrink–stretch process and stress/strain were calculated using plaque zero-load geometries. Caution should be exercised when comparing results from models with different model assumptions.

4.2. Predicting PVI Changes Based on Accurate and Reliable OCT-Based Data

It has been mentioned that IVUS resolution is not enough to quantify thin plaque cap thickness and plaque progression, meanwhile the thin cap thickness and plaque wall thickness changes in follow-up are normally smaller than the IVUS resolution. Those limitations are the reason behind IVUS-based vulnerable plaque progression and vulnerability prediction results possibly being subjected to large errors. Using IVUS data and generalized linear mixed model (GLMM) prediction method, Wang L et al. reported that an optimal combination predictor achieved AUC = 0.629 in predicting wall thickness increase and AUC = 0.845 in predicting plaque area increase [37]. Multi-modality data combining IVUS and OCT ensures accurate cap thickness quantification, C-PVI assessment and further mechanical stress/strain calculations. These improvements made accurate and reliable plaque morphology assessment and PVI predictions possible. In this work, using IO data, the mean AUC of five PVIs is 0.853, showing a superior prediction ability. Guo et al. also reported similar findings using least squares support vector machine (SVM) prediction methods that the ability of IO/OCT-based vulnerability predictions were improved compared with IVUS-based predictions (Accuracy: 0.838 vs. 0.786) [13]. Moreover, by observing different PVIs, the accuracy and AUC value of prediction results from IO slices are both higher than 75% (see Table 3). The accuracy of all the single predictors was more than 70% for C-PVI, while PB and LA had higher prediction accuracies (86.5% and 77.8%), which confirms the general reliability of the prediction based on IO data. It is worth recognizing that fusion OCT and IVUS were used in this study to provide precise measurements of fibrous cap thickness. This data set warrants more accurate mechanical and plaque vulnerability quantification and prediction accuracy. However, it is difficult to have patients to agree to simultaneously undergo OCT, IVUS, and coronary angiography at baseline and follow-up, and thus limiting the patient data set here. For prediction methods, Wang L et al. compared various prediction methods, among which the prediction accuracy of RF was the highest, superior to SVM and GLMM. The prediction accuracy of machine learning method (RF) is 5.91% higher than that of GLMM method [34]. In this paper, the above three prediction methods were all performed and compared. The prediction results from RF were selected for report since they provided best prediction accuracies.

4.3. Combining Mechanical and Morphological Predictors May Lead to Better Predictions

At present, some general scoring rules, such as CLIMA score and Burgmaier score, are often used in plaque assessment [19,38,39]. However, these scoring rules did not include mechanical forces in their assessment. It has been conjectured that those mechanical forces play an important role in plaque progression and rupture process and combining mechanical and morphological factors may lead to better prediction results. In a study using IVUS follow-up data from none patients, Wang L et al. reported that the prediction accuracy from the best morphological + mechanical combination predictor was 68.1%, 3.9% higher than that of the best morphological combination predictor (64.2%) [33]. In this paper, by using accurate multi-modality IVUS + OCT data with follow-up, for five PVIs, the combination predictors improved the prediction accuracies by 6.6%, 12.5%, 10.0%, 7.3% and 8.0% respectively, with an average improvement of 8.9%. Taking the MeanS-PVI as an example, the accuracy of the combination predictor (PA + MeanCapS) improved over that of the best single factor (PA) by 10%. That is better that the 4.01% improvement reported in work by Wang Q et al. [31]. Prediction sensitivity and specificity also had significant improvements. Our work is adding further evidence to the conjecture that combining mechanical and morphological factors may lead to better prediction results. It should be acknowledged that our sample size is small and further effort using large scale samples is needed to reach solid conclusions.

4.4. Labor Cost and Potential Implementations

The construction and simulation of a 3D thin-layered (slice) model could be finished within 10 min on a personal computer (Xeon E5-1620 v3 kernel processor (3.5 GHz)). This provides the possibility to integrate modeling with medical equipment for potential clinical implementations. Compared with 3D fluid-structure interaction (FSI) model (which requires approximately 2 weeks to construct one model), this 3D thin-layered (slice) model has the advantages of low labor cost, short construction time, ease of convergence, high accurate simulation results (relative error < 10% compared with FSI model).

It is of interest to note that only baseline data would be needed to make predictions after the model becomes well trained and validated. Indeed, follow-up data (IVUS + OCT) were only needed in this paper for verification, i.e., to verify if the predictions were indeed true. It is also needed for model training as well. If data set was large enough and the model was sufficiently trained and validated, then only baseline data would be needed to make predictions. As technology develops, OCT may have better penetration and OCT alone would be enough to construct the models and provide all values for all the predictors to make predictions.

4.5. Limitations

Some limitations of our study include the following. (a) Our sample size is still small and contained fewer slices with C-PVI = 3 and 4 (12.2% of total slices). That might account for the low sensitivity in our prediction results. Most of the patients were not in the acute progression stage of the disease, and the growth of plaque was generally slow, resulting in a small change in the condition of plaque within a year and a skewness distribution of data. Because most people were on medication, that also exacerbated to an unbalanced sample set. (b) Improving sample size and conducting our work at patient-level may help address sample imbalance, but lack of in vivo data on plaque rupture remains one of the limitations of current research. (c) Patient-specific vessel material properties were not available. Therefore, vessel material parameters from available literature were used in this study [30]. It should be noted that there is high variability of constitutive parameters among different individuals [40–42], which indeed impacts the stress/strain calculation. Guo et al. used patient-specific plaque material properties and showed that the relative errors could be 40% in stress and 123% in strain calculations if material properties obtained from ex vivo tensile testing were used [43]. (d) Thin-slice models were used in this study since they could provide better accuracy over 2D models and save model construction time compared to full 3D models [30,31]. However, they only provided plaque structure stress and strain values and did not retain flow information (for example, flow wall shear stress). Thin-slice models require much less labor to construct and could be more practical for potential clinical implementations. However, it remains true that full 3D FSI models could be a better choice for more accurate stress/strain and wall shear stress calculations. It is worth noting that the 3D thin-slice model and prediction method used in this paper are relatively more integrated, and efforts are being made to automate and streamline the whole process.

5. Conclusions

Since plaque vulnerability is hard to quantify, plaque cap thickness index and biomechanical stress/strain indices were defined as alternative quantitative plaque vulnerability indices (PVIs) to conduct predictive research. Accurate multi-modality IVUS + OCT data at both baseline and follow-up and machine learning methods were used to identify and validate the best predictors of changes in plaque vulnerability. The results showed that the accuracy of the combined predictors including mechanical factors was significantly better than that of the single predictors. Plaque cap thickness and cap stress and strain could be used as measurable and calculable evaluation indexes to predict plaque vulnerability change and provide a more complete early screening strategy for patients with vulnerable plaques.

Author Contributions: Conceptualization, D.T., L.W. and R.L.; methodology, R.L., D.T. and L.W.; formal analysis R.L., L.W. and X.G.; writing—original draft, R.L., D.T. and L.W.; writing—review and editing, D.T., L.W. and R.L.; data acquisition and preparation, A.M., M.M., G.S.M. and J.Z.; resources and supervision, H.S. and D.P.G.; project administration, D.T. and J.Z.; funding acquisition, D.T. and L.W. All authors have read and agreed to the published version of the manuscript.

Funding: This research was supported in part by National Sciences Foundation of China grants 11972117, 11802060; Natural Science Foundation of Jiangsu Province under grant number BK20180352; a Jiangsu Province Science and Technology Agency under grant number BE2016785; Fundamental Research Funds for the Central Universities and Zhishan Young Scholars Fund administrated by Southeast University (grant number 2242021R41123).

Institutional Review Board Statement: Not applicable.

Informed Consent Statement: Not applicable.

Data Availability Statement: Data are available on request. Data cannot be made publicly available for ethical or legal reasons (public availability would compromise patient privacy).

Conflicts of Interest: The authors declare no conflict of interest.

References

1. Gupta, R.; Wood, D. Primary prevention of ischaemic heart disease: Populations, individuals, and health professionals. *Lancet* **2019**, *394*, 685–696. [CrossRef] [PubMed]
2. Stary, H.C.; Chandler, A.B.; Dinsmore, R.E.; Fuster, V.; Glagov, S.; Insull, W., Jr.; Rosenfeld, M.E.; Schwartz, C.J.; Wagner, W.D.; Wissler, R.W. A Definition of Advanced Types of Atherosclerotic Lesions and a Histological Classification of Atherosclerosis. *Circulation* **1995**, *92*, 1355–1374. [CrossRef]
3. Virmani, R.; Kolodgie, F.D.; Burke, A.P.; Farb, A.; Schwartz, S.M. Schwartz. Lessons From Sudden Coronary Death A Comprehensive Morphological Classification Scheme for Atherosclerotic Lesions. *Arterioscler. Thromb. Vasc. Biol.* **2000**, *20*, 1262–1275. [CrossRef] [PubMed]
4. Tang, D.; Yang, C.; Zheng, J.; Woodard, P.K.; Saffitz, J.E.; Petruccelli, J.D.; Sicard, G.A.; Yuan, C. Local Maximal Stress Hypothesis and Computational Plaque Vulnerability Index for Atherosclerotic Plaque Assessment. *Ann. Biomed. Eng.* **2005**, *33*, 1789–1801. [CrossRef]
5. Goncalves, I.; Sun, J.; Tengryd, C.; Nitulescu, M.; Persson, A.F.; Nilsson, J.; Edsfeldt, A. Plaque Vulnerability Index Predicts Cardiovascular Events: A Histological Study of an Endarterectomy Cohort. *J. Am. Heart Assoc.* **2021**, *10*, e021038. [CrossRef] [PubMed]
6. Wang, L.; Zheng, J.; Maehara, A.; Yang, C.; Billiar, K.L.; Wu, Z.; Bach, R.; Muccigrosso, D.; Mintz, G.S.; Tang, D. Morphological and Stress Vulnerability Indices for Human Coronary Plaques and Their Correlations with Cap Thickness and Lipid Percent: An IVUS-Based Fluid-Structure Interaction Multi-patient Study. *PLoS Comput. Biol.* **2015**, *11*, e1004652. [CrossRef] [PubMed]
7. Yang, C.; Canton, G.; Yuan, C.; Ferguson, M.; Hatsukami, T.S.; Tang, D. Advanced human carotid plaque progression correlates positively with flow shear stress using follow-up scan data: An in vivo MRI multi-patient 3D FSI study. *J. Biomech.* **2010**, *43*, 2530–2538. [CrossRef]
8. Kume, T.; Akasaka, T.; Kawamoto, T.; Okura, H.; Watanabe, N.; Toyota, E.; Neishi, Y.; Sukmawan, R.; Sadahira, Y.; Yoshida, K. Measurement of the thickness of the fibrous cap by optical coherence tomography. *Am. Heart J.* **2006**, *152*, 755.e1–755.e4. [CrossRef]
9. Kini, A.S.; Vengrenyuk, Y.; Yoshimura, T.; Matsumura, M.; Pena, J.; Baber, U.; Moreno, P.; Mehran, R.; Maehara, A.; Sharma, S.; et al. Fibrous Cap Thickness by Optical Coherence Tomography In Vivo. *J. Am. Coll. Cardiol.* **2017**, *69*, 644–657. [CrossRef]
10. Liu, X.; He, W.; Hong, X.; Li, D.; Chen, Z.; Wang, Y.; Chen, Z.; Luan, Y.; Zhang, W. New insights into fibrous cap thickness of vulnerable plaques assessed by optical coherence tomography. *BMC Cardiovasc. Disord.* **2022**, *22*, 484. [CrossRef]
11. Reith, S.; Battermann, S.; Hoffmann, R.; Marx, N.; Burgmaier, M. Optical coherence tomography derived differences of plaque characteristics in coronary culprit lesions between type 2 diabetic patients with and without acute coronary syndrome. *Catheter. Cardiovasc. Interv.* **2014**, *84*, 700–707. [CrossRef]
12. Sawada, T.; Shite, J.; Garcia-Garcia, H.M.; Shinke, T.; Watanabe, S.; Otake, H.; Matsumoto, D.; Tanino, Y.; Ogasawara, D.; Kawamori, H.; et al. Feasibility of combined use of intravascular ultrasound radiofrequency data analysis and optical coherence tomography for detecting thin-cap fibroatheroma. *Eur. Heart J.* **2008**, *29*, 1136–1146. [CrossRef]
13. Guo, X.; Giddens, D.P.; Molony, D.; Yang, C.; Samady, H.; Zheng, J.; Matsumura, M.; Mintz, G.S.; Maehara, A.; Wang, L.; et al. A Multi-Modality Image-Based FSI Modeling Approach for Prediction of Coronary Plaque Progression Using IVUS and OCT Data with Follow-Up. *J. Biomech. Eng.* **2019**, *141*, 0910031–0910039. [CrossRef]
14. Bourantas, C.V.; Räber, L.; Sakellarios, A.; Ueki, Y.; Zanchin, T.; Koskinas, K.C.; Yamaji, K.; Taniwaki, M.; Heg, D.; Radu, M.D.; et al. Utility of Multimodality Intravascular Imaging and the Local Hemodynamic Forces to Predict Atherosclerotic Disease Progression. *JACC Cardiovasc. Imaging* **2020**, *13*, 1021–1032. [CrossRef] [PubMed]
15. Gijsen, F.; Katagiri, Y.; Barlis, P.; Bourantas, C.; Collet, C.; Coskun, U.; Daemen, J.; Dijkstra, J.; Edelman, E.; Evans, P.; et al. Expert recommendations on the assessment of wall shear stress in human coronary arteries: Existing methodologies, technical considerations, and clinical applications. *Eur. Heart J.* **2019**, *40*, 3421–3433. [CrossRef]
16. Costopoulos, C.; Timmins, L.H.; Huang, Y.; Hung, O.Y.; Molony, D.S.; Brown, A.J.; Davis, E.L.; Teng, Z.; Gillard, J.H.; Samady, H.; et al. Impact of combined plaque structural stress and wall shear stress on coronary plaque progression, regression, and changes in composition. *Eur. Heart J.* **2019**, *40*, 1411–1422. [CrossRef] [PubMed]
17. Costopoulos, C.; Maehara, A.; Huang, Y.; Brown, A.J.; Gillard, J.H.; Teng, Z.; Stone, G.W.; Bennett, M.R. Plaque Structural Stress. 2019. Heterogeneity of Plaque Structural Stress Is Increased in Plaques Leading to MACE. *JACC Cardiovasc. Imaging* **2020**, *13*, 1206–1218. [CrossRef] [PubMed]
18. Milzi, A.; Lemma, E.D.; Dettori, R.; Burgmaier, K.; Marx, N.; Reith, S.; Burgmaier, M. Coronary plaque composition influences biomechanical stress and predicts plaque rupture in a morpho-mechanic OCT analysis. *eLife* **2021**, *10*, e64020. [CrossRef] [PubMed]
19. Schaar, J.A.; de Korte, C.; Mastik, F.; Strijder, C.; Pasterkamp, G.; Boersma, E.; Serruys, P.W.; van der Steen, A.F. Characterizing vulnerable plaque features with intravascular elastography. *Circulation* **2003**, *108*, 2636–2641. [CrossRef]
20. Zhang, L.; Liu, Y.; Zhang, P.F.; Zhao, Y.X.; Ji, X.P.; Lu, X.T.; Chen, W.Q.; Liu, C.X.; Zhang, C.; Zhang, Y. Peak radial and circumferential strain measured by velocity vector imaging is a novel index for detecting vulnerable plaques in a rabbit model of atherosclerosis. *Atherosclerosis* **2010**, *211*, 146–152. [CrossRef]

1. Majdouline, Y.; Ohayon, J.; Keshavarz-Motamed, Z.; Cardinal, M.-H.R.; Garcia, D.; Allard, L.; Lerouge, S.; Arsenault, F.; Soulez, G.; Cloutier, G. Endovascular shear strain elastography for the detection and characterization of the severity of atherosclerotic plaques: In vitro validation and in vivo evaluation. *Ultrasound Med. Biol.* **2014**, *40*, 890–903. [CrossRef] [PubMed]
2. Khan, A.A.; Sikdar, S.; Hatsukami, T.; Cebral, J.; Jones, M.; Huston, J.; Howard, G.; Lal, B.K. Noninvasive characterization of carotid plaque strain. *J. Vasc. Surg.* **2017**, *65*, 1653–1663. [CrossRef] [PubMed]
3. Samady, H.; Eshtehardi, P.; McDaniel, M.C.; Suo, J.; Dhawan, S.S.; Maynard, C.; Timmins, L.H.; Quyyumi, A.A.; Giddens, D.P. Coronary artery wall shear stress is associated with progression and transformation of atherosclerotic plaque and arterial remodeling in patients with coronary artery disease. *Circulation* **2011**, *124*, 779–788. [CrossRef]
4. Stone, G.W.; Maehara, A.; Lansky, A.J.; de Bruyne, B.; Cristea, E.; Mintz, G.S.; Mehran, R.; McPherson, J.; Farhat, N.; Marso, S.P.; et al. A prospective natural-history study of coronary atherosclerosis. *N. Engl. J. Med.* **2011**, *364*, 226–235. [CrossRef]
5. Sakellarios, A.I.; Pezoulas, V.C.; Bourantas, C.; Naka, K.K.; Michalis, L.K.; Serruys, P.W.; Stone, G.; Garcia-Garcia, H.M.; Fotiadis, D.I. Prediction of atherosclerotic disease progression combining computational modelling with machine learning. *Annu. Int. Conf. IEEE Eng. Med. Biol. Soc.* **2020**, *2020*, 2760–2763. [CrossRef]
6. D'Ascenzo, F.; De Filippo, O.; Gallone, G.; Mittone, G.; Deriu, M.A.; Iannaccone, M.; Ariza-Solé, A.; Liebetrau, C.; Manzano-Fernández, S.; Quadri, G.; et al. Machine learning-based prediction of adverse events following an acute coronary syndrome (PRAISE): A modelling study of pooled datasets. *Lancet* **2021**, *397*, 199–207. [CrossRef]
7. Lin, S.; Li, Z.; Fu, B.; Chen, S.; Li, X.; Wang, Y.; Wang, X.; Lv, B.; Xu, B.; Song, X.; et al. Feasibility of using deep learning to detect coronary artery disease based on facial photo. *Eur. Heart J.* **2020**, *41*, 4400–4411. [CrossRef] [PubMed]
8. Lv, R.; Maehara, A.; Matsumura, M.; Wang, L.; Zhang, C.; Huang, M.; Guo, X.; Samady, H.; Giddens, D.P.; Zheng, J.; et al. Using Optical Coherence Tomography and Intravascular Ultrasound Imaging to Quantify Coronary Plaque Cap Stress/Strain and Progression: A Follow-Up Study Using 3D Thin-Layer Models. *Front. Bioeng. Biotechnol.* **2021**, *9*, 713525. [CrossRef]
9. Lansky, A.J.; Dangas, G.; Mehran, R.; Desai, K.J.; Stone, G.W.; Leon, M.B.; Mintz, G.S.; Waksman, R.; Wu, H.; Fahy, M. Quantitative angiographic methods for appropriate end-point analysis, edge-effect evaluation, and prediction of recurrent restenosis after coronary brachytherapy with gamma irradiation. *J. Am. Coll. Cardiol.* **2002**, *39*, 274–280. [CrossRef]
10. Tang, D.; Kamm, R.D.; Yang, C.; Zheng, J.; Canton, G.; Bach, R.; Huang, X.; Hatsukami, T.S.; Zhu, J.; Ma, G.; et al. Image-based modeling for better understanding and assessment of atherosclerotic plaque progression and vulnerability: Data, modeling, validation, uncertainty and predictions. *J. Biomech.* **2014**, *47*, 834–846. [CrossRef]
11. Wang, Q.; Tang, D.; Wang, L.; Canton, G.; Wu, Z.; Hatsukami, T.S.; Billiar, K.L.; Yuan, C. Combining morphological and biomechanical factors for optimal carotid plaque progression prediction: An MRI-based follow-up study using 3D thin-layer models. *Int. J. Cardiol.* **2019**, *293*, 266–271. [CrossRef]
12. Huang, X.; Yang, C.; Zheng, J.; Bach, R.; Muccigrosso, D.; Woodard, P.K.; Tang, D. 3D MRI-based multicomponent thin layer structure only plaque models for atherosclerotic plaques. *J. Biomech.* **2016**, *49*, 2726–2733. [CrossRef] [PubMed]
13. Wang, L.; Tang, D.; Maehara, A.; Molony, D.; Zheng, J.; Samady, H.; Wu, Z.; Lu, W.; Zhu, J.; Ma, G.; et al. Multi-factor decision-making strategy for better coronary plaque burden increase prediction: A patient-specific 3D FSI study using IVUS follow-up data. *Biomech. Model. Mechanobiol.* **2019**, *18*, 1269–1280. [CrossRef] [PubMed]
14. Wang, L.; Tang, D.; Maehara, A.; Wu, Z.; Yang, C.; Muccigrosso, D.; Matsumura, M.; Zheng, J.; Bach, R.; Billiar, K.L.; et al. Using intravascular ultrasound image-based fluid-structure interaction models and machine learning methods to predict human coronary plaque vulnerability change. *Comput. Methods Biomech. Biomed. Eng.* **2020**, *23*, 1267–1276. [CrossRef] [PubMed]
15. Mortensen, M.B.; Dzaye, O.; Steffensen, F.H.; Bøtker, H.E.; Jensen, J.M.; Sand, N.P.R.; Kragholm, K.H.; Sørensen, H.T.; Leipsic, J.; Mæng, M.; et al. Impact of Plaque Burden Versus Stenosis on Ischemic Events in Patients With Coronary Atherosclerosis. *J. Am. Coll. Cardiol.* **2020**, *76*, 2803–2813. [CrossRef] [PubMed]
16. Rodriguez-Granillo, G.A.; García-García, H.M.; Mc Fadden, E.P.; Valgimigli, M.; Aoki, J.; de Feyter, P.; Serruys, P.W. In Vivo Intravascular Ultrasound-Derived Thin-Cap Fibroatheroma Detection Using Ultrasound Radiofrequency Data Analysis. *J. Am. Coll. Cardiol.* **2005**, *46*, 2038–2042. [CrossRef] [PubMed]
17. Wang, L.; Tang, D.; Maehara, A.; Wu, Z.; Yang, C.; Muccigrosso, D.; Zheng, J.; Bach, R.; Billiar, K.L.; Mintz, G.S. Fluid-Structure Interaction Models Based on Patient-Specific IVUS at Baseline and Follow-Up for Prediction of Coronary Plaque Progression by Morphological and Biomechanical Factors: A Preliminary Study. *J. Biomech.* **2018**, *68*, 43–50. [CrossRef]
18. Prati, F.; Romagnoli, E.; Gatto, L.; La Manna, A.; Burzotta, F.; Ozaki, Y.; Marco, V.; Boi, A.; Fineschi, M.; Fabbiocchi, F.; et al. Relationship between coronary plaque morphology of the left anteriordescending artery and 12 months clinical outcome: The CLIMA study. *Eur. Heart J.* **2020**, *41*, 383–391. [CrossRef]
19. Burgmaier, M.; Hellmich, M.; Marx, N.; Reith, S. A score to quantify coronary plaque vulnerability in high-risk patients with type 2 diabetes: An optical coherence tomography study. *Cardiovasc. Diabetol.* **2014**, *13*, 117. [CrossRef]
20. Giudici, A.; Li, Y.; Yasmin; Cleary, S.; Connolly, K.; McEniery, C.; Wilkinson, I.B.; Khir, A.W. Time-course of the human thoracic aorta ageing process assessed using uniaxial mechanical testing and constitutive modelling. *J. Mech. Behav. Biomed. Mater.* **2022**, *134*, 105339. [CrossRef]
21. Jadidi, M.; Habibnezhad, M.; Anttila, E.; Maleckis, K.; Desyatova, A.; MacTaggart, J.; Kamenskiy, A. Mechanical and structural changes in human thoracic aortas with age. *Acta Biomater.* **2020**, *103*, 172–188. [CrossRef] [PubMed]

42. Holzapfel, G.A.; Sommer, G.; Gasser, C.T.; Regitnig, P. Determination of layer-specific mechanical properties of human coronary arteries with nonatherosclerotic intimal thickening and related constitutive modeling. *Am. J. Physiol.* **2005**, *289*, H2048–H2058. [CrossRef] [PubMed]
43. Guo, X.; Zhu, J.; Maehara, A.; Monoly, D.; Samady, H.; Wang, L.; Billiar, K.L.; Zheng, J.; Yang, C.; Mintz, G.S.; et al. Quantify patient-specific coronary material property and its impact on stress/strain calculations using in vivo IVUS data and 3D FSI models: A pilot study. *Biomech. Model. Mechanobiol.* **2017**, *16*, 333–344. [CrossRef] [PubMed]

Disclaimer/Publisher's Note: The statements, opinions and data contained in all publications are solely those of the individual author(s) and contributor(s) and not of MDPI and/or the editor(s). MDPI and/or the editor(s) disclaim responsibility for any injury to people or property resulting from any ideas, methods, instructions or products referred to in the content.

Article

Human Coronary Plaque Optical Coherence Tomography Image Repairing, Multilayer Segmentation and Impact on Plaque Stress/Strain Calculations

Mengde Huang [1], Akiko Maehara [2], Dalin Tang [1,3,*], Jian Zhu [4,*], Liang Wang [1], Rui Lv [1], Yanwen Zhu [1], Xiaoguo Zhang [4], Mitsuaki Matsumura [2], Lijuan Chen [4], Genshan Ma [4,*] and Gary S. Mintz [2]

1 School of Biological Science and Medical Engineering, Southeast University, Nanjing 210096, China
2 The Cardiovascular Research Foundation, Columbia University, New York, NY 10019, USA
3 Mathematical Sciences Department, Worcester Polytechnic Institute, Worcester, MA 01609, USA
4 Department of Cardiology, Zhongda Hospital, Southeast University, Nanjing 210009, China
* Correspondence: dtang@wpi.edu (D.T.); njzhujian@163.com (J.Z.); magenshan@hotmail.com (G.M.); Tel.: +1-508-831-5332 (D.T.)

Abstract: Coronary vessel layer structure may have a considerable impact on plaque stress/strain calculations. Most current plaque models use single-layer vessel structures due to the lack of available multilayer segmentation techniques. In this paper, an automatic multilayer segmentation and repair method was developed to segment coronary optical coherence tomography (OCT) images to obtain multilayer vessel geometries for biomechanical model construction. Intravascular OCT data were acquired from six patients (one male; mean age: 70.0) using a protocol approved by the local institutional review board with informed consent obtained. A total of 436 OCT slices were selected in this study. Manually segmented data were used as the gold standard for method development and validation. The edge detection method and cubic spline surface fitting were applied to detect and repair the internal elastic membrane (IEM), external elastic membrane (EEM) and adventitia–periadventitia interface (ADV). The mean errors of automatic contours compared to manually segmented contours were 1.40%, 4.34% and 6.97%, respectively. The single-layer mean plaque stress value from lumen was 117.91 kPa, 10.79% lower than that from three-layer models (132.33 kPa). On the adventitia, the single-layer mean plaque stress value was 50.46 kPa, 156.28% higher than that from three-layer models (19.74 kPa). The proposed segmentation technique may have wide applications in vulnerable plaque research.

Keywords: coronary; vulnerable plaque; coronary plaque models; multilayer vessel geometry

1. Introduction

Intravascular optical coherence tomography (OCT) is a new imaging modality that has been rapidly developing in recent years. It provides unprecedented resolution up to 10 μm, compared to 150–200 μm by intravascular ultrasound (IVUS). It is not only proven to be reliable and widely used in more and more clinical settings [1,2], but also brings advances and breakthroughs to vulnerable plaque research using biomechanics, allowing for more accurate cap thickness quantification and the construction of more realistic models. Nonetheless, the biggest obstacle of OCT is its low penetration depth. That is why some researchers have developed methods to combine IVUS and OCT together to obtain the complete plaque geometry, with IVUS providing the entire vessel wall geometry and OCT providing high-accuracy near-lumen plaque features, especially cap thickness, inflammations and erosion [3,4].

Despite the penetration limitation of OCT, more and more researchers have exploited the abundant information that OCT provides. OCT consensus illustrates the inaccuracy of the statement that OCT has a constant penetration depth [5]. The fact is, OCT penetration depth ranges from 0.1 mm to 2.0 mm and is heavily tissue-dependent. Particularly for high-attenuation tissues such as lipid-rich plaques, light is obscured. For lower-attenuation tissues such as fibrous and calcium, OCT can see deeper structures clearly. A study also showed that more than 180 degrees of external elastic lamina can be recognized in 95% of OCT slices [6]. The enormous visible vessel segments can be taken advantage of to repair the entire vessel wall.

It is well known that arteries have a three-layer structure: intima, media and adventitia. OCT can actually discriminate the three layers based on image intensity variations [6]. Different layers have different pixel intensities and generate a "light–dark–light" structure in OCT images. It would be a considerable advance if a multilayer vessel wall structure could be obtained from OCT images and used in vulnerable plaque research [7–9]. Different layers also have different mechanical properties [10–13]. The multilayer plaque structure will have a significant impact on plaque stress/strain distributions when multilayer models are used for calculation [14].

Recent studies have tried different methods to segment the coronary plaque and vessel wall from OCT images. Athanasiou et al. introduced an automatic segmentation method, which is able to segment four different tissue types in coronary plaque OCT images: calcium (CA), lipid tissue (LT), fibrous tissue (FT) and mixed tissue (MT) [15]. They also applied the edge detection algorithm and ellipse fitting to identify the internal elastic membrane and estimate its bounding area. The method was updated to 3D space using a linear elastic spring mesh method to fully segment the diseased segments for the first time [16]. Zahnd et al. used an original front propagation scheme depending on grayscale gradient information to segment intima, media and adventitia simultaneously in healthy vessel segments [17]. Kafieh et al. introduced a method using a coarse-grained diffusion map for the layer segmentation of retinal OCT images, which has shown robustness even in low contrast and poor layer-to-layer gradient images [18]. Other researchers also put effort into the automatic characterization of OCT plaques using artificial intelligence [19,20].

To the best of our knowledge, the above techniques have not been used to repair three-layered vessel walls simultaneously with subsequent multilayer biomechanical model construction. Three-layer vessel wall models are rarely considered in current modeling research due to the lack of usable segmented multilayer vessel image data.

In this paper, an automatic multilayer segmentation and repair method was developed to segment coronary OCT images to obtain multilayer vessel geometries for biomechanical model construction. Manually segmented data were used as the gold standard for automatic segmentation method development and validation. The edge detection method and cubic spline surface fitting were applied to detect and repair the internal elastic membrane (IEM), external elastic membrane (EEM) and adventitia–periadventitia interface (ADV). The segmented and repaired vessel slices were then used to construct 3D thin-slice models to demonstrate the impact of the multilayer vessel structure on the plaque stress/strain calculation.

2. Materials and Methods

2.1. Data Acquisition and Processing

Five existing de-identified intravascular optical coherence tomography (OCT) data sets for patients ($n = 5$) with coronary heart diseases were obtained from Cardiovascular Research Foundation (CRF). One additional patient OCT data set was acquired from Southeast University Affiliated Zhongda Hospital using protocol approved by Southeast University Zhongda Hospital Institutional Review Board (approval code 2019ZDKYSB046) with informed consent obtained. A total of 436 OCT slices from 6 patients (1 male; mean age: 70.0) were used in this study, with demographic data shown in Table 1. OCT images were acquired with ILUMIEN OPTIS System and Dragonfly JP Imaging Catheter (St. Jude Medical, Westford, MA, USA). The spatial resolution of the acquired OCT images was 4.5 μm. Slices with poor image quality were removed from this research.

Table 1. Patient demographic data. BP: blood pressure; RCA: right coronary artery; LCX: left circumflex artery; LAD: left anterior descending artery; HT: hypertension; DM: diabetes mellitus; HL: hyperlipoproteinemia.

Patient	Age	Sex	Vessel Segment	BP (mmHg)	Number of Slices	Comorbidities
P1	80	F	RCA	138/71	75	HT DM
P2	65	F	RCA	149/63	90	DM
P3	74	F	RCA	151/62	76	HT DM HL
P4	62	F	RCA	117/79	75	HL
P5	72	M	LCX	143/80	60	HT DM HL
P6	67	F	LAD	113/60	60	Not available

2.2. Multilayer Automatic Segmentation

Multilayer vessel wall manual segmentation was performed by trained experts using ImageJ 1.52v software, and served as the gold standard for automatic segmentation method development and validation. Figure 1a gives a flow chart showing the main steps of multilayer automatic segmentation and surface-repairing process using codes based on MATLAB (MATLAB R2021a, MathWorks, Natick, MA, USA). A sample slice showing definitions of lumen, three layers and three boundary contours was given. Details can also be split into three parts as follows.

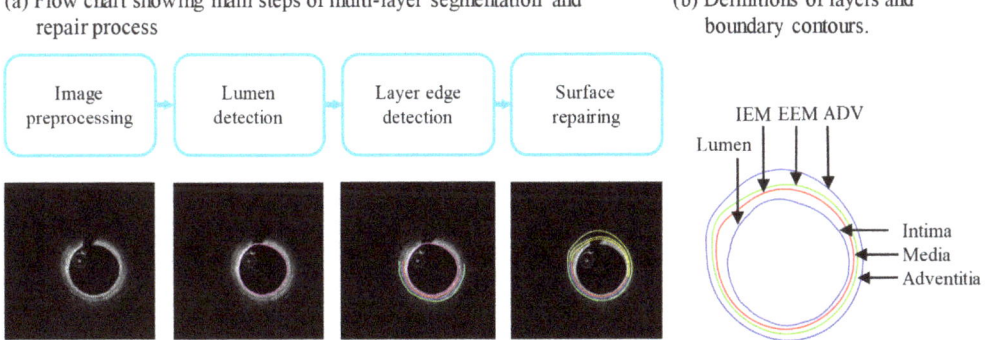

Figure 1. (a) Flow chart showing main steps of automatic multilayer segmentation and repair; (b) a sample slice showing definitions of lumen, three layers and three boundary contours.

Part A. Image preprocessing: Intensities of original OCT images were adjusted by changing the window width to increase the contrast of three layers so that intensity gradients between layers were increased and layer boundaries could be better identified. Guidewire artifacts were removed following the method in [16]. Images of one pullback were stacked in polar coordinates to prepare for subsequent segmentations.

Part B. Lumen detection: Lumen was detected using Otsu's thresholding method in each slice [21]. The threshold was given by maximizing between-class variance and minimizing in-class variance, and was then used in the binary classification of vessel tissues and lumen. Image morphological manipulations, using a square structure element whose width was 4 pixels, were performed successively to erase small noises and jump points in the image. Plaque OCT images are often spotty due to irregular plaque morphologies and scattered plaque tissue components and noises and jump points caused by guidewire attachment or residual blood. Small spotty pieces should be removed so that the segmented contours can be used for biomechanical model constructions. Contour smoothing was performed for all slices using a moving average method with a bandwidth of 50 pixels. Figure 2 shows the original and smoothed contours of a sample slice with bandwidths of

20, 50 and 100 pixels. The figure shows that 20 pixels were not enough and that smoothing using 50 pixels was sufficient for modeling use.

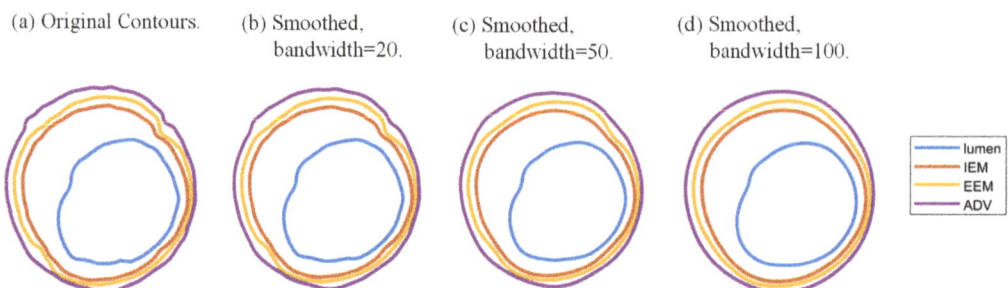

Figure 2. Contours of a sample slice using moving average smoothing with smoothing bandwidths of 20, 50 and 100 pixels.

Part C. Layer edge detection: The graphical basis of detecting different vessel layers is that intima, media and adventitia have different optical properties and pixel intensities in OCT images, and image intensities change most dramatically at layer boundaries. Figure 3a,b show the "light–dark–light" three-layer structure in a healthy vessel slice. Figure 3c,d demonstrate that intensity and intensity gradient had clear patterns in radial direction. From lumen to vessel out-boundary, intensity followed a trend of going up and down twice. Intensity gradient had a similar trend, forming two peaks and two valleys, each representing the boundary between lumen and intima, intima and media, media and adventitia, adventitia and other peripheral tissues. The boundary between intima and media is called internal elastic membrane (IEM), which is a thin membrane mainly composed of elastin. The boundary between media and adventitia is called external elastic membrane (EEM). The boundary between adventitia and other peripheral tissues is called adventitia–periadventitia interface (ADV).

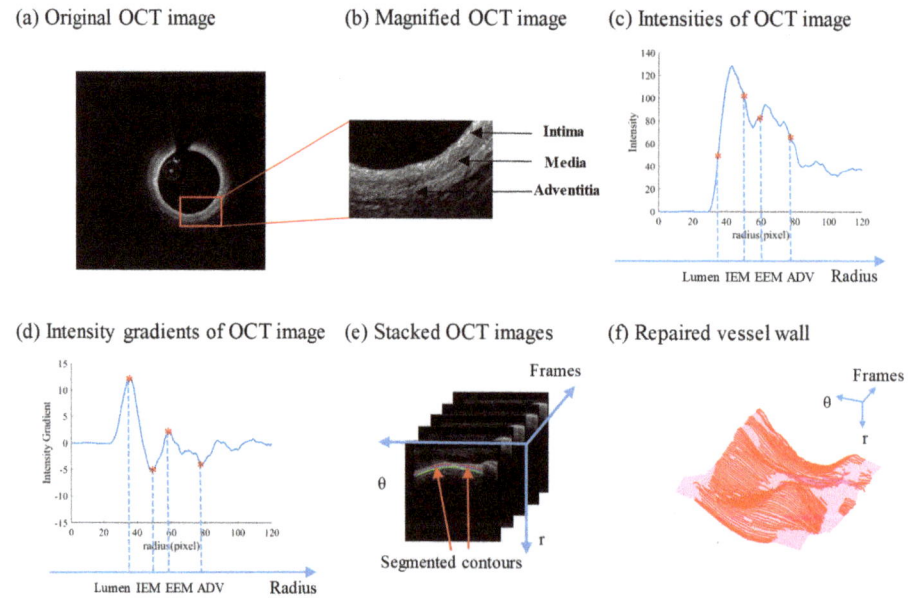

Figure 3. Schematic diagram of OCT multilayer segmentation and repairing. (**a**) Original OCT image;

(b) magnified OCT image. A "light–dark–light" three-layer structure can be clearly seen; (c) radial intensities of OCT image; (d) radial intensity gradient of OCT image. Gradient reaches its peak or valley at layer boundary; (e) OCT images and the segmented contours were stacked in polar coordinates; (f) the whole vessel wall was repaired using known contour segments. Red: contour segments. Magenta: repaired surface.

The four boundaries were identified by edge detection based on Canny method, which searches out true edges in large gradient positions amongst large noise [22]. All OCT images were firstly flattened relative to lumen to better align pixels in the same radial depth and increase the efficiency of edge detection [16]. Then, the double thresholds of Canny method and the radius of Gaussian smoothing were continually adjusted by a grid search program to determine the optimal parameters for edge detection. IEM and ADV were detected by the valleys of first derivative; EEM was detected by the peak of first derivative. The whole three-layer structure, as well as the middle of media, were detected by the peak of second derivative, implying a rough candidate region to detect edges. Of all the detected edges, manual selection was performed in the first slice of a pullback to choose the true (seed) edges, and the rest were considered noise edges. Under the assumption that edges between continuous slices do not change dramatically, edges close enough to the seed edges radially in the next slice were specified as the true edges. The remaining slices were performed in the same manner until layers of the whole pullback were segmented.

2.3. Surface Repairing

The segmented contours were reverse-flattened first relative to lumen [16]. In 3D polar coordinate space, contours of three layers were stacked and formed three surfaces with holes, as Figure 3e,f showed. Detected contours represent visible vessel inner wall, while the holes represent the invisible parts. The surfaces were then repaired by cubic spline-fitting method to obtain the parts of the vessel wall missing in OCT image [23].

After the repair, the complete surfaces were smoothed and transformed from polar coordinate system to Cartesian coordinate system. Contours of three layers are now complete in all the slices, including those obscured by lipid or other tissues.

2.4. Multilayer 3D Thin-Slice Models

Three-dimensional (3D) thin-slice models were constructed for 10 selected slices from one patient using automatically segmented slices obtained from our programs. Both multilayer and single-layer models were constructed to show plaque stress/strain results and the impact of three-layer segmentation on plaque stress/strain calculations. Since OCT data were acquired under in vivo conditions when the vessel was axially stretched and under in vivo pressure, a 5% axial shrink–stretch and a circumferential pre-shrink process were performed to obtain in vivo slice morphology [4]. Vessel tissues were assumed to be hyperelastic, anisotropic, nearly incompressible, and homogeneous. Lipid core was assumed to be hyperelastic, isotropic and nearly incompressible. Modified Mooney–Rivlin material models were used to describe the material properties of vessel tissues, including isotropic and anisotropic parts. The strain–energy density functions for tissue material properties are given below:

$$W_{iso} = c_1(I_1 - 3) + c_2(I_2 - 3) + D_1[exp(D_2(I_1 - 3)) - 1] \quad (1)$$

$$W_{aniso} = W_{iso} + \frac{K_1}{K_2}\left\{exp\left[K_2(I_4 - 1)^2\right] - 1\right\} \quad (2)$$

where $I_1 = \Sigma(C_{ii})$, $I_2 = \frac{1}{2}[I_1^2 - C_{ij}C_{ij}]$, I_1 and I_2 are the first and second invariants of right Cauchy–Green deformation tensor $C = [C_{ij}] = F^T F$, $F = [F_{ij}] = [\partial x_i / \partial a_j]$; (x_i) is current positionl (a_j) is original position; $I_4 = \lambda_\theta^2 \cos^2\varphi + \lambda_z^2 \sin^2\varphi$, where λ_θ, λ_z are the principal stretches associated with circumferential and axial direction and φ is the angle between the fiber reinforcement and the circumferential direction in individual layers; c_1, c_2, D_1, D_2, K_1 and K_2 are material parameters. Parameter values in the literature were used in this paper: Lipid: $c_1 = 0.5$ kPa, $c_2 = 0$ kPa, $D_1 = 0.5$ kPa, $D_2 = 1.5$; intima: $c_1 = -262.76$ kPa, $c_2 = 22.9$ kPa, $D_1 = 125.9$ kPa, $D_2 = 2.0$, $K_1 = 7.19$ kPa, $K_2 = 23.5$; media: $c_1 = -5$ kPa, $c_2 = -20$ kPa, $D_1 = 20$ kPa, $D_2 = 2.8$, $K_1 = 168$ kPa, $K_2 = 57$, $\varphi = 24.9°$; adventitia: $c_1 = 6.16$ kPa, $c_2 = 0$ kPa, $D_1 = 0.03$ kPa, $D_2 = 30$, $K_1 = 10$ kPa, $K_2 = 54$, $\varphi = 75.3°$ [11,13,24,25]. Uniaxial axial and circumferential stress–stretch plots for intima [24], media [13] and adventitia [11] are given by Figure 4.

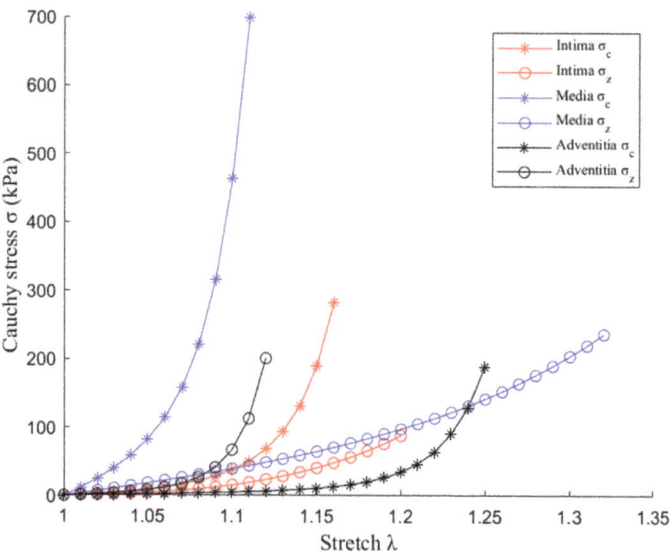

Figure 4. Stress–stretch curves of three layers obtained from uniaxial mechanical testing. σ_c: Circumferential stress; σ_z: axial stress. References: intima [24], media [13] and adventitia [11].

The thin-slice models were solved by a finite-element software ADINA 9.6 (Adina R & D, Watertown, MA, USA) following our established procedures [24]. Because stress/strain are tensors, maximum principal stress and maximum principal strain (called stress and strain from here on, respectively) were chosen as their scale representatives for stress/strain comparisons.

2.5. Data Extraction and Analysis

Since plaque slices may have irregular and nonuniform geometries, each slice was divided into 4 quarters, with each quarter containing 25 evenly spaced nodal points on the lumen. Each lumen nodal point was connected to a corresponding point on vessel wall using a piecewise equal-step method to deal with irregular nonuniform plaque morphologies [26]. Figure 5 gives an illustration for the definition of layer thickness of the three layers. Specifically, intima thickness was defined as the length of the line segment connecting lumen and IEM. Media and adventitia thicknesses were defined similarly. Layer thickness data were extracted from the 100 nodal points for each slice (total 436 slices from 6 patients) to compare their differences between automatic and manual segmentations. Plaque stress and strain data were also extracted from those nodal points of their corresponding thin-slice models for analysis.

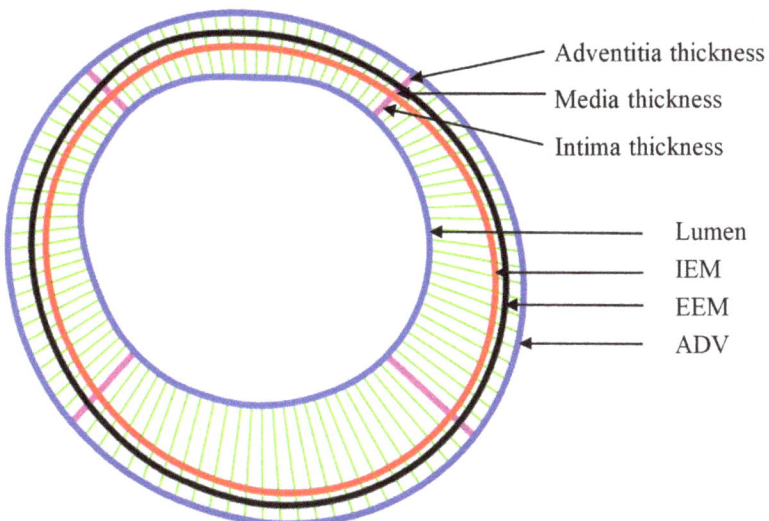

Figure 5. Schematic plot demonstrating the piecewise equal-step method for three-layer thickness and the quarter-dividing method. IEM: intima elastic membrane; EEM: external elastic membrane; ADV: adventitia–periadventitia interface.

Layer thickness data of the three layers of all slices were stored in matrix $T_q(i,j,k)$, where $q = 1, 2, 3$ represents intima, media and adventitia, respectively; i is the point-index for the 100 points in a given slice; j is the slice-index for the slices for a given patient; and k is the patient index for a given patient ($k = 1, \ldots, 6$). Equation (3) calculates the slice mean thickness of q-layer ($q = 1, 2, 3$) of j-th slice from Patient k. Equation (4) calculates the patient mean thickness of q-layer ($q = 1, 2, 3$) of all slices from Patient k. Equation (5) calculates the mean thickness of q-layer ($q = 1, 2, 3$) for all slices from all patients.

$$\text{Slice mean thickness for } q-\text{layer} = \frac{1}{100} \sum_{i=1}^{100} T_q(i,j,k), \ j \text{ and } k \text{ fixed}; \qquad (3)$$

$$\text{Patient mean thickness for } q-\text{layer} = \frac{1}{100} \times \frac{1}{m} \times \sum_{j=1}^{m} \sum_{i=1}^{100} T_q(i,j,k), \ k \text{ fixed}; \qquad (4)$$

$$\text{Mean thickness of } q-\text{layer for all patients} = \frac{1}{100} \times \frac{1}{m} \times \frac{1}{n} \times \sum_{k=1}^{n} \sum_{j=1}^{m} \sum_{i=1}^{100} T_q(i,j,k) \qquad (5)$$

Thickness error was defined as the relative error between automatic contour thickness, represented by the matrix $T_q^a(i,j,k)$, and manual contour thickness, represented by matrix $T_q^m(i,j,k)$. Equation (6) calculates the thickness error of q-layer $Error_q(j,k)$ ($q = 1, 2, 3$) of j-th slice of Patient k. Equation (7) calculates the thickness error of q-layer $Error_q(k)$ ($q = 1, 2, 3$) of Patient k. Equation (8) calculates the thickness error of q-layer $Error_q$ ($q = 1, 2, 3$) for all patients.

$$Error_q(j,k) = \frac{\frac{1}{100} \times \sum_{i=1}^{100} \left(T_q^a(i,j,k) - T_q^m(i,j,k) \right)}{\frac{1}{100} \times \sum_{i=1}^{100} T_q^m(i,j,k)} \times 100\%, \ j \text{ and } k \text{ fixed}; \qquad (6)$$

$$Error_q(k) = \frac{1}{m} \sum_{j=1}^{m} Error_q(j,k), \ k \text{ fixed}; \qquad (7)$$

$$Error_q = \frac{1}{n} Error_q(k) \tag{8}$$

3. Results

3.1. Comparison of Layer Thickness between Automatic and Manual Segmentations

Table 2 gives patient thickness values for automatic and manual segmentations from the six patients and errors of automatic segmentations compared to the gold standard (manual segmentation). Errors were calculated using Formulas (6)–(8) with point-to-point differences. Slice-averaged errors with slice standard deviations for each patient are demonstrated in Table 2. Patient-averaged errors with standard deviations are given after P6. Figure 6 gives 10 sample OCT slices selected from one patient, showing their manual and automatic contour differences. It shows that manual and automatic contours were very close. Layer thickness varied greatly from layer to layer and from patient to patient. Intima (which is really the thickened intima, or plaque) had the largest thickness and variance, with an average thickness of 0.6464 ± 0.2222 mm. Intima thickness of P3 (0.8183 mm) was more than twice of that of P4 (0.3963 mm), showing large patient variation. Mean layer thicknesses of media and adventitia were 0.2426 ± 0.0596 mm and 0.2234 ± 0.0587 mm, respectively. Automatic contours derived from our program showed a similar layer thickness compared to manual contours, with 0.6240 ± 0.2174 mm, 0.2290 ± 0.0519 mm and 0.2324 ± 0.0477 mm each for intima, media and adventitia. The relative errors of three layers were mostly less than 10%, with a mean error of −1.40 ± 8.13%, −4.34 ± 11.17% and 6.97 ± 12.00% for intima, media and adventitia, respectively. Negative errors indicated that thicknesses of automatic contours were smaller than those of manual contours. Intima had the smallest error, while media and adventitia had a slightly larger error and variance. This is expected, since accuracy for intima should be better than that for media and adventitia due to OCT penetration limitation. An alternative explanation could be that the absolute placement of the relevant contours is equally accurate/inaccurate for all layers, but because the intima is the thickest layer in these patients, the relative errors (computed by Equation (6)) ended up being smallest. Media thickness tended to be underestimated, while adventitia thickness tended to be overestimated.

Table 2. Summary of layer thickness values and slice-averaged errors of 6 patients with slice standard deviations.

Patient	Intima (mm)			Media (mm)		
	Auto	Manual	Error	Auto	Manual	Error
P1	0.6298 ± 0.0948	0.6661 ± 0.1009	−4.82 ± 4.30%	0.2585 ± 0.0455	0.2655 ± 0.0335	−5.27 ± 5.20%
P2	0.7262 ± 0.2575	0.7794 ± 0.2346	−4.09 ± 5.71%	0.2662 ± 0.0270	0.2880 ± 0.0203	−7.38 ± 7.58%
P3	0.7763 ± 0.2151	0.8183 ± 0.1945	−5.04 ± 7.21%	0.2613 ± 0.0215	0.2871 ± 0.0241	−8.76 ± 5.24%
P4	0.4268 ± 0.1478	0.3963 ± 0.1416	9.00 ± 5.11%	0.2146 ± 0.0472	0.2386 ± 0.0565	−9.30 ± 4.11%
P5	0.6439 ± 0.0935	0.6246 ± 0.0989	4.37 ± 5.13%	0.1934 ± 0.0102	0.1845 ± 0.0237	7.03 ± 14.65%
P6	0.4973 ± 0.1740	0.5390 ± 0.1674	−7.25 ± 6.40%	0.1493 ± 0.0082	0.1526 ± 0.0279	1.80 ± 16.93%
Patient-Averaged Mean ± SD	0.6240 ± 0.2174	0.6464 ± 0.2222	−1.40 ± 8.13%	0.2290 ± 0.0519	0.2426 ± 0.0596	−4.34 ± 11.17%

Patient	Adventitia (mm)			Total Vessel (mm)		
	auto	manual	error	auto	manual	error
P1	0.2429 ± 0.0325	0.2151 ± 0.0319	13.49 ± 5.55%	1.1312 ± 0.1227	1.1467 ± 0.1225	−1.32 ± 3.24%
P2	0.2377 ± 0.0451	0.2231 ± 0.0563	8.96 ± 10.87%	1.2301 ± 0.2968	1.2904 ± 0.2713	−5.29 ± 4.55%
P3	0.2217 ± 0.0441	0.2073 ± 0.0429	8.63 ± 11.00%	1.2593 ± 0.2115	1.3127 ± 0.2028	−4.16 ± 3.94%
P4	0.2097 ± 0.0465	0.2037 ± 0.0641	7.76 ± 9.87%	0.8510 ± 0.2310	0.8386 ± 0.2409	−1.91 ± 3.21%
P5	0.2745 ± 0.0402	0.2994 ± 0.0428	−5.64 ± 16.01%	1.1119 ± 0.0857	1.1085 ± 0.1003	0.53 ± 4.70%
P6	0.2112 ± 0.0531	0.2036 ± 0.0531	5.37 ± 9.19%	0.8578 ± 0.1952	0.8952 ± 0.1794	−4.49 ± 5.50%
Patient-Averaged Mean ± SD	0.2324 ± 0.0477	0.2234 ± 0.0587	6.97 ± 12.00%	1.0855 ± 0.2650	1.1124 ± 0.2711	−2.26 ± 5.00%

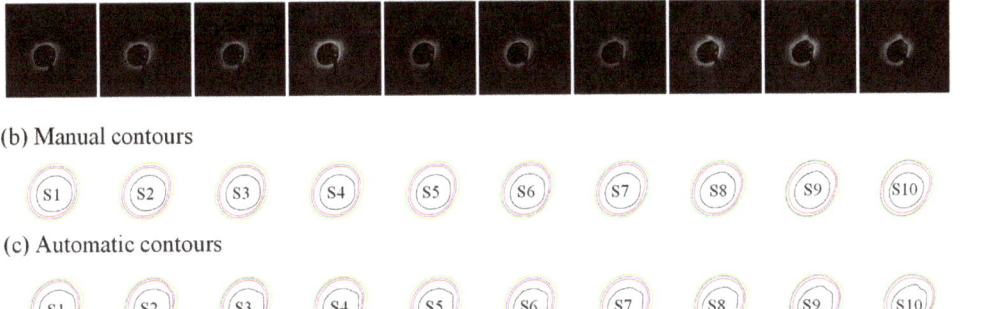

Figure 6. Sample manual and automatic contours derived from OCT images. (**a**) Original OCT images; (**b**) manual contours; (**c**) automatic contours.

Although the mean errors are somewhat informative, they likely obscure local/regional errors that may have been substantially higher (as evidenced, in part, by the standard deviation values). Given that local vessel wall and plaque characteristics are so important for clinical assessment and prognosis, the 90th percentile of the absolute value of the errors for each patient/layer are given by Table 3. The errors were based on slice-averaged results. Pointwise errors could be a little larger.

Table 3. The 90th percentile of the absolute thickness errors for 6 patients.

Patients	IEM (mm)	EEM (mm)	ADV (mm)
P1	9.78%	12.84%	18.08%
P2	11.65%	13.16%	20.34%
P3	17.68%	13.19%	20.00%
P4	14.90%	13.06%	19.33%
P5	11.47%	32.58%	29.12%
P6	14.57%	28.42%	16.54%

3.2. Point-to-Point Manual and Automatic Contour Distances of Lumen, IEM, EEM and ADV

Table 4 gives distances between corresponding manual and automatic contours (using point-to-point calculation) to show automatic contour locations relative to their corresponding manual contours. Negative values mean that automatic contours had smaller radii than manual contours in polar coordinates. Distance and standard deviation of lumen contours were the smallest, with an average distance of -0.0081 ± 0.0310 mm. The distances of IEM, EEM and ADV contours were -0.0279 ± 0.0539, -0.0689 ± 0.0563 and -0.0153 ± 0.0356 mm, respectively. Table 4 could also lead to a possible explanation for why media and adventitia thickness errors tended to have opposite signs in Table 2. For P1–P3, EEM contours had large negative values, which meant that the automatic method may be consistently placing the EEM contour slightly closer to IEM. For P4, the IEM contour errors were positive and led to a negative media thickness error.

Table 4. Point-to-point manual and automatic contour distances of lumen, IEM, EEM and ADV.

Patient	Lumen (mm)	IEM (mm)	EEM (mm)	ADV (mm)
P1	-0.0037 ± 0.0421	-0.1075 ± 0.0660	-0.1423 ± 0.0790	0.0147 ± 0.0626
P2	0.0140 ± 0.0279	-0.0392 ± 0.0453	-0.1164 ± 0.1021	-0.0690 ± 0.0668
P3	0.0160 ± 0.0312	-0.0259 ± 0.0479	-0.0942 ± 0.0777	-0.0302 ± 0.0449

Table 4. Cont.

Patient	Lumen (mm)	IEM (mm)	EEM (mm)	ADV (mm)
P4	0.0377 ± 0.0412	0.0613 ± 0.1438	−0.0104 ± 0.1346	0.0117 ± 0.1147
P5	−0.0547 ± 0.0484	−0.0213 ± 0.0360	−0.0123 ± 0.0304	−0.0372 ± 0.0571
P6	0.0177 ± 0.0547	−0.0348 ± 0.0542	−0.0381 ± 0.0655	0.0184 ± 0.0859
Mean ± SD	−0.0081 ± 0.0310	−0.0279 ± 0.0539	−0.0689 ± 0.0563	−0.0153 ± 0.0356

3.3. Impact of Multilayer Segmentation on Plaque Stress/Strain Calculations

Table 5 provided mean stress values of the 10 slices from three-layer and single-layer models and their differences using three-layer values as baseline values. Negative percentage means that the single-layer model provided an underestimate for the stress/strain value(s) calculated. The single-layer mean plaque stress value from lumen was 117.91 ± 5.55 kPa, 10.79% lower than that from the three-layer models (132.33 ± 8.40 kPa). However, on the out-boundary (adventitia), the single-layer mean plaque stress value was 50.46 ± 37.20 kPa, 156.28% higher than that from the three-layer models (19.74 ± 1.78 kPa).

Table 5. Summary of stress difference between multilayer and single-layer models.

Slice	Lumen			Out Boundary		
	Multilayer (kPa)	Single-Layer (kPa)	Error	Multilayer (kPa)	Single-Layer (kPa)	Error
1	134.07	115.97	−13.50%	20.49	48.09	134.69%
2	138.01	120.51	−12.68%	21.97	51.38	133.87%
3	114.53	105.99	−7.46%	16.57	42.31	155.33%
4	140.41	121.70	−13.33%	21.91	53.86	145.78%
5	135.68	119.49	−11.93%	20.05	51.69	157.84%
6	134.47	119.16	−11.39%	18.91	50.80	168.59%
7	138.08	122.93	−10.97%	20.17	53.35	164.51%
8	136.50	122.96	−9.92%	20.39	54.05	165.14%
9	131.47	119.76	−8.91%	19.80	52.31	164.16%
10	120.03	110.68	−7.79%	17.13	46.74	172.94%
Mean ± SD	132.33 ± 8.40	117.91 ± 5.55	−10.79 ± 2.20%	19.74 ± 1.78	50.46 ± 37.20	156.28 ± 13.82%

Table 6 gives mean strain values of the 10 slices from three-layer and single-layer models and their differences. The single-layer mean plaque strain value from lumen was 0.1916 ± 0.0034, 4.88% lower than that from the three-layer models (0.2015 ± 0.0050). On the out-boundary (adventitia), the single-layer mean plaque strain value was 0.1064 ± 0.0058, 13.40% lower than that from three-layer models (0.1228 ± 0.0066).

Table 6. Summary of strain difference between multilayer and single-layer models.

Slice	Lumen			Out-Boundary		
	Multilayer	Single-Layer	Error	Multilayer	Single-Layer	Error
1	0.2022	0.1895	−6.26%	0.1202	0.1027	−14.56%
2	0.2039	0.1922	−5.76%	0.1248	0.1075	−13.88%
3	0.1903	0.1843	−3.14%	0.1087	0.0941	−13.41%
4	0.2046	0.1925	−5.91%	0.1296	0.1116	−13.86%
5	0.2032	0.1921	−5.46%	0.1267	0.1095	−13.60%
6	0.2036	0.1929	−5.25%	0.1254	0.1087	−13.36%
7	0.2058	0.1954	−5.08%	0.1277	0.1113	−12.83%

Table 6. Cont.

Slice	Lumen			Out-Boundary		
	Multilayer	Single-Layer	Error	Multilayer	Single-Layer	Error
8	0.2047	0.1953	−4.58%	0.1272	0.1113	−12.45%
9	0.2019	0.1938	−4.02%	0.1236	0.1078	−12.79%
10	0.1949	0.1883	−3.37%	0.1146	0.0994	−13.24%
Mean ± SD	0.2015 ± 0.0050	0.1916 ± 0.0034	−4.88 ± 1.07%	0.1228 ± 0.0066	0.1064 ± 0.0058	−13.40 ± 0.62%

Figure 7 demonstrates the stress/strain distributions from three-layer and single-layer models using a sample slice. It shows that the maximum plaque stress from the three-layer model was 26% higher than that from the single-layer model (226.05 kPa vs. 178.91 kPa), while the maximum strain values from both models were almost identical (0.199 vs. 0.198). The cap stress/strain values did not show much difference.

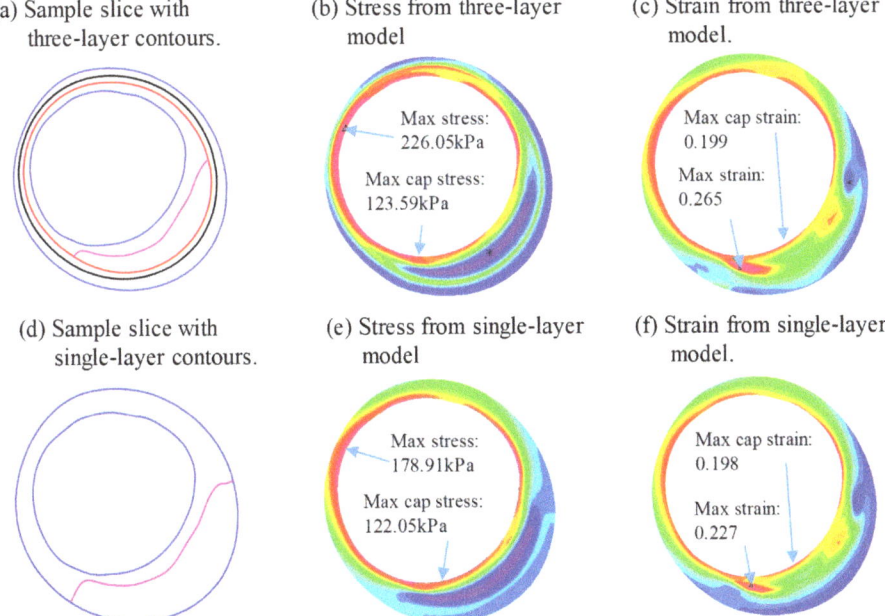

Figure 7. Comparison of plaque stress/strain distributions from three-layer and single-layer models using a sample slice. (**a**) Sample slice with three-layer contours; (**b**) stress from three-layer model; (**c**) strain from three-layer model; (**d**) sample slice with single-layer contours; (**e**) stress from single-layer model; (**f**) strain from single-layer model.

4. Discussion

4.1. Multilayer Automatic Coronary Plaque OCT Segmentation, Repairing and Its Significance to Vulnerable Plaque Research

Mechanical forces play an important role in coronary plaque initiation, progression and its eventual rupture, which often leads to critical clinical events such as heart attack and acute coronary syndrome (ACS). Accurate calculation of plaque stress/strain is of vital importance to vulnerable plaque research, including plaque progression prediction, vulnerability assessment, plaque rupture prediction and patient diagnosis, management and treatment plan optimization. Plaque stress/strain calculations depend on plaque morphology, components, pressure conditions and plaque tissue material properties. Most current plaque models use single-layer vessel structures, primarily due to image modality resolution limitations. The development and acceptance of OCT in clinical practice provide the possibility of obtaining multilayer vessel geometries. The multilayer automatic coronary plaque OCT segmentation and repairing technique proposed in this paper make it possible to use three-layer models to improve the accuracy for plaque stress/strain calculations. It will have a considerable impact on plaque progression and vulnerability predictions.

Automation is also one of the contributions of this work. The manual annotation of OCT vessel wall is extremely laborious, especially multilayer annotation. As different layer boundaries are close to each other and are hard to discriminate by naked eyes, it takes huge time and labor cost to annotate multilayers, and also brings in accidental errors. We demonstrated that multilayer thicknesses were precisely quantified using our automated process around the whole vessel wall with relative errors of 1.40%, 4.34% and 6.97% (for intima, media and adventitia, respectively) compared to manual segmented contours (gold standard), respectively.

4.2. Limitations

Some limitations of this study include: (a) Small patient size—this is a pilot study, and a larger-scale study is needed to further validate the feasibility of this method and bring it into clinical practice. (b) The method is not able to automatically characterize other plaque components such as lipid and calcification. Plaque components were manually annotated in this study. More advanced technologies such as artificial intelligence and neural network should be integrated to achieve entire automatic OCT segmentation. (c) Three-dimensional thin-slice models were used for model construction efficiency. Full 3D models should be used for better accuracy. (d) Tissue material properties used parameter values from the literature since patient-specific material properties were not available. (e) Layered plaques have different layer patterns, which is confusing for edge detection. This is a major error source [27–29]. (f) Some artifacts (such as sudden axial removal of OCT guidewire because of heart motion during acquisition) can also lead to spatial discontinuity in sequential slices, resulting in the wrong region to select true edges. (g) The whole analysis process for a patient, including automatic segmentation, repairing and thin-slice modeling, needs about 3–4 h. We will continue to try to shorten the analysis time and get closer to clinical implementations.

4.3. Future Challenges and Directions

More precise and higher-quality coronary plaque images are essential to vessel reconstruction and modeling. With the development of biomedical technologies, more advanced imaging modalities, such as a dual catheter combining both IVUS and OCT, are becoming available, providing better vessel images [30]. In terms of modeling, quantification of vessel material property has always been a pain point, and more accurate and patient-specific material parameters are needed. There is also a balance between model accuracy and model construction cost that we need to keep. Automation of modeling and data processing are needed for clinical implementation. A multilayer fluid–structure interaction (FSI) model is desirable to have and has great potential for studying the biomechanical behavior of different layers considering both solids and fluids. However, labor cost is a big concern. The automation of the 3D thin-slice model can greatly shorten the distance between laboratory experiments and clinical practice.

Author Contributions: Conceptualization, D.T., L.W. and M.H.; methodology, M.H., D.T. and L.W.; formal analysis, M.H., L.W., R.L. and X.Z.; writing—original draft preparation, M.H. and D.T.; writing—review and editing, D.T., L.W. and M.H.; data acquisition and preparation, A.M., M.M., G.S.M., G.M., J.Z., R.L., Y.Z., L.C. and X.Z.; project administration, D.T. and J.Z.; funding acquisition, D.T. and L.W. All authors have read and agreed to the published version of the manuscript.

Funding: This research was supported in part by National Sciences Foundation of China grants 11972117, 11802060; Natural Science Foundation of Jiangsu Province under grant number BK20180352; a Jiangsu Province Science and Technology Agency under grant number BE2016785; Fundamental Research Funds for the Central Universities and Zhishan Young Scholars Fund administrated by Southeast University (grant number 2242021R41123).

Institutional Review Board Statement: This study used 6 patient data. Five existing de-identified intravascular optical coherence tomography (OCT) data sets for patients ($n = 5$) with coronary heart diseases were obtained from Cardiovascular Research Foundation (CRF). IRB approval for research involving existing de-identified data is NOT required. One additional patient OCT data set was acquired from Southeast University Affiliated Zhongda Hospital using protocol approved by Southeast University Zhongda Hospital Institutional Review Board (approval code 2019ZDKYSB046) with informed consent obtained. The study was conducted according to the guidelines of the Declaration of Helsinki, and approved by the Institutional Review Board at Southeast University Zhongda Hospital.

Informed Consent Statement: Five existing de-identified intravascular optical coherence tomography (OCT) data sets do not require ethical review. One patient OCT data set was acquired from Southeast University Affiliated Zhongda Hospital using protocol approved by Southeast University Zhongda Hospital Institutional Review Board (approval code 2019ZDKYSB046) with informed consent obtained.

Data Availability Statement: Data are available on request. Data cannot be made publicly available for ethical or legal reasons (public availability would compromise patient privacy).

Conflicts of Interest: The authors declare no conflict of interest.

References

1. Gutiérrez-Chico, J.L.; Alegría-Barrero, E.; Teijeiro-Mestre, R.; Chan, P.H.; Tsujioka, H.; De Silva, R.; Viceconte, N.; Lindsay, A.; Patterson, T.; Foin, N.; et al. Optical coherence tomography: From research to practice. *Eur. Hear. J. Cardiovasc. Imaging* **2012**, *13*, 370–384. [CrossRef] [PubMed]
2. Kubo, T.; Akasaka, T.; Shite, J.; Suzuki, T.; Uemura, S.; Yu, B.; Kozuma, K.; Kitabata, H.; Shinke, T.; Habara, M.; et al. OCT Compared With IVUS in a Coronary Lesion Assessment. *JACC Cardiovasc. Imaging* **2013**, *6*, 1095–1104. [CrossRef] [PubMed]
3. Guo, X.; Giddens, D.P.; Molony, D.; Yang, C.; Samady, H.; Zheng, J.; Matsumura, M.; Mintz, G.S.; Maehara, A.; Wang, L.; et al. A Multimodality Image-Based Fluid–Structure Interaction Modeling Approach for Prediction of Coronary Plaque Progression Using IVUS and Optical Coherence Tomography Data With Follow-Up. *J. Biomech. Eng.* **2019**, *141*, 091003. [CrossRef] [PubMed]
4. Lv, R.; Maehara, A.; Matsumura, M.; Wang, L.; Zhang, C.; Huang, M.; Guo, X.; Samady, H.; Giddens, D.P.; Zheng, J.; et al. Using Optical Coherence Tomography and Intravascular Ultrasound Imaging to Quantify Coronary Plaque Cap Stress/Strain and Progression: A Follow-Up Study Using 3D Thin-Layer Models. *Front. Bioeng. Biotechnol.* **2021**, *9*, 713525. [CrossRef] [PubMed]
5. Tearney, G.J.; Regar, E.; Akasaka, T.; Adriaenssens, T.; Barlis, P.; Bezerra, H.G.; Bouma, B.; Bruining, N.; Cho, J.-M.; Chowdhary, S.; et al. Consensus Standards for Acquisition, Measurement, and Reporting of Intravascular Optical Coherence Tomography Studies: A Report From the International Working Group for Intravascular Optical Coherence Tomography Standardization and Validation. *J. Am. Coll. Cardiol.* **2012**, *59*, 1058–1072. [CrossRef] [PubMed]
6. Ali, Z.A.; Maehara, A.; Généreux, P.; Shlofmitz, R.A.; Fabbiocchi, F.; Nazif, T.M.; Guagliumi, G.; Meraj, P.M.; Alfonso, F.; Samady, H.; et al. Optical coherence tomography compared with intravascular ultrasound and with angiography to guide coronary stent implantation (ILUMIEN III: OPTIMIZE PCI): A randomised controlled trial. *Lancet* **2016**, *388*, 2618–2628. [CrossRef]
7. Eikendal, A.L.; Groenewegen, K.A.; Anderson, T.J.; Britton, A.R.; Engström, G.; Evans, G.W.; de Graaf, J.; Grobbee, D.E.; Hedblad, B.; Holewijn, S.; et al. Common Carotid Intima-Media Thickness Relates to Cardiovascular Events in Adults Aged <45 Years. *Hypertension* **2015**, *65*, 707–713. [CrossRef]
8. Fernández-Alvarez, V.; Sánchez, M.L.; Alvarez, F.L.; Nieto, C.S.; Mäkitie, A.A.; Olsen, K.D.; Ferlito, A. Evaluation of Intima-Media Thickness and Arterial Stiffness as Early Ultrasound Biomarkers of Carotid Artery Atherosclerosis. *Cardiol. Ther.* **2022**, *11*, 231–247. [CrossRef]
9. Pahkala, K.; Heinonen, O.J.; Simell, O.; Viikari, J.S.; Rönnemaa, T.; Niinikoski, H.; Raitakari, O.T. Association of Physical Activity With Vascular Endothelial Function and Intima-Media Thickness. *Circulation* **2011**, *124*, 1956–1963. [CrossRef]
10. Holzapfel, G.A.; Sommer, G.; Regitnig, P. Anisotropic Mechanical Properties of Tissue Components in Human Atherosclerotic Plaques. *J. Biomech. Eng.* **2004**, *126*, 657–665. [CrossRef]
11. Holzapfel, G.A.; Sommer, G.; Gasser, C.T.; Regitnig, P. Determination of layer-specific mechanical properties of human coronary arteries with nonatherosclerotic intimal thickening and related constitutive modeling. *Am. J. Physiol. Circ. Physiol.* **2005**, *289*, H2048–H2058. [CrossRef] [PubMed]

12. Teng, Z.; Zhang, Y.; Huang, Y.; Feng, J.; Yuan, J.; Lu, Q.; Sutcliffe, M.P.; Brown, A.J.; Jing, Z.; Gillard, J.H. Material properties of components in human carotid atherosclerotic plaques: A uniaxial extension study. *Acta Biomater.* **2014**, *10*, 5055–5063. [CrossRef] [PubMed]
13. Holzapfel, G.A.; Mulvihill, J.J.; Cunnane, E.M.; Walsh, M.T. Computational approaches for analyzing the mechanics of atherosclerotic plaques: A review. *J. Biomech.* **2014**, *47*, 859–869. [CrossRef] [PubMed]
14. Huang, J.; Yang, F.; Gutiérrez-Chico, J.L.; Xu, T.; Wu, J.; Wang, L.; Lv, R.; Lai, Y.; Liu, X.; Onuma, Y.; et al. Optical Coherence Tomography-Derived Changes in Plaque Structural Stress Over the Cardiac Cycle: A New Method for Plaque Biomechanical Assessment. *Front. Cardiovasc. Med.* **2021**, *8*, 715995. [CrossRef] [PubMed]
15. Athanasiou, L.S.; Bourantas, C.V.; Rigas, G.; Sakellarios, A.; Exarchos, T.P.; Siogkas, P.K.; Ricciardi, A.; Naka, K.; Papafaklis, M.; Michalis, L.K.; et al. Methodology for fully automated segmentation and plaque characterization in intracoronary optical coherence tomography images. *J. Biomed. Opt.* **2014**, *19*, 026009. [CrossRef]
16. Olender, M.L.; Athanasiou, L.S.; Hernandez, J.M.D.L.T.; Ben-Assa, E.; Nezami, F.R.; Edelman, E.R. A Mechanical Approach for Smooth Surface Fitting to Delineate Vessel Walls in Optical Coherence Tomography Images. *IEEE Trans. Med. Imaging* **2018**, *38*, 1384–1397. [CrossRef] [PubMed]
17. Zahnd, G.; Hoogendoorn, A.; Combaret, N.; Karanasos, A.; Péry, E.; Sarry, L.; Motreff, P.; Niessen, W.; Regar, E.; van Soest, G.; et al. Contour segmentation of the intima, media, and adventitia layers in intracoronary OCT images: Application to fully automatic detection of healthy wall regions. *Int. J. Comput. Assist. Radiol. Surg.* **2017**, *12*, 1923–1936. [CrossRef]
18. Kafieh, R.; Rabbani, H.; Abramoff, M.D.; Sonka, M. Intra-retinal layer segmentation of 3D optical coherence tomography using coarse grained diffusion map. *Med. Image Anal.* **2013**, *17*, 907–928. [CrossRef]
19. Chu, M.; Jia, H.; Gutiérrez-Chico, J.L.; Maehara, A.; Ali, Z.A.; Zeng, X.; He, L.; Zhao, C.; Matsumura, M.; Wu, P.; et al. Artificial intelligence and optical coherence tomography for the automatic characterisation of human atherosclerotic plaques. *EuroIntervention* **2021**, *17*, 41–50. [CrossRef]
20. Zhang, C.; Li, H.; Guo, X.; Molony, D.; Guo, X.; Samady, H.; Giddens, D.P.; Athanasiou, L.; Nie, R.; Cao, J.; et al. Convolution Neural Networks and Support Vector Machines for Automatic Segmentation of Intracoronary Optical Coherence Tomography. *Mol. Cell. Biomech.* **2019**, *16*, 153–161. [CrossRef]
21. Otsu, N. A Threshold Selection Method from Gray-Scale Histograms. *IEEE Trans. Syst. Man Cybern.* **1979**, *9*, 62–66. [CrossRef]
22. Canny, J. A Computational Approach to Edge Detection. *IEEE Trans. Pattern Anal. Mach. Intell.* **1986**, *8*, 679–698. [CrossRef] [PubMed]
23. Schurer, F.; Cheney, E. On Interpolating Cubic Splines with Equally-Spaced Nodes 1. *Indag. Math. Proc.* **1968**, *71*, 517–524. [CrossRef]
24. Yang, C.; Bach, R.G.; Zheng, J.; Naqa, I.E.; Woodard, P.K.; Teng, Z.; Billiar, K.; Tang, D. In Vivo IVUS-Based 3-D Fluid–Structure Interaction Models With Cyclic Bending and Anisotropic Vessel Properties for Human Atherosclerotic Coronary Plaque Mechanical Analysis. *IEEE Trans. Biomed. Eng.* **2009**, *56*, 2420–2428. [CrossRef] [PubMed]
25. Kural, M.H.; Cai, M.; Tang, D.; Gwyther, T.; Zheng, J.; Billiar, K.L. Planar biaxial characterization of diseased human coronary and carotid arteries for computational modeling. *J. Biomech.* **2012**, *45*, 790–798. [CrossRef]
26. Wang, Q.; Tang, D.; Wang, L.; Canton, G.; Wu, Z.; Hatsukami, T.S.; Billiar, K.L.; Yuan, C. Combining morphological and biomechanical factors for optimal carotid plaque progression prediction: An MRI-based follow-up study using 3D thin-layer models. *Int. J. Cardiol.* **2019**, *293*, 266–271. [CrossRef]
27. Fracassi, F.; Crea, F.; Sugiyama, T.; Yamamoto, E.; Uemura, S.; Vergallo, R.; Porto, I.; Lee, H.; Fujimoto, J.; Fuster, V.; et al. Healed Culprit Plaques in Patients With Acute Coronary Syndromes. *J. Am. Coll. Cardiol.* **2019**, *73*, 2253–2263. [CrossRef]
28. Shimokado, A.; Matsuo, Y.; Kubo, T.; Nishiguchi, T.; Taruya, A.; Teraguchi, I.; Shiono, Y.; Orii, M.; Tanimoto, T.; Yamano, T.; et al. In vivo optical coherence tomography imaging and histopathology of healed coronary plaques. *Atherosclerosis* **2018**, *275*, 35–42. [CrossRef]
29. Thondapu, V.; Mamon, C.; Poon, E.K.W.; Kurihara, O.; Kim, H.O.; Russo, M.; Araki, M.; Shinohara, H.; Yamamoto, E.; Dijkstra, J.; et al. High spatial endothelial shear stress gradient independently predicts site of acute coronary plaque rupture and erosion. *Cardiovasc. Res.* **2020**, *117*, 1974–1985. [CrossRef]
30. Terashima, M.; Kaneda, H.; Honda, Y.; Shimura, T.; Kodama, A.; Habara, M.; Suzuki, T. Current status of hybrid intravascular ultrasound and optical coherence tomography catheter for coronary imaging and percutaneous coronary intervention. *J. Cardiol.* **2021**, *77*, 435–443. [CrossRef]

Article

Experimental Study of the Propagation Process of Dissection Using an Aortic Silicone Phantom

Qing-Zhuo Chi [1,†], Yang-Yang Ge [2,†], Zhen Cao [1], Li-Li Long [1], Li-Zhong Mu [1], Ying He [1,*] and Yong Luan [2]

1. Key Laboratory of Ocean Energy Utilization and Energy Conservation of Ministry of Education, Dalian University of Technology, Dalian 116024, China
2. Department of Anesthesiology, The First Affiliated Hospital of Dalian Medical University, Dalian 116011, China
* Correspondence: heying@dlut.edu.cn; Tel.: +86-150-0451-2987
† These authors contributed equally to this work.

Abstract: Background: The mortality of acute aortic dissection (AD) can reach 65~70%. However, it is challenging to follow the progress of AD formation. The purpose of this work was to observe the process of dissection development using a novel tear-embedded silicone phantom. Methods: Silicone phantoms were fabricated by embedding a torn area and primary tear feature on the inner layer. CT scanning and laser lightening were conducted to observe the variations in thickness and volume of the true lumen (TL) and false lumen (FL) during development. Results: The model with a larger interlayer adhesion damage required a lower pressure to trigger the development of dissection. At the initiation stage of dissection, the volume of TL increased by 25.5%, accompanied by a 19.5% enlargement of tear size. The force analysis based on the change of tear size verified the deduction of the process of interlaminar separation from the earlier studies. Conclusions: The primary tear and the weakening adhesion of the vessel layers are key factors in AD development, suggesting that some forms of primary damage to the arterial wall, in particular, the lumen morphology of vessels with straight inner lumen, should be considered as early risk predictors of AD.

Keywords: 3D printing; silicone phantom; CT scanning; aortic dissection; interlayer adhesion damage

1. Introduction

Acute aortic syndrome is a group of interrelated life-threatening diseases that includes aortic dissection (AD), intramural hematoma, and penetrating atherosclerotic ulcer. The disease of the aortic wall develops rapidly and can present as chest tightness and chest or back pain. The incidence of acute aortic syndrome ranges from approximately 3.5 to 6.0 per 100,000 people per year worldwide but more than 27 per 100,000 in older adults [1–3]. The most common acute aortic syndrome is acute AD, accounting for approximately 85–95% of all reported cases [4–6].

AD arises from a primary tear in the aortic intima, which can frequently be found in vascular segments that are exposed to abnormal stress or shear stress. Abnormal changes in the physiological pulsating flow environment are one of important factors that induce primary tearing. Elongation of the ascending aorta can give rise to a region that is affected by low and oscillating wall shear stress (WSS) [7,8]. Such abnormal changes in the flow environment can lead to a state of atherogenic phenotype for the endothelial layer [9–11]. Histological studies have also shown that the fibrin, laminin, and fibronectin in the arterial wall face dissolution under the long-term increase the WSS in the progress of adaptive dilation remodeling of the vascular diameter, which may cause the interlayer adhesion damage and elastic fiber degeneration [12–15]. At the same time, hypertension is associated with the hypertrophy of the arterial wall [16], leading to increased wall stiffness [17]. In addition, hypertension has a long-term effect on the Inner and one-third depth of the external media [18], where AD primarily occurs [19].

It is seen from the previous research that vascular wall damage can be characterized by penetrating wounds of the inner aortic wall and delamination between aortic layers. Several experimental studies have shown that much higher pressure is needed to induce dissection on healthy aortic tissue, of which the triggering pressure of dissection can reach as high as 600 mmHg. In contrast, even much lower blood pressure can cause the continuing development of tears if an inner aortic wound exists [20–22]. It can be seen that the trigger pressure of the dissection significantly depends on different types of vascular wall damage.

In vivo experiments with large animals have great research significance for investigating AD formation mechanisms. Tang et al. and Wang et al. achieved a 75% success rate for inducing type B dissection in canine samples by injecting adrenaline and making an initial tear on the arterial wall [23,24]. In the experiment of Guo et al., the initial tear feature was a transverse circumferential notch that covered one-third of the proximal descending aorta [25]. The delamination between aortic layers, termed the "entry pocket", was also incised on the aortic circumferential direction to induce dissection development. However, due to the complexity of animal experiments and the difficulty of control, it is challenging to observe the specific development process.

In vitro studies that use compliance-matched idealized aortic phantoms may be more controllable to implement and cause less animal suffering. Coupling the in vitro and zero-dimensional model, Rudenick et al. and Soudah et al. analyzed the influence of tear size and hemodynamic changes on the pressure and flow rate in the true lumen (TL) and false lumen (FL) [26,27]. Birjiniuk et al. characterized the relationship between pressure and flow rate in the false lumen and carried out suggestions for the stratification of deterioration risk in patients with type B AD [28,29]. Brunet et al. presented a novel experimental method to observe the propagation of the arterial dissection using a porcine carotid artery, where a longitudinal notch was created from the inside. It was concluded from their experiment that high shear stress in the crack tip is a possible trigger for the propagation of the dissection [30]. However, another opinion is that the alternating stress of stretching and contracting acting on the crack tip is far more crucial than the shear stress [31,32].

Despite many advances in the comprehension of the underlying mechanisms of aortic dissection in the last two decades, there is still a lack of work concerning the mechanism of tear propagation due to its complexity and nonlinearity. The advantage of the in vitro dissection model is that the deformation of the primary tear and the early stage of dissection propagation can be precisely investigated. Nevertheless, to date, fabrication of a developable dissection aortic phantom has not been realized because of the complexity of making primary tears in the inner layer [28,33,34]. Our previously proposed brush-spin-coating method can be further developed for making a vascular model with an initial inner layer tear since the coating method can mimic the laminated structure and, therefore, can be used for fabricating the aortic phantoms with dissection features [35].

This work aimed to develop a method for fabricating silicone models with dissection features inside. In vitro experimental research was also carried out to analyze the delamination behaviors. The overall structure of the paper is as follows. Section 2.1 discusses the characteristic of AD in human aortic tissue. Section 2.2 describes the processing method of the silicone phantom model with an initial tear in detail. In Sections 2.3 and 2.4, the experimental setup and the observing method are introduced step by step. The results of the tube pressure measurement computed tomographic (CT) scanning and laser imaging measurement throughout the development of AD are presented in Sections 3.1 and 3.2. In Section 4, the possible key factors inducing the dissection are analyzed, and conclusions are drawn in Section 5.

2. Materials and Methods

2.1. Key Features of the Tear of AD in Human Aortic Tissue

The structural features of the primary tear for the idealized phantom were modeled by observing a piece of human aortic tissue and the CT images of the same patient. The tissue was obtained from a 52-year-old man and soaked in formalin for 48 h. The patient gave written informed consent. Figure 1 shows the layout of aortic tissue from the intima side. The aortic tissue has been cut through the middle of the tear (section A) and 10 mm distal to the tear (section B) to observe the tissue morphology.

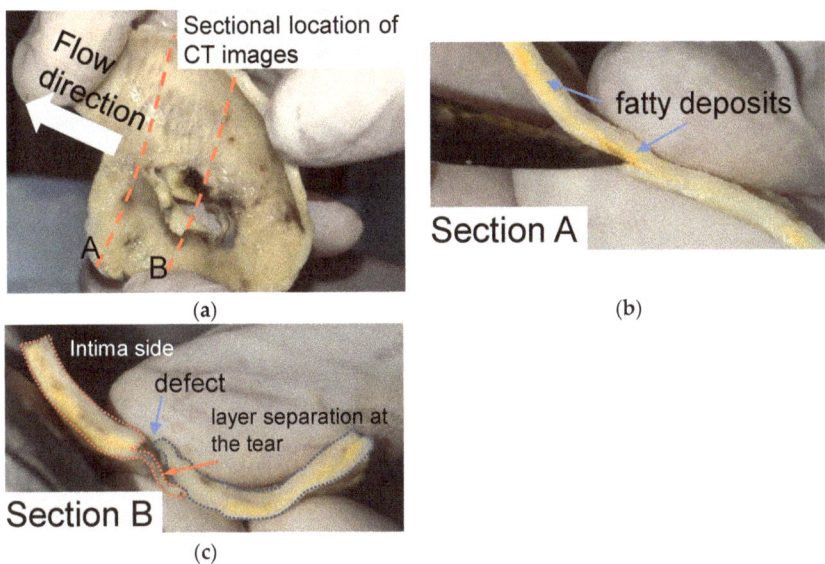

Figure 1. Anatomic structure of the aortic tissue and schematic diagram of the tissue at the breach. (**a**) A piece of formalin-soaked aortic tissue and its two cut positions. (**b**) Yellow fatty deposits showing an atherosclerotic layer on section A. (**c**) Vascular defect and layer separation around the tear on section B.

Pronounced delamination was observed in section A of the aortic tissue (Figure 1b). The yellow loose connective tissue between the media around the ostia of the rupture was identified as accumulated adipose tissue via pathological examination. As shown in Figure 1c, a defect was found on the intima side of the tissue, and a thin film of vascular tissue covered the defect. The thin vascular film extended only about 7 mm and then detached from the blood vessel, leaving a wound in the tissue. It should be noted that the aortic tissue was the dissection flap, which contained only the aortic intima and part of the media.

The aortic tissue observed in section B (Figure 1c) provides the basic anatomic structure for creating a primary tear on a silicone phantom. AD's two critical anatomic features are the defect located on the intima layer and the separation begins from the defect between the layers. Accordingly, for a silicone phantom, the primary tear can be composed of a tear in the inner wall and an area of a tiny pocket between the silicone layers.

Based on the observation of human aortic tissue, four critical parameters were considered in the modeling to represent the dissection: the thickness of the TL and the FL, the area of the inner tear, and the area of the pocket-shaped layer separation. The manufacturing steps of tear-embedded silicone phantoms include the fabrication of an inner layer with a tear, the outer layer, and a tiny pocket between the inner and outer layers to form the primary tear region.

2.2. Fabrication Method of Tear-Embedded Silicone Phantom

Brush-spin-coating was carried out to obtain a target thickness of 1.6 mm for the inner layer. In this study, two kinds of silicone materials were selected for the coating of the models, which were HY-E620 (Hong Ye E620, Shen Zhen Hong Ye Jie Technology Co., Ltd., Shenzhen, China) and DSA-7055 (DSA 7055, Guangdong Doneson New Materials, Guangdong, China). The mechanical properties of the two silica gels were measured and compared with the human aortic tissue obtained after surgery for aortic dissection. The aortic tissue of the patient was removed during the treatment of type A AD, then directly stored in a low-temperature environment of 4 °C in phosphate-buffered saline. Informed consent and ethical approval were obtained as well. In order to obtain more comparable results, patient-specific aortic arch silicone phantom models were made and then sliced into tension test strips at the corresponding positions of the obtained aortic tissue sample. The results of the direct tension test on the phantom samples and aortic tissue samples are shown in Figure 2. It can be seen that the tension test profile of DSA 7055 was closer to that of the aortic tissue when the strain rate was more significant than 1.5.

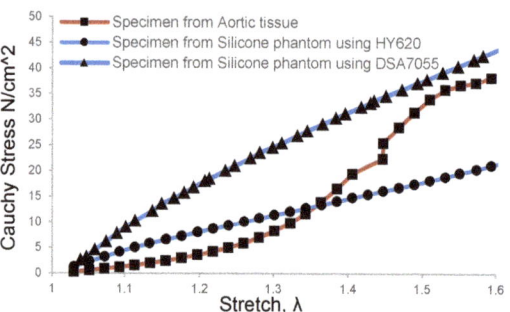

Figure 2. Tension test results of the aortic tissue sample and silicone phantom were processed with the brush-spin-coating method at the same size and position.

A recent ex vivo study on blood vessel expansion showed that pressure increments from 10 mmHg to 49 mmHg and 93 mmHg caused a circumferential strain rate of 1.32 and 1.50, respectively [30]. This means that the in vivo blood vessels underwent about 1.5 times the strain rate in a healthy state since the mean blood pressure of humans is 93.3 mmHg. As depicted in Figure 2, the elastic modulus of the silicone phantom DSA 7055 gradually approached the vascular tissue when the strain rate was about 1.5 rather than that of phantom E620. Meanwhile, the elastic modulus of our phantom was 0.72 MPa, which is slightly larger than that of normal tissue and approaching the value of elder or hypertensive patients [36,37]. This implies that the material can mimic the elder's aortic tissue, which can be more easily developed into dissection. Thus the tear-embedded phantom model was made of silicone material DSA7055. The refractive index of DSA 7055 was 1.41, and its dynamic viscosity was approximately 3000 cps, which needed 90 min of curing and baking at 45 °C on a 3D-printed water-soluble cylinder model.

The manufacturing process of a phantom with developable dissection features began when the inner layer's coating was finished. Figure 3a shows the schematic diagram of the water-soluble inner core and the coated inner layer. A small circular piece of the inner layer was removed from the inner core, as depicted in Figure 3b. In cutting off the silicone sheet, a circular blade is important to avoid additional injury to the inner layer. From the side view, an empty region on the inner core's surface was left after this operation. The circular empty area and the silicone tear were then coated with a release agent (Figure 3c), and the silicone tear was returned to its original height to perform the final coating of the outer layer (Figure 3d). After the silicone phantom was completely cured, the inner core was dissolved to leave the inner and outer silicone layers, which remained glued together at this stage. Subsequently, the circular area coated with the release agent could be easily removed from the inside. From Figure 3e, it is seen that the torn area was a thinner part that was composed of only the outer silicone layer. The silicone model's thickness was similar to that of an aortic ulcer or porcine model of AD [25].

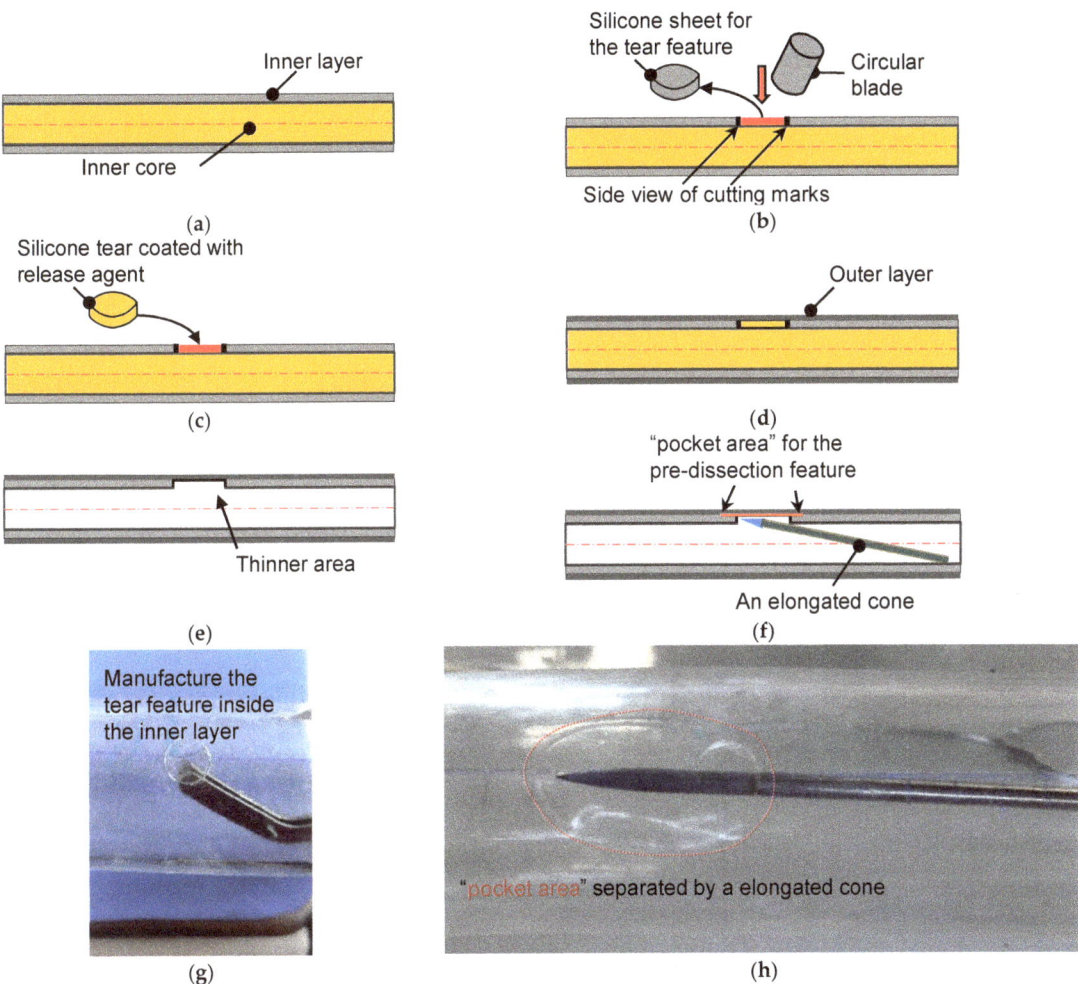

Figure 3. The schematic diagram for the fabrication of the aortic dissection phantom. (**a**) Side view of the inner core and its inner coating layer. (**b**) A circular blade cut a circular torn area. (**c**) The silicone sheet from the inner core was coated with a release agent and returned to the original position. (**d**) Coating of the outer layer on the inner layer with a targeted thickness. (**e**) Silicone tube with a torn area after the dissolution of the inner core. (**f**) Making pocket-shaped delamination between the inner and outer layers. (**g**) The actual operation is to remove the silicone sheet. (**h**) Creation of a pocket area between the layers by a cone probe.

The pocket-shaped slight delamination can be created using an elongated probe with a conical head (Figure 3f). Figure 3g,h show the pictures for creating the initial and pocket-shape delamination in an idealized aortic phantom. Two silicone models with different sizes of pocket areas were produced, among which the small dissection (SD) was an elliptical area with major and minor axes of 16 mm and 12 mm, respectively. Meanwhile, the principal axes of the elliptical area of the large dissection (LD) were 20 mm, but the short axes were the same as that of the small ones. The high-transparency silicone material, named DSA-7055, was adopted for the fabrication. In order to capture the specific process of tear development in the axial direction, laser irradiation of the middle plane of the model and video filming of the entire process of tearing in the top view were used. Pre-experiments have shown that laser irradiation is prone to reflections and refraction, which affects the quantitative study of features near the specific tear. Therefore, an LD model was used for the laser experiment, and an SD model was used for the CT scanning experiment. In addition, five more silicone models of the same tear size but without the interlayer delamination feature of the primary tear were fabricated as the control group. Due to limitations in CT room booking, all control models in this study were performed in a laser experiment.

2.3. Experimental Setup of the Laser-Enhanced Observation

The in vitro experiment to investigate dissection in the idealized aortic phantom was conducted in a steady state. Figure 4 shows the essential elements of the in vitro experimental circulation. The experimental device contained three main parts: a pressure sensor and recording instrument, a laser-enhanced observation system, and a pipeline circulation with a dead-end for pressurization (Figure 4). The injected working medium was water at 20 degrees Celsius. Air exhaust components were used to eliminate all bubbles for the initial experimental condition. A pressure transmitter, which converted the pressure data into a 4–20 mA current signal, was recorded in real-time at a refresh rate of 0.1 s. Alterations in the morphology of the idealized aortic phantom and in the pressure were documented with cameras to ensure time-synchronized output.

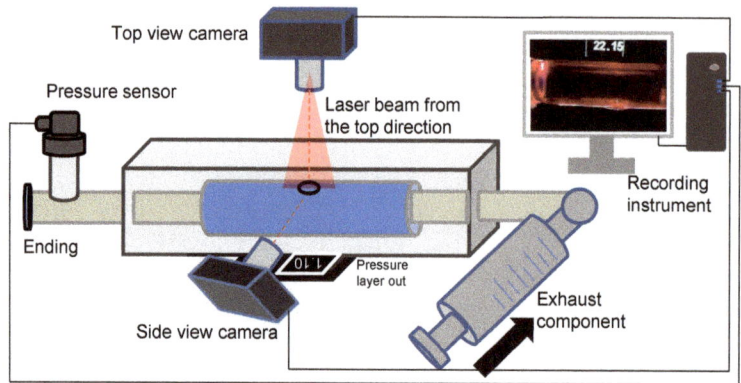

Figure 4. Schematic diagram of an experimental platform for laser-enhanced observation.

The specific steps of laser-enhanced observation are as follows. At the beginning of the experiment, all bubbles were eliminated. Then, the pressure was adjusted to 2.5 kPa by twisting the screw. Subsequently, the recording was turned on, and the screw was manually rotated. The pressure readings in the tube lumen gradually increased during the twisting process, which was also recorded in the camera footage. The pressurization process then stopped when the experimenter observed the development of an interlayer or a rupture at the interlayer. The recording process continued until the development of the tube wall interlayer was stable again, or a rupture had occurred.

2.4. Experimental Setup of the CT Imaging

CT scanning was conducted to investigate the volume and wall thickness alteration during the development of wall dissection. The experimental pipeline was settled on the CT scanning equipment as depicted in Figure 5 (X-ray equipment for CT; SOMATOM Force, Siemens Healthcare GmbH, Erlangen, Germany). The scope for scanning was targeted at the silicone phantom. The direction of the scan coincided with the phantom's axial direction, and 15 mL iodixanol of contrast agent was added to 1000 mL of water and then filled into the pipeline. The in-plane accuracy of the CT images was 0.1563 × 0.1563 mm, and the interval between the in-plane sections was 1 mm. The tear-embedded model for scanning was an SD with a smaller pocket area around the internal tear.

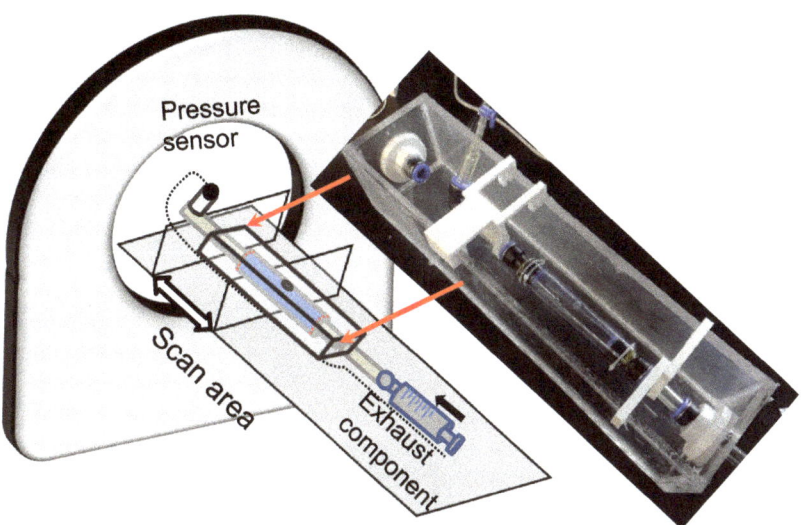

Figure 5. Schematic diagram of the experiment with CT scanning to capture the development of dissection.

In pressurization, the measurements could only be obtained before flow injection and CT scanning to avoid injury to the operations. Experimental pressurization and CT data extraction were performed alternately. The experimenter entered the CT control room to increase the pressure and quickly left the room while another operator immediately started the CT scan. If the CT scan revealed wall damage developing into a dissecting structure, then a continuous CT scan was maintained until the termination of the dissection process.

3. Results
3.1. Deformation of the FL and TL during the Development of Dissection

Pressurization experiments were performed for normal silicone phantoms and phantoms with initial damages. When the normal silicone tube was pressurized over 30 kPa, all the models were ruptured. In contrast, pressurization delaminated the tube models containing the feature of the primary tear into dissecting models.

The laser-enhanced observation was used to track the interfaces of TL and FL during the dissection development. Figure 6a shows four side views at different pressures. The first pressure state of 19.05 kPa was a moment in the gradual pressurization process. Pressure state 22.80 kPa was the pressure when the dissection started to develop. It is possible to qualitatively distinguish the increase of the tear at the 22.80 kPa from that of the pressure of 19.05 kPa.

There was no significant change in the line pressure from 22.80 kPa to 22.62 kPa within 3 s of the dissection starting to develop. However, the process underwent a dramatic increase in the volume of the FL. When the FL development process of the dissection was finished, the pressure of the line was stabilized at 15.67 kPa.

In order to observe the deformation of FL and TL during the pressurization, the temporal variation of tube pressure and deformation for the outer and inner layers are plotted in Figure 6b. The temporal deformations of the outer and inner layer are expressed using the variations of distance BC and AC indexed in Figure 6a, where A refers to the point of the edge of the internal tear, B is the intersection point of the transverse dashed line and the outer interface, and C is 3 px away from the largest-deformation point of B.

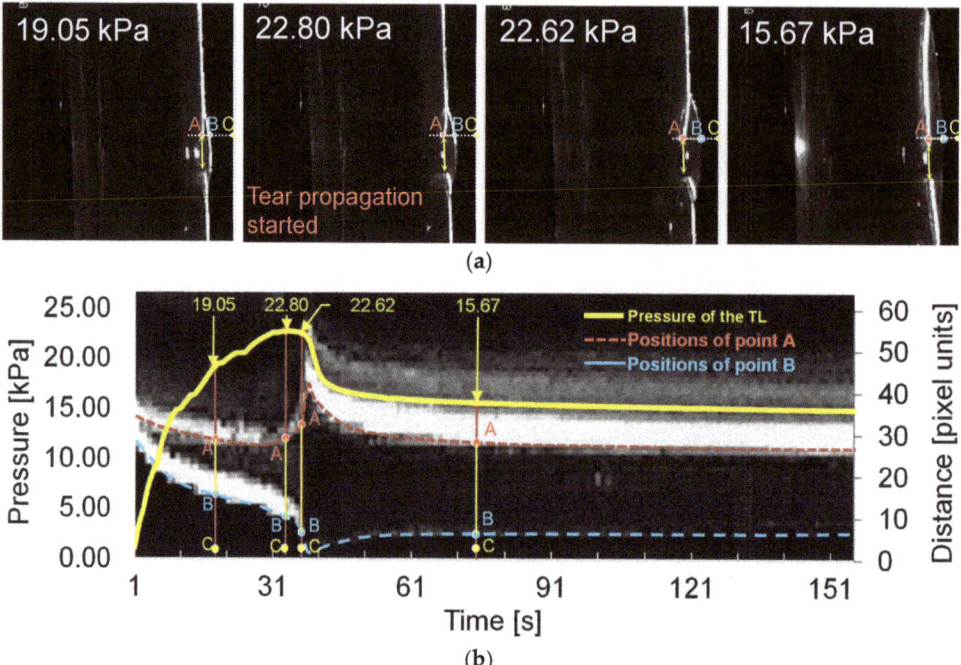

Figure 6. Main results of the laser-enhanced observation experiment. (**a**) Positions of the inner and outer layer during the pressurization. (**b**) Pressure recording and sequential images of the flap and outer layer positions during the pressurization. The left axis of Figure 6b is the pressure of the experiment, and the right side is the height scale of the picture, which is in pixels. The four experimental moments in Figure 6a are marked in Figure 6b using yellow arrows. In temporal latitude, point A constitutes the change in position of the inner layer of the membrane during the experiment, indicated by the red dashed line. The blue dashed line indicates the outer layer of the model.

It can be seen that the distance of AC gradually declined before the pressure reached 22.80 kPa. As the pressure reached 22.80 kPa, the distance of AC increased severely while the pressure remained at almost the same value. After 3 s, the distance of AC decreased fast, accompanied by a rapid drop of pressure. When the pressure dropped to 15.67 kPa, at the time of 81 s, the variations of AC became stable. The variation of AC indicates that there was a dramatic inward movement of the flap to TL during the development of FL.

In contrast, the distance of BC decreased rapidly before the pressure reached 22.62 kPa, suggesting that the FL volume increased quickly with the pressurization. After the FL reached its largest size, the distance of BC increased slightly and reached a stable state gradually, suggesting that the development of FL had finished.

3.2. Volume and Wall Thickness Variations during the Progress of Dissection

Pressurization and CT scanning were performed alternately until the experimenters discovered that the tear development process had begun. Figure 7a shows the time points for the data collection and timing for CT scanning operations. At the beginning of the experiment, 2.5 kPa was the pressure, so a set of CT scans were performed. The pressurization and scanning were alternated until the pressurization reached 32.5 kPa (36 times of scanning as depicted in Figure 7a), when the experimenter discovered that interlayer delamination had developed (the first sacan of 10 times of scanning as depicted in Figure 7a). After the delamination began to develop, the experiment entered the observation phase, and only CT scans were performed without further pressurization.

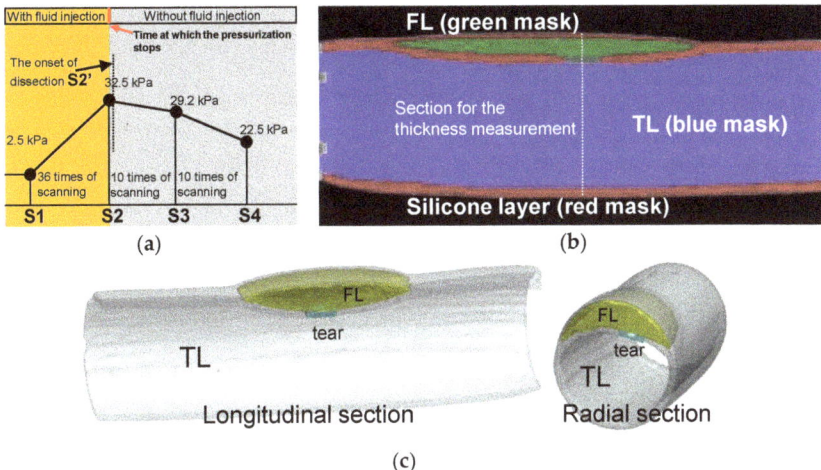

Figure 7. Experimental process of CT scanning and image reconstruction. (**a**) Pressure setting and the protocol of the CT scanning. (**b**) Reconstructed model of the longitudinal section from CT images at time point S2. (**c**) 3D reconstructed model of the longitudinal and cross-sectional view at time point S2′.

For data processing, this study positioned the initial pressure as S1 and the last CT scan before the onset of the dissection as S2. The CT that detected the development of the delamination was designated as S2′, one of the subsequent CTs during the development of the dissection was positioned as S3, and the final development status was positioned as S4. CT scan experiments were carried out under steady-state stepwise pressurization condition, which was the same as that in the laser tracking. Using the ScanIP module in the Simpleware software package (Simpleware Ltd., Exeter, UK), the volume of the TL and FL and the wall thicknesses of the silicone phantom were measured. Figure 7b gives a sectional reconstruction model at time point S3. Similar reconstruction models were obtained for all CT images at every scanning time point from S1 to S4. Figure 7c shows the reconstruction model at time point S2′ as dissection started to develop.

Regarding the error estimation of the experiment, the CT scanning was carried out three times at each scanning moment. However, FL was rapidly formed and increased at time point S2′. The scanning at time point S2′ was taken only once. For time points S1, S2, S3, and S4 (Figure 7a), three sets of CT data at each scanning point were selected for model reconstruction. On the other hand, the CT images at the moment S2′ were processed and constructed three times to avoid user errors.

The sequential data of CT imaging were used to measure the thickness, and the three-dimensional model was used for the volume estimation. Figure 8 presents the images of CT scanning in the longitudinal direction at different time points in Figure 7a. It is observed from Figure 8a–d that the initiation and propagation of AD were the same as the observations in the laser lightning experiment. A small tear was observed in the inner layer from Figure 8a at the beginning of pressurization corresponding to the time point S1. At time point S2, a slightly enlarged FL can be observed from Figure 8b. The lumen's radial views for three different positions, A-A′, B-B′, and C-C′, at the scanning moments of S3 and S4 are depicted in Figure 8c,d. The distance between the sections was set to 10 mm. The thickness of TL and FL, the volume of TL and FL, and the length of tear size were measured according to the cross-sectional images of the B-B′ position (B-B′ in Figure 8c).

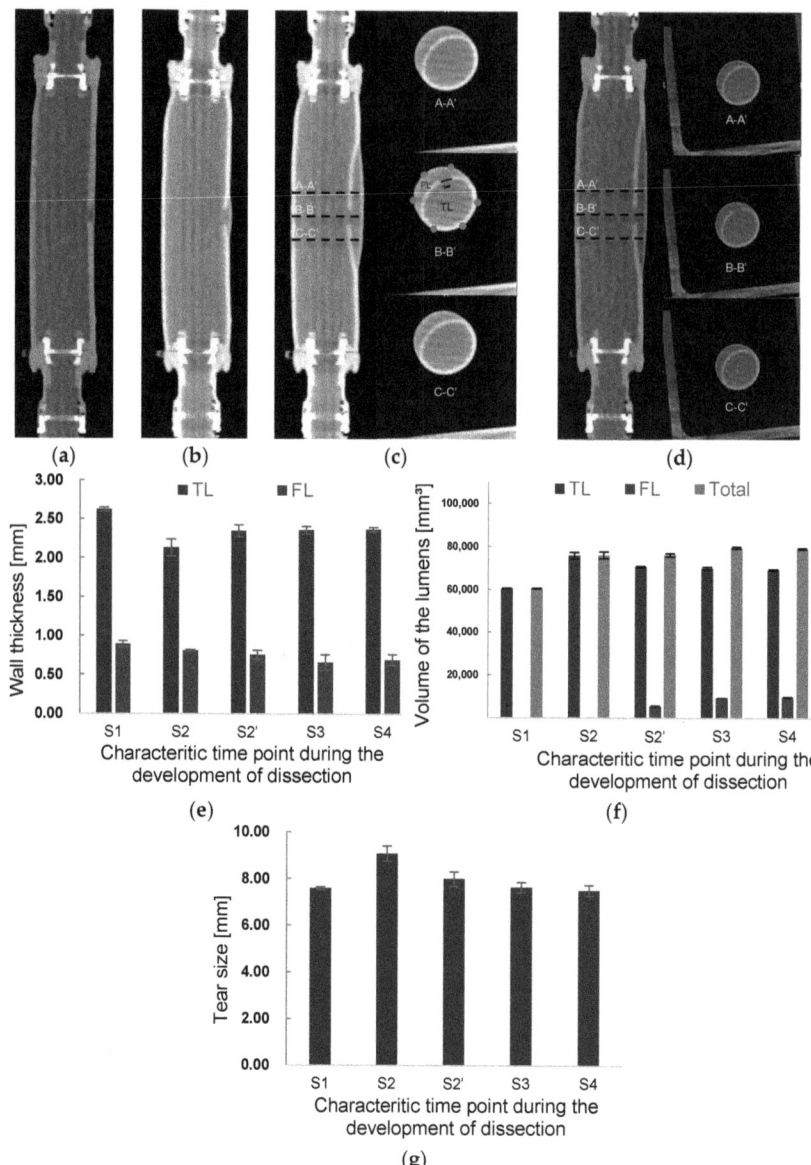

Figure 8. Axial CT images of at scanning time points (**a**) S1 and (**b**) S2. Axial CT images and the images of the cross sections A-A', B-B', and C-C' at a scanning time point (**c**) S3 and (**d**). The distances between A-A', B-B', and C-C' are 10 mm. Changes in the (**e**) wall thickness and (**f**) volume of TL, FL, and in total during the development onset of dissection. (**g**) Alteration of tear length during the development of dissection.

Figure 8e shows the measured wall thickness at each scanning point. At the initial scanning time S1, the normal and tear wall thickness was 2.6 and 0.9 mm, respectively. In contrast, the average wall thickness of TL and FL at the moment S2 was 18.7% and 9.1% thinner than those of S1, respectively. Correspondingly, the tear size was enlarged by 19.5% in the process of increasing TL.

At time point S2′, as shown in Figure 8f, the volume of FL increased considerably, being 25-times larger than that at S2 due to the sudden delamination of layers. Meanwhile, the volume of TL was reduced by about 6.9%. The volume of the phantom at time point S2′ still remained the same as that at S2. The diameter of the tear decreased by 11.9%, approaching the original size. The wall thickness of TL recovered to 92% of the original one after decreasing at time point S2.

In addition, the dissection further spread along the axial direction after the S2 time point, resulting in a 78.6% increase in FL volume and a small decrease of TL at the S3 time point. Correspondingly, the wall thickness of FL decreased, but TL wall thickness remained almost the same as that at the previous time point S2′. The expansion of FL stopped at the time S4. Except the slight augment of FL, the TL volume, wall thickness of the TL and FL, and tear size remained nearly the same as those at the previous time stage S3, indicating that the propagation of dissection had ended.

4. Discussion

4.1. Experimental Biomechanical Study for Aortic Dissection

Many studies have been conducted to quantify the strength of the aorta by means of stretching, bulge inflation, the peeling test, the trouser test, the direct tension test, and the in-plane shear test [38]. Iliopoulos et al. and Manopoulos et al. found that the circumferential tensile failure stresses in the anterior, posterior, right, and left lateral were not significantly different between the control, aneurysm, and dissection groups [39–41]. However, in the axial direction, there were significant differences in tensile failure stresses between the various types of samples, with the right lateral being higher than the anterior and the posterior side in control samples. At the same time, the left and right lateral were higher than the anterior side in the aneurysm samples, and the right lateral was higher than the left in the dissected samples.

Although uniaxial tensile tests provide valuable insight into the strength properties of aortic tissue, unidirectional tensile tests are limiting in representing in vivo loading conditions. Therefore, some studies have attempted to measure the mechanical properties of the aorta by means of bulge inflation tests. Sugita et al. reported the longitudinal weakest of the normal aorta under the bulge inflation test [42], which is close to the tensile test data, but the crack formation and crack development aspects of aneurysmal tissue still need further study [43]. Through bulge inflation experiments, Pearson et al. found that the fracture elongation of ascending aortic samples was significantly greater compared with isthmus samples [44].

Brunet et al. carried out a series of studies that were closer to physiological realism than the bulge inflation experiments. Brunet et al. designed a custom tension inflation device that fit into an X-ray microscopic imaging setup. The X-ray tomography device allowed observation of the wall structure and incision behavior during carotid artery inflation. In the following pressurization experiment of a vascular lumen with an intima notch [30], when the pressure was continuously inflated to 687 mmHg (91.6 kPa), widening and deepening of the notch were observed. However, the vascular still had difficulty developing a dissection structure even though the experimental pressure was increased to approximately 687 mmHg. This may be related to the form of the intima notch [30].

A reasonable feature of the tear may be one of the critical factors in whether the tearing process can be triggered under experimental conditions. During the formation phase of aortic dissection, a long-term abnormal hemodynamic environment may lead to wall damage developing gradually. As a result, by the time the dissection is triggered, the mechanical properties of the vessel wall may already be damaged, and then pressure incensement is often considered an essential trigger for the onset of dissection. Taking into account how the in vivo experiments of dissection were produced in the previous study by Guo et al. [25] and the dissected tissue samples obtained in this study, the key features of the tear were designed. At the same time, the previous study proposed the corresponding method of producing the wall injury in the form of tearing and pocket interlaminar separation. In the following experimental research, the tear model developed into a sandwich structure during pressurization, suggesting that further exploration of injury forms may help to improve the physiological realism of dissection-triggered experiments.

4.2. Analysis of the Forces during the Development of Dissection

The mechanism of the interlayer forces during tear development has been described by previous studies [31,32], but there is a lack of experimental evidence. The present study, although differing in material properties from aortic tissue, may provide an analytical basis for the overall mechanical behavior of dissection development.

The open area around the intima tear is directly associated with whether the dissection can be triggered. The mechanisms may be explained by the force balance depicted in Figure 9, where Fp, Fr, and Fe are the tube pressure, adhesion force of the interlayer, and elastic tension force of the silicone layers, respectively. At the early stage of pressurization (Figure 9a), the silicone wall expanded with pressure. When the experimental pressure was inflated to 32.5 kPa, as shown in Figure 9b, a small dome area was formed (red region in Figure 9b) above the internal tear due to a thinner pipe wall at the outer side of FL. More fluid inside the lumen may flow into the small dome, breaking the force balance on the flap's two sides and the edge point A.

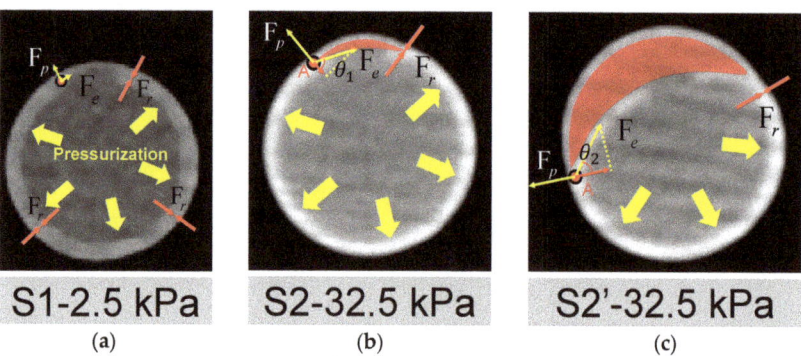

Figure 9. Analysis of mechanisms for the initiation and development of AD. Deformation of the cross-section of the silicone phantom and the forces acted on the wall (**a**) at the time point S1; (**b**) at the time point S2; (**c**) at the time point S2′. Fp, Fr and Fe are the pressure of the tube, the adhesion force of the silicone layers and the elastic tension of the inner silicone layer, respectively. The FL area is indicated by the red shading. A is the point where TL and FL are still stuck together. Fe and Fp act on point A, indicated by the yellow arrow. The projection of Fe in the opposite direction of Fp is the force that may cause interlayer detachment to start developing at point A.

The force of the inward pulling effect at point A is the component of the Fe opposite to Fp, as shown in Figure 9b,c. When the pressure increased, it led the combined force of Fe and Fp to be more significant than the resistance to tearing Fr at Point A. Then, the dissection began to develop, corresponding to the period of S2 to S2′. As can be seen in Figures 8f and 9c, during this period, the dissection was mainly in the circumferential direction, which confirms the viewpoint of Mikich and Qiao et al. [31,32].

The influence of the size of the pocket area between layers can be explained by the analysis of force balance acting on the edge point A as well. The increase in pocket area made the flap flatten. At this condition, the angle θ at point A will increase, as depicted in Figure 9c, which means a lower pressure increment can trigger the dissection.

The tension force Fe is related to the thickness of the flap and maybe even the shape of the local vascular morphology. If the flap thickness increases, then the Fe will increase. As previous studies have reported, a thicker flap may lead to a more severe dissection level [22,25]. In addition, the properties of the aortic flap in chronic AD may also need to be further considered because the stiffening of the dissected flap is common in AD patients [45]. In addition, from Figure 8c,d, it is seen that there was no significant difference for the edge of FL in the circumferential direction at S2' and S3, both of which ended at the half circle of the cross-section. However, the FL developed in an axial direction considerably at S3 time point, implying that the development of AD may first occur in the circumferential direction and end while reaching a new force balance [46]. The feature of developing first in the circumferential direction is most likely related to the radius of curvature of the vessel wall. This should interest investigators because some common areas of proximal tear, such as the lateral aspect of the descending aorta, the aortic arch, and the proximal area of the brachiocephalic artery, have a far more complex radius of curvature feature than the straight tube features in this experiment.

The present study also suggests that the "straining" pattern of the intimal wall may be a promising screening indicator for aortic dissection risk. It is likely that minor vascular injuries or even small tears in the intimal wall of the aorta exist prior to the onset of dissection, but such tears are difficult to detect with conventional imaging techniques [47,48]. In this study, the strained state of the wall (affected by tension force Fe) can assist in clinical screening for the early stage of AD. For example, if the CT angiographic findings of an intermural hematoma occurred, the aortic vessel also showed morphological features of a wall being straightened, as shown in Figure 9 of this study. At the same time, its outer wall also experienced more significant distension than the adjacent location. Therefore, based on the present experimental findings, the risk of dissection needs to be taken seriously.

This dissected phantom has good clinical translational benefit prospects. Traditional dissected models usually make the FL into a channel. The present study allows for structural changes in the dissection from injury to the dissected lumen and, therefore, more closely resembles the actual process of AD. It is also possible to quickly test for blood flow and pressure changes within the interstitial layer. Therefore, this model's most direct clinical translation is to be used by interventionalists as a teaching tool for clamping stent placement. At this stage, the technique is available for personalized models, as it is depicted in Figure 10. Clinicians can perform a series of studies on this model regarding stent anchorage zones and techniques involving in vivo and ex vivo stent fenestration.

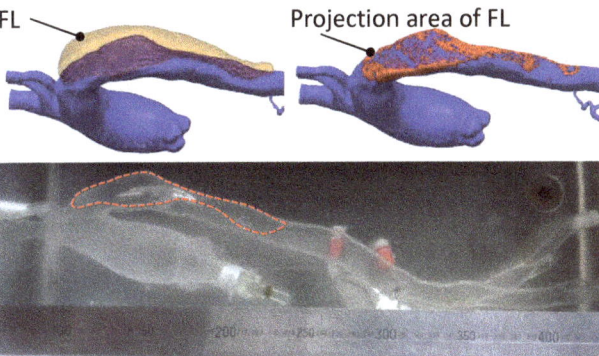

Figure 10. Personalized aortic silicone phantom with interlaminar damage structures.

4.3. Limitations

Despite the detailed observation of the dissection propagation process in the study, several shortcomings should be investigated further. The material's elastic modulus did not show the nonlinearity of actual aortic tissue. A phantom with a nonlinear elastic modulus or anisotropic mechanical properties may be fabricated by coating silica gel fibers on the model's surface. Another noticed limitation of this study is the dissection-triggering condition. In order to facilitate the observation of results, stepwise steady-state pressurization boundary conditions have been employed in most studies on the interlayer triggering of AD [20,30]. In future work, AD development in various pulsatile pressure conditions should be carried out to see what kind of pulsatile pattern is the most dangerous for development.

For simplicity, we employed water as the injected fluid in the two observation experiments. In a further study, glycerol and water should be used to obtain a fluid with the same viscosity as the blood. At the same time, the current research ignores the effect of personalized aortic structure on a primary tear, of which tortuosity, curvature, or tapering of the aorta may influence the sizes and positions of the primary tear [8,49]. By coupling personalized aortic structure fabrication techniques with this study [35], more realistic AD structures can be achieved, where thoracic endovascular repair (TEVAR) may be simulated.

In terms of subsequent development of the method, the intermural hematoma model is also one of the directions. Hypertension and aortic intermural hematoma are not uncommon in the advanced age group. In this model, the intermural hematoma aortic phantom model can be obtained by reducing the tear characteristics or using highly permeable silicone instead of tear fabrication. Future studies of tear triggering and hematoma volume on wall morphology using the intermural hematoma phantom model, and even the assessment of high-risk hematoma location, will be carried out gradually based on the technical solutions of this study.

5. Conclusions

We fabricated an in vitro silicone model with primary tear and adhesion damage, which developed a dissection via a stepwise pressurization experiment. The experimental approach allows for clearly tracking the initiation and propagation process of dissection.

At the initiation stage, only the volume of TL increased with the pressurization. There were no significant size changes for the primary interlayer damage. At the propagation stage, a severe increase of FL volume was observed, accompanied by rapid inward movement of the flap and the shrinking of the tear, leading to the decline of pressure in the silicone phantom. Meanwhile, the thickness of the FL gradually decreased as the dissection developed, which increased the risk of FL rupture. The size of the primary pocket-shaped tear significantly influenced the pressure required to trigger dissection. The larger the primary pocket-shaped area, the lower the pressure required.

The experimental results indicate that the primary tear and the weakening adhesion of the vessel layers are key factors in AD development. These two features are also frequently observed in patients with the acute aortic syndrome. This study suggests that some forms of primary damage to the arterial wall, such as aortic ulceration with aortic hematoma, which weakens the interlayer adhesion of the aorta, should be considered as early risk predictors of AD. This type of dissected model will have good prospects for clinical translation in areas such as interventional surgery rehearsal.

6. Patents

Qing-Zhuo Chi, Li-Zhong Mu, Ying He, and Zhen Cao. Manufacturing method of personalized extracorporeal interlayer physical model. China Invention Patent, CN112669687A.4[P], 16 April 2021 (Licensed).

Author Contributions: Conceptualization, Y.H., Q.-Z.C., Y.-Y.G., L.-Z.M. and Y.L.; methodology, Q.-Z.C., Z.C., L.-L.L. and L.-Z.M.; software, Q.-Z.C.; validation, Q.-Z.C. and Z.C.; resources, Y.H.; data curation, Q.-Z.C.; writing—original draft preparation, Q.-Z.C.; writing—review and editing, Y.H.; visualization, Y.-Y.G.; supervision, Y.H.; project administration, Y.H.; funding acquisition, Y.H. All authors have read and agreed to the published version of the manuscript.

Funding: This research was funded by the National Natural Science Foundation of China (Nos. 51976026), Fundamental Research Funds of Central Universities of China (No. DUT22YG206), and the Fundamental Research Funds for the Central Universities (DUT20GJ203).

Institutional Review Board Statement: The study was conducted in accordance with the Declaration of Helsinki, and approved by the Institutional Review Board (or Ethics Committee) of Dalian University of Technology (protocol code: 2021-083, date of approval: 5 November 2021) and by Dalian Medical University (protocol code: PJ-KS-KY-2021-187, date of approval: 23 August 2021).

Informed Consent Statement: Informed consent was obtained from all subjects involved in the study.

Data Availability Statement: Not applicable.

Conflicts of Interest: Authors declare no conflict of interest.

References

1. Clouse, W.D.; Hallett, J.W., Jr.; Schaff, H.V.; Spittell, P.C.; Rowland, C.M.; Ilstrup, D.M.; Melton, L.J., 3rd. Acute aortic dissection: Population-based incidence compared with degenerative aortic aneurysm rupture. *Mayo Clin. Proc.* **2004**, *79*, 176–180. [CrossRef] [PubMed]
2. Howard, D.P.; Banerjee, A.; Fairhead, J.F.; Perkins, J.; Silver, L.E.; Rothwell, P.M.; Oxford Vascular, S. Population-based study of incidence and outcome of acute aortic dissection and premorbid risk factor control: 10-year results from the Oxford Vascular Study. *Circulation* **2013**, *127*, 2031–2037. [CrossRef] [PubMed]
3. Olsson, C.; Thelin, S.; Stahle, E.; Ekbom, A.; Granath, F. Thoracic aortic aneurysm and dissection: Increasing prevalence and improved outcomes reported in a nationwide population-based study of more than 14,000 cases from 1987 to 2002. *Circulation* **2006**, *114*, 2611–2618. [CrossRef] [PubMed]
4. Harris, K.M.; Braverman, A.C.; Eagle, K.A.; Woznicki, E.M.; Pyeritz, R.E.; Myrmel, T.; Peterson, M.D.; Voehringer, M.; Fattori, R.; Januzzi, J.L.; et al. Acute Aortic Intramural Hematoma An Analysis From the International Registry of Acute Aortic Dissection. *Circulation* **2012**, *126*, S91–S96. [CrossRef] [PubMed]
5. Ganaha, F.; Miller, D.C.; Sugimoto, K.; Do, Y.S.; Minamiguchi, H.; Saito, H.; Mitchell, R.S.; Dake, M.D. Prognosis of aortic intramural hematoma with and without penetrating atherosclerotic ulcer: A clinical and radiological analysis. *Circulation* **2002**, *106*, 342–348. [CrossRef]
6. Evangelista, A.; Mukherjee, D.; Mehta, R.H.; O'Gara, P.T.; Fattori, R.; Cooper, J.V.; Smith, D.E.; Oh, J.K.; Hutchison, S.; Sechtem, U.; et al. Acute intramural hematoma of the aorta: A mystery in evolution. *Circulation* **2005**, *111*, 1063–1070. [CrossRef]
7. Sugawara, J.; Hayashi, K.; Yokoi, T.; Tanaka, H. Age-associated elongation of the ascending aorta in adults. *JACC Cardiovasc. Imaging* **2008**, *1*, 739–748. [CrossRef]
8. Chi, Q.Z.; Chen, H.M.; Mu, L.Z.; He, Y.; Luan, Y. Haemodynamic Analysis of the Relationship between the Morphological Alterations of the Ascending Aorta and the Type A Aortic-Dissection Disease. *Fluid Dyn. Mater. Proc.* **2021**, *17*, 721–743. [CrossRef]
9. Chiu, J.J.; Chien, S. Effects of disturbed flow on vascular endothelium: Pathophysiological basis and clinical perspectives. *Physiol. Rev.* **2011**, *91*, 327–387. [CrossRef]
10. Gimbrone, M.A., Jr.; Garcia-Cardena, G. Endothelial Cell Dysfunction and the Pathobiology of Atherosclerosis. *Circ. Res.* **2016**, *118*, 620–636. [CrossRef]
11. Malek, A.M.; Alper, S.L.; Izumo, S. Hemodynamic shear stress and its role in atherosclerosis. *JAMA* **1999**, *282*, 2035–2042. [CrossRef]
12. Tronc, F.; Mallat, Z.; Lehoux, S.; Wassef, M.; Esposito, B.; Tedgui, A. Role of matrix metalloproteinases in blood flow-induced arterial enlargement: Interaction with NO. *Arterioscler. Thromb. Vasc. Biol.* **2000**, *20*, E120–E126. [CrossRef]
13. Dolan, J.M.; Sim, F.J.; Meng, H.; Kolega, J. Endothelial cells express a unique transcriptional profile under very high wall shear stress known to induce expansive arterial remodeling. *Am. J. Physiol. Cell Physiol.* **2012**, *302*, C1109–C1118. [CrossRef]
14. Guzzardi, D.G.; Barker, A.J.; van Ooij, P.; Malaisrie, S.C.; Puthumana, J.J.; Belke, D.D.; Mewhort, H.E.; Svystonyuk, D.A.; Kang, S.; Verma, S.; et al. Valve-Related Hemodynamics Mediate Human Bicuspid Aortopathy: Insights From Wall Shear Stress Mapping. *J. Am. Coll Cardiol.* **2015**, *66*, 892–900. [CrossRef]
15. Metaxa, E.; Tremmel, M.; Natarajan, S.K.; Xiang, J.; Paluch, R.A.; Mandelbaum, M.; Siddiqui, A.H.; Kolega, J.; Mocco, J.; Meng, H. Characterization of Critical Hemodynamics Contributing to Aneurysmal Remodeling at the Basilar Terminus in a Rabbit Model. *Stroke* **2010**, *41*, 1774–1782. [CrossRef]
16. Hayashi, K.; Makino, A.; Kakoi, D. Remodeling of arterial wall: Response to changes in both blood flow and blood pressure. *J. Mech. Behav. Biomed. Mater.* **2018**, *77*, 475–484. [CrossRef]
17. Sehgel, N.L.; Sun, Z.; Hong, Z.; Hunter, W.C.; Hill, M.A.; Vatner, D.E.; Vatner, S.F.; Meininger, G.A. Augmented vascular smooth muscle cell stiffness and adhesion when hypertension is superimposed on aging. *Hypertension* **2015**, *65*, 370–377. [CrossRef]
18. Angouras, D.; Sokolis, D.P.; Dosios, T.; Kostomitsopoulos, N.; Boudoulas, H.; Skalkeas, G.; Karayannacos, P.E. Effect of impaired vasa vasorum flow on the structure and mechanics of the thoracic aorta: Implications for the pathogenesis of aortic dissection. *Eur. J. Cardiothorac. Surg.* **2000**, *17*, 468–473. [CrossRef]
19. Osada, H.; Kyogoku, M.; Ishidou, M.; Morishima, M.; Nakajima, H. Aortic dissection in the outer third of the media: What is the role of the vasa vasorum in the triggering process? *Eur. J. Cardiothorac. Surg.* **2013**, *43*, e82–e88. [CrossRef]
20. Carson, M.W.; Roach, M.R. The strength of the aortic media and its role in the propagation of aortic dissection. *J. Biomech.* **1990**, *23*, 579–588. [CrossRef]

21. Tiessen, I.M.; Roach, M.R. Factors in the initiation and propagation of aortic dissections in human autopsy aortas. *J. Biomech. Eng.* **1993**, *115*, 123–125. [CrossRef] [PubMed]
22. Tam, A.S.M.; Sapp, M.C.; Roach, M.R. The effect of tear depth on the propagation of aortic dissections in isolated porcine thoracic aorta. *J. Biomech.* **1998**, *31*, 673–676. [CrossRef] [PubMed]
23. Tang, J.; Wang, Y.; Hang, W.; Fu, W.; Jing, Z. Controllable and uncontrollable Stanford type B aortic dissection in canine models. *Eur. Surg. Res.* **2010**, *44*, 179–184. [CrossRef] [PubMed]
24. Wang, L.X.; Wang, Y.Q.; Guo, D.Q.; Jiang, J.H.; Zhang, J.; Cui, J.S.; Fu, W.G. An experimental model of Stanford type B aortic dissection with intravenous epinephrine injection. *Kaohsiung J. Med. Sci.* **2013**, *29*, 194–199. [CrossRef]
25. Guo, B.; Dong, Z.; Pirola, S.; Liu, Y.; Menichini, C.; Xu, X.Y.; Guo, D.; Fu, W. Dissection level within aortic wall layers is associated with propagation of type B aortic dissection: A swine model study. *Eur. J. Vasc. Endovasc. Surg.* **2019**, *58*, 415–425. [CrossRef]
26. Rudenick, P.A.; Bijnens, B.H.; Garcia-Dorado, D.; Evangelista, A. An in vitro phantom study on the influence of tear size and configuration on the hemodynamics of the lumina in chronic type B aortic dissections. *J. Vasc. Surg.* **2013**, *57*, 464–474.e465. [CrossRef]
27. Soudah, E.; Rudenick, P.; Bordone, M.; Bijnens, B.; Garcia-Dorado, D.; Evangelista, A.; Onate, E. Validation of numerical flow simulations against in vitro phantom measurements in different type B aortic dissection scenarios. *Comput. Methods Biomech. Biomed. Eng.* **2015**, *18*, 805–815. [CrossRef]
28. Birjiniuk, J.; Timmins, L.H.; Young, M.; Leshnower, B.G.; Oshinski, J.N.; Ku, D.N.; Veeraswamy, R.K. Pulsatile Flow Leads to Intimal Flap Motion and Flow Reversal in an In Vitro Model of Type B Aortic Dissection. *Cardiovasc. Eng. Technol.* **2017**, *8*, 378–389. [CrossRef]
29. Birjiniuk, J.; Veeraswamy, R.K.; Oshinski, J.N.; Ku, D.N. Intermediate fenestrations reduce flow reversal in a silicone model of Stanford Type B aortic dissection. *J. Biomech.* **2019**, *93*, 101–110. [CrossRef]
30. Brunet, J.; Pierrat, B.; Adrien, J.; Maire, E.; Curt, N.; Badel, P. A Novel Method for In Vitro 3D Imaging of Dissecting Pressurized Arterial Segments Using X-Ray Microtomography. *Exp. Mech.* **2021**, *61*, 147–157. [CrossRef]
31. Mikich, B. Dissection of the aorta: A new approach. *Heart* **2003**, *89*, 6–8. [CrossRef]
32. Qiao, A.K.; Li, X.Y.; Zhang, H.J. Mechanism of the initiation, propagation and treatment of aortic dissection. *J. Beijing Univ. Technol.* **2007**, *33*, 959–963.
33. Shiraishi, Y.; Yambe, T.; Narracott, A.J.; Yamada, A.; Morita, R.; Qian, Y.; Hanzawa, K. Modeling Approach for An Aortic Dissection with Endovascular Stenting. *Annu. Int. Conf. IEEE Eng. Med. Biol. Soc.* **2020**, *2020*, 5008–5011. [CrossRef]
34. Marconi, S.; Lanzarone, E.; De Beaufort, H.; Conti, M.; Trimarchi, S.; Auricchio, F. A novel insight into the role of entry tears in type B aortic dissection: Pressure measurements in an in vitro model. *Int. J. Artif. Organs* **2017**, *40*, 563–574. [CrossRef]
35. Chi, Q.Z.; Mu, L.Z.; He, Y.; Luan, Y.; Jing, Y.C. A brush–spin–coating method for fabricating in vitro patient-specific vascular models by coupling 3D-printing. *Cardiovasc. Eng. Technol.* **2021**, *12*, 200–214. [CrossRef]
36. Roccabianca, S.; Figueroa, C.A.; Tellides, G.; Humphrey, J.D. Quantification of regional differences in aortic stiffness in the aging human. *J. Mech. Behav. Biomed. Mater* **2014**, *29*, 618–634. [CrossRef]
37. De Beaufort, H.W.L.; Ferrara, A.; Conti, M.; Moll, F.L.; van Herwaarden, J.A.; Figueroa, C.A.; Bismuth, J.; Auricchio, F.; Trimarchi, S. Comparative Analysis of Porcine and Human Thoracic Aortic Stiffness. *Eur. J. Vasc. Endovasc. Surg.* **2018**, *55*, 560–566. [CrossRef]
38. Sherifova, S.; Holzapfel, G.A. Biomechanics of aortic wall failure with a focus on dissection and aneurysm: A review. *Acta Biomater.* **2019**, *99*, 1–17. [CrossRef]
39. Dimitrios, C.I.; Rejar, P.D.; Eleftherios, P.K.; Despina, P.; George, D.S.; Konstantinos, T.; Christodoulos, S.; Dimitrios, P.S. Regional and directional variations in the mechanical properties of ascending thoracic aortic aneurysms. *Med. Eng. Phys.* **2009**, *31*, 1–9. [CrossRef]
40. Iliopoulos, D.C.; Kritharis, E.P.; Giagini, A.T.; Papadodima, S.A.; Sokolis, D.P. Ascending thoracic aortic aneurysms are associated with compositional remodeling and vessel stiffening but not weakening in age-matched subjects. *J. Thorac. Cardiovasc. Surg.* **2009**, *137*, 101–109. [CrossRef]
41. Manopoulos, C.; Karathanasis, I.; Kouerinis, I.; Angouras, D.C.; Lazaris, A.; Tsangaris, S.; Sokolis, D.P. Identification of regional/layer differences in failure properties and thickness as important biomechanical factors responsible for the initiation of aortic dissections. *J. Biomech.* **2018**, *80*, 102–110. [CrossRef] [PubMed]
42. Sugita, S.; Matsumoto, T.; Ohashi, T.; Kumagai, K.; Akimoto, H.; Tabayashi, K.; Sato, M. Evaluation of Rupture Properties of Thoracic Aortic Aneurysms in a Pressure-Imposed Test for Rupture Risk Estimation. *Cardiovasc. Eng. Technol.* **2011**, *3*, 41–51. [CrossRef]
43. Kim, J.H.; Avril, S.; Duprey, A.; Favre, J.P. Experimental characterization of rupture in human aortic aneurysms using a full-field measurement technique. *Biomech. Model Mechanobiol.* **2012**, *11*, 841–853. [CrossRef] [PubMed]
44. Pearson, R.; Philips, N.; Hancock, R.; Hashim, S.; Field, M.; Richens, D.; McNally, D. Regional wall mechanics and blunt traumatic aortic rupture at the isthmus. *Eur. J. Cardiothorac. Surg.* **2008**, *34*, 616–622. [CrossRef]
45. Karmonik, C.; Duran, C.; Shah, D.J.; Anaya-Ayala, J.E.; Davies, M.G.; Lumsden, A.B.; Bismuth, J. Preliminary findings in quantification of changes in septal motion during follow-up of type B aortic dissections. *J. Vasc. Surg.* **2012**, *55*, 1419–1426.e1411. [CrossRef]
46. Rolf-Pissarczyk, M.; Li, K.W.; Fleischmann, D.; Holzapfel, G.A. A discrete approach for modeling degraded elastic fibers in aortic dissection. *Comput. Method Appl. M* **2021**, *373*, 113511. [CrossRef]

7. Li, Y.; Zhang, N.; Xu, S.; Fan, Z.; Zhu, J.; Huang, L.; Chen, D.; Sun, Z.; Sun, L. Acute type A aortic intramural hematoma and type A aortic dissection: Correlation between the intimal tear features and pathogenesis. *Quant. Imaging Med. Surgery* **2020**, *10*, 1504–1514. [CrossRef]
8. Kitai, T.; Kaji, S.; Yamamuro, A.; Tani, T.; Kinoshita, M.; Ehara, N.; Kobori, A.; Kim, K.; Kita, T.; Furukawa, Y. Detection of intimal defect by 64-row multidetector computed tomography in patients with acute aortic intramural hematoma. *Circulation* **2011**, *124*, 174–178. [CrossRef]
9. Chi, Q.; He, Y.; Luan, Y.; Qin, K.; Mu, L. Numerical analysis of wall shear stress in ascending aorta before tearing in type A aortic dissection. *Comp. Biol. Med.* **2017**, *89*, 236–247. [CrossRef]

Journal of Functional Biomaterials

Article

Lingering Dynamics of Type 2 Diabetes Mellitus Red Blood Cells in Retinal Arteriolar Bifurcations

Lili Long [1], Huimin Chen [1], Ying He [1,*], Lizhong Mu [1] and Yong Luan [2]

1. Key Laboratory of Ocean Energy Utilization and Energy Conservation of Ministry of Education, School of Energy and Power Engineering, Dalian University of Technology, Dalian 116000, China
2. Department of Anesthesiology, The First Affiliated Hospital of Dalian Medical University, Dalian 116011, China
* Correspondence: heying@dlut.edu.cn

Abstract: It has been proven that the deformability of red blood cells (RBC) is reduced owing to changes in mechanical properties, such as diabetes mellitus and hypertension. To probe the effects of RBC morphological and physical parameters on the flow field in bifurcated arterioles, three types of RBC models with various degrees of biconcave shapes were built based on the in vitro experimental data. The dynamic behaviors of the RBCs in shear flow were simulated to validate the feasibility of the finite element-Arbitrary Lagrangian–Eulerian method with a moving mesh. The influences of the shear rate and viscosity ratios on RBC motions were investigated. The motion of RBCs in arteriolar bifurcations was further simulated. Abnormal variations in the morphological and physical parameters of RBCs may lead to diminished tank-tread motion and enhanced tumbling motion in shear flow. Moreover, abnormal RBC variations can result in slower RBC motion at the bifurcation with a longer transmit time and greater flow resistance, which may further cause inadequate local oxygen supply. These findings would provide useful insights into the microvascular complications in diabetes mellitus.

Keywords: diabetes mellitus erythrocyte; retinal vessel; fluid–structure interaction; lingering time

Citation: Long, L.; Chen, H.; He, Y.; Mu, L.; Luan, Y. Lingering Dynamics of Type 2 Diabetes Mellitus Red Blood Cells in Retinal Arteriolar Bifurcations. *J. Funct. Biomater.* **2022**, *13*, 205. https://doi.org/10.3390/jfb13040205

Academic Editor: Vassilios Sikavitsas

Received: 12 September 2022
Accepted: 22 October 2022
Published: 27 October 2022

Publisher's Note: MDPI stays neutral with regard to jurisdictional claims in published maps and institutional affiliations.

Copyright: © 2022 by the authors. Licensee MDPI, Basel, Switzerland. This article is an open access article distributed under the terms and conditions of the Creative Commons Attribution (CC BY) license (https://creativecommons.org/licenses/by/4.0/).

1. Introduction

Diseases related to microcirculation, such as diabetes mellitus (DM), hypertension, and atherosclerosis, are major health problems in modern society. Retinopathy is one of the most common complications of diabetes mellitus (DM). Retinal vasculature is readily displayed in vivo and can be directly observed and characterized by non-invasive means. Health assessment of the retinal vasculature is a potential biomarker of pathological processes in the eyes and other organs [1].

Microvasculature is one of the basic components of microcirculation, and its size is comparable to that of a red blood cell (RBC). RBCs are the main suspension component of blood, functioning as transporters of oxygen and carbon dioxide between the lungs and peripheral tissues. RBCs are composed of a viscous fluid and an elastic membrane encapsulating the viscous fluid (hemoglobin solution). The RBC membrane is linked by a combination of a phospholipid bilayer and a cytoskeleton (spectrin network) [2,3], with the composite properties of a phospholipid bilayer and a protein network determining the RBC biconcave shape, resulting in membrane elasticity and biomechanical properties [4] as well as controlling RBC deformability [5]. Normal RBC deformability is a key determinant of proper blood flow and function in microcirculation [6].

Several medical studies have demonstrated that, compared to healthy RBCs, the concave depth, diameter, and deformation index of RBCs in DM decreased, while the stiffness and viscosity increased [7]. Ciasca et al. [8] measured the viscoelasticity of RBCs on a nanoscale and found increased viscosity and stiffness of diabetic RBC membranes compared to healthy RBCs. Jin et al. [9] and AlSahi et al. [10] studied the effect of DM on

RBC morphology by applying atomic force microscopy and fluorescence spectroscopy, and pointed out that excessive glucose in the blood would lead to a swollen RBC shape. In addition, excessive glucose causes RBCs aggregation, which eventually leads to increased viscosity and slow the mobility of RBCs. Consequently, examining the effects of RBC morphological and physical parameters on the flow field may enhance the basic understanding of RBC behavior under healthy and DM conditions.

Fluid–structure interaction (FSI) is the most ideal method for studying the mechanism of RBC motion in addition to in vitro experiments. Pozrikidis et al. [11] in 1995 developed a simulation of capsule deformation in shear flow using the boundary integral method. Wei et al. [12] applied the immersed boundary-lattice Boltzmann method to study RBC motion and deformation in two-dimensional capillaries and calculated NO transport properties across the RBC membrane. In a recent study, Balogh et al. [13] employed the immersed-boundary method to present multiple RBCs motions in near-physiological three-dimensional microvascular networks. Most studies on RBCs have ignored the thickness of RBC membranes and treated RBCs as elastic capsules wrapped in a thin shell with no thickness. Abnormal changes in the morphology and mechanical properties of RBC can adversely affect RBC deformation.

Therefore, we considered the RBC membrane thickness and used classical shear flow to validate the feasibility of the finite element-Arbitrary Lagrangian–Eulerian (ALE) algorithm for calculating the RBC motion. Subsequently, we calculated the motion of single and multiple RBCs in retinal bifurcating arterioles. Meanwhile, the pressure drop at the bifurcation and the lingering time of the RBC at the bifurcation were quantified and analyzed.

2. Materials and Methods

2.1. Retinal Vessel Model

The circulatory system plays a principal role in transporting blood, which contains oxygen and nutrients indispensable for the growth and maintenance of the body, to the immediate vicinity of the tissues of the organs [14]. The branching vascular parameters of retinal arterioles can be described by the power law:

$$d_0^k = d_1^k + d_2^k \tag{1}$$

where k is the vessel index, d_0 is the parent vessel diameter, and d_1, d_2 are the child branch vessel diameters, respectively. The branching power law was derived by Murray and Zamir according to the principle of minimal work of the arterial system [15,16]. Several studies have reported k values varying between 2 and 3 [17–19]. Here, the k value was taken as 2.1 [20].

The asymmetry index is used to describe the relationship between the child vessels:

$$\psi = d_2/d_1 \tag{2}$$

Diameter ratio:

$$\psi_1 = d_1/d_0 \tag{3}$$

$$\psi_2 = d_2/d_0 \tag{4}$$

Bifurcation angle in accordance with the minimum pumping power and volume:

$$\cos\theta_1 = \frac{\psi_1^{-4} + 1 + \psi^4}{2\psi_1^{-2}} \tag{5}$$

$$\cos\theta_2 = \frac{\psi_1^{-4} + \psi^4 - 1}{2\psi^2 \psi_1^{-2}} \tag{6}$$

Length of the retinal vessels:

$$l = 7.4(d_i/2)^{1.15}, i = 0, 1, 2 \tag{7}$$

The diameter, length, and bifurcation angle of each segment of the child vessel can be calculated, given the diameter and asymmetry ratio of the parent vessel. In this study, the parent vessel diameter d_0 is set to 9.5 μm [21], and the asymmetry index ψ is 1, which means that the child vessels are perfectly symmetrical.

2.2. RBC Model

An RBC is considered a viscous fluid wrapped by a viscoelastic membrane with a biconcave shape. The model equation is as follows [22]:

$$\left(x^2 + y^2 + a^2\right)^2 - 4a^2x^2 = b^4 \tag{8}$$

where a and b are geometric parameters related to the RBC diameter d and height h [23]:

$$a^2 = d^2 - h^2/8, b^2 = d^2 + h^2/8 \tag{9}$$

Here, the RBC membrane thickness was quoted from in vitro experimental studies, setting as 54 nm [24,25]. We define three types of RBCs with different degrees of biconcave shapes based on in vitro experimental studies: the healthy red blood cell (H-RBC), pre-diabetic red blood cell (P-RBC), and diabetic red blood cell (D-RBC), whereas for a detailed description of the RBC model parameters, we refer to [26].

2.3. Governing Equations

Blood is an incompressible fluid, and its flow equation is described by the Navier–Stokes equation:

$$\rho_{fluid}\left[\frac{\partial u_{fluid}}{\partial t} + \left(u_{fluid} \times \nabla\right)u_{fluid}\right] = \nabla \times \left[-pI + \mu(\nabla u_{fluid} + (\nabla u_{fluid})^T)\right] \tag{10}$$

$$\nabla \times u_{fluid} = 0 \tag{11}$$

where ρ_{fluid} is the plasma density, u_{fluid} is the plasma velocity vector, p is pressure, and μ is viscosity. Plasma density and cytoplasmic density were chosen to have the same magnitude, 1060 kg/m^3. The viscosity ratio λ of cytoplasm viscosity and plasma viscosity was defined as $\lambda = \mu_{cytoplasm}/\mu_{plasma}$, and the physiological viscosity ratio λ_c was 5 [27].

The RBC membrane is isotropic and viscous, and the behavior of the RBC fluid interaction can be described by the elastic dynamic equation:

$$\rho_{solid}\frac{\partial^2 U_{solid}}{\partial t^2} = \nabla\sigma + F_V \tag{12}$$

where ρ_{solid} is the RBC membrane density, U_{solid} is the RBC membrane displacement vector, σ is the stress tensor of the RBC membrane, and F_V is the force per unit volume of RBCs. The stress–strain relationship is expressed as follows:

$$\sigma = G\varepsilon + \eta\frac{d\varepsilon}{dt} \tag{13}$$

where G represents the RBC membrane shear modulus and $G = E/(2(1+\nu))$. Notably, E and ν are Young's modulus and Poisson's ratio of the RBC membrane, respectively. η is the viscosity of the RBC membranes. ε is the strain tensor of the RBC membrane, and its equation is as follows:

$$\varepsilon(t) = \frac{\sigma_0}{G}(1 - e^{-t/\tau}) \tag{14}$$

where σ_0 denotes the initial strain tensor of the RBC membrane. τ is the relaxation time of the RBC membrane, $\tau = \eta/G$.

The interfaces between the RBC membrane, plasma, and cytoplasm are FSI boundaries. Two-way coupling is captured along the FSI boundary by the fluid applying forces on the RBC membrane, and the RBC membrane displacement imposes a moving wall boundary condition on the fluid. These conditions can be expressed as:

$$\sigma \times n = \Gamma \times n \quad (15)$$

$$u_{fluid} = u_{soild} \quad (16)$$

$$u_{soild} = \frac{\partial U_{solid}}{\partial t} \quad (17)$$

where Γ is the force of the fluid acting on the RBC membrane, $\Gamma = \left[-pI + \mu(\nabla u_{fluid} + (\nabla u_{fluid})^T)\right]$, n is the normal vector at the fluid–structure boundary, and u_{solid} refers to the velocity vector of the RBC membrane.

The Neo-Hookean model of hyper-elastic material was chosen for the RBC membrane [28,29], and the Kelvin–Voigt model was adopted to realize the viscoelasticity of the RBC membrane. The RBC membrane density is set to 1090 kg/m³ [30]. The parameters used in the model are shown in Table 1.

Table 1. Values for parameters used in the model.

Symbol	Value, Units	Description
d_0	9.5 μm	parent vessel diameter
k	2.36	vessel index
ψ	1	asymmetry index
ρ_{fluid}	1060 kg/m³	plasma density and cytoplasmic density
ρ_{solid}	1090 kg/m³	RBC membrane density
λ_c	5	physiological viscosity ratio
λ	1, 3, 5, 7, 9	viscosity ratio
v	50, 100, 150 1/s	shear flow rate

2.4. Boundary Conditions

The numerical simulations were carried out by the commercial software COMSOL Multiphysics 5.6 on AMD EPYC 7452 64-Core @2.35GHz computer node processor. COMSOL Multiphysics is a largescale finite element analysis software for multi-physics simulation, which is widely used in various fields of scientific research and engineering project calculation. A fully developed velocity profile was set at the inlet, and the mean velocity was set to 0.26 cm/s, which refers to a similar vessel size [21,31]. The outlet pressure was set to 0 Pa. A moving mesh model was used to track the fluid deformation. Meshes are performed using an ALE mesh, where the Eulerian mesh traces the fluid, and the Lagrangian mesh describes the RBC membrane. RBCs undergo large deformations during movement, resulting in mesh distortion. Therefore, an automatic remeshing technology was adopted in the calculation when the mesh quality fells below a specified value of 0.2, and the calculation continued after the solver remeshed. The meshes of the RBC membrane were particularly refined. The numerical results are considered mesh-independent when the difference in the pressure drop between two consecutive simulations is less than 2%. Figure 1 shows the flow chart of numerical simulation framework in this study.

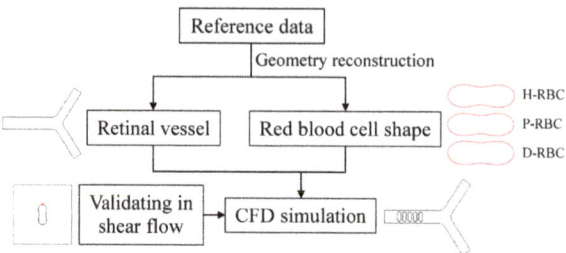

Figure 1. Numerical simulation framework.

The values of RBC morphology and blood properties used were taken from the published literature, as summarized in Table 2.

Table 2. Parameters used in the numerical simulation.

Parameters	H-RBC	P-RBC	D-RBC
Diameter (μm)	6.91 [26]	6.82 [26]	6.89 [26]
Thickness (μm)	2.36 [26]	2.4 [26]	2.68 [26]
Concave depth (μm)	0.43 [26]	0.37 [26]	0.143 [26]
μ_{plasma} (Pa·s)	0.00128 [32]	0.00131	0.00133 [32]
E (kPa)	1820 [8]	2020	2520 [8]
τ (s)	0.1 [33,34]	0.08 [35]	0.07 [35]

3. Results

In the present study, we simulated the dynamic behavior of H-RBC, P-RBC, and D-RBC in shear flow based on a finite element ALE method with a moving mesh, to validate the feasibility of calculating RBC motion and explore the influence of shear rate v and viscosity ratios λ on RBC motion. Subsequently, we simulated RBC motion in a retinal bifurcating arteriole to examine the effects of RBC morphology and physical parameters on the hemodynamics of the bifurcating vessels.

3.1. Dynamic Behavior of RBC in Shear Flow

RBC in shear flow has identified two major dynamic behaviors: tank-tread (TT) motion and tumbling (TB) motion (Figure 2). It can be observed that the TT motion of the marker point and TB motion in D-RBC are the fastest. In this study, we focused on the form of RBC motion at different viscosity ratios and different shear rates and compared the differences among the three groups of RBC motion. Based on the previous studies, we simulated RBC motion in shear flow at shear rates in $v = 50, 100, 150$ 1/s corresponding to viscosity ratios $\lambda = 1, 3, 5, 7, 9$ [36,37]. The shear flow region was 20 μm × 20 μm, and the RBC was located at the center. The upper and lower walls impose opposite velocities but the same magnitude to form the shear flow, and the left and right sides are period boundaries.

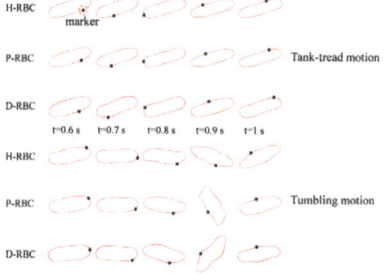

Figure 2. Tank-tread (TT) motion (**top**) and tumbling (TB) motion (**bottom**).

Under the shear force, when $\lambda \prec \lambda_c$, the RBC begins to rotate, after rotating to a certain inclination angle, it maintains a stable state. The RBC membrane undergoes TT motion. The concave point of RBC gradually disappears, and its shape changes irregularly, Subsequently, it stretches along the inclination direction [38]. Figure 3a provides the inclination angle of three groups of RBCs in shear flow when $\lambda = 1, 3$ at $v = 50\ 1/s$. When $\lambda \geq \lambda_c$, the RBC only undergoes TT motion in shear flow. Notably, with λ increasing slowly, the inclination angle increases [39–41]. This agrees with the previous studies [39] that the form of RBC motion is highly dependent on λ. Figure 3b shows the angle trajectory of the marker point on RBC membrane over time at $v = 50\ 1/s$. Notably, it can be found that lower λ has a short period, D-RBC moves a little faster than P-RBC and H-RBC. Figure 3c shows TT frequency of three groups of RBCs increases linearly with increasing v [35]. The slope of D-RBC is the largest, followed by P-RBC.

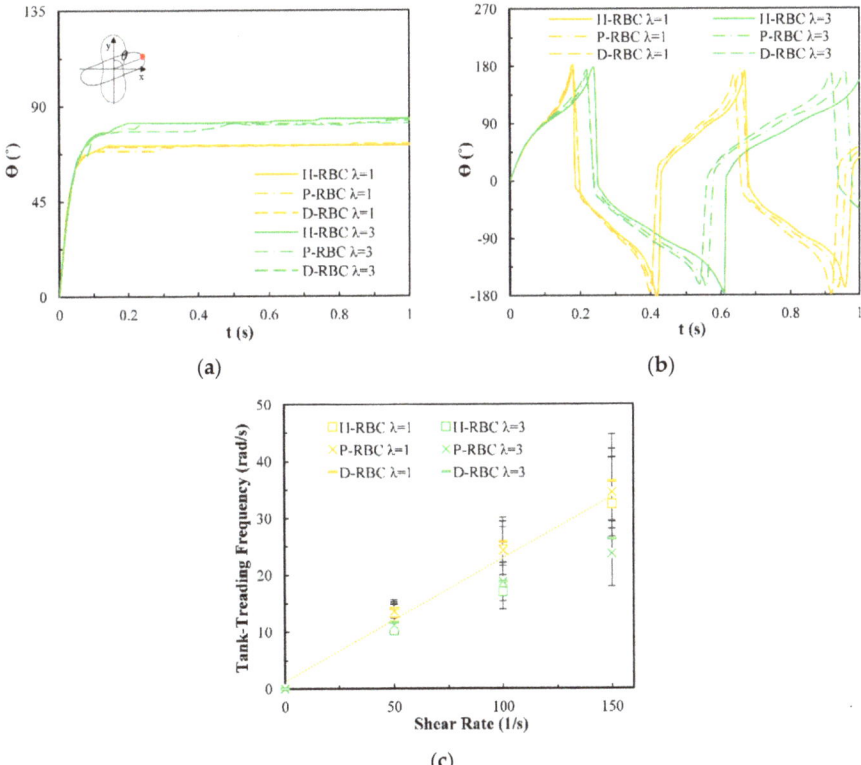

Figure 3. TT motion for $\lambda = 1, 3$ at $v = 50\ 1/s$. (a) Inclination angle in shear flow $\lambda = 1, 3$ of RBC; (b) trajectories of the marked point on the RBC membrane over time; (c) functional dependence of RBC TT frequency with respect to v.

Figure 4a shows the variation in the TB angle with time at $v = 50\ 1/s$. During $\lambda \geq \lambda_c$, the RBC mainly undergo TB motion. The rotation speed of the D-RBC was greater than that of the P-RBC and H-RBC. Higher λ has a faster TB speed and shorter TB period for all RBCs. We counted TB frequency at $v = 50, 100, 150\ 1/s$, as shown in Figure 4b. The TB frequency increased linearly with increasing v. In contrast to TT motion, greater λ has a greater TB frequency. The TB frequency in D-RBC was higher than that in H-RBC and P-RBC. It can be clearly seen that the TB frequency of the D-RBC is the largest at any λ value. This is expected because the RBC shape and physical parameters have changed significantly, which makes a big difference to RBC motion.

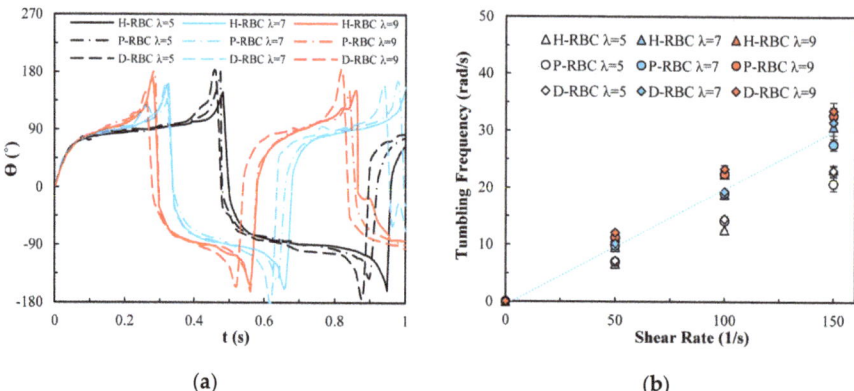

Figure 4. TB motion for $\lambda = 5, 7, 9$ at $v = 50$ 1/s. (**a**) TB angle of RBC at $\lambda = 5, 7, 9$ over time; (**b**) functional dependence of TB frequency with respect to v.

3.2. Dynamics of an RBC in Flowing through a Retinal Bifurcating Vessel

In this section, we studied the dynamics and rheology of three groups of RBCs through the retinal bifurcating vessels at $\lambda = \lambda_c = 5$, as shown in Figure 5. The RBC is released along the center of the vessel and gradually adopts the shape of a parachute in the parent vessel. After flowing into a child vessel with a diameter smaller than that of RBC, the RBC stretches to flow in an elongated bullet shape. As the RBC approaches the bifurcation, it immediately folds in the opposite direction, and we have observed that the RBC lingers for some time at the bifurcation.

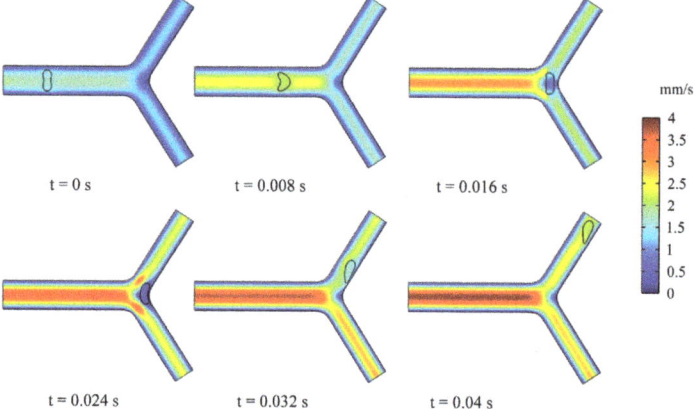

Figure 5. Snapshots of an RBC motion and deformation at different moments in the retinal bifurcating vessel.

To examine the influence of RBCs on the flow field, we extracted the velocity distributions at moments t1 and t2 of the a-a section near the bifurcation of the vessel. Figure 6 presents the velocity profile distribution of the a-a section at t1 and t2 moments (t1 < t2). The velocity profile in D-RBC is blunt compared to the other groups at the t1 moment shown in Figure 6a, while after RBC has passed through the a-a section, the velocity profiles in H-RBC and P-RBC are blunt compared to those in D-RBC, as shown in Figure 6b. This indicates that the RBC shape and physical parameters have little effect on the flow field during a single RBC moving. However, the redistribution of RBC may affect the flow field after approaching the bifurcation.

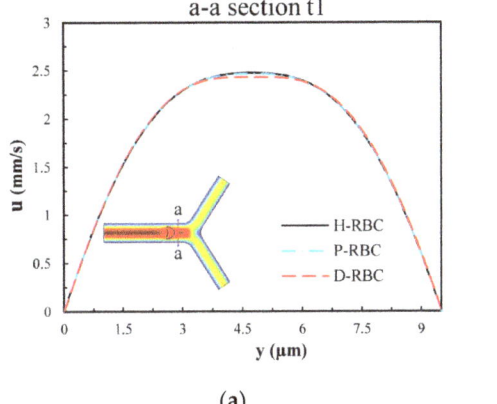
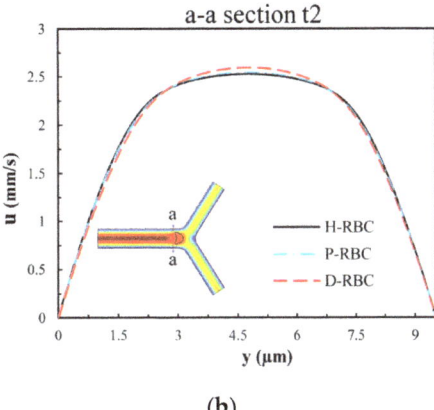

Figure 6. The velocity profile distribution of the a-a section: (**a**) when the RBC does not pass through the a-a section but close to it at the moment, t1; (**b**) when the RBC passes through the a-a section at the moment, t2.

To take lingering into account, we refer to the definition of the lingering phenomenon of RBCs at the bifurcation in reference [42], indicating that the RBC lingers at the bifurcation if RBC velocity decreases severely near the bifurcation. The velocity of the three groups of RBCs increased slowly in the parent vessel, and close to the bifurcation, the RBC velocity decreased rapidly for some time, as shown in Figure 7a. The D-RBC velocity decreased more at the bifurcation as well as at the child vessel. Therefore, we withdrew the periods when the RBC velocity was less than 30 μm/s to quantify the lingering time of RBCs in each group, as shown in Figure 7b. The P-RBC lingering time was slightly longer than H-RBC, and the D-RBC lingering time was the longest. Thus, we predict that the biconcave shape of RBC decreases and the viscosity and stiffness of RBC increases, which may increase the resistance of vessel.

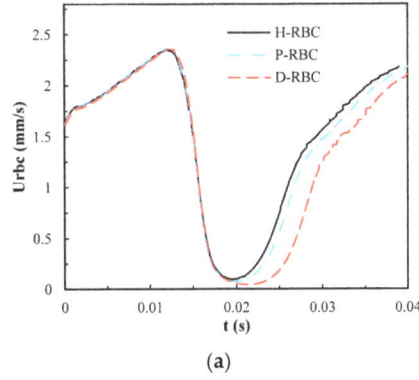
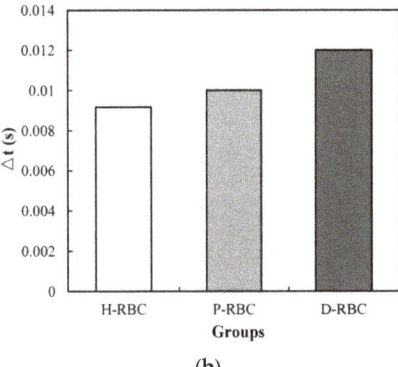

Figure 7. (**a**) The RBC velocity distribution in the retinal bifurcating vessel; (**b**) RBC lingering time in the retinal bifurcating vessel.

Due to the lingering of RBCs at the bifurcation, we quantify the pressure drop changes during an RBC moving in the retinal bifurcating vessel over time, as shown in Figure 8. The pressure drop of three groups of RBCs is basically the same when the RBC moves in the parent vessel, whereas the pressure drop increases sharply when the RBC closes the bifurcation. Consequently, the pressure drop remains slightly increasing for a period of time at the bifurcation, when the RBC flows out of the bifurcation, and the pressure drop decreases slowly and increases slightly in the child vessel. The pressure drop in the D-RBC is greater than that in the H-RBC and P-RBC during RBC motion, which means that the D-RBC leads to a higher resistance at the bifurcation during motion.

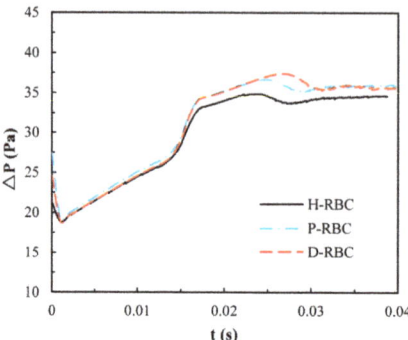

Figure 8. The pressure drop in the retinal bifurcating vessel.

3.3. Dynamics of Multi-RBCs in Flowing through a Retinal Bifurcating Vessel

To examine the effect of multiple RBCs motions on the flow field, we simulated the dynamic behavior of five RBCs in the retinal bifurcating vessel at $\lambda = \lambda_c = 5$. All RBCs were placed in the center of the parent vessel at an intercellular distance of 4 µm, as shown in Figure 9. When the RBCs reached the bifurcation, the preceding RBC was observed to undergo deformation. It is worth noting that the deformation of the preceding RBC affects the deformation of the following RBC, and the preceding RBC stays at the bifurcation for a long time, which can cause the next RBC to stack at the bifurcation and block the entrance of a child vessel. The deflection of the mass center determines which child branches the next RBC enters.

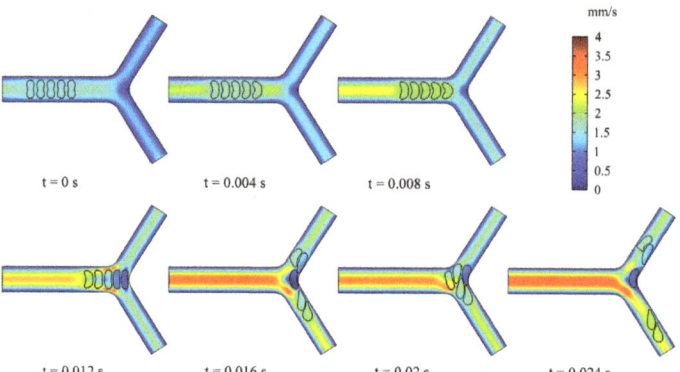

Figure 9. Snapshots of multi-RBCs motion and deformation at different moments in the retinal bifurcating vessel.

Figure 10 presents the velocity distribution of the b-b section at t1 and t2 moments (t1 < t2). The velocity profile is blunted owing to the approach of the RBCs at the t1 moment in Figure 10a, while there is essentially no difference in the velocity distribution among the three groups of RBCs. Figure 10b shows the velocity distributions of section b-b when all RBCs have just passed through the b-b section at the moment, t2. For the H-RBCs, the velocity profile has two peaks, which are indicative of a relatively equal distribution of H-RBCs at the bifurcation. We predicted that this would result in a more homogenous distribution of the RBC flux in the child vessel. Conversely, the P-RBCs and D-RBCs have enhanced aggregation at the bifurcation, causing the velocity profile to be inclined toward the wall. This indicates that there is a more uneven RBC distribution in DM and hyperglycemia at the bifurcation, which may cause unequal redistribution of RBCs in the child vessel, resulting in large heterogeneity in the vessel hematocrit and a reduction in the average capillary discharge hematocrit.

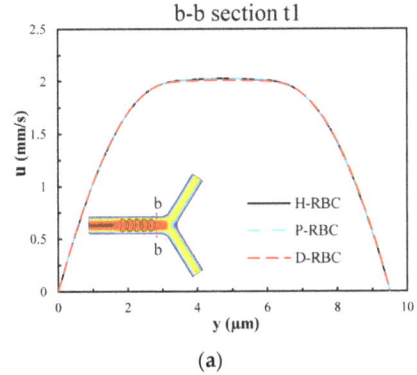
(a)

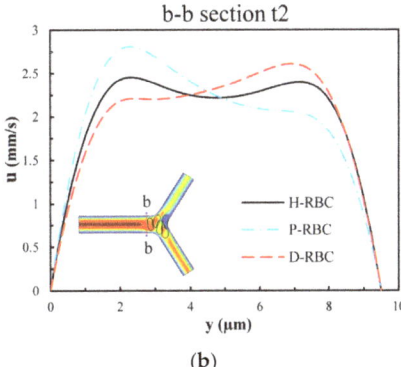
(b)

Figure 10. The velocity distribution of the b-b section: (**a**) when RBCs do not pass through the b-b section but close to it at the t1 moment; (**b**) when RBCs have just passed through the b-b section at the t2 moment.

Previously we observed that one RBC at the bifurcation causes an increase in pressure drop, whereas one RBC is far from the physiological situation. Therefore, in this study, we quantified the change in pressure drop when multiple RBCs move in the vessel. Figure 11 shows the distribution of the pressure drop in the vessel during RBC motion. The pressure drop in the parent was essentially the same for all the groups. When RBCs flowed into the bifurcation, the pressure drop increased sharply and then decreased sharply at the bifurcation. The largest pressure drop for D-RBCs is 167.67 Pa at the bifurcation, while that for H-RBCs is 155.82 Pa, implying that in a specific vessel, D-RBC causes a higher flow resistance compared to H-RBC.

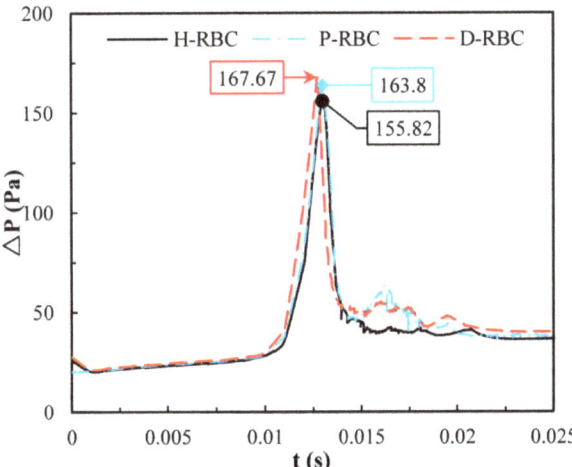

Figure 11. The pressure drop in the retinal bifurcating vessel.

4. Discussion

First, we analyzed the effects of λ and v in shear flow on the dynamic behavior of RBC. When $\lambda \prec \lambda_c$, RBCs perform TT motion, and when $\lambda \geq \lambda_c$, RBCs mainly perform

TB motion. We found that the TT motion and TB motion speed of the D-RBC are greater than those of the H-RBC, which is consistent with the results of previous experiments [43]. The TT motion perimeter becomes longer as the RBC is stretched in shear flow, while the biconcave shape and diameter of the D-RBC also decreased to some extent, which led to a slightly greater TT frequency in the D-RBC than in the other two groups. In addition, the slowing down of the TT motion with an increasing viscosity ratio is due to the increased viscosity in the cytoplasm, which slows down the rotation of the cytoplasm and motion of the RBC membrane [44].

As the cytoplasmic viscosity increases further, a shift from the TT motion to TB motion can be induced. This is due to the fact that as the viscosity increases, the transfer of shear torque to the membrane becomes increasingly difficult, at which point the RBC behaves like a near rigid solid and shows TB motion [45]. The larger the λ, the faster the TB frequency. It is closer to a rigid particle, which leads to a higher TB frequency as the viscosity of the RBC increases. It is possible that increased viscosity makes it less likely to flow, and a decrease in the RBC deformability leads to a significant increase in the TB motion [46,47].

Second, we analyzed the velocity, pressure drop, and lingering time of the RBC in the retinal bifurcating vessel. Our simulations revealed that the transit time of the H-RBC throughout the retinal bifurcating vessel was lesser than that of the D-RBC and P-RBC. This difference mainly occurs near the bifurcation and downstream of the bifurcation. Peduzzi et al. [48] showed that blood rheology disorders can cause capillary and post-capillary small venous stagnation in patients with DM and that stagnation is a major cause of blindness in patients with DM. The increased lingering time of RBC at the bifurcation decreases the blood flow in the retinal vessels, and the stagnation of microcirculatory blood flow leads to local hypoxia and lactic acidosis, which in turn causes microvascular damage [49]. Related studies have shown that a concomitant increase in blood viscosity caused by hyperglycemia [50] and increased RBC viscosity in DM leads to decreased RBC deformability and mobility, which increases the intrinsic resistance to blood flow in the vasculature [51,52]. The transit time of RBC increases with increased blood glucose [53], indicating that the retinal vasculature is susceptible to hyperglycemic injury, and the presence of hyperglycemia has given importance by patients.

In addition, we found that the local pressure drop at the bifurcation increased and the overall pressure drop in the D-RBC and P-RBC flowing in the vessel was greater than that in the H-RBC. This is due to the lingering of RBC at the bifurcation, which leads to a decrease in flow into the child vessel and an increased pressure near the entrance of the child vessel. Therefore, a sharp rise in the pressure drop near the bifurcation, leads to a significant increase in the resistance to flow, which is consistent with the findings of Peter Balogh [13]. In the DM state, increased glycosylation, cholesterol, and oxidative stress may lead to swelling of RBCs with impaired deformability [54]. Thus, the combination of the reduced biconcave shape of the D-RBC and physical parameters leads to a higher local pressure drop than that in the H-RBC. Several qualitative studies have been performed to study the effects of RBC physical parameters on hemodynamics. We additionally investigated the effect of RBC shape on hemodynamics (the results are not yet published). Keeping other parameters constant, the reduced biconcave shape of RBC still increases the transit time and flow resistance, indicating that the hemodynamic parameters are highly sensitive to RBC morphology and that the alteration of RBC morphology is a potential marker of microvascular disease.

Finally, we analyzed the velocity and pressure drop of multiple RBCs in a bifurcating retinal vessel. The pressure drop induced by D-RBCs at the bifurcation is significantly greater than that induced by H-RBCs, indicating that D-RBCs induce greater resistance to the flow in retinal vessels, which may reflect more pronounced hemodynamic disturbances in retinal vessels in DM. The enhanced aggregation of D-RBCs and P-RBCs at the bifurcation, increased lingering time, and decreased RBC deformability resulted in a dramatic increase in local pressure drop, and created greater flow resistance, which is the result of a combination of reduced RBC biconcave shape, increased viscosity, and RBC membrane stiffness. This suggests that the DM state is more prone to vascular occlusion, leading to local hypoxia. Kiani et al. [55] simulated flowing deformable disc-shaped particles and found an order-of-magnitude increase in the pressure gradient at a bifurcation. P-RBCs also caused a certain degree of rising pressure drop relative to H-RBCs. Previous studies have shown that a hyperglycemic state can induce the accumulation of oxidative stress, which promotes cellular damage and the development of DM complications [56,57]. As the blood glucose level increases, RBC aggregation is also enhanced [53]. Vekasi et al. [58] noted that chronic hyperglycemia may lead to disturbances in blood rheology, leading to further microcirculatory disturbances. D-RBCs exhibit an increased adhesion to the vascular endothelium at bifurcation [59], and RBCs and endothelial cells may act synergistically to severely impede retinal perfusion and trigger retinopathy. Previous studies have shown that a significantly higher number of endothelial cells at bifurcation occurs in patients with DM than in other patients [60]. This is due to the altered structure and function of RBCs, which can increase their aggregation and adhesion to endothelial cells [61]. Therefore, a decrease in the biconcave shape of RBC and an increase in the viscosity and RBC membrane stiffness may exacerbate the process of disease.

In this study, although the thickness of the RBC membrane was taken into account, the effect of the actual composition of the RBC membrane on its movement was not thoroughly considered. In addition, due to the limitation of computing resources, the calculated retinal blood vessel is a symmetric idealized model, and the real microvessel network with the development of the disease has not been established. In the 3D modeling work of biomechanics of D-RBC by Chang et al. using spring network model [35], the computed frequency of the angular trajectory for the marked particle of the cell membrane is larger than ours, implying that the overall two-dimensional model can solely represent part of the RBC behaviors in microvessel network. In the future, a three-dimensional RBC continuum model should be developed in the next work. A foreseeable extension of the work would be to investigate the RBC movement and vasodilation function in the real retinal vascular network model extracted from available imaging data, so that more clinical supports can be used for comparison.

5. Conclusions

First, hemodynamic calculations were performed for H-RBCs, P-RBCs, and D-RBCs in shear flow, and the effects of RBC morphology and physical parameters on RBC dynamic behavior were investigated. The TT motion, TB motion, and linear relationship between TT frequency, TB frequency, and shear rate are consistent with the results of previous studies, indicating that the finite element method can successfully predict the complex dynamics of RBC in the shear flow under different periods. We found that abnormal variations in the morphological and physical parameters of RBCs may lead to diminished tank-tread motion and enhanced tumbling motion in shear flow.

Second, single and multiple RBC motions were simulated in the retinal bifurcation vessels to compare changes in hemodynamic parameters of H-RBCs, P-RBCs, and D-RBCs in the retinal bifurcation vessels. The results show that as DM progresses, the RBC lingering time becomes longer at the bifurcation, resulting in sharply increasing local pressure and a greater resistance to flow, leading to longer transit times of RBCs throughout the vessel, and which may promote further development of DM, and provoke microcirculatory complications.

Author Contributions: Conceptualization: Y.H. and L.L.; Investigation: L.L., H.C., Y.H., L.M. and Y.L.; Writing—original draft preparation: L.L., Y.H. and H.C.; Writing—review and editing: L.L., Y.H. and H.C.; Supervision: Y.H. All authors have read and agreed to the published version of the manuscript.

Funding: This work is supported by the National Natural Science Foundation of China (No. 51976026) and the Fundamental Research Funds of Central Universities of China (No. DUT22YG206).

Institutional Review Board Statement: The Institutional Review Board of Dalian University of Technology (DUTSEPE221007-01) approved the protocol of this study.

Informed Consent Statement: Not applicable.

Data Availability Statement: Correspondence and requests for materials should be addressed to L.L

Conflicts of Interest: The authors declare no conflict of interest.

References

1. Ikram, M.K.; Cheung, C.Y.; Lorenzi, M.; Klein, R.; Jones, T.L.Z.; Wong, T.Y.; Retinal, N.J.W. Retinal Vascular Caliber as a Biomarker for Diabetes Microvascular Complications. *Diabetes Care* **2013**, *36*, 750–759. [CrossRef]
2. Kavdia, M.; Popel, A.S. Venular endothelium-derived NO can affect paired arteriole: A computational model. *Am. J. Physiol. Heart Circ. Physiol.* **2006**, *290*, H716. [CrossRef]
3. Peng, Z.; Li, X.; Pivkin, I.V.; Dao, M.; Suresh, S. Lipid bilayer and cytoskeletal interactions in a red blood cell. *Proc. Natl. Acad. Sci. USA* **2013**, *110*, 13356–13361. [CrossRef] [PubMed]
4. Diez-Silva, M.; Dao, M.; Han, J.Y.; Lim, C.T.; Suresh, S. Shape and Biomechanical Characteristics of Human Red Blood Cells in Health and Disease. *MRS Bull.* **2010**, *35*, 382–388. [CrossRef] [PubMed]
5. Mohandas, N.; Evans, E. Mechanical Properties of the Red Cell Membrane in Relation to Molecular Structure and Genetic Defects. *Annu. Rev. Biophys. Biomol. Struct.* **1994**, *23*, 787–818. [CrossRef] [PubMed]
6. Brown, C.D.; Ghali, H.S.; Zhao, Z.H.; Thomas, L.L.; Friedman, E.A. Association of reduced red blood cell deformability and diabetic nephropathy—Reply. *Kidney Int.* **2005**, *67*, 2066–2067. [CrossRef]
7. Loyola-Leyva, A.; Loyola-Rodríguez, J.; Atzori, M.; Micron, F.G.J. Morphological changes in erythrocytes of people with type 2 diabetes mellitus evaluated with atomic force microscopy: A brief review. *Micron* **2017**, *105*, 11–17. [CrossRef] [PubMed]
8. Ciasca, G.; Papi, M.; Di Claudio, S.; Chiarpotto, M.; Palmieri, V.; Maulucci, G.; Nocca, G.; Rossi, C.; De Spirito, M. Mapping viscoelastic properties of healthy and pathological red blood cells at the nanoscale level. *Nanoscale* **2015**, *7*, 17030–17037. [CrossRef]
9. Jin, H.; Xing, X.B.; Zhao, H.X.; Chen, Y.; Huang, X.; Ma, S.Y.; Ye, H.Y.; Cai, J.Y. Detection of erythrocytes influenced by aging and type 2 diabetes using atomic force microscope. *Biochem. Biophys. Res. Commun.* **2010**, *391*, 1698–1702. [CrossRef]
10. AlSalhi, M.S.; Devanesan, S.; AlZahrani, K.E.; AlShebly, M.; Al-Qahtani, F.; Farhat, K.; Masilamani, V. Impact of Diabetes Mellitus on Human Erythrocytes: Atomic Force Microscopy and Spectral Investigations. *Int. J. Environ. Res. Public Health* **2018**, *15*, 2368. [CrossRef]
11. Pozrikidis, C. Finite Deformation of Liquid Capsules Enclosed by Elastic Membranes in Simple Shear-Flow. *J. Fluid Mech.* **1995**, *297*, 123–152. [CrossRef]
12. Wei, Y.; Mu, L.; Tang, Y.; Shen, Z.; He, Y. Computational analysis of nitric oxide biotransport in a microvessel influenced by red blood cells. *Microvasc. Res.* **2019**, *125*, 103878. [CrossRef] [PubMed]
13. Balogh, P.; Bagchi, P. Direct Numerical Simulation of Cellular-Scale Blood Flow in 3D Microvascular Networks. *Biophys. J.* **2017**, *113*, 2815–2826. [CrossRef] [PubMed]
14. Takahashi, T. (Ed.) A Theoretical Model for the Microcirculatory Network. In *Microcirculation in Fractal Branching Networks*; Springer: Tokyo, Japan, 2014; pp. 25–45.
15. Murray, C. The Physiological Principle of Minimum Work: I. The Vascular System and the Cost of Blood Volume. *Proc. Natl. Acad. Sci. USA* **1926**, *12*, 207–214. [CrossRef]
16. Zamir, M. Nonsymmetrical bifurcations in arterial branching. *J. Gen. Physiol.* **1978**, *72*, 837. [CrossRef]
17. Iberall, A.S. Anatomy and steady flow characteristics of the arterial system with an introduction to its pulsatile characteristics. *Math. Biosci.* **1967**, *1*, 375–395. [CrossRef]
18. Zamir, M. On fractal properties of arterial trees. *J. Theor. Biol.* **1999**, *197*, 517–526. [CrossRef]
19. Karch, R.; Neumann, F.; Neumann, M.; Schreiner, W. Staged growth of optimized arterial model trees. *Ann. Biomed. Eng.* **2000**, *28*, 495–511. [CrossRef]
20. Luo, T.; Gast, T.J.; Vermeer, T.J.; Burns, S.A. Retinal Vascular Branching in Healthy and Diabetic Subjects. *Investig. Ophth. Vis. Sci.* **2017**, *58*, 2685–2694. [CrossRef]
21. Takahashi, T. Oxygen Consumption by Vascular Walls in the Retinal Vasculature. In *Microcirculation in Fractal Branching Networks*; Springer: Tokyo, Japan, 2014; pp. 47–69.
22. Hochmuth, R.M.; Waugh, R.E. Erythrocyte-Membrane Elasticity and Viscosity. *Annu. Rev. Physiol.* **1987**, *49*, 209–219. [CrossRef]
23. Vayo, H.W.; Shibata, M.K. Numerical Results on Red-Blood-Cell Geometry. *Jpn. J. Physiol.* **1982**, *32*, 891–894. [CrossRef] [PubMed]
24. Nans, A.; Mohandas, N.; Stokes, D.L. Native Ultrastructure of the Red Cell Cytoskeleton by Cryo-Electron Tomography. *Biophys. J.* **2011**, *101*, 2341–2350. [CrossRef] [PubMed]
25. Heinrich, V.; Ritchie, K.; Mohandas, N.; Evans, E. Elastic thickness compressibilty of the red cell membrane. *Biophys. J.* **2001**, *81*, 1452–1463. [CrossRef]
26. Loyola-Leyva, A.; Loyola-Rodriguez, J.P.; Teran-Figueroa, Y.; Camacho-Lopez, S.; Gonzalez, F.J.; Barquera, S. Application of atomic force microscopy to assess erythrocytes morphology in early stages of diabetes. A pilot study. *Micron* **2021**, *141*, 102982. [CrossRef]

27. De Haan, M.; Zavodszky, G.; Azizi, V.; Hoekstra, A.G. Numerical Investigation of the Effects of Red Blood Cell Cytoplasmic Viscosity Contrasts on Single Cell and Bulk Transport Behaviour. *Appl. Sci.* **2018**, *8*, 1616. [CrossRef]
28. Bagchi, P.; Johnson, P.C.; Popel, A.S. Computational fluid dynamic simulation of aggregation of deformable cells in a shear flow. *J. Biomech. Eng.* **2005**, *127*, 1070–1080. [CrossRef] [PubMed]
29. Zhang, J.; Johnson, P.C.; Popel, A.S. Effects of erythrocyte deformability and aggregation on the cell free layer and apparent viscosity of microscopic blood flows. *Microvasc. Res.* **2009**, *77*, 265–272. [CrossRef] [PubMed]
30. Mohandas, N.; Groner, W. Cell-Membrane and Volume Changes during Red-Cell Development and Aging. *Ann. N. Y. Acad. Sci.* **1989**, *554*, 217–224. [CrossRef]
31. Tsukada, K.; Sekizuka, E.; Oshio, C.; Minamitani, H. Direct measurement of erythrocyte deformability in diabetes mellitus with a transparent microchannel capillary model and high-speed video camera system. *Microvasc. Res.* **2001**, *61*, 231–239. [CrossRef]
32. Biro, K.; Feher, G.; Vekasi, J.; Kenyeres, P.; Toth, K.; Koltai, K. Hemorheological Parameters in Diabetic Patients: Role of Glucose Lowering Therapies. *Metabolites* **2021**, *11*, 806. [CrossRef]
33. Baskurt, O.K.; Meiselman, H.J. Determination of red blood cell shape recovery time constant in a Couette system by the analysis of light reflectance and ektacytometry. *Biorheology* **1996**, *33*, 489–503. [CrossRef] [PubMed]
34. Hochmuth, R.M.; Worthy, P.R.; Evans, E.A. Red-Cell Extensional Recovery and the Determination of Membrane Viscosity. *Biophys. J.* **1979**, *26*, 101–114. [CrossRef]
35. Chang, H.Y.; Li, X.J.; Karniadakis, G.E. Modeling of Biomechanics and Biorheology of Red Blood Cells in Type 2 Diabetes Mellitus. *Biophys. J.* **2017**, *113*, 481–490. [CrossRef] [PubMed]
36. Fischer, T.M. Tank-tread frequency of the red cell membrane: Dependence on the viscosity of the suspending medium. *Biophys. J.* **2007**, *93*, 2553–2561. [CrossRef]
37. Shen, Z.Y.; Farutin, A.; Thiebaud, M.; Misbah, C. Interaction and rheology of vesicle suspensions in confined shear flow. *Phys. Rev. Fluids* **2017**, *2*, 103101. [CrossRef]
38. Fischer, T.; Schmidschonbein, H. Tank Tread Motion of Red-Cell Membranes in Viscometric Flow—Behavior of Intracellular and Extracellular Markers (with Film). *Blood Cells* **1977**, *3*, 351–365.
39. Fischer, T.M.; Korzeniewski, R. Angle of Inclination of Tank-Treading Red Cells: Dependence on Shear Rate and Suspending Medium. *Biophys. J.* **2015**, *108*, 1352–1360. [CrossRef]
40. Noguchi, H. Swinging and synchronized rotations of red blood cells in simple shear flow. *Phys. Rev. E* **2009**, *80*, 021902. [CrossRef]
41. Sugihara, M. Motion and deformation of a red blood cell in a shear flow: A two-dimensional simulation of the wall effect. *Biorheology* **1985**, *22*, 1–19. [CrossRef]
42. Kihm, A.; Quint, S.; Laschke, M.W.; Menger, M.D.; John, T.; Kaestner, L.; Wagner, C. Lingering Dynamics in Microvascular Blood Flow. *Biophys. J.* **2021**, *120*, 432–439. [CrossRef]
43. Mazzanti, L.; Faloia, E.; Rabini, R.A.; Staffolani, R.; Kantar, A.; Fiorini, R.; Swoboda, B.; Depirro, R.; Bertoli, E. Diabetes-Mellitus Induces Red-Blood-Cell Plasma-Membrane Alterations Possibly Affecting the Aging Process. *Clin. Biochem.* **1992**, *25*, 41–46. [CrossRef]
44. Dodson, W.R.; Dimitrakopoulos, P. Tank-treading of swollen erythrocytes in shear flows. *Phys. Rev. E* **2012**, *85*, 021922. [CrossRef] [PubMed]
45. Sui, Y.; Chew, Y.T.; Roy, P.; Cheng, Y.P.; Low, H.T. Dynamic motion of red blood cells in simple shear flow. *Phys. Fluids* **2008**, *20*, 112106. [CrossRef]
46. Chien, S.; Usami, S.; Dellenback, R.J.; Gregersen, M.I. Shear-Dependent Deformation of Erythrocytes in Rheology of Human Blood. *Am. J. Physiol.* **1970**, *219*, 136–142. [CrossRef] [PubMed]
47. Forsyth, A.M.; Wan, J.D.; Ristenpart, W.D.; Stone, H.A. The dynamic behavior of chemically "stiffened" red blood cells in microchannel flows. *Microvasc. Res.* **2010**, *80*, 37–43. [CrossRef] [PubMed]
48. Peduzzi, M.; Melli, M.; Fonda, S.; Codeluppi, L.; Guerrieri, F. Comparative-Evaluation of Blood-Viscosity in Diabetic-Retinopathy. *Int. Ophthalmol.* **1984**, *7*, 15–19. [CrossRef] [PubMed]
49. Grindle, C.F.; Buskard, N.A.; Newman, D.L. Hyperviscosity retinopathy. A scientific approach to therapy. *Trans. Ophthalmol. Soc. U K (1962)* **1976**, *96*, 216–219.
50. Fornal, M.; Lekka, M.; Pyka-Fosciak, G.; Lebed, K.; Grodzicki, T.; Wizner, B.; Styczen, J. Erythrocyte stiffness in diabetes mellitus studied with atomic force microscope. *Clin. Hemorheol. Microcirc.* **2006**, *35*, 273–276.
51. Tamariz, L.J.; Young, J.H.; Pankow, J.S.; Yeh, H.C.; Schmidt, M.I.; Astor, B.; Brancati, F.L. Blood viscosity and hematocrit as risk factors for type 2 diabetes mellitus: The atherosclerosis risk in communities (ARIC) study. *Am. J. Epidemiol.* **2008**, *168*, 1153. [CrossRef]
52. Lowe, G.D.O.; Lee, A.J.; Rumley, A.; Price, J.F.; Fowkes, F.G.R. Blood viscosity and risk of cardiovascular events: The Edinburgh Artery Study. *Brit. J. Haematol.* **1997**, *96*, 168–173. [CrossRef]
53. Babu, N.; Singh, M. Influence of hyperglycemia on aggregation, deformability and shape parameters of erythrocytes. *Clin. Hemorheol. Microcirc.* **2004**, *31*, 273–280. [PubMed]
54. Pelikanova, T.; Kohout, M.; Valek, J.; Base, J.; Stefka, Z. Fatty acid composition of serum lipids and erythrocyte membranes in type 2 (non-insulin-dependent) diabetic men. *Metabolism* **1991**, *40*, 175–180. [CrossRef]
55. Kiani, M.F.; Cokelet, G.R. Additional Pressure-Drop at a Bifurcation Due To the Passage of Flexible Disks in a Large-Scale Model. *J. Biomech. Eng.* **1994**, *116*, 497–501. [CrossRef]

56. Fiorentino, T.V.; Prioletta, A.; Zuo, P.; Folli, F. Hyperglycemia-induced Oxidative Stress and its Role in Diabetes Mellitus Related Cardiovascular Diseases. *Curr. Pharm. Des.* **2013**, *19*, 5695–5703. [CrossRef] [PubMed]
57. Amiya, E. Interaction of hyperlipidemia and reactive oxygen species: Insights from the lipid-raft platform. *World J. Cardiol.* **2016**, *8*, 689–694. [CrossRef] [PubMed]
58. Vekasi, J.; Marton, Z.; Kesmarky, G.; Cser, A.; Russai, R.; Horvath, B. Hemorheological alterations in patients with diabetic retinopathy. *Clin. Hemorheol. Microcirc.* **2001**, *24*, 59–64. [PubMed]
59. Wautier, J.L. Blood cell—Vessel wall interactions. *Clin. Hemorheol. Microcirc.* **1992**, *12*, 55–58. [CrossRef]
60. Cho, Y.I.; Mooney, M.P.; Cho, D.J. Hemorheological Disorders in Diabetes Mellitus. *J. Diabetes Sci. Technol.* **2008**, *2*, 1130. [CrossRef] [PubMed]
61. Moon, J.S.; Kim, J.H.; Kim, J.H.; Park, I.R.; Lee, J.H.; Kim, H.J.; Lee, J.; Kim, Y.K.; Yoon, J.S.; Won, K.C.; et al. Impaired RBC deformability is associated with diabetic retinopathy in patients with type 2 diabetes. *Diabetes Metab.* **2016**, *42*, 448–452. [CrossRef]

Quasi-Static Mechanical Properties and Continuum Constitutive Model of the Thyroid Gland

Peng Su [1], Chao Yue [1], Likun Cui [2], Qinjian Zhang [1], Baoguo Liu [2,*] and Tian Liu [1,*]

1 College of Mechanical and Electrical Engineering, Beijing Information Science and Technology University, Beijing 110192, China
2 Key Laboratory of Carcinogenesis and Translational Research (Ministry of Education/Beijing), Department of Head & Neck Surgery, Peking University Cancer Hospital & Institute, Beijing 100142, China
* Correspondence: lbg29@163.com (B.L.); liutian@bistu.edu.cn (T.L.); Tel.: +86-010-13910799925 (B.L.); +86-010-82426933 (T.L.)

Abstract: The purpose of this study is to obtain the digital twin parameters of the thyroid gland and to build a constitutional model of the thyroid gland based on continuum mechanics, which will lay the foundation for the establishment of a surgical training system for the thyroid surgery robot and the development of the digital twin of the thyroid gland. First, thyroid parenchyma was obtained from fresh porcine thyroid tissue and subjected to quasi-static unconfined uniaxial compression tests using a biomechanical test platform with two strain rates ($0.005\ \text{s}^{-1}$ and $0.05\ \text{s}^{-1}$) and two loading orientations (perpendicular to the thyroid surface and parallel to the thyroid surface). Based on this, a tensile thyroid model was established to simulate the stretching process by using the finite element method. The thyroid stretching test was carried out under the same parameters to verify the validity of the hyperelastic constitutive model. The quasi-static mechanical property parameters of the thyroid tissue were obtained by a quasi-static unconstrained uniaxial compression test, and a constitutional model that can describe the quasi-static mechanical properties of thyroid tissue was proposed based on the principle of continuum media mechanics, which is of great value for the establishment of a surgical training system for the head and neck surgery robot and for the development of the thyroid digital twin.

Keywords: thyroid; biomechanics; constitutive model; hyperelasticity

1. Introduction

Thyroid cancer is a malignant tumor with a high clinical incidence, and its incidence has been increasing year by year in recent years. Surgery is a common treatment modality [1]. With the development of surgical robotics, more and more surgical robots are used in thyroid surgery; however, traditional surgical robot training has the challenges of a high cost; long lead time; high risk; difficult quantitative evaluation; and high reliance on cadavers, silicone models, and animals [2,3]. Supported by virtual reality, artificial intelligence, human–computer interaction, digital twin, modern medicine, and other technologies, establishing virtual operating area models, carrying out virtual surgery training, realizing digital simulation of the soft tissue structure and function from microscopic to macroscopic, and complete presentation of intraoperative tissue and organ morphology, rheological models are the key breakthroughs for solving the above problems.

Virtual surgical training, based on interaction and feedback, has significant advantages over traditional surgical robot training, which relies on physical objects, in terms of reducing risk costs and avoiding ethical and moral issues [4]. To establish soft tissue digital twins in virtual surgical training systems, it is urgent to understand the real mechanical properties of tissues and organs, as well as to provide accurate material constitutive models. Therefore, it is of great importance to fully understand the quasi-static mechanical properties of the thyroid gland and to develop corresponding material constitutive models.

Yamada H. [5] and Thibault L.E. et al. [6] earlier focused on the mechanical properties of soft tissue organs and studied the mechanical properties of organ soft tissues using rabbits and primates, respectively. In the last two decades, surgical robots have developed more rapidly, and their safety assessment has received more and more attention and more experimental tools have been used. Several researchers have investigated the mechanical properties of soft tissues using compression tests [7–9], tensile tests [10,11], and indentation tests [12,13], and there are some differences in the results obtained under different testing conditions or experimental specifications. In addition, some indirect measurement methods based on imaging techniques, such as magnetic resonance elastography (MRE) [14] and ultrasound imaging [15], have also been applied to study the mechanical properties of soft tissues. Most of these direct or indirect testing methods are for soft tissues with a larger morphology, such as liver, muscle, and tracheal hose, and mechanical properties studies on the thyroid gland have not been reported.

Biomechanical modeling of soft tissues is one of the applications of rheological models. The functional relationship between the stress tensor and strain tensor is usually referred to as constitutive model, which is a mathematical model reflecting the macroscopic properties of the material. Zahra et al. [16] proposed a hyperelastic constitutive model to describe the behavior of soft tissue (as an isotropic homogeneous material) based on continuum media mechanics. Yang et al. [17] proposed a porous hyperelastic model using the shear wave elastography (SWE) technique and, based on this constitutive model, derived the relationship between the wave velocity and the solid matrix deformation generated by the parameters of the constitutive model and the internal pressure. There are less experimental data and studies on the mechanical properties of thyroid tissues in previous studies, and there is no consultable constitutive model of the thyroid that can correspond to the experimental results.

In this paper, we investigated the quasi-static mechanical properties of porcine thyroid tissue using a biomechanical test platform equipped with Nano25 high-precision sensors, and designed quasi-static unconfined uniaxial compression tests with strain rates of $0.005\ \text{s}^{-1}$ and $0.05\ \text{s}^{-1}$, and loading directions perpendicular to and parallel to the thyroid surface. The effects of different strain rates and loading directions on the mechanical properties of porcine thyroid tissue were investigated. Based on this, a constitutive model was developed to describe the quasi-static mechanical properties of the porcine thyroid gland based on the theory of a hyperelastic constitutive model of a rubber-like material.

2. Materials and Methods

2.1. Specimen Preparation

Out of respect and adherence to ethical norms, animal tissues and organs are usually used as substitutes to approximate the mechanical properties of human tissues and organs, and this result will be influenced by the animal species [18], and the reference significance of choosing animals with mechanical properties closer to those of human tissues and organs for the study will be greater. Numerous scholars have tested the mechanical properties of the soft tissues of pigs and then studied the mechanical properties of human soft tissues [9,19,20]. Therefore, in this paper, the porcine thyroid gland was chosen as a substitute for studying the mechanical properties of the human thyroid gland.

The thyroid gland used in the experiment was obtained from 10 adult Landrace pigs, and the whole neck was cut and separated after slaughter and transported to the laboratory at a constant temperature of 4 °C. In order to maintain the moisture of the organ tissues, saline was sprayed regularly during the preparation of specimens and experiments after delivery to the laboratory; meanwhile, to minimize the effect of post-mortem time on tissue mechanics, each experiment was controlled within 8 h after slaughter.

Isolation of the thyroid gland from the neck of a pig is a prerequisite for specimen preparation, a process that requires the experimenter to be able to accurately identify the thyroid gland to be isolated from the complex tissue structure of the pig neck. The location of the porcine thyroid gland is similar to that of the human body and is located at the junction of the larynx and trachea, approximately anterior to the second to fourth tracheal cartilage rings. The porcine thyroid samples in the experiment were taken from the parenchymal portion of the thyroid gland. The thyroid gland was first identified as described above, and was detached from the cervical trachea and surrounding tissues and muscles with a scalpel. A total of 10 thyroid glands were isolated, each weighing approximately 50 g. Then, a rectangular sample strip with a thickness of about 6 mm was cut out using a double-row tool, and then annular drilling tools were used to drill a circular sample strip of the thyroid gland with a diameter of about 10 mm in two directions: perpendicular to the thyroid surface and parallel to the thyroid surface, respectively, as shown in Figure 1. The prepared specimen was in the form of a cylinder with a diameter of 10 mm and a height of 6 mm. Large defects in the sample should be avoided during the preparation process, such as the inclusion of blood vessels, apparent unevenness of the upper and lower surfaces, and other parts. The original dimensions of the thyroid sample strips were measured using digital vernier calipers to measure their length, width, and thickness several times, and the average value was taken as the initial dimensions of the sample strips.

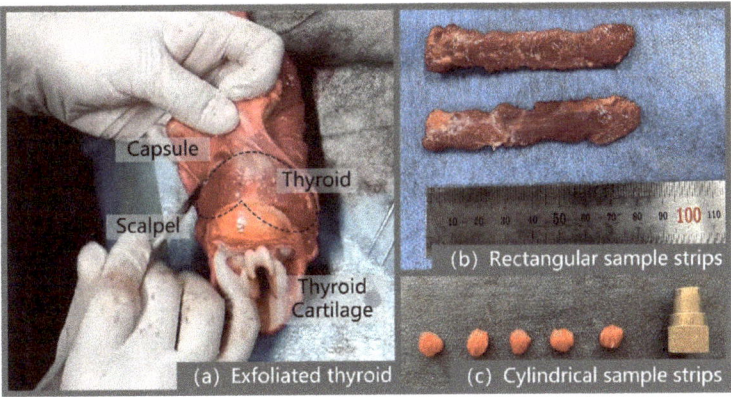

Figure 1. Diagram of the process for preparing thyroid specimens.

2.2. Uniaxial Compression Experiment

In the study of soft tissue biomechanical properties, multi-axial tensile and compression experiments are often designed. In the article, an unrestricted compression test under quasi-static conditions was designed to study the quasi-static mechanical properties of the thyroid gland. The reason for this is that considering the small size of the thyroid gland, it was difficult to prepare suitable samples for multiaxial stretching. To try to compensate for the shortcomings of the uniaxial experiments, the article prepared experimental samples from perpendicular to the surface of the porcine thyroid gland and parallel to the surface of the porcine thyroid gland, respectively, and the samples were taken from the parenchymal part of the porcine thyroid gland, avoiding the blood vessels and uneven parts of the thyroid gland as much as possible. The unconfined compression experimental method is one of the common means to obtain the mechanical response of soft tissues [18], which usually uses a compression platform to compress soft tissue samples with a certain regular shape, and the contact surface between the platform and the soft tissue sample is in a free state during the compression process, and ensures that the soft tissue sample can be deformed freely along the radial direction as much as possible.

The thyroid gland quasi-static compression experiments were performed on a built biomechanical test platform, as shown in Figure 2, which was equipped with a Nano25 high-precision six-dimensional force/moment transducer with a high accuracy and high stiffness, and the experimental strain rate was set at $0.005\ \text{s}^{-1}$ and $0.05\ \text{s}^{-1}$, in accordance with the requirements of the quasi-static tests. As the elasticity of the biological soft tissues, such as the thyroid gland, mainly comes from changes in entropy, and there is no unique natural state for it [8], in order to achieve a relatively stable state for the samples, the prepared samples were pretreated before conducting the experiments, i.e., 10 load–unload cycles of compression were performed with a compression load of 3 N each time, and the samples were left to stand for 200 s after the pretreatment as the recovery time of the samples. After the recovery phase, the thyroid sample was placed in the center of the compression lower platform, and the compression table was controlled to compress the sample at 0.03 mm/s until the strain was greater than 25%, and the parameters of pressure, displacement and time during this process were collected at 125 Hz. Ten sets of experiments were conducted for both samples (vertical thyroid surface and parallel thyroid surface) according to the above test method. During the above experiments, a small amount of saline was dropped on the sample surface at regular intervals in order to avoid excessive water loss from the sample. In addition, to meet the needs of unrestricted compression, a layer of vegetable oil (vegetable oil processed from GM soybeans with a viscosity (E020 °C) of 8.5) was brushed on the surface of the two indenters in contact with the specimen.

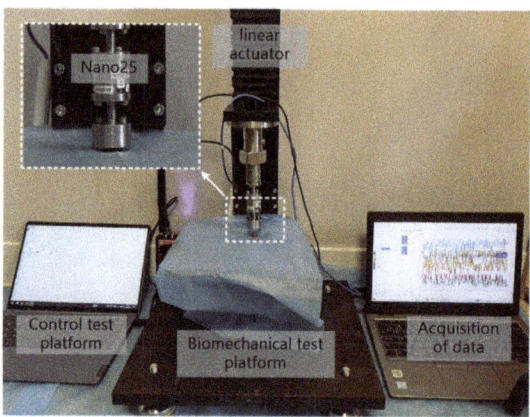

Figure 2. Uniaxial compression test of the thyroid gland.

2.3. Hyperelastic Modeling

A constitutive model based on strain energy functions is commonly used to describe the hyperelastic properties of rubber-like materials, and several commonly used strain energy density functions are provided in some finite element software, such as the Mooney–Rivlin model [21,22], the Yeoh model [23,24], the Ogden model [25], and the Neo-Hookean model [26]. Based on these commonly used strain energy density functions, or with appropriate modifications, some researchers have developed constitutive models describing the hyperelastic characteristics of various types of biological soft tissues and other materials [27–29], and the article will also develop a constitutive model for hyperelasticity of the thyroid tissue under low strain rate conditions by choosing an appropriate strain energy density function.

According to the finite deformation theory, a material point is located at X before deformation and it is located at x after deformation, then the deformation gradient is

$$F = \partial x / \partial X \tag{1}$$

In the case of finite deformation, the deformation of the material can be described by the right Cauchy–Green deformation tensor C. The right Cauchy–Green tensor C is expressed by the following equation

$$C = F^T F \tag{2}$$

The effect of tissue anisotropy on the model cannot be ignored in the modeling of biological soft tissues. Giannokostas et al. [30] included the influence of the anisotropy of the parts of the arteries in the modeling of blood vessels, which greatly improved the accuracy of the model; in the modeling of organs such as the liver, kidney, and spleen [8,9,29], they are usually considered as an isotropic material to study. In the article, it is considered that the thyroid gland is a gland with internal capillaries, capsules, and other structures, similar to organs in terms of structure and somewhat different from blood vessels. In order to simplify the model as much as possible and reduce the computational effort of the model, the thyroid gland is approximated as an isotropic material.

The three invariants, I_1, I_2 and I_3, of the Cauchy–Green deformation tensor C for an isotropic material are denoted as

$$\begin{cases} I_1 = \text{tr}(C) = \lambda_1^2 + \lambda_2^2 + \lambda_3^2 \\ I_2 = \frac{1}{2}[I_1^2 - \text{tr}(C^2)] = \lambda_1^2\lambda_2^2 + \lambda_2^2\lambda_3^2 + \lambda_3^2\lambda_1^2 \\ I_3 = \det(C) = \lambda_1^2\lambda_2^2\lambda_3^2 \end{cases} \quad (3)$$

where λ_1, λ_2, and λ_3, denote the three principal elongations. When the material is incompressible, $I_3 = 1$. For incompressible isotropic materials, the strain energy density function is usually a function of the strain invariants I_1 and I_2.

The Yeoh model [23] is a cubic strain energy function, which is able to describe the materials in which the shear modulus changes with deformation, and the parameters obtained from the experimental fitting of some simple deformation can predict the mechanical behavior of other deformation cases, and the applicable deformation range is also wide enough to simulate large deformations. The model contains only the invariant variable I_1, and the strain energy density function is

$$W = C_{10}(I_1 - 3) + C_{20}(I_1 - 3)^2 + C_{30}(I_1 - 3)^3 \quad (4)$$

where, C_{10}, C_{20}, and C_{30} are the material parameters.

For incompressible hyperelastic materials, the principal Cauchy stress is usually determined by the following equation

$$\sigma_i^e = \lambda_i \frac{\partial W}{\partial \lambda_i} - p^e (i = 1, 2, 3) \quad (5)$$

where p^e is the hydrostatic pressure of the hyperelastic material.

Substituting Equations (3) and (4) into Equation (5), the principal Cauchy stress can be expressed as

$$\sigma_i^e = 2\lambda_i^2\left[C_{10} + 2C_{20}(I_1 - 3) + 3C_{30}(I_1 - 3)^2\right] - p^e \quad (6)$$

For the uniaxial compression test, let $\lambda_1 = \lambda$ denote the elongation in the loading direction and $\sigma_1^e = \sigma^e$ denote the first principal Cauchy stress, under the assumption of incompressibility, we have

$$\begin{cases} \lambda_2 = \lambda_3 = \lambda^{-1/2} \\ \sigma_2^e = \sigma_3^e = 0 \end{cases} \quad (7)$$

Substituting the above equation into Equation (6), we have

$$\sigma^e = 2\lambda^2\left[C_{10} + 2C_{20}(I_1 - 3) + 3C_{30}(I_1 - 3)^2\right] - p^e \quad (8)$$

$$0 = 2\lambda^{-1}\left[C_{10} + 2C_{20}(I_1 - 3) + 3C_{30}(I_1 - 3)^2\right] - p^e \quad (9)$$

Combining the above two equations with Equations (3) and (7) yields

$$\sigma^e = 2\left(\lambda^2 - \lambda^{-1}\right)\left[C_{10} + 2C_{20}\left(\lambda^2 + 2\lambda^{-1} - 3\right) + 3C_{30}\left(\lambda^2 + 2\lambda^{-1} - 3\right)^2\right] \quad (10)$$

In the above equation, the principal elongation λ can be expressed by $\lambda = 1 + \varepsilon$, and ε is the engineering strain; the principal Cauchy stress σ^e can be expressed by $\sigma^e = T^e \lambda$, and T^e is the engineering stress.

The engineering stress–strain curve obtained from the quasi-static compression test with a strain rate of $0.005\ \text{s}^{-1}$ was selected for the fitting, and parameters C_{10}, C_{20}, and C_{30} were obtained.

3. Results and Discussion

3.1. Quasi-Static Mechanical Properties

In the quasi-static compression test of the porcine thyroid, the directly obtained data are the loading force, loading velocity, time, etc. The corresponding calculations using the directly obtained raw data allow for the analysis of stress–strain, stress–elongation, stress–time, strain–time, and other relationships. The relevant parameters are calculated as follows

$$\sigma = \frac{F}{A} \quad (11)$$

$$\varepsilon = \frac{L - L_0}{L_0} = \frac{\Delta L}{L_0} \quad (12)$$

where σ is the stress in MPa; F is the loading force applied to the soft tissue collected by the transducer in N; A is the initial cross-sectional area of the sample in mm²; ε is the strain, dimensionless; L is the length of the specimen after stretching in mm; and L_0 is the initial specimen length in mm.

The resulting stress–strain curves were obtained for porcine thyroid tissue at two strain rates ($0.005\ \text{s}^{-1}$ and $0.05\ \text{s}^{-1}$) and two loading directions (perpendicular to the thyroid surface and parallel to the thyroid surface). As the samples had some individual variability, the average stress at the same strain in the six sets of experiments was taken to make a stress/strain graph, as shown in Figure 3. It can be seen from the figure that the mechanical properties of the thyroid tissue did not change significantly when the strain rate increased from $0.005\ \text{s}^{-1}$ to $0.05\ \text{s}^{-1}$; the average stress–strain curves of the thyroid tissue were also very close for both loading directions, which indicates that the thyroid tissue can be considered as an isotropic material as far as the specimens used in the tests are concerned. Similar conclusions were reached by Farhana [8] and Umale et al. [9] in their studies on the mechanical properties of soft tissues such as the liver, kidney, and spleen.

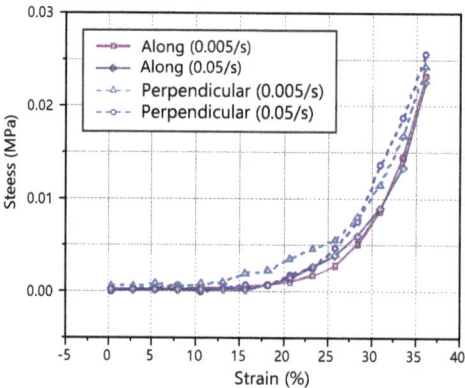

Figure 3. Stress–strain curve of the thyroid quasi-static compression test.

The average stress–strain diagram obtained from the compression test data of porcine thyroid tissue is shown in Figure 4, and the quasi-static mechanical properties of the thyroid were analyzed based on this diagram. The figure shows that the stress–strain diagram under quasi-static showed a concave-upward nonlinear characteristic, the stress amplitude was low in the initial stage, and when the strain exceeded 30%, the thyroid tissue was gradually compacted and the stress increased rapidly. The compression experimental process was divided into three stages: the first stage was the small deformation stage (segment OA in the figure), in which the stress–strain curve varied approximately linearly (slope k_1) and the stress increased slowly with the increasing strain; the second stage was the nonlinear stage (segment AB in the figure), in which the stress–strain curve showed a nonlinear variation; the third stage was the large deformation stage (segment BC in the figure), in which the stress–strain curve again showed a linear variation (slope k_2) and the stress increased rapidly with the increasing strain.

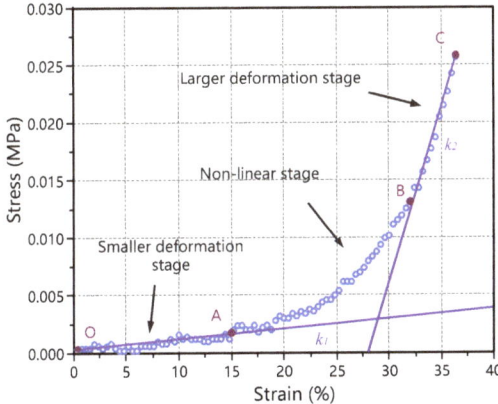

Figure 4. Analysis of the mechanical properties of the thyroid gland in pigs.

In a material stress–strain curve, the slope of the linear phase represents the Young's modulus. For thyroid tissue, the slope of the stress–strain curve in the small deformation phase (k_1) can be used as the Young's modulus in the small deformation phase of the thyroid (E_1), and the slope of the stress–strain curve in the large deformation phase (k_2) can be used as the Young's modulus in the large deformation phase of the thyroid (E_2). It is of interest that in the large deformation stage, a certain degree of damage occurred inside the thyroid tissue, thus this stage contained the plastic phase of the thyroid tissue deformation process. According to the average stress–strain diagram, the Young's modulus of the thyroid tissue in the small deformation stage and the Young's modulus in the large deformation stage could be obtained, and the calculated results are shown in Table 1.

Table 1. Results of calculating Young's modulus of the porcine thyroid.

Test Group	E_1 (MPa)	E_2 (MPa)
1	2.238×10^{-5}	3.430×10^{-3}
2	1.696×10^{-5}	3.590×10^{-3}
3	2.459×10^{-5}	2.680×10^{-3}
4	2.670×10^{-5}	2.950×10^{-3}
5	2.362×10^{-5}	2.740×10^{-3}
6	1.972×10^{-5}	3.260×10^{-3}
Mean	2.233×10^{-5}	3.108×10^{-3}
Variance	1.230×10^{-11}	1.405×10^{-7}

E_1 Young's modulus in the small deformation phase of the thyroid; E_2 Young's modulus in the large deformation phase of the thyroid.

3.2. Hyperelastic Constitutive Model

The hyperelastic model fitting function provided by ABAQUS finite element software was used to fit the quasi-static stress–strain curve of thyroid tissue with several common hyperelastic models (Mooney–Rivlin, Ogden, Neo-Hooke, and Yeoh), and the fitting results are shown in Figure 5 and the model parameters are shown in Table 2. The fitting results showed that the Mooney–Rivlin and Neo-Hooke models differed significantly from the experimental results when describing large strains (strains exceeding 25%), and the five-parameter Ogden and Yeoh models could better describe the quasi-static mechanical properties of the thyroid tissue. Considering the minimization of the model parameters, the Yeoh model was chosen to describe it. based on the experimental data of the compression test, the coefficients of the Yeoh model could be determined, i.e., $C_{10} = 1.9 \times 10^{-3}$ MPa, $C_{20} = -2.3 \times 10^{-3}$ MPa and $C_{30} = 0.04$ MPa. Therefore, the Yeoh hyperelastic constitutive model of the thyroid tissue could be described as follows

$$W = 1.9 \times 10^{-3}(I_1 - 3) - 2.3 \times 10^{-3}(I_1 - 3)^2 + 0.04(I_1 - 3)^3 \tag{13}$$

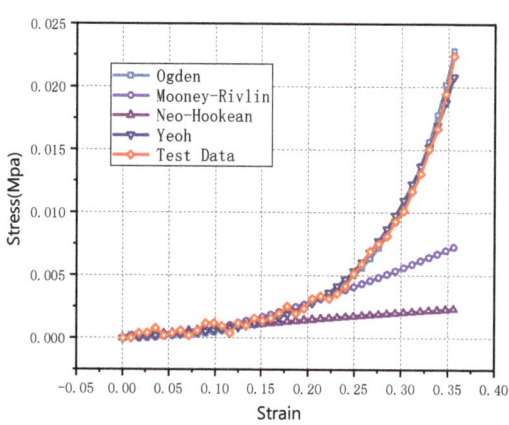

Figure 5. Fitting of the thyroid stress–strain curve to the constitutional model.

Table 2. Hyperelasticity constitutive model fitting parameters.

	C_{10}/MPa	C_{20}/MPa	C_{30}/MPa	A_1/MPa	A_2/MPa	R^2
N-H	2.357	-	-	-	-	0.564
M-R	1.7×10^{-2}	1.7×10^{-2}	-	-	-	0.661
Ogden	-0.305	0.297	1.619	10.267	-25	0.996
Yeoh	1.9×10^{-3}	-2.3×10^{-3}	0.04	-	-	0.999

In the table, C_{10}, C_{20}, C_{30}, A_1 and A_2 are the material parameters; R^2 is the correlation coefficient; and "-" represents no parameter value.

This model parameter will provide an important material basis for the simulation of thyroid gland deformation and damage assessment during the interaction of surgical instruments, which is important for the safety assessment of thyroid surgery robots.

Su et al. [31] performed viscoelastic modeling of biological soft tissues using the following viscoelastic model

$$E(t) = \sum_{i=1}^{n} E_i e^{-t/\tau_i} + E_{n+1} \tag{14}$$

where E_i and τ_i are the model parameters; n is the number of the Prony-series; E_{n+1} is the static elastic modulus (i.e., the long-term relaxation modulus), and

$$E_{n+1} = \frac{\sigma_\infty}{\varepsilon_0} \tag{15}$$

where σ_∞ is residual stress and ε_0 is the constant strain.

The model describes the viscoelasticity of the material more accurately and its parameters are easier to obtain by numerical fitting based on stress relaxation tests. To obtain the parameters of this model, we designed the thyroid stress relaxation test: the sample preparation was similar to the compression experiment in the form of a cylinder with a diameter of 10 mm and a height of 6 mm. The indenter in the controlled experimental platform (shown in Figure 2) compressed the sample at a speed of 200 mm/min to 0.75 times the original length, stopped compression, and fixed the indenter position for 1000 s. Six tests were performed, and the statistics of the initial and final stresses are shown in Table 3.

Table 3. Statistics on stress in stress-relaxation tests.

Test Group	Initial Stress (MPa)	Last Stress (MPa)
1	2.646×10^{-2}	3.738×10^{-3}
2	2.854×10^{-2}	4.012×10^{-3}
3	1.762×10^{-2}	3.532×10^{-3}
4	2.931×10^{-2}	4.216×10^{-3}
5	1.912×10^{-2}	2.974×10^{-3}
6	2.491×10^{-2}	3.685×10^{-3}
Mean	2.433×10^{-2}	3.693×10^{-3}
Variance	2.39×10^{-5}	1.84×10^{-7}

It was found that the thyroid stress relaxation was accurately reflected when $i = 4$. The mathematical expression was a relatively simple viscoelastic model. The Maxwell viscoelastic model was used to fit the thyroid stress relaxation curve, as shown in Figure 6. The model can be described as

$$G(t) = \sum_{i=1}^{4} E_i e^{-t/\tau_i} + E_5 \tag{16}$$

where E_i is relaxation modulus (unit: MPa), and τ_i is relaxation time (unit: s), expressed by

$$(E_1, E_2, E_3, E_4, E_5) = (0.91, 0.82, 0.76, 0.49, 0.51)$$

And

$$(\tau_1, \tau_2, \tau_3, \tau_4) = \left(8.83, 88.68, 784.29, 2.89 \times 10^3\right)$$

The results show that the behavior of the thyroid under progressive stress relaxation compression conditions consisted of an immediate stiff response, a transient relaxation phase, and a steady-state stage. In the final steady-state stage, the steady-state values reflected the residual stresses inside the thyroid gland.

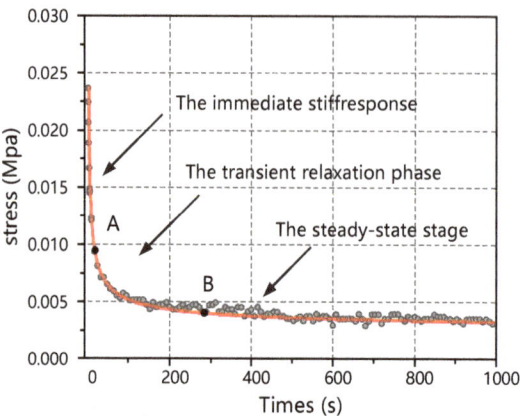

Figure 6. Test results corresponding to Equation (16).

3.3. Verification by Stretching Thyroid Specimens

Uniaxial tensile tests were performed on the porcine thyroid gland using a biomechanical test rig equipped with a Nano25 high-precision sensor, as shown in Figure 7. The test subject was the porcine thyroid gland, and the specimen was prepared as a rectangular sample strip, about 80 mm × 15 mm × 5 mm, as shown in Figure 1b, which should be left for two hours after preparation to eliminate the effect of internal stress. The use of suitable jigs in tensile experiments is one of the important factors affecting the experimental results. The current jig is not fully suitable for mechanical measurement experiments of viscoelastic materials. First, the existing jig often uses two metal clamping blocks with a certain surface roughness for clamping by threaded screwing. Secondly, there are two ways to realize the automatic centering function of the existing clamps. One is to use the restoring force of the spring and the other is to use the one-way clamping block movement. The former is not adjustable due to the non-adjustable elasticity of the spring, resulting in the clamping force of the clamps not being able to be adjusted and it is not universal for biological materials; the latter is not sufficient to maintain the force due to the one-way movement, the material can easily slip or fall off prematurely, and the stroke of the clamping block is small, which is not suitable for the clamping of large size materials. The latter is not universally applicable to biological materials because of the unidirectional motion, insufficient holding force, easy material slippage or early dislodgement, and small travel of the clamping block. Thirdly, the existing jig that can adjust the size of the clamping force lacks a synchronous locking device, and the force on the clamping block will inevitably shift during the mechanical test experiment, resulting in a change in the clamping force and the phenomenon of the material slipping or falling off prematurely. To address the above situation, a tensile test jig with automatic alignment was designed, as shown in Figure 7b. The spiral structure with the same pitch and opposite rotation direction was used to realize the synchronous anisotropic motion of the two clamping blocks, and to then realize the automatic alignment function of the jig. A device for a synchronous locking function was also added to the jig, which could lock the clamping blocks synchronously after the jig completed the clamping action. A special 3D-printed auxiliary part was designed between the jig and the specimen. The special structure of the auxiliary part could increase the friction coefficient between the biomaterial and the jig to prevent the biomaterial from slipping or falling off prematurely during the experiment. During the test, the fixture stretched the specimen at a speed of 2 mm/s until fracture, and 10 tests were conducted separately, and the relevant data such as the loading force, loading speed, and time were recorded, and the stress and strain were calculated according to Equations (11) and (12).

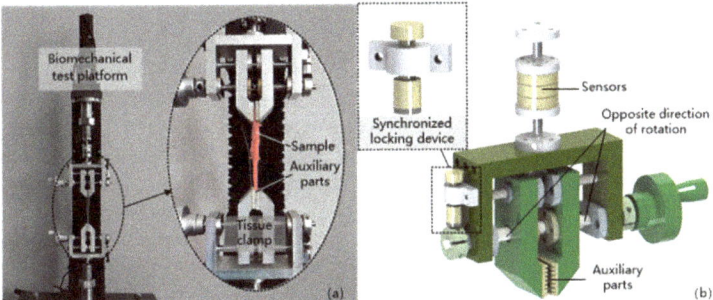

Figure 7. Thyroid tensile test. (**a**) Test platform and sample clamping (**b**) Tensile test jig.

Then, based on the above thyroid hyperelastic model, the stretching thyroid process was simulated in an ABAQUS environment. It provides a quantitative judgment basis for the determination of the thyroid deformation. The correctness of the developed hyperelastic constitutive model was verified.

- Modeling. The stretched sample model was created in ABAQUS according to the structure and dimensions (80 mm × 15 mm × 5 mm) of the thyroid sample prepared in Figure 1b.
- Properties. The thyroid material properties were defined by the hyperelastic constitutive model established above.
- Loading. The lower end of the fixed thyroid-like strip is shown in Figure 8a. The upper end of the sample was stretched at a rate of 2 mm/s.

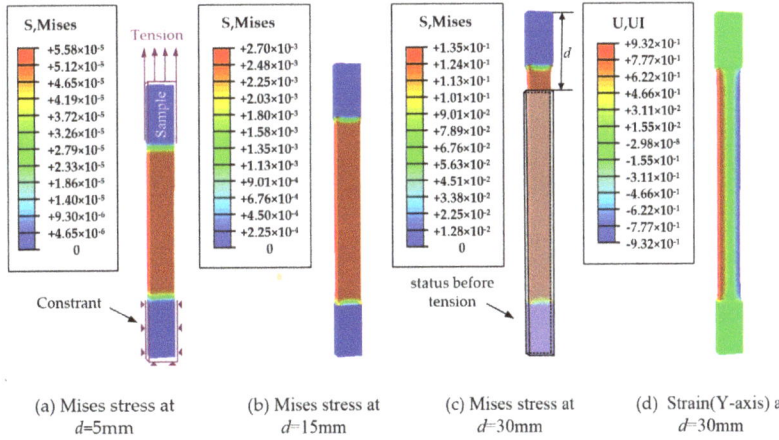

(**a**) Mises stress at d=5mm (**b**) Mises stress at d=15mm (**c**) Mises stress at d=30mm (**d**) Strain(Y-axis) at d=30mm

Figure 8. Stress–strain clouds of the tensile thyroid specimens. (**a**–**c**) show the stress clouds for tensile displacements of 5 mm, 15 mm, and 30 mm, respectively. (**d**) indicates the elastic strain cloud in the y-axis direction when the tensile displacement is 15 mm. In (**c**), the gray wireframe depicts the state before being stretched, where d indicates the displacement during extrusion.

In the simulated stretching thyroid simulation, the stress–strain cloud diagram of the sample is shown in Figure 8. At a stretching displacement of 5 mm, a small deformation of the thyroid sample occurred, as shown in Figure 8a, and the elastic strain in the y-axis direction was not obvious; as the stretching proceeded, the thyroid sample was gradually elongated, as shown in Figure 8b,c, and the elastic strain in the y-axis direction gradually increased. The elastic strain in the y-axis direction at a tensile displacement of 15 mm is shown in Figure 8d. Of course, there were some differences between the simulation clouds and the experimental results, because there were many factors affecting thyroid deformation, such as local deformation and uneven force. Therefore, due to the limitation of parameter setting, the simulation can only simulate the thyroid deformation under ideal conditions. The results of the uniaxial tensile test and simulated tensile test are shown in Figure 8. Obviously, in the stress–strain curves, the curves obtained from both the test and the simulation showed obvious hyperelastic characteristics. Compared with the experimental results, the slope of the stress–strain curve obtained from the simulation was larger and varied more significantly (in the large deformation phase).

In Figure 9, it can be seen that there is a certain difference between the simulation results and the experimental results, and in order to express the error between the simulation and experimental values, the absolute error Δ and the relative error η between them are defined as follows

$$\Delta = A - L \tag{17}$$

$$\eta = \Delta/L \tag{18}$$

where A is the simulation value and L is the test value. The calculated results are shown in Table 4.

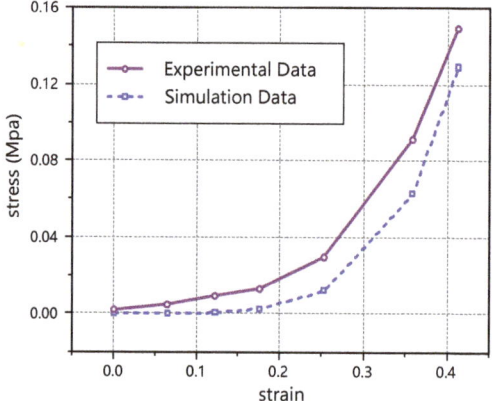

Figure 9. Comparison of the tensile test and simulated stress–strain.

Table 4. Error between the simulation and test values.

Strain	Stress (MPa)		Absolute Error (MPa)	Relative Error (%)
	Simulation Data	Experimental Data		
0	0	0	0	0
0.064	5.576×10^{-5}	6.650×10^{-5}	-1.074×10^{-5}	16.16
0.121	6.477×10^{-4}	8.300×10^{-4}	-1.823×10^{-4}	21.97
0.175	2.690×10^{-3}	4.301×10^{-3}	-1.611×10^{-3}	37.46
0.253	1.227×10^{-2}	2.077×10^{-2}	-0.0085	40.92
0.388	6.298×10^{-2}	7.414×10^{-2}	-0.0112	15.05
0.413	0.1294	0.1489	-0.0196	13.13
Mean error				20.67

The experimental data are the mean value and the simulation data are the fitting value.

The reasons for the errors are as follows. First, the sampling process of the sample strips would produce errors. As a result of the characteristics of the soft tissue, there was an error in the size of the sample in the test and the simulation model, and even if the sample was taken to avoid large blood vessels and uneven areas as much as possible, it was still difficult to avoid the existence of capillaries and other structures in the sample, which could lead to local deformation and uneven force in the test, while the simulation was performed under ideal conditions, ignoring the influence of this situation. The second is that the load setting in the simulation was not exactly the same as the actual test. In the simulation, the clamped part of the sample (the upper and lower ends of the sample) was assumed to be rigid, i.e., the clamped part was assumed not to deform in the tensile part, but the deformation of this part could not be avoided during the test. The third is the existence of extrusion in the clamping part. During the test, in order to avoid excessive water loss of the sample, saline would be applied to the sample at regular intervals, and the soft tissue itself also contained water, which would squeeze out the water in the sample during the clamping process, and the structure of the squeezed part would be changed, and this complex process is not simulated in the simulation, which will also lead to errors between the simulation results and the test results. Therefore, due to the above limitations, the simulation can only simulate the thyroid deformation under ideal conditions.

4. Conclusions

Uniaxial compression tests were performed on fresh porcine thyroid tissue at quasi-static strain rates of 0.005 s^{-1} and 0.05 s^{-1} and two loading directions (perpendicular to the thyroid surface and parallel to the thyroid surface) using a biomechanical test platform equipped with Nano25 high-precision sensors to investigate the effects of the strain rate and loading direction on the mechanical properties of the porcine thyroid tissue. The results showed that the porcine thyroid tissue did not exhibit significant strain rate effects in the low strain rate range of $0.005 \text{ s}^{-1} \sim 0.05 \text{ s}^{-1}$, and the loading direction had no significant effect on the mechanical properties of the porcine thyroid tissue, which can be regarded as an isotropic material. The mechanical properties of porcine thyroid were also studied under quasi-static conditions, and it was found that there was no significant difference in Young's modulus at each stage under two strain rates and loading directions, and the average Young's modulus was calculated to be about 2.233×10^{-5} MPa for the small deformation stage and 3.108×10^{-3} MPa for the large deformation stage of the porcine thyroid.

In order to describe the mechanical properties of porcine thyroid tissue at a quasi-static, it was considered as a non-compressible isotropic hyperelastic material, and the Yeoh strain energy density function was used to develop a hyperelastic constitutive model of porcine thyroid tissue at quasi-static low strain rate, and the three parameters of the function were obtained by fitting: $C_{10} = 1.9 \times 10^{-3}$ MPa, $C_{20} = -2.3 \times 10^{-3}$ MPa, $C_{30} = 0.04$ MPa. On the basis of this hyperelastic constitutive model, the tensile test simulation was performed on the thyroid specimens. The stress–strain curves obtained from the simulation were generally consistent with the experimental results, but some differences could be neglected because there were often some factors that could not be simulated in the actual test, such as local deformation and uneven stresses.

The quasi-static mechanical properties of the thyroid gland obtained from uniaxial unconstrained compression tests and the developed hyperelastic constitutive model are suitable for digital modeling of thyroid materials. Based on the results of this study, we are evaluating the possibility of simulating thyroid surgery and using it in a surgical training system. With the further development of virtual reality technology, we will explore methods to realize virtual thyroid surgery training by combining robotics, finite elements, and digital twins.

Author Contributions: Methodology, P.S; Formal analysis, T.L.; Investigation, Q.Z.; Resources, P.S. and B.L.; Data curation, L.C.; Writing—original draft, C.Y.; Writing—review & editing, T.L.; Project administration, P.S. and B.L. All authors have read and agreed to the published version of the manuscript.

Funding: This study is funded by the National Key R&D Program of China (grant no. 2019YFC0119200), the National Natural Science Foundation of China (grant no. 52005045), and the Natural Science Foundation of Beijing Municipality (grant no. L192018).

Institutional Review Board Statement: Not applicable.

Informed Consent Statement: Not applicable.

Data Availability Statement: Not applicable.

Acknowledgments: The authors extend their gratitude to the doctors of the Beijing Cancer Hospital for their experience and advice.

Conflicts of Interest: There are no conflict of interest in this study.

References

1. Rosko, A.J.; Gay, B.L.; Reyes-Gastelum, D.; Hamilton, A.S.; Ward, K.C.; Haymart, M.R. Surgeons' Attitudes on Total Thyroidectomy vs Lobectomy for Management of Papillary Thyroid Microcarcinoma. *JAMA Otolaryngol. Head Neck Surg.* **2021**, *147*, 667–669. [CrossRef] [PubMed]
2. Oquendo, Y.A.; Chua, Z.; Coad, M.M.; Nisky, I.; Jarc, A.M.; Wren, S.M.; Lendvay, T.S.; Okamura, A.M. Robot-Assisted Surgical Training Over Several Days in a Virtual Surgical Environment with Divergent and Convergent Force Fields. *Proc. Hamlyn Symp. Med. Robot.* **2021**, 81–82.
3. Vinck, E.E.; Smood, B.; Barros, L.; Palmen, M. Robotic cardiac surgery training during residency: Preparing residents for the inevitable future. *Laparosc. Endosc. Robot. Surg.* **2022**, *5*, 75–77. [CrossRef]
4. Ujala, Z.; Zarafshan, Z. Surgical resident training in Pakistan and benefits of simulation based training. *JPMA. J. Pak. Med. Assoc.* **2020**, *70*, 904–908.
5. *Strength of Biological Materials*; Williams & Wilkins: Philadelphia, PA, USA, 1970; p. 582.
6. Thibault, L.E.; Gennarelli, T.A. Biomechanics of Diffuse Brain Injuries. In Proceedings of the 10th International Technical Conference on Experimental Safety Vehicles (ESV), Oxford, UK, 1–4 July 1985; p. 185.
7. Duma, S.M.; Stitzel, J.D.; Hardy, W.N.; Sparks, J.L. *Characterizing the Biomechanical Response of the Liver*; Virginia Polytechnic Institute and State University: Blacksburg, VA, USA, 2010.
8. Pervin, F.; Chen, W.W.; Weerasooriya, T. Dynamic compressive response of bovine liver tissues. *J. Mech. Behav. Biomed. Mater.* **2011**, *4*, 76–84. [CrossRef] [PubMed]
9. Umale, S.; Deck, C.; Bourdet, N.; Dhumane, P.; Soler, L. Jacques Marescaux, Remy Willinger. Experimental mechanical characterization of abdominal organs: Liver, kidney & spleen. *J. Mech. Behav. Biomed. Mater.* **2013**, *17*, 22–33. [PubMed]
10. Brunon, A.; Bruyere-Garnier, K.; Coret, M. Mechanical characterization of liver capsule through uniaxial quasi-static tensile tests until failure. *J. Biomech.* **2010**, *43*, 2221–2227. [CrossRef]
11. Kemper, A.R.; Santago, A.C.; Stitzel, J.D.; Sparks, J.L.; Duma, S.M. Biomechanical response of human spleen in tensile loading. *J. Biomech.* **2012**, *45*, 348–355. [CrossRef]
12. Yarpuzlu, B.; Ayyildiz, M.; Tok, O.E.; Aktas, R.G.; Basdogan, C. Correlation between the mechanical and histological properties of liver tissue. *J. Mech. Behav. Biomed. Mater.* **2014**, *29*, 403–416. [CrossRef]
13. Kugler, M.; Hostettler, A.; Soler, L.; Borzacchiello, D.; Chinesta, F.; George, D.; Rémond, Y. Numerical simulation and identification of macroscopic vascularised liver behaviour: Case of indentation tests. *Biomed. Mater. Eng.* **2017**, *28*, S107. [CrossRef]
14. Brinker, S.; Klatt, D. Demonstration of concurrent tensile testing and magnetic resonance elastography. *J. Mech. Behav. Biomed. Mater.* **2016**, *63*, 232–243. [CrossRef] [PubMed]
15. Lin, H.; Shen, Y.; Chen, X.; Zhu, Y.; Zheng, Y.; Zhang, X.; Guo, Y.; Wang, T.; Chen, S. Viscoelastic properties of normal rat liver measured by ultrasound elastography: Comparison with oscillatory rheometry. *Biorheology* **2016**, *53*, 193–207. [CrossRef] [PubMed]
16. Matin, Z.; Ghahfarokhi; Zand, M.M.; Tehrani, M.S.; Wendland, B.R.; Dargazany, R. A visco-hyperelastic constitutive model of short- and long-term viscous effects on isotropic soft tissues. *Proc. Inst. Mech. Eng. Part C J. Mech. Eng. Sci.* **2019**, *234*, 3–17. [CrossRef]
17. Zheng, Y.; Jiang, Y.; Cao, Y. A porohyperviscoelastic model for the shear wave elastography of the liver. *J. Mech. Phys. Solids* **2021**, *150*, 104339. [CrossRef]
18. Mattei, G.; Ahluwalia, A. Sample, testing and analysis variables affecting liver mechanical properties: A review. *Acta Biomater.* **2016**, *45*, 60–71. [CrossRef] [PubMed]
19. Pervin, F.; Chen, W.W. Effect of inter-species, gender, and breeding on the mechanical behavior of brain tissue. *Neuroimage* **2011**, *54*, S98–S102. [CrossRef]
20. Wen, Y.; Zhang, T.; Yan, W.; Chen, Y.; Wang, G. Mechanical response of porcine hind leg muscles under dynamic tensile loading. *J. Mech. Behav. Biomed. Mater.* **2020**, *115*, 104279. [CrossRef]
21. Fan, L.; Yao, J.; Wang, L.; Xu, D.; Tang, D. Optimization of Left Ventricle Pace Maker Location Using Echo-Based Fluid-Structure Interaction Models. *Front. Physiol.* **2022**, *13*, 215. [CrossRef]
22. Zhang, L.; Li, Z.H.; Ma, X.Q. Parametric characterization of the Mooney-Rivlin hyperelastic intrinsic model of rubber. *Noise Vib. Control* **2018**, *A02*, 4.
23. Yeoh, O.H. Some Forms of the Strain Energy Function for Rubber. *Rubber Chem. Technol.* **2012**, *66*, 754–771. [CrossRef]
24. Cheng, K.E.; Wang, F.; Cao, Z.; Kong, H.; Zhang, J.; Fan, Y. A continuous medium intrinsic model for skeletal muscle at variable strain rates. *Med. Biomech.* **2021**, *36*, 7.
25. Ogden, R.W. Large Deformation Isotropic Elasticity-On the Correlation of Theory and Experiment for Incompressible Rubberlike Solids. *Proc. R. Soc. A Math.* **1972**, *326*, 26. [CrossRef]

26. Stasio, L.D.; Liu, Y.; Moran, B. Large deformation near a crack tip in a fiber-reinforced neo-Hookean sheet with discrete and continuous distributions of fiber orientations—ScienceDirect. *Theor. Appl. Fract. Mech.* **2021**, *114*, 103020. [CrossRef]
27. Chu, H.Y.; Sun, D.M.; Chen, Q.; Cai, L.G. Establishment of hyperelastic-viscoplastic model and structural dynamic response of rubber. *J. Beijing Univ. Technol.* **2019**, *45*, 10.
28. Guan, G.Y.; Meng, Z.G.; Xie, L.; Peng, N. Evaluation of the fitting effect of the hyperelasticity constitutive model of polyurethane rubber. *Q. J. Mech.* **2021**, *42*, 10.
29. Estermann, S.J.; Pahr, D.H.; Reisinger, A. Hyperelastic and viscoelastic characterisation of hepatic tissue under uniaxial tension in time and frequency domain. *J. Mech. Behav. Biomed. Mater.* **2020**, *112*, 104038. [CrossRef]
30. Giannokostas, K.; Dimakopoulos, Y.; Tsamopoulos, J. Shear stress and intravascular pressure effects on vascular dynamics: Two-phase blood flow in elastic microvessels accounting for the passive stresses. *Biomech. Model. Mechanobiol.* **2022**, *21*, 1659–1684. [CrossRef]
31. Su, P.; Yang, Y.; Xiao, J.; Song, Y. Corneal hyper-viscoelastic model: Derivations, experiments, and simulations. *Acta Bioeng. Biomech.* **2015**, *17*, 73–84.

Article

Analysis of the Effect of Thickness on the Performance of Polymeric Heart Valves

Jingyuan Zhou [1], Yijing Li [2], Tao Li [2], Xiaobao Tian [1], Yan Xiong [2,*] and Yu Chen [1,*]

1 Department of Applied Mechanics, Sichuan University, Chengdu 610065, China
2 College of Mechanical Engineering, Sichuan University, Chengdu 610065, China
* Correspondence: xy@scu.edu.cn (Y.X.); yu_chen@scu.edu.cn (Y.C.)

Abstract: Polymeric heart valves (PHVs) are a promising and more affordable alternative to mechanical heart valves (MHVs) and bioprosthetic heart valves (BHVs). Materials with good durability and biocompatibility used for PHVs have always been the research focus in the field of prosthetic heart valves for many years, and leaflet thickness is a major design parameter for PHVs. The study aims to discuss the relationship between material properties and valve thickness, provided that the basic functions of PHVs are qualified. The fluid–structure interaction (FSI) approach was employed to obtain a more reliable solution of the effective orifice area (EOA), regurgitant fraction (RF), and stress and strain distribution of the valves with different thicknesses under three materials: Carbothane PC−3585A, xSIBS and SIBS−CNTs. This study demonstrates that the smaller elastic modulus of Carbothane PC−3585A allowed for a thicker valve (>0.3 mm) to be produced, while for materials with an elastic modulus higher than that of xSIBS (2.8 MPa), a thickness less than 0.2 mm would be a good attempt to meet the RF standard. What is more, when the elastic modulus is higher than 23.9 MPa, the thickness of the PHV is recommended to be 0.1–0.15 mm. Reducing the RF is one of the directions of PHV optimization in the future. Reducing the thickness and improving other design parameters are reliable means to reduce the RF for materials with high and low elastic modulus, respectively.

Keywords: fluid–structure interaction (FSI); polymeric heart valves (PHVs); thickness; strain; stress

Citation: Zhou, J.; Li, Y.; Li, T.; Tian, X.; Xiong, Y.; Chen, Y. Analysis of the Effect of Thickness on the Performance of Polymeric Heart Valves. *J. Funct. Biomater.* **2023**, *14*, 309. https://doi.org/10.3390/jfb14060309

Academic Editor: Ilia Fishbein

Received: 4 April 2023
Revised: 17 May 2023
Accepted: 26 May 2023
Published: 1 June 2023

Copyright: © 2023 by the authors. Licensee MDPI, Basel, Switzerland. This article is an open access article distributed under the terms and conditions of the Creative Commons Attribution (CC BY) license (https://creativecommons.org/licenses/by/4.0/).

1. Introduction

Prosthetic heart valves are employed to replace the diseased native valve as a treatment for severe aortic valve (AV) disease. Mechanical heart valves (MHVs) and bioprosthetic heart valves (BHVs) are two main prostheses that have been utilized therapeutically; however, they are susceptible to thrombosis and structural valve degeneration (SVD), respectively [1,2]. It was found that the mechanical properties of polymer can easily be formulated to match that of native tissue and the geometry of it can be designed to produce physiological flow [3]. Consequently, polymeric heart valves (PHVs) are expected to overcome the shortcomings of MHVs and BHVs.

Materials with good durability and biocompatibility have always been the research focus in the field of PHVs for many years. Polysiloxanes, polytetrafluoroethylene (PTFE) and polyurethane (PUs) are the earliest materials applied for developing PHVs, which have good biocompatibility, hemodynamic properties and viscoelasticity, respectively [4,5]. Still, as time goes on, many PHVs made from them have been tested and failed because of thrombosis, calcification, hydrolysis, etc. [6]. Therefore, the challenges of durability remain. Novel polymers with better performance are obtained by adjusting the microstructure of materials with insufficient mechanical properties or reinforcing with filler. xSIBS and hydrogen−bonding−enhanced supramolecular hydrogels are typical cases of the former, and both of them have shown promising in vitro results [7–9]. Aerogel is widely used in medicine due to its attractive structural characteristics [10]. Macroscale composites have focused on reinforcing the elastomeric leaflets with macroscale fibers, such as Dacron

(PET) and PTFE [11,12]. However, no commercially–viable devices have been developed using macroscopic composite materials so far. Nanocomposites have a better application foreground because of their superior mechanical properties, biocompatibility and simpler valve manufacturing process [3,4,13], such materials include polyhedral oligomeric silsesquioxane poly(carbonate–urea) urethane (POSS–PCU), polyvinyl alcohol (PVA) hydrogels reinforced with bacterial cellulose (BC) [14], the integration of graphene oxide (FGO) nanomaterials and PCUs [15], and the addition of carbon nanotubes (CNTs) to polymers with acceptable biocompatibility and mechanical properties [16,17]. Except for POSS–PCU that has undergone in vitro testing [18], the other three nanocomposites are still in the early stages of development.

The valve design is another crucial factor influencing the performance and lifetime of PHVs. Bending is a primary deformation mode of the leaflets. According to Euler–Bernoulli beam theory, bending stiffness is proportional to the cube of leaflet thickness [3]. Therefore, leaflet thickness is a major design parameter for PHVs. Previous studies qualitatively provided information on the relationship between leaflet modulus and thickness. The results show that the modulus and thickness are positively and negatively related to low stress, respectively [19]. Additionally, low modulus materials are subjected to higher strains than high modulus materials at the same stress level and are therefore more susceptible to creep and, ultimately, failure [20].

The performances of some PHVs made from novel materials have been tested in vitro hydrodynamic assessments. These include the Polynova xSIBS valve with variable thickness [7,8], the POSS–PCU valve with three leaflet thicknesses (0.1, 0.15 and 0.2 mm) [18], the PHV of silk fibroin fiber membranes (ISF) combined with poly (ethylene glycol) diacrylate (PEGDA) hydrogels with a thickness of 0.4 mm [21] and the PHV made from hyaluronan–enhanced linear–low–density polyethylene (LLDPE–HA) with a thickness of 0.08 mm [22], etc. What is more, the performances of PHVs are also well suited for measuring with the fluid–structure interaction (FSI) simulations due to the interplay between structure and fluid. The PHV constructed from bionate thermoplastic polycarbonate urethane with 0.127 mm thickness [23] and the PHV made from poly(styrene–ethylene–propylene–styrene) (SEPS) block copolymers with 0.20 mm thickness [24] were simulated where the flow rate was defined for the ventricular outflow rate. In addition, PHVs made from PU (0.16 mm) [25] and block co–polymer (0.3 mm) [26], as well as the Polynova xSIBS valve with variable thickness [7], were tested with the pressure boundary condition applied to the inlet section.

Due to the significant impact of the material and thickness of the PHV on its performance, it is necessary to adapt the thickness to the material to achieve optimal valve performance. However, whether it was testing the functionalities of PHVs made of novel polymers or focusing on improving design to enhance the performances of PHVs, few studies have mentioned how the thickness was selected. The former concerns mainly one material; therefore, its results have little reference significance for other materials. If the latter lacks consideration for thickness, the performance of the PHV obtained by optimizing other design parameters is probably not optimal.

The aim of the current work was to investigate the appropriate thickness range of PHVs that can meet basic functions. It is assumed that the appropriate thickness should enable PHVs to satisfy clinical requirements for the effective orifice area (EOA) and the regurgitant fraction (RF), as well as ensure that the stress and strain on the leaflets are small and evenly distributed. In this study, the FSI approach was employed for two cardiac cycles to obtain the performances of PHVs of three materials with different thicknesses. The selected materials consist of Carbothane PC–3585A, xSIBS and SIBS with higher molecular weight (Mn~70,000 g mol^{-1}) reinforced by 1 wt% of CNTs (SIBS–CNTs). Carbothane PC–3585A is a kind of medical–grade thermoplastic polyurethane, it is widely used in the fields of biomedical applications [27]. xSIBS and SIBS–CNTs are the most promising developments in recent years [4,17]. SIBS–CNTs have enhanced biocompatibility and increased strength compared to neat SIBS [17], which will contribute to improving long–term durability and fatigue resistance, particularly in high–stress environments such as the cardiac cycle. The proposed thickness range supplies an option for the testing and optimization of PHVs with enhanced efficiency.

2. Materials and Methods

2.1. Geometry and Mesh Generation

The 3D geometry of the valve and blood flow volume were created in Rhino 7.0 (Robert McNeel & Assoc, Seattle, WA, USA). Both the structure and fluid were meshed using ANSA 21.0.1 (BETA CAE Systems, Root, Switzerland).

The geometry of the model was designed to resemble a typical prosthetic heart valve with leaflets in an almost closed configuration, based on the design of commercially available Sapien XT valve (Edwards Life Sciences, Irvine, CA, USA), setting the diameter to 23 mm and the height to 14.3 mm of all valves [28]. Leaflets were generated by the attachment curve, the free edge and the belly curve. The free edge was swept along the belly curve to intersect with the sinus at the attachment curve. We rotated one leaflet around the vertical axis to create three identical leaflets to generate the final valve model; each leaflet occupied 120° in the cross−section.

The geometry of the aortic root in the current model was based on the measurements of aortic root in adults [29]. To move the boundary conditions away from the regions of interest, two straight−tube extensions with lengths of 10 and 20 mm were added to the upstream and downstream, respectively. The physiological significance of these lengths is that the upstream boundary was within the left ventricle, while the downstream boundary did not reach the aortic arch. Additionally, the two extensions have smaller influence on the results because the pressure drops in these extensions are obviously smaller than the transvalvular pressure difference of the valve [30]. The structural and fluid parts used for this study are shown in Figure 1. Dimension of the various design parameters are given in Table 1.

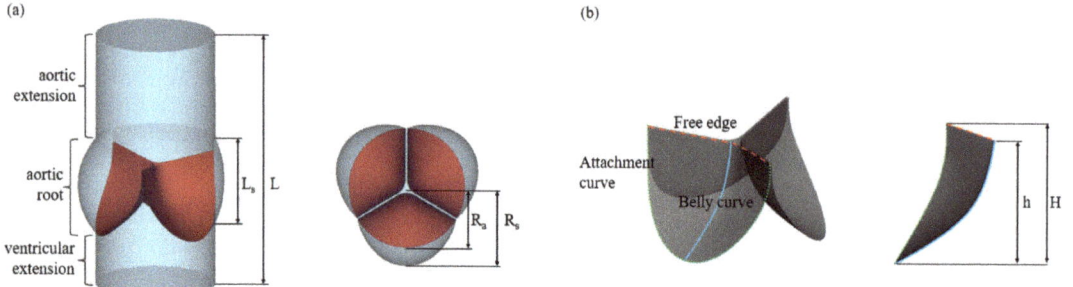

Figure 1. The geometry design of the model using the modeling design parameters. (**a**) Fluid field and (**b**) PHV.

Table 1. Dimensions of design parameters used in modeling the AV.

	H (mm)			h (mm)	
Structure	14.3			11.3	
	R_a (mm)	R_s (mm)		L_s (mm)	L (mm)
Fluid	11.5	15.0		17.0	47.0

H: valve height, h: leaflet coaptation height, R_a: radius of aorta, R_s: radius of sinus, L_s: sinus length, L: model length.

For spatial discretization, the valve was discretized into 17,130 quadrilateral fully integrated shell elements. Five integration points were distributed through shell thickness. The fluid domain was discretized into 529,152 tetrahedral Eulerian elements with maximum characteristic length of 1.0 mm (mean characteristic length of 0.60 mm). Finer resolution of 0.15 mm was used in the region surrounding the interface, which had the same geometry as the structure, and the attachment nodes of the interface were shared with the elements at the base of the sinus. The discretized solution fields for blood flow zone and valve leaflet structure are shown in Figure 2.

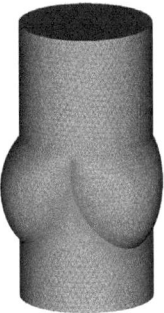 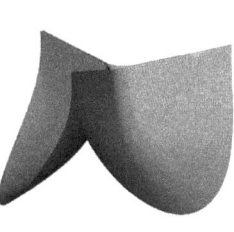

Figure 2. The discretization of fluid domain and PHV.

2.2. FSI Approach

FSI simulations with two−way coupling were applied to consider the interaction between the blood and PHVs. In order to limit the numerical instabilities caused by the approximate density of blood and PHVs, a strongly coupled strategy was implemented by the incompressible flow solver (ICFD) and the implicit mechanics solver in the LS−DYNA, Release 13.0 (Ansys, Canonsburg, PA, USA).

For FSI simulations, ICFD uses an Arbitrary Lagrangian−Eulerian (ALE) approach for mesh movement and models the interaction between the structural and fluid parts; the interfaces between the solid and the fluid are Lagrangian and deform with the structure. The solid and fluid geometry must match at the interface but not necessarily the meshes. Therefore, ICFD can automatically re−mesh to keep an acceptable mesh quality to support large deformation rate of the leaflets. Further, interaction detection coefficient (IDC, IDC = 0.25) was set to ensure FSI interaction.

2.3. Boundary Conditions

The time−dependent physiological pressure difference between the left ventricle and the ascending aorta [31] was applied to the inlet (Figure 3a) and the outlet was set to zero pressure. A constant time step of 0.1 ms was set in the simulations. One cardiac cycle (0.8 s) was discretized in 8000 time steps. Two complete cardiac cycles were simulated to achieve convergence and eliminate the effect of sudden start during 1st cardiac cycle. All results were extracted from the 2nd cycle.

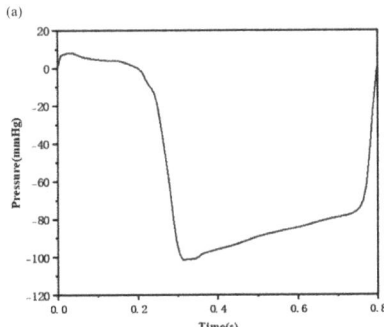 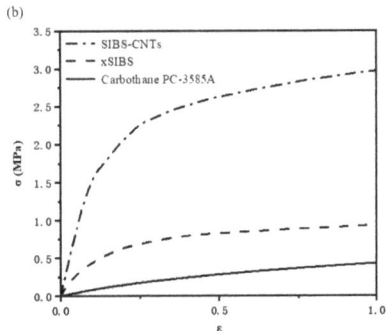

Figure 3. (**a**) The time−dependent physiological pressure difference between the left ventricle and the ascending aorta and (**b**) Engineering stress–strain curves of three materials.

During the closing phase, the three leaflets experience significant contact with one another. Similar to previous studies, a penalty-based contact formulation and frictionless conditions were adopted here [32]. For two segments on each side of the contact interface that were overlapping and penetrating, a consistent nodal force assembly taking into account the individual shape functions of the segments was performed. When the three leaflets experienced significant contact with one another, both surface penetrations and edge-to-edge penetrations were checked.

For all models, the aortic wall boundaries were assumed rigid and no-slip condition was employed at the wall-blood interface. Surfaces of valve leaflets in contact with blood flow were considered as fluid-structure interfaces, across which loads and displacements were transferred. All nodes along the attachment curves were considered to be fixed in the FSI simulations.

2.4. Material Properties and Thickness of PHVs

The blood is assumed to be Newtonian and incompressible, with constant dynamic viscosity of 0.004 Pa·s and a density of 1050 kg/m^3 [33].

The selected polymers were modeled with 1st-order Ogden hyperelastic material models, which express the strain energy (W) by principal stretches (λ_1):

$$W = \frac{\mu_1}{\alpha_1}(\lambda_1^{\alpha_1} - 3) + \frac{1}{d_1}(J-1)^2 \tag{1}$$

where λ_1 is the principal stretch ratio, μ_1 and α_1 are material constants, and d_1 is incompressible parameter.

Based on the uniaxial tensile test results from previous research [17,34,35], the engineering stress-strain curves of the materials were used to solve material parameters of three materials according to the requirements in LS-DYNA [36]. Figure 3b shows the engineering stress-strain curves of three materials. The elastic modulus of xSIBS and SIBS-CNTs is 2.8 MPa and 23.9 MPa, respectively [17,37]. Because of the lack of data on the elastic modulus of Carbothane PC-3585A subjected to uniaxial tensile tests under the same conditions as [35], considering that the true stress-strain curve of Carbothane PC-3585A is almost linearly elastic within 40% strain, which is sufficient for PHV deformation, the elastic modulus of Carbothane PC-3585A is assumed to be 1.0 MPa in this study.

Taking the available application of Carbothane PC-3585A in PHV [35,38] into account, the thickness range of PHVs made from Carbothane PC-3585A was set to 0.25 mm, 0.30 mm and 0.35 mm. A variable thickness design has extensive applications in the investigations on PHVs made from xSIBS [7,34,37,39,40]. According to the design, the thickness range of PHVs made from xSIBS was set to 0.15 mm, 0.20 mm and 0.25 mm. Materials with an elastic modulus greater than 20 MPa, similar to SIBS-CNTs, were used to produce PHVs with a thickness of less than 0.15 mm for testing [18,23], while too small thickness may cause the PHV to prolapse [19]. Therefore, the thickness of SIBS-CNTs was set to 0.10 mm, 0.15 mm and 0.2 mm.

3. Results

3.1. Valve Performance Parameters

In order to assess the quantification of valve stenosis, the effective orifice area (EOA) was used as a measurement of the effective jet area during left ventricular ejection of the cardiac cycle [41]. The EOA was calculated using Equation (2) [42]:

$$\text{EOA} = \frac{q_{vRMS}}{51.6\sqrt{\frac{\Delta p}{\rho}}} \tag{2}$$

where q_{vRMS} (cm^3/s) is the root mean square of the forward flow rate during the positive differential pressure period; Δp (mmHg) is the average pressure difference measured during the positive differential pressure period; and ρ (g/cm^3) is the density of the fluid.

According to the boundary conditions shown in Figure 3a, $\Delta p = 4.79$ mmHg, which was close to that in the healthy young condition [42].

Regurgitant fraction (RF) evaluates leakage after valve closure; Equation (3) from ISO 5840–3:2021 [43] was applied to calculate the RF:

$$RF = \frac{V_C + V_L}{V_F} \times 100\% \qquad (3)$$

where V_c is the closing volume, V_L is the leakage volume and V_F is the forward volume.

The EOA and RF calculated by Equations (2) and (3), respectively, are shown in Figure 4. PHVs made from Carbothane PC–3585A obtained larger EOA and smaller RF. The EOA and RF of the thinnest PHV (0.15 mm) of xSIBS were equivalent to those of the thickest PHV (0.35 mm) made from Carbothane PC–3585A. More significant changes were observed in the results of SIBS–CNTs. The PHV made from SIBS–CNTs with a thickness of 0.20 mm had the smallest V_F (21.31 mL), as shown in Table 2. As the thickness reduced to 0.15 mm, the V_F rapidly increased to 52.39 mL; still, large regurgitation made its RF relatively high. In addition, satisfactory results were achieved when the thickness was reduced to 0.10 mm.

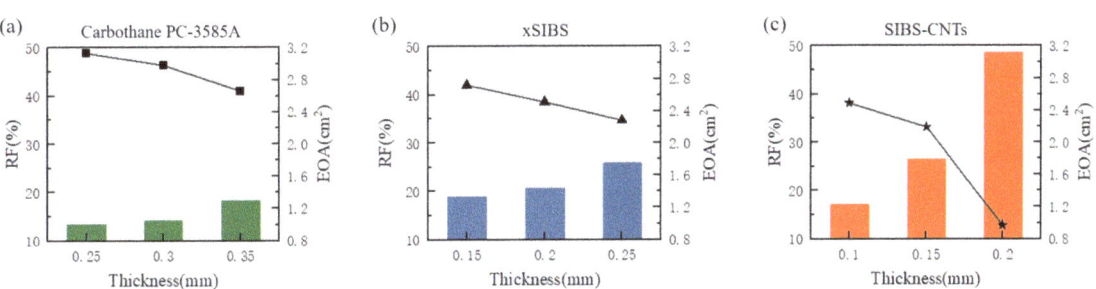

Figure 4. EOA (bars, left Y–axis units) and RF (points and connecting lines, right Y–axis units) of (**a**) Carbothane PC–3585A, (**b**) xSIBS and (**c**) SIBS–CNTs.

Table 2. The cardiac output (CO) and regurgitation of all models.

	Carbothane PC–3585A			xSIBS			SIBS–CNTs		
t (mm)	0.25	0.30	0.35	0.15	0.20	0.25	0.10	0.15	0.20
$V_c + V_L$ (mL)	6.81	10.75	12.32	12.94	12.76	14.33	10.46	13.89	10.36
V_F (mL)	80.34	76.03	67.01	68.23	61.72	55.49	61.00	52.39	21.31

3.2. Valve Dynamics

Figure 5 shows von Mises stresses distribution on the leaflets. The von Mises stresses were used to quantify the stress field in the leaflets because the yield behavior of polymers is often described by modified versions of the von Mises criterion [44]. Similar to previous analyses [45–47], the maximum in–plane principal Green–Lagrange strain (MIPE) on the aortic side of the leaflets, as shown in Figure 6, was applied to include contributions from both stretching and bending [45].

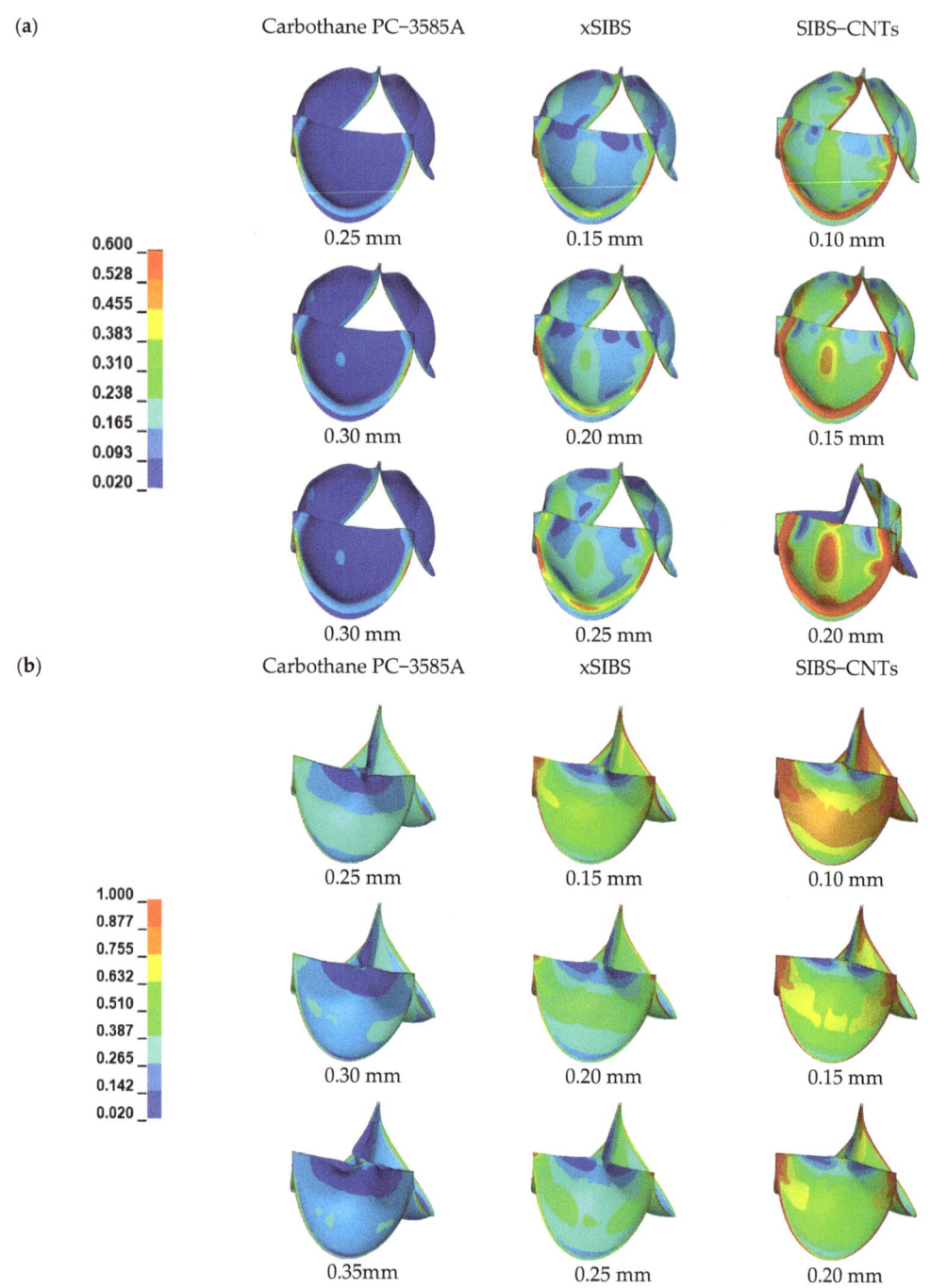

Figure 5. von−Mises stresses (MPa) distribution of the three material models. (**a**) At the moment of fully opening and (**b**) at the moment of fully closing. Note that the color scales are different for fully−opened and fully−closed results.

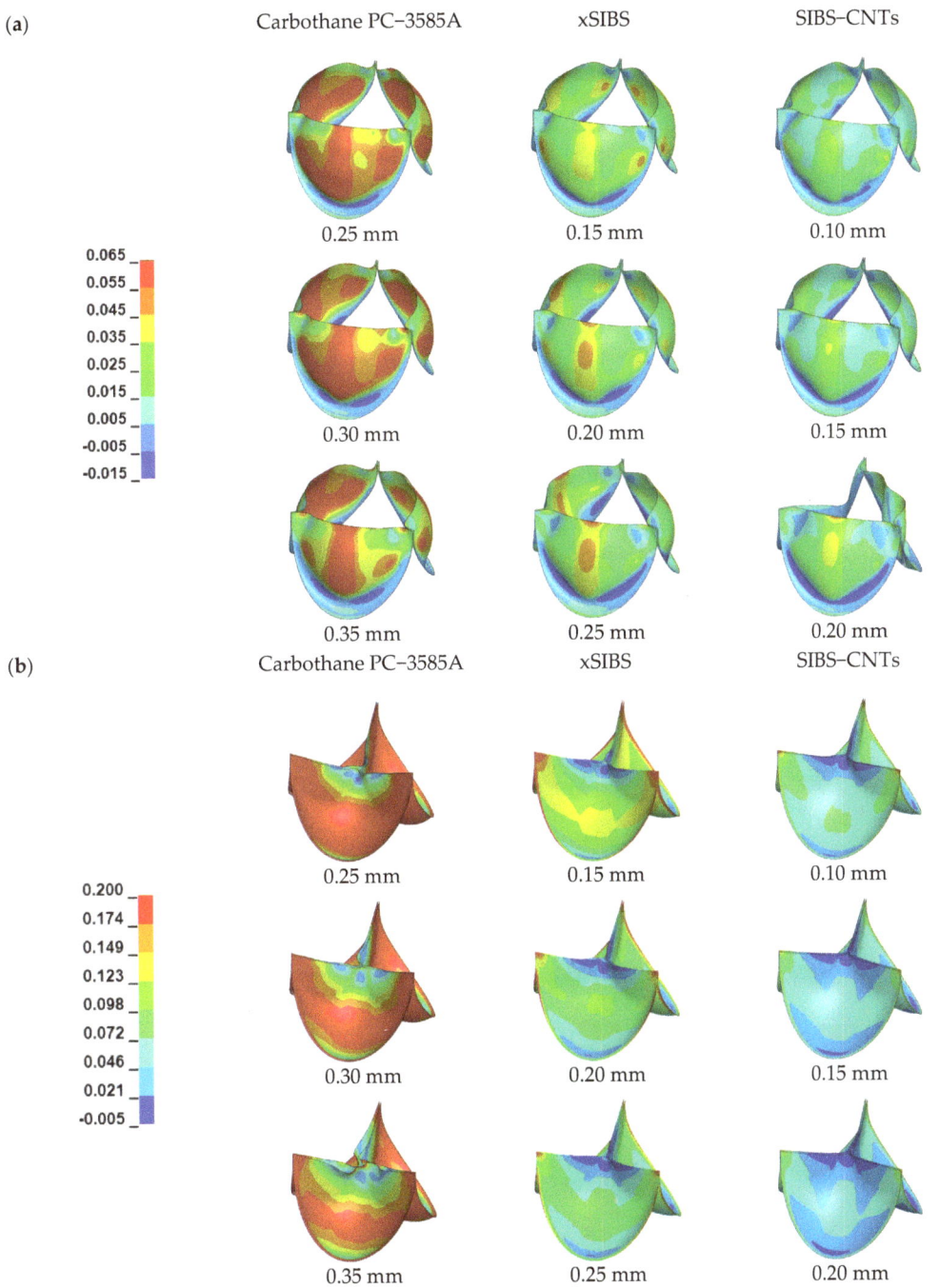

Figure 6. MIPE distributions of the three material models. (**a**) At the moment of fully opening and (**b**) at the moment of fully closing. Note that the color scales are different for fully-opened and fully-closed results.

Figure 5 depicts that the stress concentrations distributed along the attachment curve and at commissure points were negatively correlated with the thickness. The valves made from Carbothane PC−3585A had the lowest stresses during the entire cardiac cycle. The valve (0.10 mm) made from SIBS−CNTs had the maximum stress value of all the models, which was 2.15 MPa after the valves were completely closed.

Figure 6 shows that the valves made from Carbothane PC−3585A had strains significantly higher than the valves made from other materials, especially at the fully−closed stage. The same distribution of strains was observed for xSIBS and SIBS−CNTs. That is, at the fully−opened stage, as thickness increased, the strain in the middle of the valve belly became more obvious and the high strain on both sides of the belly came to disappear; at the same time, the low strain area at the bottom of the belly expanded. After the valve was fully closed, high strains along the attachment curve and at commissure points were reduced with the increase in thickness, which also made the strain distribution on the valve more uniform.

4. Discussion

With the same PHV design and boundary conditions, it can be observed from Figure 4 that EOA and RF were negatively and positively related to the thickness, respectively, with the same material, this result ties well with the test results of POSS−PCU [18] that are shown in Table 3 and it is more obvious with the increase in elastic modulus. According to the performance requirements for PHVs with a diameter of 23 mm in ISO 5840−3:2021: RF < 20%, EOA > 1.25 cm^2. PHVs made from materials that have a smaller elastic modulus (<2.8 MPa) can achieve the EOA and RF that met the standard within the thickness of 0.15–0.20 mm. This is especially true for materials with a higher elastic modulus, otherwise it will be difficult to obtain qualified performance parameters, such as the PHV of SEPS [48], which had a smaller geometric orifice area (the GOA shown in Table 3) under the same flow rate of 4 L/min as the xSIBS−PHV (0.25 mm) calculated in this study.

The thickness of 0.20 mm seriously hampered the complete opening of the PHV for SIBS−CNTs with an elastic modulus of 23.9 MPa. Fortunately, the decrease in thickness rapidly increased the CO, significantly enlarged EOA and reduced RF. However, it should be noted that reducing the thickness to improve the performance of PHVs may be more suitable for materials with a high elastic modulus, as they have better resistance to the leaflet flutter caused by reduced thickness [49]. On balance, materials with an elastic modulus higher than 23.9 MPa can be used to make PHVs with a thickness of less than 0.15 mm to achieve satisfactory EOA and RF.

Compared to previous studies, the RF in this study seems to be higher. On the one hand, the same physiological pressure pulse was applied as the pressure boundary condition for all simulations. The greater the elastic modulus of the material, the greater the pressure differential that should be applied to ensure the complete closure of the valve [19]. Therefore, the high RF shown in Figure 4 can indicate that the normal physiological pressure difference is not sufficient to completely close PHVs of xSIBS with a thickness of 0.25 mm and PHVs of SIBS−CNTs with a thickness greater than 0.15 mm. PEGDA−ISF demonstrated a similar situation, it had a bigger elastic modulus (4.54 ± 0.43 MPa) than xSIBS while the PHV (0.4 mm) made from PEGDA−ISF showed a qualified RF (14.5%), this may be caused by the larger pressure difference (Figure 5c of [21]) in the diastole of the experiment than the 100 mmHg used here. On the other hand, in vitro observations of PHVs made from xSIBS indicate that the RF was decreased during long−term experiments [7], while it is difficult for FSI to observe such an obvious phenomenon in multiple cycles because of its high calculation cost [50].

Although higher stresses were observed from the valves made from xSIBS and SIBS−CNTs, the thicknesses set for them were smaller than the thickness range of Carbothane PC−3585A; hence, it is reasonable to observe greater stresses by the joint effect of thinner thicknesses and a bigger elastic modulus.

A qualified EOA and RF can be obtained easier in the valve made from Carbothane PC−3585A even if the thickness of the valve was large, however, the strain of such valves was significantly higher; the strain accumulation due to high strain is related to creep and valve failure [20,49]. This is consistent with previous finite element results, that is, the strain of the valve increases as the elastic modulus of its material decreases [19]. However, it should be noted that strains changed obviously with the elastic material. At the fully–closed stage, it is observed from Figure 6 that the maximum strain of valves decreased from 40.1% to 20.6% only by raising the elastic modulus from 1 MPa (Carbothane PC−3585A) to 2.8 MPa (xSIBS) at the same valve thickness (0.25 mm). Smaller and uniformly distributed strains can be observed in SIBS–CNTs; even in the thinnest PHV (0.1 mm), its maximum strain did not exceed 20%.

In addition, the results show that the design of the PHV is also a key element in valve performance. On the one hand, after the closure of the valve (0.35 mm) made from Carbothane PC−3585A, the observed unnecessary distortion from Figure 7 indicated that there is a relationship between thickness and the length of the free edge. In this case, reducing the length of the free edge appropriately may allow more complete closure, thereby reducing the regurgitation. On the other hand, the RF of valves made of xSIBS (Figure 4) is higher than that of the PolyNova Valve, a novel xSIBS polymer−based valve which has a special design as shown in [7]. Apart from changing the radial cross−sectional profile of the leaflets from uniform to variable thickness, the coaptation height is equal to the valve height. Previous literature suggests that decreasing the ratio of valve height to coaptation height increases the coaptation area (CA) of the leaflets significantly, and a large CA reduces the possibility of aortic regurgitation [51]. Therefore, the design of the PolyNova Valve may explain why the RF obtained through in vitro experiment [7] is lower than the RF shown in Figure 4.

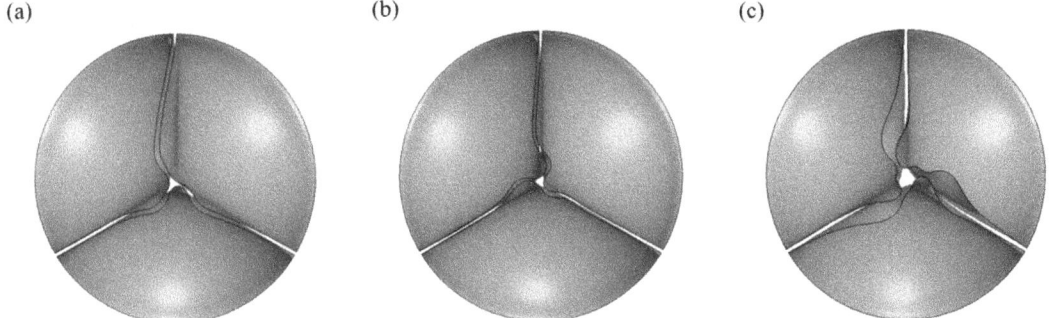

Figure 7. The top view of PHVs of Carbothane PC−3585A after the PHVs were completely closed. (a) 0.25 mm, (b) 0.30 mm and (c) 0.35 mm.

Table 3. Available data on material properties and performance assessment of PHVs in recent years.

Material	Elastic Modulus (E)	Thickness	Test Standard	Results	References
POSS–PCU	E = 26.2 MPa	0.10 mm	In vitro hydrodynamic assessment: ISO 5840−3	EOA = 3.34 cm², RF = 4.68%	[18]
	E = 23.0 MPa	0.15 mm		EOA = 3.13 cm², RF = 10.77%	
	E = 15.9 MPa	0.20 mm		EOA = 2.69 cm², RF = 12.34% (10 cycles)	
xSIBS	E = 2.8 MPa	Various thicknesses (0.15–0.25 mm)	In vitro hydrodynamic assessment: ISO 5840−3	EOA = 1.8 ± 0.04 cm², RF = 7.5 ± 1.0% (400 million cycles)	[7]
				EOA = 1.7 cm², RF = 16.7% (10 cycles)	[8]
PEGDA–ISF	E = 4.54 ± 0.43 MPa	0.4 mm	In vitro hydrodynamic assessment: ISO 5840−3	EOA = 2.30 cm², RF = 14.5% (10 cycles)	[21]
LLDPE–HA *	E > 76.49 ± 1.86 MPa	0.08 mm	In vitro hydrodynamic assessment: a pulsatile flow driving the PHV, 1 s is one cycle.	EOA= 2.08 ± 0.04 cm², RF = 11.23 ± 0.55 (60 cycles)	[22]

Table 3. Cont.

Material	Elastic Modulus (E)	Thickness	Test Standard	Results	References
PU	E = 3.67 MPa	0.15 mm	In vitro hydrodynamic assessment: a pulsatile flow driving the PHV, 1 s is one cycle.	GOA = 3.90 cm^2 (24 cycles)	[50]
Bionate thermoplastic polycarbonate urethane	E = 8 MPa	0.16 mm	FSI: Pressure difference (600 bpm) (inlet)	GOA = 2.83 cm^2 (3 cycles)	[25]
	E = 23.93 MPa	0.127 mm	FSI: low rate that was equal to 4.5 L/min of CO (inlet) and a mean arterial pressure of 100 mmHg (outlet)	GOA = 3.31 cm^2, RF = 5.64% (4 cycles)	[23]
SEPS	E = 3.2 MPa	0.20 mm	FSI: flow rate (inlet) and zero pressure (outlet)	GOA = 2.24 cm^2 (0.4 s)	[24]
		Two leaflets with a thickness of 0.36 mm and one leaflet of 0.39 mm	FSI: pressure difference corresponded to a flow rate of 4 L/min	GOA = 1.81 cm^2 (2 cycles)	[48]

* The description of HA content is vague in [22], but the addition of HA will cause the elastic modulus of LLDPE to be at least higher than 76.49 ± 1.86 MPa according to [52].

Limitation

This paper solely discussed the relationship between thickness and the elastic modulus of materials based on TAVI Sapien−XT leaflet design. The impact of other designs on the performance of the valve has not been taken into account. From the results, it can be seen that the TAVI Sapien−XT leaflet design used in this study was not entirely suitable for materials with high elastic modulus. Although how to improve the design for such materials has been discussed, and the conclusions were obtained by comparing the results with other studies shown in Table 3, which should reduce the impact of a single design to some extent, the limitations brought about by the design need to be further pointed out.

Furthermore, due to the limited scope of this work, further tests must be performed in vitro and in vivo with PHVs to assess how the thickness affects the long−term durability of the device. For example, PHVs have shown excellent potential in TAVR procedures [7,8,16,53], which requires the valve to be crimped at a low diameter for the purpose of catheter insertion [54]. It is obvious that the process is closely related to the material and thickness of the PHVs. Therefore, it is necessary to conduct crimping experiments on the PHVs to evaluate the effectiveness and stability of the material under high crimping strain conditions.

Finally, isotropic nonlinear hyperelastic material was assumed for the mechanical properties of the leaflets, more comprehensive results can be obtained by considering anisotropic materials.

5. Conclusions

This study investigated the relationship between valve thickness and material properties using FSI simulations. Three polymer materials, Carbothane PC−3585A, XSIBS and SIBS−CNTs were selected. The smaller elastic modulus of Carbothane PC−3585A allowed for a thicker valve (>0.3 mm) to be produced, while for materials with an elastic modulus higher than that of xSIBS (2.8 MPa), a thickness less than 0.2 mm would be a good attempt to meet the RF standard. What is more, when the elastic modulus is higher than 23.9 MPa, the recommended thickness of the PHV is 0.1−0.15 mm. Reducing the RF is one of the directions of PHV optimization in the future. Materials with a high elastic modulus can resist high strain, so it is advisable to reduce backflow by decreasing the thickness. However, considering that a reduction in the elastic modulus of the material will significantly increase the strain on the PHV, for materials with a small elastic modulus, adjusting other design optimization parameters based on an appropriate thickness is more secure.

Author Contributions: Conceptualization, J.Z. and Y.C.; methodology, J.Z. and T.L.; software, J.Z., Y.L. and T.L.; validation, X.T.; formal analysis, Y.L.; investigation, J.Z. and T.L.; resources, Y.C.; data curation, J.Z. and T.L.; writing—original draft preparation, J.Z.; writing—review and editing, Y.X.; visualization, Y.L.; supervision, Y.C.; project administration, Y.C.; funding acquisition, Y.C. All authors have read and agreed to the published version of the manuscript.

Funding: This study was funded by the National Natural Science Research Foundation of China (grant number: 12172239).

Data Availability Statement: Not applicable.

Conflicts of Interest: The authors declare no conflict of interest.

References

1. Dvir, D.; Bourguignon, T.; Otto, C.M.; Hahn, R.T.; Rosenhek, R.; Webb, J.G.; Treede, H.; Sarano, M.E.; Feldman, T.; Wijeysundera, H.; et al. Standardized Definition of Structural Valve Degeneration for Surgical and Transcatheter Bioprosthetic Aortic Valves. *Circulation* **2018**, *137*, 388–399. [CrossRef]
2. Iung, B.; Rodés-Cabau, J. The optimal management of anti-thrombotic therapy after valve replacement: Certainties and uncertainties. *Eur. Hear. J.* **2014**, *35*, 2942–2949. [CrossRef]
3. Li, R.L.; Russ, J.; Paschalides, C.; Ferrari, G.; Waisman, H.; Kysar, J.W.; Kalfa, D. Mechanical considerations for polymeric heart valve development: Biomechanics, materials, design and manufacturing. *Biomaterials* **2019**, *225*, 119493. [CrossRef]
4. Rezvova, M.A.; Klyshnikov, K.Y.; Gritskevich, A.A.; Ovcharenko, E.A. Polymeric Heart Valves Will Displace Mechanical and Tissue Heart Valves: A New Era for the Medical Devices. *Int. J. Mol. Sci.* **2023**, *24*, 3963. [CrossRef]
5. Ghanbari, H.; Viatge, H.; Kidane, A.G.; Burriesci, G.; Tavakoli, M.; Seifalian, A.M. Polymeric heart valves: New materials, emerging hopes. *Trends Biotechnol.* **2009**, *27*, 359–367. [CrossRef]
6. Nazir, R. Collagen–hyaluronic acid based interpenetrating polymer networks as tissue engineered heart valve. *Mater. Sci. Technol.* **2016**, *32*, 871–882. [CrossRef]
7. Rotman, O.M.; Kovarovic, B.; Bianchi, M.; Slepian, M.J.; Bluestein, D. In Vitro Durability and Stability Testing of a Novel Polymeric Transcatheter Aortic Valve. *ASAIO J.* **2020**, *66*, 190–198. [CrossRef]
8. Rotman, O.M.; Kovarovic, B.; Chiu, W.-C.; Bianchi, M.; Marom, G.; Slepian, M.J.; Bluestein, D. Novel Polymeric Valve for Transcatheter Aortic Valve Replacement Applications: In Vitro Hemodynamic Study. *Ann. Biomed. Eng.* **2019**, *47*, 113–125. [CrossRef]
9. Wu, J.; Wu, Z.; Zeng, H.; Liu, D.; Ji, Z.; Xu, X.; Jia, X.; Jiang, P.; Fan, Z.; Wang, X.; et al. Biomechanically Compatible Hydrogel Bioprosthetic Valves. *Chem. Mater.* **2022**, *34*, 6129–6141. [CrossRef]
10. Yahya, E.B.; Amirul, A.A.; H.P.S., A.K.; Olaiya, N.G.; Iqbal, M.O.; Jummaat, F.; A.K., A.S.; Adnan, A.S. Insights into the Role of Biopolymer Aerogel Scaffolds in Tissue Engineering and Regenerative Medicine. *Polymers* **2021**, *13*, 1612. [CrossRef]
11. Wang, Q.; McGoron, A.J.; Bianco, R.; Kato, Y.; Pinchuk, L.; Schoephoerster, R.T. In-vivo assessment of a novel polymer (SIBS) trileaflet heart valve. *J. Hear. Valve Dis.* **2010**, *19*, 499–505.
12. Kütting, M.; Roggenkamp, J.; Urban, U.; Schmitz-Rode, T.; Steinseifer, U. Polyurethane heart valves: Past, present and future. *Expert Rev. Med Devices* **2011**, *8*, 227–233. [CrossRef]
13. Udayakumar, G.P.; Muthusamy, S.; Selvaganesh, B.; Sivarajasekar, N.; Rambabu, K.; Banat, F.; Sivamani, S.; Sivakumar, N.; Hosseini-Bandegharaei, A.; Show, P.L. Biopolymers and composites: Properties, characterization and their applications in food, medical and pharmaceutical industries. *J. Environ. Chem. Eng.* **2021**, *9*, 105322. [CrossRef]
14. Mohammadi, H. Nanocomposite biomaterial mimicking aortic heart valve leaflet mechanical behaviour. *Proc. Inst. Mech. Eng. Part H J. Eng. Med.* **2011**, *225*, 718–722. [CrossRef]
15. Ovcharenko, E.A.; Seifalian, A.; Rezvova, M.A.; Klyshnikov, K.Y.; Glushkova, T.V.; Akenteva, T.N.; Antonova, L.V.; Velikanova, E.A.; Chernonosova, V.S.; Shevelev, G.Y.; et al. A New Nanocomposite Copolymer Based On Functionalised Graphene Oxide for Development of Heart Valves. *Sci. Rep.* **2020**, *10*, 1–14. [CrossRef]
16. Rozeik, M.M.; Wheatley, D.J.; Gourlay, T. Investigating the Suitability of Carbon Nanotube Reinforced Polymer in Transcatheter Valve Applications. *Cardiovasc. Eng. Technol.* **2017**, *8*, 357–367. [CrossRef]
17. Rezvova, M.A.; Nikishau, P.A.; Makarevich, M.I.; Glushkova, T.V.; Klyshnikov, K.Y.; Akentieva, T.N.; Efimova, O.S.; Nikitin, A.P.; Malysheva, V.Y.; Matveeva, V.G.; et al. Biomaterials Based on Carbon Nanotube Nanocomposites of Poly(styrene-b-isobutylene-b-styrene): The Effect of Nanotube Content on the Mechanical Properties, Biocompatibility and Hemocompatibility. *Nanomaterials* **2022**, *12*, 733. [CrossRef]
18. Rahmani, B.; Tzamtzis, S.; Ghanbari, H.; Burriesci, G.; Seifalian, A.M. Manufacturing and hydrodynamic assessment of a novel aortic valve made of a new nanocomposite polymer. *J. Biomech.* **2012**, *45*, 1205–1211. [CrossRef]
19. Thornton, M.A.; Howard, I.C.; Patterson, E.A. Three-dimensional stress analysis of polypropylene leaflets for prosthetic heart valves. *Med Eng. Phys.* **1997**, *19*, 588–597. [CrossRef]
20. Bernacca, G.M.; O'connor, B.; Williams, D.F.; Wheatley, D.J. Hydrodynamic function of polyurethane prosthetic heart valves: Influences of Young's modulus and leaflet thickness. *Biomaterials* **2002**, *23*, 45–50. [CrossRef]

21. Guo, F.; Liu, C.; Han, R.; Lu, Q.; Bai, Y.; Yang, R.; Niu, D.; Zhang, X. Bio-inspired anisotropic polymeric heart valves exhibiting valve-like mechanical and hemodynamic behavior. *Sci. China Mater.* **2019**, *63*, 629–643. [CrossRef]
22. Heitkemper, M.; Hatoum, H.; Dasi, L.P. In vitro hemodynamic assessment of a novel polymeric transcatheter aortic valve. *J. Mech. Behav. Biomed. Mater.* **2019**, *98*, 163–171. [CrossRef]
23. Gharaie, S.H.; Mosadegh, B.; Morsi, Y. In Vitro Validation of a Numerical Simulation of Leaflet Kinematics in a Polymeric Aortic Valve Under Physiological Conditions. *Cardiovasc. Eng. Technol.* **2018**, *9*, 42–52. [CrossRef]
24. Ghanbari, J.; Dehparvar, A.; Zakeri, A. Design and Analysis of Prosthetic Heart Valves and Assessing the Effects of Leaflet Design on the Mechanical Attributes of the Valves. *Front. Mech. Eng.* **2022**, *8*, 2. [CrossRef]
25. Wu, W.; Pott, D.; Mazza, B.; Sironi, T.; Dordoni, E.; Chiastra, C.; Petrini, L.; Pennati, G.; Dubini, G.; Steinseifer, U.; et al. Fluid–Structure Interaction Model of a Percutaneous Aortic Valve: Comparison with an In Vitro Test and Feasibility Study in a Patient-Specific Case. *Ann. Biomed. Eng.* **2016**, *44*, 590–603. [CrossRef]
26. Luraghi, G. Insights from experimental and FSI simulation of a polymeric heart valve. In Proceedings of the ESB-ITA Thematic Conference 2016, Palermo, Italy, 8–9 September 2016.
27. Thiebes, A.L.; Kelly, N.; Sweeney, C.A.; McGrath, D.J.; Clauser, J.; Kurtenbach, K.; Gesche, V.N.; Chen, W.; Kok, R.J.; Steinseifer, U.; et al. PulmoStent: In Vitro to In Vivo Evaluation of a Tissue Engineered Endobronchial Stent. *Ann. Biomed. Eng.* **2017**, *45*, 873–883. [CrossRef]
28. Toggweiler, S.; Wood, D.A.; Rodés-Cabau, J.; Kapadia, S.; Willson, A.B.; Ye, J.; Cheung, A.; Leipsic, J.; Binder, R.K.; Gurvitch, R.; et al. Transcatheter Valve-In-Valve Implantation for Failed Balloon-Expandable Transcatheter Aortic Valves. *JACC Cardiovasc. Interv.* **2012**, *5*, 571–577. [CrossRef]
29. Roman, M.J.; Devereux, R.B.; Kramer-Fox, R.; O'Loughlin, J. Two-dimensional echocardiographic aortic root dimensions in normal children and adults. *Am. J. Cardiol.* **1989**, *64*, 507–512. [CrossRef]
30. Marom, G.; Haj-Ali, R.; Raanani, E.; Schäfers, H.-J.; Rosenfeld, M. A fluid–structure interaction model of the aortic valve with coaptation and compliant aortic root. *Med Biol. Eng. Comput.* **2012**, *50*, 173–182. [CrossRef]
31. Sturla, F.; Votta, E.; Stevanella, M.; Conti, C.A.; Redaelli, A. Impact of modeling fluid–structure interaction in the computational analysis of aortic root biomechanics. *Med Eng. Phys.* **2013**, *35*, 1721–1730. [CrossRef]
32. Cao, K.; Sucosky, P. Computational comparison of regional stress and deformation characteristics in tricuspid and bicuspid aortic valve leaflets. *Int. J. Numer. Methods Biomed. Eng.* **2017**, *33*, e02798. [CrossRef]
33. Wei, Z.A.; Sonntag, S.J.; Toma, M.; Singh-Gryzbon, S.; Sun, W. Computational Fluid Dynamics Assessment Associated with Transcatheter Heart Valve Prostheses: A Position Paper of the ISO Working Group. *Cardiovasc. Eng. Technol.* **2018**, *9*, 289–299. [CrossRef]
34. Claiborne, T.; Xenos, M.; Sheriff, J.; Chiu, W.-C.; Soares, J.; Alemu, Y.; Gupta, S.; Judex, S.; Slepian, M.J.; Bluestein, D. Toward Optimization of a Novel Trileaflet Polymeric Prosthetic Heart Valve via Device Thrombogenicity Emulation. *ASAIO J.* **2013**, *59*, 275–283. [CrossRef]
35. Gulbulak, U.; Ertas, A.; Baturalp, T.B.; Pavelka, T. The effect of fundamental curves on geometric orifice and coaptation areas of polymeric heart valves. *J. Mech. Behav. Biomed. Mater.* **2020**, *112*, 104039. [CrossRef]
36. *LS-DYNA_Manual_Volume_II_R13 LSTC*; Livermore Software Technology: Canonsburg, PA, USA, 2021.
37. Piatti, F.; Sturla, F.; Marom, G.; Sheriff, J.; Claiborne, T.E.; Slepian, M.J.; Redaelli, A.; Bluestein, D. Hemodynamic and thrombogenic analysis of a trileaflet polymeric valve using a fluid–structure interaction approach. *J. Biomech.* **2015**, *48*, 3641–3649. [CrossRef]
38. Gulbulak, U.; Gecgel, O.; Ertas, A. A deep learning application to approximate the geometric orifice and coaptation areas of the polymeric heart valves under time—Varying transvalvular pressure. *J. Mech. Behav. Biomed. Mater.* **2021**, *117*, 104371. [CrossRef]
39. Claiborne, T.E.; Sheriff, J.; Kuetting, M.; Steinseifer, U.; Slepian, M.J.; Bluestein, D. In Vitro Evaluation of a Novel Hemodynamically Optimized Trileaflet Polymeric Prosthetic Heart Valve. *J. Biomech. Eng.* **2013**, *135*, 021021–0210218. [CrossRef]
40. Ghosh, R.P.; Marom, G.; Rotman, O.M.; Slepian, M.J.; Prabhakar, S.; Horner, M.; Bluestein, D. Comparative Fluid–Structure Interaction Analysis of Polymeric Transcatheter and Surgical Aortic Valves' Hemodynamics and Structural Mechanics. *J. Biomech. Eng.* **2018**, *140*, 121002. [CrossRef]
41. Dasi, L.P.; Simon, H.A.; Sucosky, P.; Yoganathan, A.P. Fluid Mechanics of Artificial Heart Valves. *Clin. Exp. Pharmacol. Physiol.* **2009**, *36*, 225–237. [CrossRef]
42. Tango, A.M.; Ducci, A.; Burriesci, G. In silico study of the ageing effect upon aortic valves. *J. Fluids Struct.* **2021**, *103*, 103258. [CrossRef]
43. ISO 5840-3:2021; Cardiovascular implants-Cardiac valve prostheses-Part 3:Heart valve substitutes implanted by transcatheter techniques. The International Organization for Standardization: Geneva, Switzerland, 2021.
44. Caddell, R.; Raghava, R.; Atkins, A. Pressure dependent yield criteria for polymers. *Mater. Sci. Eng.* **1974**, *13*, 113–120. [CrossRef]
45. Fan, R.; Bayoumi, A.S.; Chen, P.; Hobson, C.M.; Wagner, W.R.; Mayer, J.E., Jr.; Sacks, M.S. Optimal elastomeric scaffold leaflet shape for pulmonary heart valve leaflet replacement. *J. Biomech.* **2013**, *46*, 662–669. [CrossRef]
46. Hsu, M.-C.; Kamensky, D.; Bazilevs, Y.; Sacks, M.S.; Hughes, T.J.R. Fluid–structure interaction analysis of bioprosthetic heart valves: Significance of arterial wall deformation. *Comput. Mech.* **2014**, *54*, 1055–1071. [CrossRef]
47. Hsu, M.-C.; Kamensky, D.; Xu, F.; Kiendl, J.; Wang, C.; Wu, M.C.H.; Mineroff, J.; Reali, A.; Bazilevs, Y.; Sacks, M.S. Dynamic and fluid–structure interaction simulations of bioprosthetic heart valves using parametric design with T-splines and Fung-type material models. *Comput. Mech.* **2015**, *55*, 1211–1225. [CrossRef]

28. Luraghi, G.; Wu, W.; De Gaetano, F.; Matas, J.F.R.; Moggridge, G.D.; Serrani, M.; Stasiak, J.; Costantino, M.L.; Migliavacca, F. Evaluation of an aortic valve prosthesis: Fluid-structure interaction or structural simulation? *J. Biomech.* **2017**, *58*, 45–51. [CrossRef]
29. Johnson, E.L.; Wu, M.C.H.; Xu, F.; Wiese, N.M.; Rajanna, M.R.; Herrema, A.J.; Ganapathysubramanian, B.; Hughes, T.J.R.; Sacks, M.S.; Hsu, M.-C. Thinner biological tissues induce leaflet flutter in aortic heart valve replacements. *Proc. Natl. Acad. Sci. USA* **2020**, *117*, 19007–19016. [CrossRef]
30. Sigüenza, J.; Pott, D.; Mendez, S.; Sonntag, S.J.; Steinseifer, U.; Nicoud, F.; Kaufmann, T.A.S. Fluid-structure interaction of a pulsatile flow with an aortic valve model: A combined experimental and numerical study. *Int. J. Numer. Methods Biomed. Eng.* **2018**, *34*, e2945. [CrossRef]
31. Xu, F.; Morganti, S.; Zakerzadeh, R.; Kamensky, D.; Auricchio, F.; Reali, A.; Hughes, T.J.R.; Sacks, M.S.; Hsu, M.-C. A framework for designing patient-specific bioprosthetic heart valves using immersogeometric fluid-structure interaction analysis. *Int. J. Numer. Methods Biomed. Eng.* **2018**, *34*, e2938. [CrossRef]
32. Prawel, D.A.; Dean, H.; Forleo, M.; Lewis, N.; Gangwish, J.; Popat, K.C.; Dasi, L.P.; James, S.P. Hemocompatibility and Hemodynamics of Novel Hyaluronan–Polyethylene Materials for Flexible Heart Valve Leaflets. *Cardiovasc. Eng. Technol.* **2014**, *5*, 70–81. [CrossRef]
33. Rahmani, B.; Tzamtzis, S.; Sheridan, R.; Mullen, M.J.; Yap, J.; Seifalian, A.M.; Burriesci, G. In Vitro Hydrodynamic Assessment of a New Transcatheter Heart Valve Concept (the TRISKELE). *J. Cardiovasc. Transl. Res.* **2017**, *10*, 104–115. [CrossRef]
34. Khoffi, F.; Heim, F. Mechanical degradation of biological heart valve tissue induced by low diameter crimping: An early assessment. *J. Mech. Behav. Biomed. Mater.* **2015**, *44*, 71–75. [CrossRef]

Disclaimer/Publisher's Note: The statements, opinions and data contained in all publications are solely those of the individual author(s) and contributor(s) and not of MDPI and/or the editor(s). MDPI and/or the editor(s) disclaim responsibility for any injury to people or property resulting from any ideas, methods, instructions or products referred to in the content.

Review

Updated Perspectives on Direct Vascular Cellular Reprogramming and Their Potential Applications in Tissue Engineered Vascular Grafts

Saneth Gavishka Sellahewa [1], Jojo Yijiao Li [1] and Qingzhong Xiao [1,2,*]

[1] William Harvey Research Institute, Faculty of Medicine and Dentistry, Queen Mary University of London, London EC1M 6BQ, UK
[2] Key Laboratory of Cardiovascular Diseases, School of Basic Medical Sciences, Guangzhou Institute of Cardiovascular Disease, The Second Affiliated Hospital, Guangzhou Medical University, Guangzhou 511436, China
* Correspondence: q.xiao@qmul.ac.uk; Tel.: +44-(0)2078826584

Abstract: Cardiovascular disease is a globally prevalent disease with far-reaching medical and socio-economic consequences. Although improvements in treatment pathways and revascularisation therapies have slowed disease progression, contemporary management fails to modulate the underlying atherosclerotic process and sustainably replace damaged arterial tissue. Direct cellular reprogramming is a rapidly evolving and innovative tissue regenerative approach that holds promise to restore functional vasculature and restore blood perfusion. The approach utilises cell plasticity to directly convert somatic cells to another cell fate without a pluripotent stage. In this narrative literature review, we comprehensively analyse and compare direct reprogramming protocols to generate endothelial cells, vascular smooth muscle cells and vascular progenitors. Specifically, we carefully examine the reprogramming factors, their molecular mechanisms, conversion efficacies and therapeutic benefits for each induced vascular cell. Attention is given to the application of these novel approaches with tissue engineered vascular grafts as a therapeutic and disease-modelling platform for cardiovascular diseases. We conclude with a discussion on the ethics of direct reprogramming, its current challenges, and future perspectives.

Keywords: cellular reprogramming; cell transdifferentiation; direct cellular lineage-conversion; tissue engineered vascular grafts; vascular regeneration; stem cells; vascular progenitor cells; smooth muscle cells; endothelial cells; atherosclerosis; cardiovascular disease

1. Introduction

Cardiovascular disease (CVD) is an increasingly prevalent cause of global morbidity and mortality and affects half a billion people worldwide [1]. Endovascular and surgical revascularisation therapies have been increasingly applied in patients with severe CVD. However, such measures are still limited and fail to recapitulate healthy arterial tissue, as evidenced by that almost 1 in 2 autologous saphenous vein grafts will experience graft failure by 10 years post-surgery due to factors such as vulnerability to arterial pressure, intimal hyperplasia, and continued atherosclerosis [2,3]. Moreover, small diameter vessels (<6 mm) commonly found in cerebral, cardiac and peripheral regions are hard to treat with limited grafts, poor surgical accessibility and subpar performance of synthetic polymer prosthetics [4].

To overcome the abovementioned limitations and present a new treatment modality, research has looked to an innovative regenerative approach known as direct reprogramming (synonyms include transdifferentiation and direct lineage-conversion) defined as "the process of inducing a desired cell fate, by converting somatic cells from one lineage to another without transitioning through an intermediate pluripotent or multipotent state" [5].

Direct reprogramming is a descendent of a large body of cellular reprogramming research which, alongside the Nobel Prize-winning work by Takahashi and Yamanaka, has shown that cell fate is not locked but can be manipulated through the ectopic expression of pluripotency factors, lineage-specific transcription factors, small molecules and non-coding RNAs [6]. The key galvanizing feature of this approach is the avoidance of pluripotency or multipotency. In contrast, induced-pluripotent stem cell (iPSC) generation requires the dedifferentiation of somatic cells to a pluripotent cell which then differentiates to the required lineage, often requiring ex vivo expansion [5]. Consequently, iPSC generation comes with multiple distinct limitations including costly, time-consuming, complex cell generation protocols, and tight regularity oversight due to the risks of tumorigenicity [7].

Multiple reprogramming approaches have emerged to generate vascular cells from somatic cells. The first protocol is using a partial reprogramming approach to generate partially iPSCs (PiPSCs). Specifically, the four pluripotency factors (OSKM; OCT3/4, SOX2, KLF4 and C-MYC) discovered by Takahashi and Yamanaka [6] is initially transfected into somatic cells for a short period (normally up to seven days) to produce PiPSCs. When placed in differentiation mediums, the PiPSCs further differentiate to the desired cellular lineage. Another common protocol uses a single or combination of lineage-specific transcription factors to induce the expression of lineage-specific genes while silencing original somatic cell genes by modulating chromatin configurations of target genes [5]. Two other approaches use small molecules to either induce a progenitor-like state [8] or activate innate immunity to form a state of epigenetic plasticity [9] which is sensitive to differentiation signals.

To provide an in-depth knowledge of the current arterial direct reprogramming landscape and future research directions, we conducted the narrative literature review by identifying literature sources from peer-reviewed journals on the PubMed and MEDLINE databases in the last 10 years (2012–2022). The key search terms used in this review included: cellular reprogramming, direct cellular reprogramming, transdifferentiation, cellular lineage-conversion, somatic cells, vascular cells, endothelial cells, smooth muscle cells, vascular progenitor cells, vascular regeneration and tissue engineered vascular grafts. The search terms were combined in various combinations with the 'AND' command to identify primary literature sources that specifically explored direct reprogramming approaches and their application in TEVGs. Any sources solely exploring iPSCs reprogramming approaches were excluded. Particularly, in this review we examine the generation of each arterial cell group (endothelial (ECs), smooth muscle (SMCs), and vascular progenitor (VPCs) cells) with an analysis of the specific reprogramming factors and their molecular pathways. We also briefly discuss how direct reprogramming strategies could be applied to tissue engineered vascular grafts (TEVG) for disease-modelling and for clinical application as vascular conduits. Finally, we conclude the review with a discussion on the ethics of direct reprogramming, its current challenges, and future perspectives.

2. Endothelial Cell Generation

Endothelial cells are found as a continuous monolayer in the tunica intima of arteries and directly interface with the bloodstream. Their roles are numerous and include the regulation of haemostasis, vascular tone, immunity, and angiogenesis. Here, we explore several strategies to directly generate functional ECs: pluripotency factors-, lineage-specific transcription factors-, innate-immune activation-, and microRNA-based reprogramming (Table 1; Figure 1).

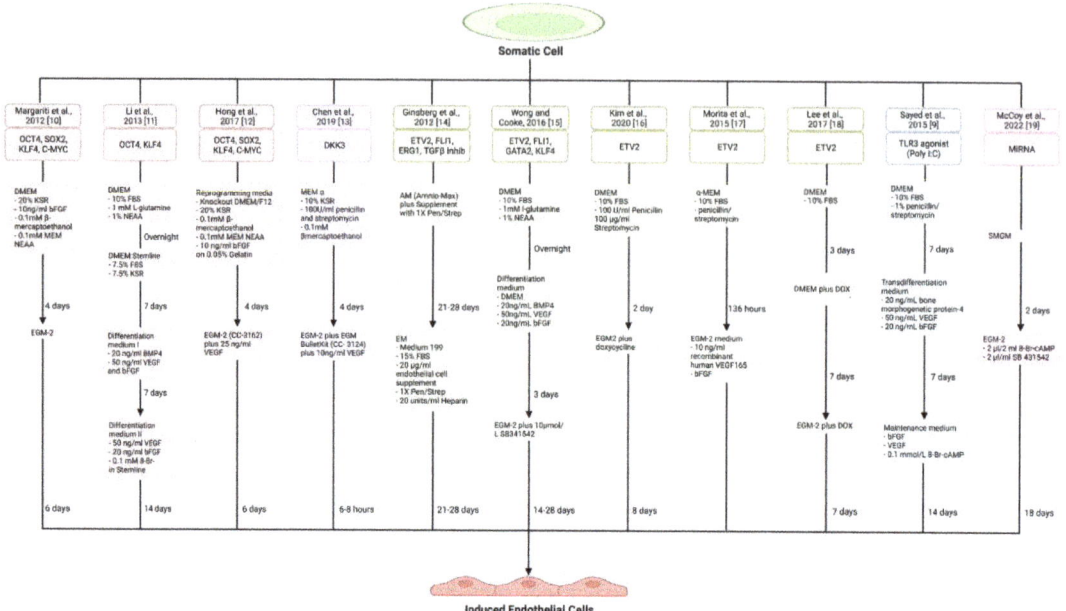

Figure 1. Schematic diagram showing key reprogramming factors, induction medium, and duration used for creating induced endothelial cells (iECs) in the different studies. Colour code: Red, Pluripotency factor-based reprogramming; Purple, DKK3-induced cellular reprogramming; Green: Lineage-specific transcription factor ETV2-based reprogramming; Blue, Innate-immune activation-induced cellular reprogramming; Pink, MicroRNA-based reprogramming. bFGF, basic Fibroblast Growth Factor; 8-Br-CAMP, 8-Bromo-cAMP; cAMP, cyclic AMP; BMP4, Bone Morphogenetic Protein 4; DKK3, Dickkopf WNT Signalling Pathway Inhibitor 3; DMEM, Dulbecco's Modified Eagle Medium; EC, Endothelial Cell; EGM-2, Endothelial Growth Medium-2; EM, Endothelial Medium; ERG1, Early Growth Response Protein 1; ETV2, ETS Variant Transcription Factor 2; FBS, Fetal Bovine Serum; FLI1, Friend Leukaemia Integration 1 Transcription Factor; GATA2, GATA-binding factor 2; KSR, Knockout Serum Replacement; MEM, Minimum Essential Medium; miRNA, MicroRNA; NEAA, Non-Essential Amino Acid; OSKM, OCT4 (Octamer-Binding Transcription Factor 4), SOX2 (SRY-Box Transcription Factor 2), KLF4 (Kruppel-Like Factor 4) and C-MYC (c-Myc proto-oncogene protein); Pen, Penicillin; Poly I:C, Polyinosinic-polycytidylic acid; SB341542, TGF-β Receptor Kinase Inhibitor; SMGM, Smooth Muscle Cell Growth Media; Strep, Streptomycin; TGFβ, Transforming Growth Factor Beta; TLR3, Toll-Like Receptor 3; VEGF, Vascular Endothelial Growth Factor.

Table 1. Generation of endothelial cells through cellular reprogramming.

Reference	Source Cells	Transcription Factors	Culture Medium	Functional Outcome	Therapeutic Potential	Signalling Pathway	Limitations
Margariti et al., 2012 [10]	Human embryonic lung fibroblasts (HELF)	OCT4, SOX2, KLF4, C-MYC	EGM2	iECs were stable and formed patent vessels when constituted onto a decellularised vessel scaffold	Hindlimb ischaemia: increased capillary number and blood perfusion	SETSIP activation which promotes EC-specific gene expression	Embryonic cell source is ethically controversial

Table 1. *Cont.*

Reference	Source Cells	Transcription Factors	Culture Medium	Functional Outcome	Therapeutic Potential	Signalling Pathway	Limitations
Li et al., 2013 [11]	Human neonatal fibroblasts	OCT4, KLF4	Differentiation medium II (50 ng/mL VEGF, 20 ng/mL bFGF, 0.1 mM 8-Br-CAMP)	Addition of 8-Br-cAMP increased trans-differentiation of fibroblasts into iECs	Murine hindlimb ischaemic model observed increased capillary number and blood perfusion	Not assessed	Conversion efficacy was low compared to studies using all 4 OSKM factors prior to sorting methods
Hong et al., 2017 [12]	Human umbilical artery smooth muscle cells	OCT4, SOX2, KLF4, C-MYC	EGM-2 (CC-3162) plus 25 ng/mL VEGF	SMCs are capable of trans-differentiation to iECs	Murine hindlimb ischaemic model observed increased capillary number and blood perfusion	OSKM upregulates VE-cadherin, HES4 and JAG1 which increases EC-specific gene expression	Lentiviral vectors possess safety risks. Viability of plasmid delivery confirmed but not explored
Chen et al., 2019 [13]	HELF	DKK3	EGM-2 plus EGM BulletKit (CC-3124) plus 10 ng/mL VEGF	iECs formed a patent monolayer in an ex vivo vascular graft	Formation of microvascular structures in vivo	Increases MET and VEGFR2, decreases miR-125a-5p and promotes Stat3	Embryonic cell source is ethically controversial
Ginsberg et al., 2012 [14]	Human amniotic fluid-derived cells	ETV2, FLI1, ERG1, TGFβ inhibition	EM (Medium 199, 15% FBS, 20 µg/mL endothelial cell supplement, 1X Pen/Strep, 20 units/mL Heparin)	iECs were generic and may hold potential for further subtype specification	In a regenerating mouse liver model, engraftment of iECs resulted in patent capillaries	Not assessed	Unsuccessfully with human postnatal cells. Use of amniotic cells is ethically controversial
Wong and Cooke, 2016 [15]	Human neonatal foreskin fibroblasts	ETV2, FLI1, GATA2, KLF4	EGM-2 plus 10 µmol/L SB341542	iECs uptake Ac-LDL and formed capillary-like networks	Not assessed	Not assessed	Therapeutic potential unknown
Kim et al., 2020 [16]	Human dermal fibroblasts	ETV2	EGM-2 plus doxycycline	iECs formed stable endothelial layers when seeded on a decellularised liver scaffold	Hindlimb ischaemia: improved angiogenic capabilities and blood perfusion	cAMP/EPAC/RAP1	iECs were not easily expandable
Morita et al., 2015 [17]	Human dermal fibroblasts	ETV2	EGM-2 medium (10 ng/mL recombinant human VEGF165, bFGF)	iECs displayed venous properties but adopted arteriole characteristics when combined with mural cells	Hindlimb ischaemia: improved angiogenic capabilities and blood perfusion	Modify DNA methylation states of EC genes	Extensive 50-day ETV2 exposure, lacking maturity, failed to induce NOS3
Lee et al., 2017 [18]	Human postnatal dermal fibroblasts	ETV2	EGM-2 plus DOX	Generation of early immature iECs, followed by matured iECs	Injection of early iECs into a murine hindlimb ischaemic model improved vessel generation and tissue perfusion	Not assessed	Early immature iECs failed to direct incorporation into host vasculature. Long timeline to cultivate mature iECs

Table 1. Cont.

Reference	Source Cells	Transcription Factors	Culture Medium	Functional Outcome	Therapeutic Potential	Signalling Pathway	Limitations
Sayed et al., 2015 [9]	Human neonatal foreskin fibroblasts	Poly I:C (TLR3 agonist)	Maintenance medium (bFGF, VEGF, 0.1 mmol/L 8-Br-cAMP)	Innate immune activation is necessary for human fibroblasts to transdifferentiate into ECs effectively	Murine hindlimb ischaemic model observed increased expansion of host vasculature, blood perfusion and decreased tissue injury	Innate immune activation, TLR3/NF-κB/iNOS, epigenetic plasticity. Metabolic switching from oxidative phosphorylation to glycolysis	Low transdifferentiation efficacy. Therapeutic potential unknown, metabolic heterogeneity in iECs
McCoy et al., 2022 [19]	Human coronary artery smooth muscle cells (CASMCs)	miRNA	EGM-2 (2 µL/2 mL 8-Br-cAMP, 2 µL/mL SB 431542)	iECs exhibit high similarity to native ECs	Quicker limb reperfusion	Upregulation of NOTCH1, JAG1, and DLL4	Other miRNA targets need to be explored further

Ac-LDL, Acetylated-Low-Density Lipoprotein; bFGF, basic Fibroblast Growth Factor; 8-Br-CAMP, 8-Bromo-cAMP; cAMP, cyclic AMP; DKK3, Dickkopf WNT Signalling Pathway Inhibitor 3; DLL4, Delta Like Canonical Notch Ligand 4; DNA, Deoxyribonucleic Acid; DOX, Doxycycline; EC, Endothelial Cell; EGM-2, Endothelial Growth Medium-2; EM, Endothelial Medium; EPAC, Exchange Proteins directly Activated by cAMP; ERG1, Early Growth Response Protein 1; ETV2, ETS Variant Transcription Factor 2; FLI1, Friend Leukaemia Integration 1 Transcription Factor; GATA2, GATA-binding factor 2; HELF, Human Embryonic Lung Fibroblasts; HES4, Hes Family BHLH Transcription Factor; iEC, induced Endothelial Cell; iNOS, inducible Nitric Oxide Synthase; JAG1, Jagged Canonical Notch Ligand 1; MET, Mesenchymal-to-Epithelial Transition; miRNA, MicroRNA; miR-125a-5p, MicroRNA-125a-5p; NF-κB, Nuclear Factor Kappa B; NOTCH1, Neurogenic Locus Notch Homolog Protein 1; NOS3, Nitric Oxide Synthase 3; OSKM, OCT4 (Octamer-Binding Transcription Factor 4), SOX2 (SRY-Box Transcription Factor 2), KLF4 (Kruppel-Like Factor 4) and C-MYC (c-Myc proto-oncogene protein); Poly I:C, Polyinosinic-polycytidylic acid; RAP1, Ras-Related Protein 1; SB341542, TGF-β Receptor Kinase Inhibitor; SETSIP, Set Like Protein; Stat3, Signal transducer and activator of transcription 3; TGFβ, Transforming Growth Factor Beta; TLR3, Toll-Like Receptor 3; VEGF, Vascular Endothelial Growth Factor; VEGFR2, Vascular Endothelial Growth Factor Receptor 2.

2.1. Pluripotency Factor-Based Reprogramming

Direct reprogramming with the Yamanaka factors to generate induced-endothelial cells (iECs) was first demonstrated by Margariti et al. [10]. 4-day lentiviral overexpression of OSKM successfully dedifferentiated human fibroblasts into a partially induced-pluripotent state which did not express pluripotency markers such as SSEA-1 and did not generate any tumours when injected into mice. With the addition of endothelial differentiation culture medium (EGM-2 media), iECs were formed displaying typical endothelial morphology and functions (Low-density lipoprotein (LDL) uptake and angiogenesis), as well as expressing a panel of endothelial-specific markers such as vascular endothelial growth factor receptor 2 (VEGFR2; also known as kinase domain receptor, KDR), endothelial nitrous oxide synthase (eNOS), and von Willebrand factor (vWF). eNOS produces nitric oxide (NO), which vasodilates arteries, controls cell growth and resists inflammatory changes such as platelet aggregation. NO production is an important indicator of endothelial function and exerts an atheroprotective effect. However, this study did not assess NO production [10].

Based on Margariti and colleagues' discoveries, it was soon shown that not all Yamanaka factors are required for transdifferentiation. Li et al. showed that OCT4 and KLF4 are sufficient for successfully reprogramming human fibroblasts into iECs, albeit with lower conversion efficacies of around 1%, improved to 3.85% with 8-Br-cAMP [11]. This conversion efficacy is significantly lower when compared to the 3-factor combinations of OCT4, SOX2 and KLF4 at 11.8% [11] and 4-factor combinations of OSKM at ~30% [10,12]. Nevertheless, sorting methods for the two-factor protocol achieved a 97% pure population of cells expressing a key endothelial marker, platelet endothelial adhesion molecule-1 (PECAM-1; also known as cluster of differentiation 31 or CD31) [11]. Using fewer factors may achieve cheaper and faster protocols while avoiding oncogenic factors like C-MYC. As technology improves and more specific markers are identified, selection and sorting methods may overcome low transdifferentiation efficacies. Indeed, such methods will need to be robust as any potentially pluripotent cell that does escape filtering can exert a teratoma risk [20].

Hong et al. showed that arterial SMCs were also amenable to the Yamanaka factor-based reprogramming approach [12]. SMCs are a valuable cell source as they are abundant and found immediately adjacent to the endothelium. Moreover, they possess significant phenotypic plasticity and share mesodermal origins and common progenitors with ECs- during embryogenesis, suggesting that their genetic and epigenetic mountains are easier to climb in the transdifferentiation process [21]. Hong and colleagues first generated CD34-positive vascular progenitors with 4 days of OSKM overexpression which was followed by differentiation in EGM-2 media and VEGF for 6 days [12]. Within their population of iECs, 33.4% of cells expressed CD31. Other key endothelial genes were upregulated (CD34, KDR, CD144, eNOS and vWF) while SMC genes were downregulated (α-SMA, SM22α, calponin, SM-MHC). Representative of functional mature ECs, the iECs took up LDL and increased expression of an inflammatory molecule, intercellular adhesion molecule 1 (ICAM-1) in response to TNF-alpha stimulation [12].

Hong and colleagues also identified several key insights into the reprogramming mechanism. The mesenchymal-to-epithelial transition (MET) occurred simultaneously with the reprogramming of SMCs to a vascular progenitor [12]. The upregulation of vascular endothelial cadherin (VE-cadherin, also known as CD144), an essential endothelial homeostasis and angiogenesis regulator, was the key event in the transition. In addition, two components within the Notch pathway, HES5 (HES family transcription factor 5) and JAG1 (Jagged canonical Notch ligand 1, a surface protein), had essential regulatory functions in differentiation. Overexpression of both proteins increased EC marker expression, while knockout studies on HES5 lowered marker expression. JAG1 increased the promoter activity of HES5 and eNOS. The Notch signalling pathway is affiliated with various cellular functions from cell growth, cell fate regulation and angiogenesis [22]. In keeping with the above findings, a recent pioneering study confirmed Notch signalling as a promoter of MET by increasing JAG1 mRNA expression through the RBP-Jκ transcription factor [23]. Moreover, JAG1 expression by ECs has been shown to propagate the development of multi-layered SMCs around an endothelial layer through lateral induction, which holds an exciting research avenue where JAG1/Notch signalling can facilitate reconstitution of an arterial wall from induced cells [24]. Furthermore, crosstalk between Notch and VEGF has been shown to promote angiogenic sprouting, vascular branching and stabilise cell-to-cell junctions [25].

Although the above studies have demonstrated no in vivo tumour formation for the respective observation periods in mice, the potential for tumorigenesis cannot be ruled out with short-term overexpression of pluripotency factors. Lentiviral delivery of the factors is rather troublesome due to the risks of insertional mutagenesis, host genetic alterations and germline transfers [7]. A safer route with plasmid delivery of OSKM was trialled by Hong et al., although a thorough analysis of the protocol is yet to be established [12]. An alternative route to the dedifferentiated, progenitor-like state that bypasses viral genetic handling was achieved with the cytokine-like protein dickkopf-3 (DKK3) [13]. Again, 4-day overexpression of adenoviral-delivered DKK3 followed by culturing in EGM-2 media supplemented with VEGF for 6 days, robustly reprogrammed human fibroblasts into functional iECs. Pluripotency marker expression (OSKM) did not change throughout the protocol. Showcasing their novel ex vivo circulation bioreactor system, the authors implanted their DKK3-reprogrammed iECs onto a decellularised aortic mouse graft, and after 5 days of culturing, observed an iEC monolayer surrounded by a multi-layered SMC wall. However, further in-depth analysis of cellular changes, marker expression patterns and EC-mural associations was not conducted.

The authors add to the growing evidence that the MET occurs when the cells move towards a progenitor-like state with VE-cadherin interactions acting as a prerequisite for further endothelial differentiation. Furthermore, the anti-angiogenic activity of microRNA (miR)-125a-5p through regulation of Stat3 (signal transducer and activator of transcription 3) was observed. Stat3 has been identified to improve the production of NO and prostacyclin (a vasodilator) and promote angiogenesis through interactions with VEGF [26]. Mimics of miR-125a-5p resulted in reduced activity of Stat3, whereas Stat3 silencing was associated with reductions in EC-specific gene expression and in vitro vascular formation. Interestingly, in the absence of Stat3, miR-125a-5p had no negative regulatory effects on EC differentiation [13].

While no in vivo studies were conducted, and the reprogramming efficacy is not yet established, this single-factor protocol by Chen and colleagues does set the foundations for safer and faster reprogramming approaches without extensive genetic manipulation. Moreover, adenoviral delivery harbours a reduced risk of host DNA integration and is unlikely to promote immune reactions to lentiviral counterparts [27]. Moreover, DDK3 is atheroprotective in ECs, which highlights a synergistic therapeutic benefit in reprogramming strategies [28].

2.2. Lineage-Specific Transcription Factors—The Advent of ETV2

ETV2 (E-twenty-six variant transcription factor 2) is well-established as a potent regulator of embryonic vascular development [29]. The factor is transiently expressed during embryogenesis, becoming virtually undetectable in postnatal ECs but significantly elevated following endothelial injury [30]. Indeed, both ETV2 knockout and prolonged exposure are associated with abnormal embryological vascular development [31,32]. Based on the above findings, Ginsberg et al. theorised that ETV2 alongside two other ETS-domain transcription family factors, FLI1 and ERG1, together with TGFβ inhibition, could induce expression of endothelial-specific genes and hence reprogramming of human midgestation lineage-committed c-Kit negative amniotic cells to iECs [14]. Indeed, transient lentiviral ETV2 expression induced an immature endothelial progenitor-like state which matured and 'locked-in' the vascular identity in response to continued FLI1 and ERG1 expression. Interestingly, they observed that stoichiometric expression of ETV2 with FLI1 and ERG1 was required to generate iECs from amniotic cells efficiently. Amniotic cell-derived iECs were functional and expandable with a similar transcriptome to human umbilical vein endothelial cells (HUVECs). Further constitutive signalling with the protein kinase AKT1 and transcription factor SOX17 has been shown to improve the activation of the global endothelial gene program [33]. However, Ginsberg and colleagues' protocol failed to reprogram human postnatal cells, and interestingly, when repeated in another study, the addition of FLI1 and ERG1 observed a lower reprogramming efficacy in murine fibroblasts [34]. Moreover, the non-autologous implications of amniocentesis-derived therapy and the amniotic cell source have spurred a debate on the potential ethical and clinical limitations.

To circumvent the drawback of amniotic cells, Wong and Cooke used three other endothelial-specific transcription factors, FLI1, GATA2 and KLF4, alongside ETV2 to reprogram human neonatal fibroblasts into ECs, achieving a conversion efficacy of 16% for CD31-positive cells which was four times higher than the 5-factor protocol of ETV2, FOXO1, KLF2, TAL1 and LMO2 by Han and colleagues. ETV2 was identified as the most potent factor for reprogramming induction as systematically removing each transcription factor except ETV2 in 3-factor combinations only resulted in minimal reductions of CD31-positive cells. Neonatal fibroblasts were reportedly easier to reprogram than adult fibroblasts due to a more fluid epigenetic state. However, a study used modified mRNA encoding ETV2, FLI1, GATA2, and KLF4 to compare differences in angiogenic behaviour between reprogrammed cells derived from neonatal fibroblasts and dermal fibroblasts in patients with peripheral artery disease [35]. While successfully demonstrating reprogramming with modified mRNA, the study showed that neonatal and patient-derived iECs exhibited the same angiogenic behaviour, although this response was inconsistent among iECs derived from different patients. Further investigations are required to delineate differences between iECs generated from neonatal and adult cells.

Three studies reported ETV2 as an independent inductor of human adult fibroblast reprogramming, each with key insights into the endothelial reprogramming mechanism [16–18]. Morita and colleagues used a doxycycline-inducible system for lentiviral delivery of ETV2 to human adult fibroblasts, which after 15 days, were sorted for CD31-expressing iECs (3.5%). After suspension in Matrigel plugs and implantation into non-obese diabetic, severe combined immune deficiency mice, their iECs formed patent vasculature, which enabled erythrocyte circulation and expressed eNOS while almost 50% of iEC vessels associated with adjacent mural cells [17]. The authors further identified high levels of endogenous FOXC2 in human fibroblasts, interacting with ectopic ETV2 through a DNA binding site known as the FOX:ETS motif. Binding to the motif recruited factors which facilitated gene expression through post-transcriptional ETV2 alteration and modification of DNA methylation states of key endothelial genes. Different from the findings reported by Ginsberg et al. [14], ETV2 sufficiently induced endogenous FLI1 and ERG expression, and ectopic expression of the latter two factors was not required for generating iECs. An optimal differentiation efficacy was achieved with an intermediate ETV2 overexpression level, whereby deviations from this level negatively affected the reprogramming efficacy [17].

However, Morita and colleagues' protocol used 15-day exposure to ETV2 with an extensive 50-day culturing period which researchers suggest may have forced the upregulation of endothelial markers without promoting cellular maturity [7]. With a similar methodology, Lee et al. failed to produce ECs from human fibroblasts and instead showed that transient ETV2 expression, independent of FOXC2, is obligatory for transdifferentiation and is reminiscent of embryogenic ETV2 expression patterns [18]. Initial lentiviral ETV2 induction generated early iECs characterised by mixed endothelial and fibroblast gene markers and low vWF and PECAM1 expression. The authors acknowledged that the residual fibroblastic features might be advantageous through synergistic paracrine modulation that matures and maintains early iECs. In a murine model, these early iECs were demonstrated to increase EC proliferation, neovascularisation and reduce limb ischaemia. However, whether the therapeutic benefits were attained through a paracrine effect or direct incorporation into existing vasculature is unclear. Early-to-late transformation of ECs was demonstrated through a 20-day transgene-free period followed by 6-day ETV2 exposure. Late-iECs observed increased silencing of fibroblastic signatures, low ETV2 and increased PECAM1 and NO production, suggesting more significant phenotypic similarities to primary mature ECs. Adding valproic acid (VPA) alongside the early-to-late iEC conversion improved the differentiation efficacy from 2% to 60%. In vivo studies demonstrated the incorporation of late-iECs into host vessels [18].

Morita et al. further observed an arterial endothelial specification with enhanced eNOS expression and mural-endothelial associations when their reprogrammed cells were constituted in Matrigel plugs containing mural cells and exposed to environmental changes such as shear stress [17]. However, without mural cell association, cells expressed venous markers suggesting a venous phenotype. In keeping with previous studies, these findings attribute the pivotal role of local microenvironments and dynamic mechanical forces in maturing and directing ECs to a distinct sub-phenotype. Further arterial specification was achieved with forskolin, a labdane diterpenoid extracted from the Coleus barbatus plant, which has shown several cardioprotective effects in early clinical studies [36,37]. Kim and colleagues first demonstrated that ETV2 activates cyclic AMP (cAMP) and exchange proteins directly activated by cAMP (EPAC) [16]. Subsequently, GTP-bound RAP1 protein is activated, which stimulates EC gene expression. Forskolin acted as a cAMP activator which promoted cAMP/EPAC/RAP1 signalling and observed a 3.2-fold increase in the number of CD31$^+$/VE-cadherin$^+$ ECs generated while promoting arterial-specific endothelial markers through Notch signalling. Forskolin treatment demonstrated further suppression of mesenchymal markers and improved neo-vascularisation in vivo compared to vehicle-treated cells. RAP1 activation by ETV2 was shown in another study to promote the development of durable lumens [38].

Solidifying the role of ETV2 in endothelial reprogramming, ETV2 was recently identified as a pioneer factor [29]. From studies on mouse embryonic fibroblasts (MEFs), the authors showed that ETV2 binds to nucleosomes and recruits BRG1, which then forms a complex acting to relax chromatin configurations of endothelial genes while recruiting other cofactors and increasing H3K27ac deposition. BRG1 maintained open chromatin formations allowing ETV2 to recruit other factors to activate endothelial gene expression. The authors also highlight the contribution of immune signalling in facilitating or repressing reprogramming based on their observation that a MEF cluster with greater inflammatory activity experienced greater upregulation of genes compared to a cluster with less inflammatory activity. Finally, ETV2 suppressed non-endothelial lineages by downregulating mesodermal genes [29].

2.3. Innate-Immune Activation—The Big Potential of Small Molecules

A research group has contributed significantly towards a paradigm shift away from pluripotency and lineage-specific transcription factors-based cellular reprogramming [9,39,40]. The researchers showed that innate immune activation is a facilitator and independent inductor of direct reprogramming. Innate immune activation via small molecules could induce a state of epigenetic plasticity, which is then amenable to differentiation with specific signals—a phenomenon termed 'transflammation' [39]. A key benefit of this approach is that viral vectors are avoided, and genetic manipulation is minimised. Through three progress studies, the research group reported three key findings: (1) innate immunity could be activated by targeting TLR3 (Toll-like receptor 3), which induces phenotypic plasticity through the (2) activation of the inducible-NOS (iNOS) signalling pathway and (3) the metabolic conversion from oxidative phosphorylation to glycolysis. Inhibition of iNOS and glycolytic switching reduced or inhibited endothelial differentiation. Through these processes, innate immune activation effectively increased DNA accessibility for reprogramming.

In their introductory study, an agonist of TLR3, polyinosinic:polycytidylic acid (polyI:C, PIC), induced a state of epigenetic plasticity, namely a global changes in epigenetic modifiers that increase the probability for an open chromatin state, in human fibroblasts which trans-differentiated into ECs in response to culture medium containing VEGF, BMP4, FGF, 8-Br-cAMP, and a TGFβ-inhibitor [9]. The resulting iECs shared similar functional and genetic characteristics to native ECs, with a conversion efficacy of 2% for CD31-expressing cells. Even though the iECs failed to incorporate with the endogenous vasculature in a peripheral artery disease murine model, a potential paracrine effect of iECs was observed to promote angiogenesis and improve tissue perfusion. The investigators further alluded to unpublished data suggesting the viability of targeting alternative receptors such as TLR4 and RIG-I (retinoic acid-inducible gene I) for innate immune activation.

The following study aimed to elucidate the cellular pathways that underline the above reprogramming process [40]. Upon agonist activity of TLR3, an extensive signalling cascade began with activation of iNOS mediated by NFκB. The end feature of the cascade through NO production or direct binding to iNOS was the destabilisation of PRC1, which maintains suppressive epigenetic marks such as the trimethylation of histone 3 at lysine 27 (H3K27me3) of the CD31 promoter region. Decreased H3K27me3 results in greater DNA accessibility for key transdifferentiation factors.

In their latest study on the topic, innate immune activation was coupled to metabolic activity with a switch from oxidative phosphorylation to glycolysis [39]. Specifically, iECs were generated by treating fibroblasts with induction medium containing Poly I:C (30 ng/mL) and EC georth medium supplemented with VEGF (50 ng/mL), bFGF (20 ng/mL), BMP4 (20 ng/mL), and 8-Br-cAMP (100 µM) for the 1st and 2nd week, respectively. The CD31$^+$ iECs were sorted with flow cytometer and expanded in EC growth medium with SB431542 (10 µM). Mechanistically, by conducting multiple biochemical experimentation including Seahorse assay, the authors found that the metabolic changes occurred through numerous steps beginning with diverting some pyruvate and citrate away from the citric acid cycle and into the cytoplasm facilitated by the upregulation of mitochondrial citrate transporters (such as Slc25a1). Increased expression of nuclear ATP citrate lyase leads to the generation of nuclear acetyl-CoA. Acetyl-CoA then increases the activity of histone acetyltransferases which support histone acetylation and configures access to genes that promote transdifferentiation. The absence of alterations in fatty acid synthesis, upregulation of glycolytic enzymes by PIC and the induction of iNOS instead of eNOS provides compelling evidence that innate immune signalling independently contributed to the metabolic shift. However, the authors did observe metabolic heterogeneity in their population, which they suggested may be attributed to the metabolic heterogeneity of the starting fibroblast population. Ultimately, the iNOS pathway and glycolytic shift indirectly increase endothelial-specific gene expression by promoting an open chromatin composition. Further investigations into the metabolic shifts and mitochondria-nuclear signalling may identify more effective transdifferentiation strategies.

2.4. MicroRNA-Based Reprogramming

MicroRNAs (miRNA/miR) are increasingly observed to hold important roles in cell fate specification [41]. MiRNA was recently used to reprogram SMCs into iECs [19]. By identifying miRNA enrichment levels between ECs and SMCs, McCoy and colleagues demonstrated that transfection of EC-enriched miR-146a-5p and 181b-5p mimics, with the converse inhibition of SMC-enriched miR-143-4p and miR-145-5p could reprogram human coronary artery SMCs (CASMCs) into iECs. Their transdifferentiation protocol took place over 20 days with initial transfection followed by culturing in media consisting of EGM-2 (with a TFGβ inhibitor and 8-Br-cAMP), sorting for ICAM-1 positive cells and a final expansion phase. The iECs shared similar transcriptional and phenotypical profiles to HUVECs but not to CASMCs. In a murine hindlimb ischaemia model, mice transplanted with iECs experienced significantly faster perfusion times (142% faster) on day-11 post-injection and fewer toe loss (a sign of prolonged ischaemia) than HUVEC transplant recipients. The authors observed increased eNOS expression, which they believe reflects their angiogenic benefits. Furthermore, the upregulation of NOTCH1, JAG1 and DLL4 provides further evidence for the involvement of Notch signalling in SMC-to-EC conversion [19].

The 4-miR cocktail protocol confirms the viability of non-viral reprogramming approaches to generate iECs that are theoretically comparable to iECs generated through pluripotency and lineage-specific factors. However, there are a few areas for further improvement. Firstly, there may be better markers for iEC sorting than ICAM-1. Mature, quiescent ECs observe very low basal expression of ICAM-1, and its role is more robustly related to endothelial activation in pathogenic pathways, including atherosclerosis [42]. ICAM-1 is also expressed in other cells, such as macrophages and lymphocytes. Vascular cell adhesion molecule 1 is a more specific inflammatory marker for ECs, while CD31 and CD144 are generally considered highly reflective of mature ECs [43]. Secondly, the reprogramming efficacy is not clear which hinders any comparisons with other reprogramming approaches. Furthermore, several other miRs that could independently or synergistically induce reprogramming. For example, miR-539 and miR-582 are associated with SMC-to-EC communication to guide SMC development around ECs, while miR-125/126 have regulatory effects on TGFβ activity [19,44]. A cocktail of curated suppressive and expressive miR elements may yield high reprogramming efficacies. Further analysis of the effect of miR on disease states like atherosclerosis and diabetes will be essential to reduce the risk of adverse effects.

3. Smooth Muscle Generation

SMCs are found in the tunica media and help the artery withstand and regulate blood pressure by constricting and relaxing. Like ECs, several studies have reported using pluripotency and lineage-specific factors to reprogramme various somatic cells into induced-vascular smooth muscle cells (iSMCs) (Table 2; Figure 2). Human embryonic lung fibroblasts (HELF) were successfully reprogrammed into iSMCs with 4-day exposure to

OSKM, followed by 4-day culturing in SMC differentiation medium [45]. The resulting iSMCs closely resembled native SMCs with the upregulation of SMC-specific markers such as SM22 (smooth muscle protein 22), calponin, SMA (smooth muscle actin-alpha), MYH11 (myosin heavy chain), SRF (serum response factor), MYOCD, and smoothelin. MYH11 and smoothelin are regarded as the best indicators of a mature SMC [46]. Other lineage-specific markers were not upregulated in the process, an important finding as some SM-specific markers are transiently expressed in other cell types like myofibroblasts. However, expression patterns of SM-specific miRs (e.g., miR-143) seen in mature SMCs and used as reprogramming factors by Mccoy et al. [19] were not investigated in this study. Nonetheless, fibroblast gene downregulation and contractile responses to KCl administration did suggest a mature and functional SMC phenotype of iSMCs. Although only 38% of cells were SM22-positive in this approach, antibiotic selection of cells with neomycin produced a pure population of SM22-positive iSMCs, as reported in this study [45].

Table 2. Generation of smooth muscle cells through cellular reprogramming.

Reference	Source Cell	Transcription Factors	Culture Medium	Functional Outcome	In Vivo Therapeutic Potential	Signalling Pathway	Limitations
Karamariti et al., 2013 [45]	HELF	OCT4, SOX2, KLF4, C-MYC	DM (MEM α, 10% FBS, 100 U/mL penicillin and streptomycin, 0.2 mM L-glutamine, 0.1 mM β-mercaptoethanol, 10 ng/mL PDGF-BB)	iVSMCs	Transplantation of iVSMCs-seeded decellularised vessel in mice increased survival	DKK3/Kremen1/ Wnt signalling	Limited to HELF, Unknown efficacy of iVSMC generation, HELF is ethically controversial
Karamariti et al., 2018 [47]	HELF	DKK3	DMEM (ATCC, 10% EmbryoMax® ES Cell Qualified FBS, 10 ng/mL LIF, 0.1 mM 2-mercaptoethanol) on a 0.04% gelatin substrate	VPCs, iVSMCs	Promotes stabilisation of atherosclerotic plaques by increasing SMCs and suppressing inflammation	DKK3/ATF6/ TGFβ1	HELF is ethically controversial.
Hirai et al., 2018 [48]	MEF and adult dermal fibroblasts	Myocd, GATA6, MEF2C	SMC medium (DMEM/F-12, 10% KSR, 2 ng/mL recombinant human TGF-β1, 10 ng/mL human PDGF-BB, 1% penicillin-streptomycin)	iVSMCs	Not assessed	Not assessed	Partially reprogrammed iVSMCs

ATF6, Activating Transcription Factor 6; DKK3, Dickkopf WNT Signalling Pathway Inhibitor 3; DMEM, Dulbecco's Modified Eagle Medium; FBS, Fetal Bovine Serum; GATA6, GATA-binding factor 6; HELF, Human Embryonic Lung Fibroblasts; iVSMCs, iPSC-derived Vascular Smooth Muscle Cells; KREMEN 1, Kringle Containing Transmembrane Protein 1; KSR, Knockout Serum Replacement; MEF, Mouse Embryonic Fibroblasts; MEF2C, Myocyte-Specific Enhancer Factor 2C; MEM α, Minimum Essential Medium α; Myocd, Myocardin; OSKM, OCT4 (Octamer-Binding Transcription Factor 4), SOX2 (SRY-Box Transcription Factor 2), KLF4 (Kruppel-Like Factor 4) and C-MYC (c-Myc proto-oncogene protein); PDGF-BB, Platelet-Derived Growth Factor-BB; SMC, Smooth Muscle Cell; TGFβ1, Transforming Growth Factor Beta 1; VPCs, Vascular Progenitor Cells.

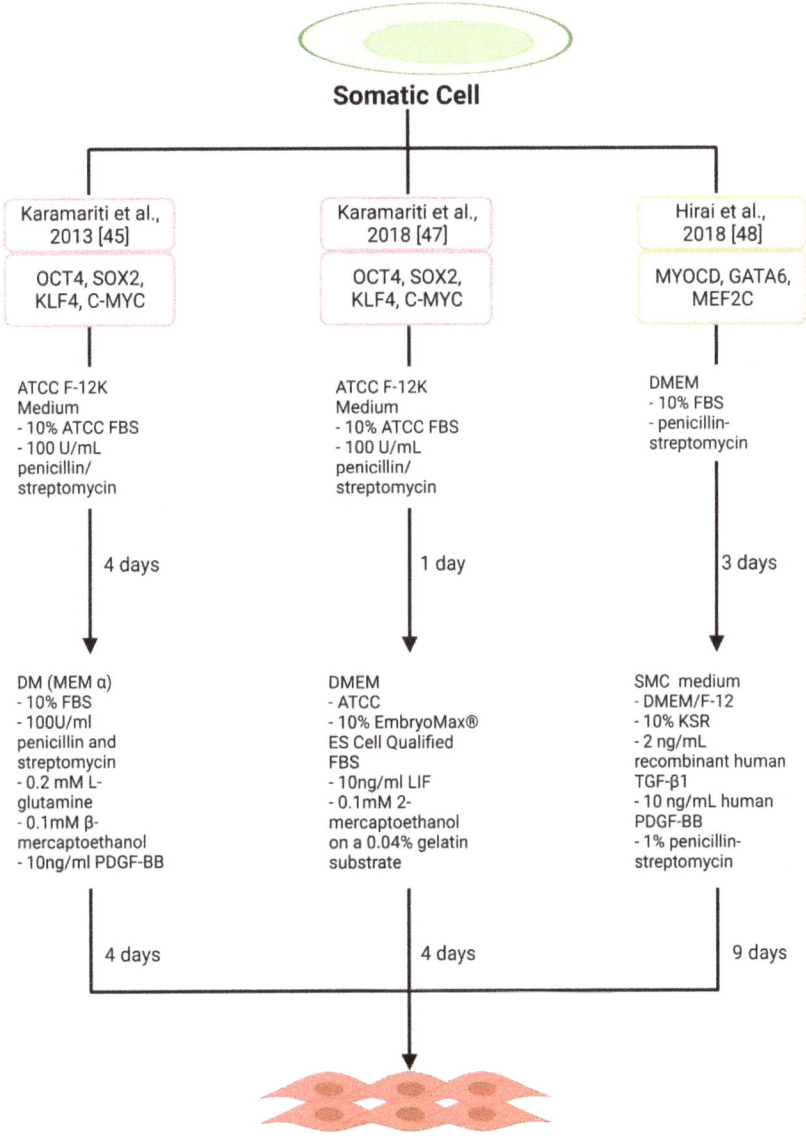

Figure 2. Schematic diagram showing key reprogramming factors, induction medium, and duration used for creating induced smooth muscle cells (iSMCs) in the different studies. Colour code: Red, Pluripotency factor-based reprogramming; Yellow, Lineage-specific transcription factors-based reprogramming. ATCC, ATCC medium; ATCC F-12K, ATCC-Kaighn's Modification of Ham's F-12 Medium; DKK3, Dickkopf WNT Signalling Pathway Inhibitor 3; DM, Differentiation Medium; DMEM, Dulbecco's Modified Eagle Medium; GATA6, GATA-binding factor 6; KSR, Knockout Serum Replacement; MEF, Mouse Embryonic Fibroblasts; MEF2C, Myocyte-Specific Enhancer Factor 2C; MEM α, Minimum Essential Medium α; MYOCD, Myocardin; OSKM, OCT4 (Octamer-Binding Transcription Factor 4), SOX2 (SRY-Box Transcription Factor 2), KLF4 (Kruppel-Like Factor 4) and C-MYC (c-Myc proto-oncogene protein); PDGF-BB, Platelet-derived growth factor-BB; SMC, Smooth Muscle Cell; TGFβ1, Transforming Growth Factor Beta 1.

Reprogramming HELFs to iSMCs identified a novel role of DKK3. During transdifferentiation, DKK3 binds to Kremen1 and activates canonical Wnt signalling, which then causes the translocation of β-catenin into the nucleus. B-catenin then enhances the transcriptional activity of SM22 [45]. Interestingly, the reprogramming protocol presented in this study appears to be specific to HELFs and OSKM, as repeating the approach on skin fibroblasts observed no significant changes in SMC marker expression [45]. From their identification of DKK3's involvement in reprogramming, Karamariti and colleagues further demonstrated adenoviral nucleofection of DKK3 as an independent inductor of reprogramming and readily transdifferentiated HELFs into iSMCs [47]. The iSMCs observed similar transcriptional and behaviour profiles to native SMCs. Analysis of the molecular pathway highlighted a different signalling cascade to OSKM-based iSMC reprogramming. DKK3 induced activity of the transcription factor ATF6 (activating transcription factor 6) which then increases transcription of TGFβ1. TGFβ1 expression increased SMC gene expression in fibroblasts. Furthermore, through several in-vitro and animal studies, the authors showed that DKK3-based reprogramming of both vascular progenitors and fibroblasts into iSMCs promotes stabilisation of atherosclerotic plaques (by suppressing inflammation and increasing SMCs)—highlighting a synergistic benefit of the protocol [47]. However, as DKK3 and TGFβ are both involved in reprogramming SMCs and ECs, the next challenge is identifying what pathways and molecules regulate each cell's fate. Moreover, further work must elucidate whether different mediums or combinations of factors could better transform skin fibroblasts into SMCs and avoid the ethical implications of obtaining HELFs and the invasive nature of lung biopsies.

The conversion of human dermal fibroblasts (HDFs), a more ethically acceptable cell source, into iSMCs was achieved through the retroviral introduction of three factors, MYOCD, GATA6 and MEF2C (collectively known as MG2) with a conversion efficacy of 80% for MYH11-positive cells [48]. MYOCD is well-known as the master regulator of smooth muscle differentiation. However, the three-factor combination only induced endogenous MYOCD in mouse embryonic fibroblasts (MEFs), which the authors speculate may reflect an MYOCD-independent reprogramming pathway in human aorta media. Active suppression of Kruppel-like transcription factors (KLFs) is an alternative explanation for the finding. Identifying the regulators of MYOCD is a topic of ongoing research. However, evidence does demonstrate the KLF family's repressive nature on MYOCD expression and their participation in the movements towards the synthetic-SMC phenotype [49,50]. Hence, high KLF5 expression in iSMCs compared to the control suggests partially reprogrammed cells. Further research needs to elucidate the role of MYOCD and its regulators in reprogramming. Suppressing KLF may facilitate the drive towards and maintenance of mature iSMCs.

4. Vascular Progenitor Cells

Several studies have reported small and rare colonies of progenitor cells found in different locations of the arterial wall, particularly the tunica adventitia [51]. These progenitors include endothelial, smooth muscle, haematopoietic, and multipotent stem/progenitor cells. For successful induction of a vascular progenitor, the final cell should exhibit high proliferative capacity, interact with the cellular matrix and readily differentiate into their respective cells. Similar to iECs and iSMCs, multiple cellular reprogramming methods have been reported to generate directly induced vascular progenitor cells (iVPCs) (Table 3; Figure 3).

Table 3. Generation of vascular progenitor cells through cellular reprogramming.

Reference	Source Cell	Transcription Factors	Culture Medium	Functional Outcome	In Vivo Therapeutic Potential	Signalling Pathway	Limitations
Kurian et al., 2013 [8]	Human neonatal and adult fibroblasts	OCT4, SOX2, KLF4, C-MYC	MIM (DMEM:F12, 15 mg mL^{-1} stem cell–grade BSA, 17.5 µg mL^{-1} human insulin, 275 µg mL^{-1} human holo-transferrin, 20 ng mL^{-1} bFGF, 50 ng mL^{-1} human VEGF-165 aa, 25 ng mL^{-1} human BMP4, 450 µM monothioglycerol, 2.25 mM L-glutamine, 2.25 mM NEAA)	CD34$^+$ angioblast-like bipotent progenitors	Forming functional blood vessels that integrated with host vasculature	Not investigated	Heterogenous cells
Zhang et al., 2017a [52]	Human adult and neonatal dermal fibroblast	OCT4, SOX2, KLF4, C-MYC	DMEM/F12 (20% KSR, 10 ng mL^{-1} bFGF, 1 mM GlutaMAX, 0.1 mM NEAA, 55 µM β-mercaptoethanol)	Induced tripotent cardiac progenitor cells (iSMCs, iECs, iCMs)	Improved cardiac function and reduced adverse cardiac remodelling	Not investigated	Teratoma risk
Zhang et al., 2016 [53]	MEF	OCT4, SOX2, KLF4, C-MYC	ieCPC basal medium plus Advanced DMEM/F12: Neural basal (1:1) (1X N2, 1X B27 without Vitamin A, 1X Glutamax, 1X NEAA, 0.05% BSA, 0.1 mM β-ME) plus BACS, (5 ng/mL BMP4, 10 ng/mL Activin A, 3 µM CHIR99021, 2 µM SU5402)	BACS as a reliable prerequisite for the effective creation and ongoing renewal of ieCPCs	Directly produce CMs, ECs, and SMCs when exposed to the infarcted heart environment in vivo	Not investigated	Translational applicability of these cells
Pham et al., 2016 [54]	Human dermal fibroblasts	ETV2	Medium 200 (5% PRP, 5 ng/mL recombinant EGF, 1 ng/mL recombinant VEGF, 20 ng/mL insulin-like growth factor, 1 µg/mL ascorbic acid, 0.2 µg/mL hydrocortisone, 22.5 µg/mL heparin, 1% antibiotic-antimycotic)	Unipotent iEPCs	Improve hindlimb ischemia	Not investigated	Venous not arterial ECs
Park et al., 2020 [55]	Mouse fibroblasts	ETV2, Fli1	VPC medium (10% FBS, 2 mmol/L L-glutamine, β-mercaptoethanol, penicillin/streptomycin, 10 ng/mL VEGF)	Self-renewal and biopotency iVPCs	Enhanced blood flow without tumour formation	Not investigated	Contamination of residual undifferentiated PSC

BACS, BMP4, Activin A, CHIR99021, SU5402; bFGF, basic Fibroblast Growth Factor; BMP4, Bone Morphogenetic Protein 4; BSA, Bovine Serum Albumin; CHIR99021, Glycogen Synthase Kinase 3 Inhibitor; CM, Cardiomyocyte; EC, Endothelial Cell; ETV2, ETS Variant Transcription Factor 2; Fli1, Friend leukaemia integration 1; iCM, induced Cardiomyocyte; iEC, induced Endothelial Cell; iEPC, induced Endothelial Progenitor Cell; iSMC, induced Smooth Muscle Cell; iVPC, induced Vascular Progenitor Cell; OSKM, OCT4 (Octamer-Binding Transcription Factor 4), SOX2 (SRY-Box Transcription Factor 2), KLF4 (Kruppel-Like Factor 4) and C-MYC (c-Myc proto-oncogene protein); MEF, Mouse Embryonic Fibroblasts; MEM, Minimal Essential Medium; MIM, Mesodermal Induction Medium; SMC, Smooth Muscle Cell; SU5402, Inhibitor of FGF, VEGF, and PDGF signaling; VEGF, Vascular Endothelial Growth Factor; VPC, Vascular Progenitor Cell.

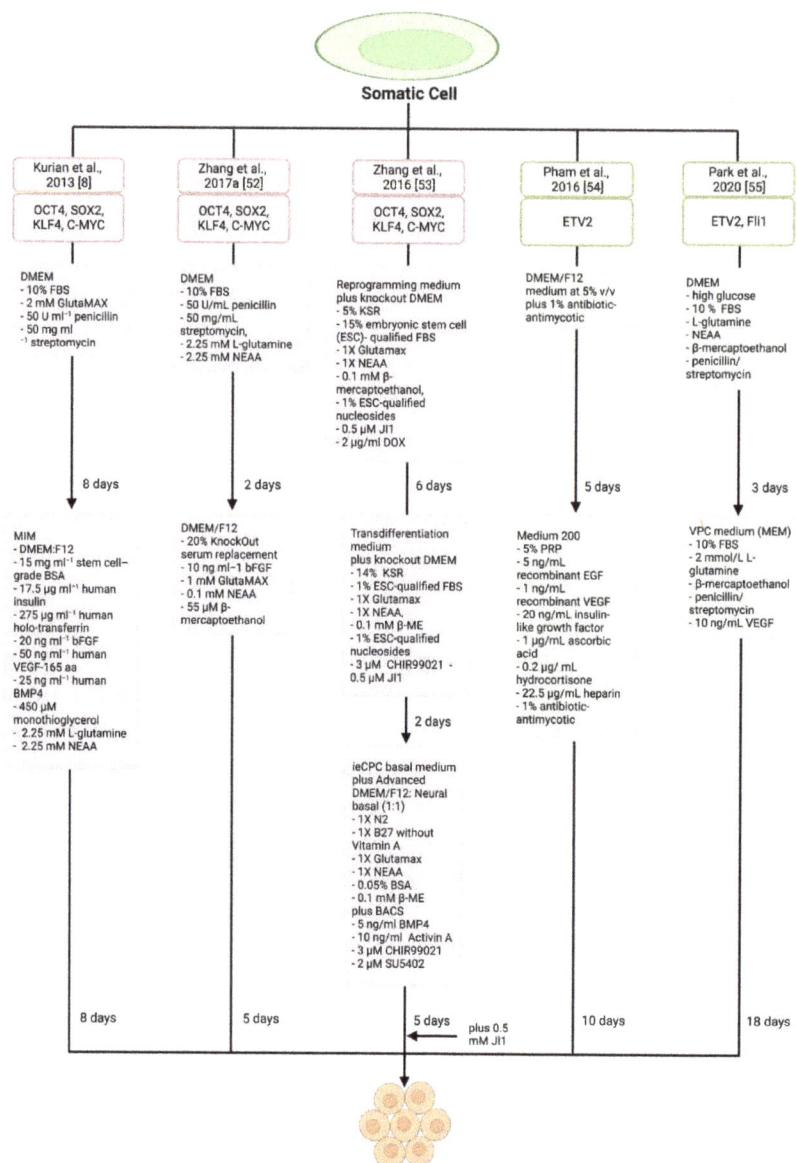

Figure 3. Schematic diagram showing key reprogramming factors, induction medium, and duration used for creating induced vascular progenitor cells (iVPCs) in the different studies. Colour code: Red, Pluripotency factor-based reprogramming; Green: Lineage-specific transcription factor ETV2-based cellular reprogramming. BACS, BMP4, Activin A, CHIR99021, SU5402; bFGF, basic Fibroblast Growth Factor; β-ME, β-mercaptoethanol; BMP4, Bone Morphogenetic Protein 4; BSA, Bovine Serum Albumin; CHIR99021, Glycogen Synthase Kinase 3 Inhibitor; DMEM, Dulbecco's Modified Eagle Medium; DOX, Doxycycline; EGF, Endothelial Growth Factor; FBS, Fetal Bovine Serum; Fli1, Friend leukaemia integration 1; JI1, Jak Inhibitor 1; KSR, Knockout Serum Replacement; OSKM, OCT4 (Octamer-Binding Transcription Factor 4), SOX2 (SRY-Box Transcription Factor 2), KLF4 (Kruppel-Like Factor 4) and C-MYC (c-Myc proto-oncogene protein); MEM, Minimal Essential Medium; MIM, Mesodermal Induction Medium; NEAA, Non-Essential Amino Acid; VEGF, Vascular Endothelial Growth Factor.

4.1. Pluripotency Factor-Based Reprogramming

Forced expression of OSKM has directly reprogrammed adult fibroblasts into two types of bipotent VPCs [8,52]. The first approach by Kurian and colleagues used 8-day expression of OSKM and 8-day incubation in a mesodermal-induction-medium to generate CD34-positive progenitor cells capable of forming functional iECs and iSMCs [8]. CD34 is a characteristic marker of haematopoietic progenitor cells [56]. The benefit of progenitor cells in regenerative therapy was highlighted when an initial conversion efficacy of 20–60% for CD34-positive cells was increased to 400–1200% through expansion methods. Genetic characterisation of terminally differentiated cells indicated heterogeneous expression of markers, with arterial, venous, and lymphatic gene expression in iEC populations, and the presence of pericyte markers in iSMC populations. The authors attribute these findings to experimental alterations, poor downregulation of fibroblast signatures or variations in epigenetic plasticity between cells. Finally, the ability to repeat the protocol with episomal delivery of OSKM showcased the viability of non-integrating delivery methods that circumvent viral integration's safety issues [8].

The second approach by Zhang et al. introduced OSKM into human adult dermal fibroblasts for 7-days to produce CD34-positive cells capable of differentiation into either iECs or induced-erythroblasts (iEBs) through culturing in endothelial and erythroblast differentiation mediums for 10-days and 4-days, respectively [52]. iEBs were identified by the erythroid marker CD235a. When assessed in vivo, new vessels from human $CD34^+$ progenitor cells readily communicated with the existing murine vessel and contained murine erythrocytes. The release of differentiation signals like VEGF in ischemic tissue and erythropoietin in circulating blood is a potential mechanism for differentiating transplanted $CD34^+$ progenitors into iECs or iEBs. Furthermore, the transcription factor SOX17 (SRY-box transcription factor 17) was identified as a 'tuneable rheostat-like switch' whereby overexpression favoured the endothelial lineage compared to depletion, which promoted the erythroblast fate [52]. Thus, SOX17 is a regulator of endothelial development and a cell-fate decider. Another key finding from this study is the increased telomerase activity upon de-differentiation, which marks a crucial therapeutic benefit where cells could undergo prolonged proliferation.

The generation of cardiac progenitor cells (iCPCs) with tripotent potential for ECs, SMCs and cardiomyocytes (CMs) was achieved through a three-stage protocol [53]. (1) Murine fibroblasts were first exposed to OSKM and JI1 (Jak inhibitor 1) for 5-days followed by (2) 2-day treatment with JI1 and an activator of canonical Wnt signalling (CHIR99021). The transition to a progenitor was completed with (3) 14-day culturing in a medium containing BMP-4, Activin A, CHIR99021 and SU5402 (an inhibitor of VEGF), collectively known as BACS. The expression of markers FLK-1 and PDGFR-α indicated the progenitor state. The terminal progenitor cells were restricted to the cardiovascular lineage, displayed significant self-renewable capacity, expanded more than 10^{10}-fold and were genetically stable for more than 18 passages. Although aimed at cardiac regeneration, this protocol is advantageous for vascular repair. For example, progenitors may support coronary artery regeneration, and the CMs can improve cardiac function. In vivo transplantation of iCPCs into mice resulted in the generation of cells of the 3 lineages, with 90% of iCPCs undergoing differentiation to form SMCs, CMs and ECs in an approximated ratio of 6:3:1. In a myocardial infarction murine model, functional improvements such as reduced scar size were observed. Whether the cells directly promoted the therapeutic benefits or indirectly through paracrine modulation is unclear and requires further investigation. A caveat with the authors' protocol was that not all iCPCs were tripotent, with 45.5%, 22.7%, and 31.8% of them being unipotent, bipotent or tripotent, respectively. It will be interesting to investigate whether variations to the reprogramming protocol could dictate which potency level and lineage(s) is derived). The mechanism underlying BACS in facilitating de-differentiation to a progenitor is another subject for future research.

4.2. ETV2

In the previous discussion of EC generation, ETV2 transduction reportedly led to the development of a progenitor-like state which was amenable to differentiation under specific cultures [14]. However, the progenitor cells needed to be better characterised, and only their lineage derivates were transplanted into mice for in vivo studies. Recently, in the absence of differentiation media, ETV2 successfully generated induced-endothelial progenitor cells (iEPCs) [54,57]. Through either lentiviral delivery or using modified mRNA, ETV2 overexpression for 14-days propagated expression of CD31-positive iEPCs, which robustly formed capillary-like networks within Matrigel plugs and improved tissue perfusion in a murine hindlimb ischaemia model. Hypoxic conditions (5% oxygen) improved the lentiviral-based reprogramming efficacy 6-fold from 1.21% under normoxia to 7.5% for CD31-positive cells with hypoxia. Notably, a much lower efficacy of 3.1% under hypoxic conditions was observed with modified ETV2 mRNA delivery. Nevertheless, following sorting, both protocols achieved almost pure CD31-positive cell populations.

A significant finding of the ETV2-based approach was using an alternative medium for culturing progenitor cells [54]. Thus far, most studies discussed have used 10% fetal bovine serum (FBS) to supplement the differentiation media. The use of FBS has several ethical implications, as some see the procedure to derive the serum from unborn foetuses as inhumane. Moreover, animal proteins may lead to adverse immune reactions and animal-to-human viral transmission when transplanted [58]. Hence, the researchers showed that platelet-rich plasma derived from a patient's peripheral blood (PRP) could be an alternative to FBS in various mediums [54]. Non-viral factor delivery and animal-free culture present a breakthrough in the safety of reprogramming. However, the investigators did not attempt a thorough genetic characterisation of cells derived from iEPCs. Moreover, the proliferation capacity of iEPCs was not assessed but did indicate that fibroblasts proliferated prior to ETV2 administration.

In another study, ETV2 and FLI1 co-expression induced bipotent VPCs (iVPCs) identified by the expression of the CD144 marker [55]. The CD144-positive iVPCs stably expanded for 25 passages with an average doubling time of 40 h. The iVPCs differentiated into functional iECs and iSMCs when cultured in respective differentiation media. In a murine hindlimb ischaemia model, transplantation of the iVPCs resulted in improved tissue perfusion. Further research should explore the molecular pathways in which ETV2 and FLI1 facilitate reprogramming to iVPCs. It is worth noting that in the above two papers, CD31 and CD144 were used to select iVPCs. However, these markers are also present on mature ECs. Hence, one may question the validity of the iVPCs phenotype. Using more specific markers or a full panel of genetic markers to confirm the cellular phenotype will ensure improved reliability and complete characterisation of generated cells.

5. Tissue Engineered Vascular Grafts

TEVGs are biodegradable, cell-seeded scaffolds which can be implanted into a host to develop over time into functional vessel-like structures [59]. The central idea with TEVGs is that the scaffold enables cell anchorage and maturation, stimulates extracellular matrix (ECM) deposition and host cell tissue repair, which can maintain the robust vascular structure after the polymer scaffold biodegrades. The value of TEVGs in direct reprogramming research is two-fold. Firstly, TEVGs can be used to investigate reprogrammed cell function within the microvascular environment. For example, induced cells could be evaluated for their response to various factors, including the mechanical strain from high-pressure pulsatile blood flow (through bioreactor systems), circulating hormones, cytokines, oxygenation, immune activity and ECM deposition. Moreover, CVD's underlying disease processes and risk factors (e.g., high blood glucose and lipid levels) can alter vascular cell function. Hence, the response of reprogrammed cells to both normal and diseased physiology is paramount for safety assessments before clinical translation. Secondly, contemporary autologous and artificial grafts fail to achieve long-term survivability and sustainability in replacing diseased vessels with small diameter (<6 mm) [60,61]. By combining direct reprogramming with innovative bioengineering materials and manufacturing methods, patient-specific and genetically augmented vascular cells could be grafted onto scaffolds with low immunogenic properties. iPSCs-seeded TEVGs have already demonstrated good integration and therapeutic benefits in animal studies, and one may sufficiently assume that the benefits of direct reprogramming compared to iPSCs may have greater advantages with TEVGs [62,63]. Moreover, while in vivo methods such as directly injecting reprogrammed cells can be therapeutically beneficial in medium to long-term treatment plans, short-term or urgent clinical requirements may be less feasible. Hence, 'off-the-shelf' or rapidly developed TEVGs will provide holistic coverage of any therapeutic requirements. TEVG is a dynamic interdisciplinary field with an overwhelming number of strategies, from improving the fundamental mechanical and chemical properties to the detailed microscopic control of surface morphology and augmentation of post-transplant thrombogenic processes [4,64–66]. Since TEVG is a complex topic outside the scope of the present view, the following section provides the reader with a brief exploration and exciting examples of where direct reprogramming can be applied to TEVGs. Herein we explore using decellularised tissue, 3D bioprinting and scaffold-based systems to generate TEVGs for research and therapeutic purposes (Figure 4).

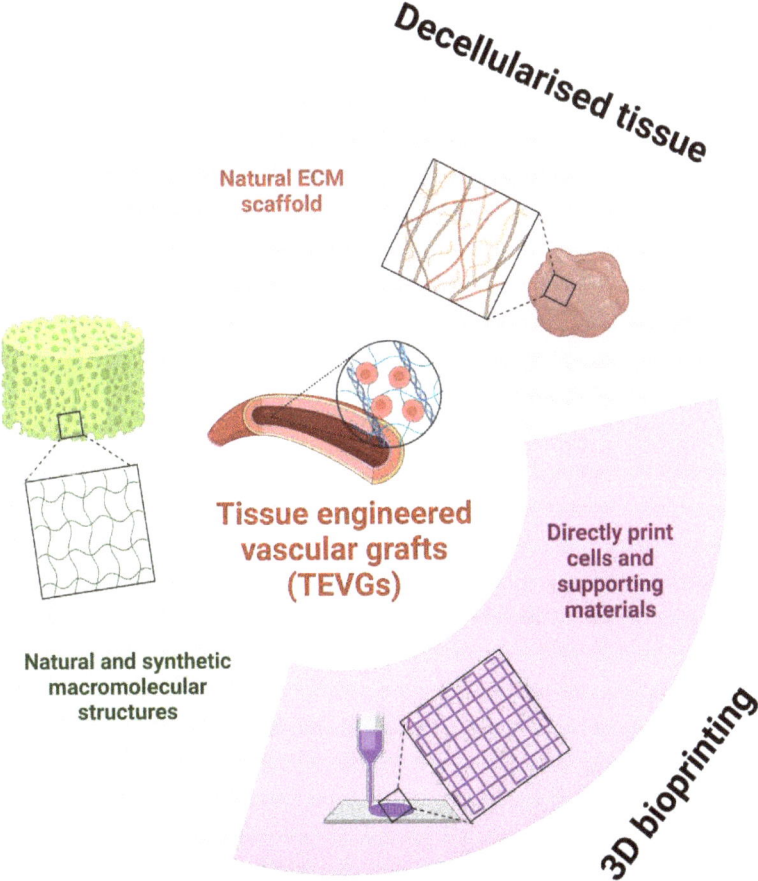

Figure 4. Tissue engineered vascular grafts (TEVGs) generated using decellularized tissue, 3D bioprinting and scaffold-based grafts, respectively.

5.1. Decellularised Tissue

Decellularised tissue grafts typically arise from animal or cadaveric vessels that undergo cell and nuclear removal to leave behind the natural ECM scaffold [65]. Decellularised grafts have several advantages: reduced host-immune responses; presence of natural biochemical and biomechanical properties; presence of bioactive substances that facilitate the migration of endogenous cells and progenitors; observe high biodegradability without the release of toxic products, and can be re-populated with patient-derived cells [67]. Disadvantages include incomplete decellularization, loss of some properties through the decellularization process, difficulty achieving complete recellularisation and mismatch between graft degradation rates and the tissue regeneration rate [67,68]. Several papers have elegantly documented the advantages and disadvantages of decellularised grafts [69–71].

Several authors have provisionally examined their directly reprogrammed cells on decellularised aortic grafts (DAG). Margariti et al. seeded their PiPSC-Ecs onto the DAG, which was followed by culturing in a bioreactor system that emulated physiological blood flow. The cells aggregated into the typical vessel morphology with Ecs in an elongated and orientated pattern with patent lumens [10]. Double-seeded PiPSCs demonstrated the ability to generate an EC monolayer and multiple SMC layers, which improved vessel stability. Hong et al. injected SMC-derived iECs on the luminal surface of the DAG with primary SMCs on the outside and observed similar results as Margariti and colleagues but further exemplified the potential for the same cell source to develop both the endothelial and smooth muscle components [12]. Karamariti et al. went further and engrafted their double-seeded PiPS-derived Ecs and SMCs into mice [45]. Their graft observed patent vasculature and a survival rate of 60% 21-days post-transplantation compared to unseeded DAG and fibroblast-seeded DAG, which experienced rupture and luminal occlusion, respectively. In an alternative to DAG, Kim et al. observed complete endothelialisation and maintenance of endothelial features on a decellularised rat liver scaffold with their iECs [16]. While the studies mentioned confirm the viability of reprogrammed cells in engineered grafts, they fail to analyse key vascular physiology markers such as contractile function and ECM formation. For example, Ji et al. observed dilatory and constriction responses to vasoactive stimuli (flow rate changes, phenylephrine and acetylcholine) in their TEVG composed of iSMCs transdifferentiated from endothelial progenitors [72]. Such methods are more conducive to evaluating cell and tissue function in 3D microenvironments than traditional 2D dish-based analysis.

One limitation of decellularised grafts is the difficulty replanting cells due to the complex ECM architecture [68]. A solution is to use solubilised decellularised grafts to create 2D coatings and 3D hydrogels to improve cell adhesion and maturation [73]. 3D hydrogels may provide greater recapitulation of the mature ECM microenvironment in vivo [74]. Jin and colleagues demonstrated that brain ECM-based 2D coatings and 3D hydrogels not only enhanced conversion efficacies of induced-neuronal cells transdifferentiated from fibroblasts (through plasmid-delivery of reprogramming factors), but further propagated cell maturation, gene expression and attained significant therapeutic benefits in animal studies [73]. The authors further observed only marginal differences between their human brain-ECM batches, which were nevertheless all conducive to successful reprogramming, highlighting that batch-to-batch variation studies for decellularized arterial grafts are imperative to understand future reproducibility. Another key takeaway is that their hydrogel can successfully overcome the low conversion efficacies experienced by non-viral delivery systems [73].

5.2. 3D Bioprinting

3D bioprinting is an emerging approach to directly print cells and supporting material (referred to as bioink) to create 3D structures with high precision and curated spatial distribution [75]. Liguori et al. used a DAG hydrogel to bioprint the tunica media of a small-diameter blood vessel [76]. The resulting tissue was densely populated and viable over their 7-day observation period. Interestingly, the hydrogel provided sufficient cues to drive spontaneous differentiation of stem cells to SMCs. Ho and Hsu showcased an approach to combine direct reprogramming and 3D bioprinting. The authors first generated a thermosensitive and waterborne polyurethane gel which contained human fibroblasts and the neural direct reprogramming inductor, FoxD3. The authors then demonstrated that the extrusion pressure generated when extruding the fibroblast and FoxD3-laden polyurethane gel through a syringe needle of a 3D bioprinter was sufficient to transfect fibroblasts with FoxD3 and consequently generate neuron-like cells with a conversion efficacy of 15.6%. Thus, this study provides a non-viral and in-situ approach for reprograming cells and synergistically generating a 3D graft [77]. The addition of biomechanical stimuli post-bioprinting can further mature and functionalise cells, as seen with bioprinted iPSC-cardiomyocytes, which attained greater sarcomere length and contractile forces after mechanical stretching [78]. Similarly, drug-releasing microspheres could be added to the bioink or post-bioprinting to provide constitutive signalling (such as EC differentiation signals) that may mature and maintain the induced-cell phenotype [79].

5.3. Scaffold-Based Grafts

Scaffold-based grafts use synthetic and natural macromolecular structures to facilitate tissue regeneration [65]. Different properties such as fibre material, fibre and pore size, mechanical stiffness and degradation rates can be tightly controlled to develop niche microenvironments to regulate cell fate and function. Sato et al. demonstrated the development of a 3D synthetic scaffold that facilitated direct reprogramming of fibroblasts into induced osteoblasts (iOBs) [80]. After transducing fibroblasts with their reprogramming inductors, the cells were placed and cultured on a fibronectin coated nanogel-cross-linked porous-freeze-dried (NanoCliP-FD). iOBs readily adhered to the scaffold due to the large pores and deposited large amounts of bone matrix with significant bone regeneration observed in vivo. The authors further state that dehydration of the gel into a matrix would improve storage and transport, with the return to a gel undertaken prior to transplantation. Controlling pore size is important for vascular cells. Smaller pores result in greater SMC populations in the lumen with more ECM deposition. If the pore is too small, cells may not readily seed on the graft [81]. In another study examining iPSC-neural crest cell seeded scaffolds, it was observed that the 3D scaffold selectively removed undifferentiated iPSCs through cell confinement which also matured induced-cells to achieve an almost 38-fold increase in cell survival compared to dissociated cells [82]. This finding is exciting and highlights the role of scaffolds to organise isolated cells, improve conversion efficacies and to reduce the risk of tumorigenesis in direct reprogramming approaches. Similarly, mesenchymal stem cells readily differentiated to SMCs with stiffer pectin hydrogel scaffolds, while the EC lineage was promoted with softer scaffolds [79]. Other scaffold materials such as poly (lactide-co-glycolide)/polyethylene glycol [83] and polylactic acid [84] have also observed remarkable success with directly reprogrammed cells.

6. The Ethics of Direct Reprogramming

The fundamental ethical advantage of direct reprogramming is the reduced focus on human embryonic stem cells (hESCs). As somatic cells are the cell source, there is no need for human blastocysts to be used or produced, which many regards as unethical due to the risk to and loss of potential life [85]. Moreover, therapeutic cloning becomes redundant if direct reprogramming can achieve the same results [86]. However, there is still a residual ethical conflict as the use of hESCs cannot be disregarded entirely. It is worth noting that direct reprogramming research is still in its infancy. Further in-depth and side-by-side comparisons of hESCs and their derivatives will need to be conducted to identify and understand the role of various transcription factors and regulators. hESCs are a natural derivation of embryogenic physiology compared to what may be seen as 'unnatural' with iPSCs and reprogramming. Therefore, hESCs will better assess how genes, transcription factors, molecules, materials, markers and mediums are employed during natural de-differentiation and differentiation processes. Thus, in the short term, hESCs are required to develop standards to compare cells and understand programming pathways. However, once differentiation and reprogramming mechanisms are fully validated, hESCs may be wholly replaced. Similarly, reprogramming approaches using amniotic or human embryonic cells will experience challenges associated with gaining informed consent and the lengthy process of gaining approval and oversight from ethical and regulatory institutions. In the meantime, the most sensible and uncontroversial sources for cellular reprogramming are adult skin fibroblasts, peripheral blood cells, urine epithelial cells, and cells from bariatric or metabolic surgery.

7. Current Challenges & Future Perspectives

The review thus far has identified exciting outcomes and outlooks for direct reprogramming approaches. However, several notable challenges remain and need to be addressed with an outlook for clinical translation. In the following section, we highlight key challenges and provide potential avenues for overcoming them.

7.1. Factor Identification and Reprogramming Efficacies

The biggest challenge for direct reprogramming is identifying a single or group of factors that efficiently propagate reprogramming. Traditional factor identification strategies have used trial-and-error methods, which are tedious, time-consuming, and slow [34]. Thus, studies have primarily used well-known pluripotency factors or master regulators such as ETV2. However, even with the use of such potent regulators, reprogramming strategies have encountered hugely varied conversion efficacies ranging from as low as 1% to >90%, which are dependent on a complex network of factors such as the specific reprogramming factor(s) used, their exposure period, molecular boosters, microenvironmental aids (e.g., shear stress and hypoxic environments) and level of inflammatory activity. While direct reprogramming strategies are still in their infancy, it is still valuable to explore a range of factors outside the well-known regulators as there may be better inductors of reprogramming.

To rapidly explore other transcription factors and combinations, several groups have developed computational prediction systems which use gene expression and regulatory network data to identify factors that promote the desired lineage conversion [87]. Such systems enable rapid screening of factors and essentially fall into two categories: transcription factor identification, which aims to identify transcription factors involved in conversion, and transcription factor perturbation which is a simulation of the effects of transcription factors on a cell [88]. Examples of the former include TRANSDIRE (TRANS-omics-based approach for DIrect REeprogramming) by Eguchi et al., which predicts pioneering factors and transcription factors for a range of cell conversion, and Mogrify by Rackham et al., which produces an eight-list rank of top transcription factors that potentially regulate the conversion from specific initial cells to final targeted cells [89,90]. Ronquist et al. demonstrated a perturbation system that further identifies the amount and induction time of the required transcription factors for a specific conversion [91]. As technology advances, computational systems may incorporate non-transcription factors to identify other small molecules and miRs capable of successfully reprogramming cells. Likewise, it is vital to explore molecules, culture conditions and mediums that can boost the reprogramming efficacy, as seen with VPA and forskolin. For further information on computational prediction algorithms and future perspectives, the reader is directed to several in-depth reviews [87,88,92,93].

7.2. Heterogeneity of Derived Cell Populations

Very few studies examined the specific subtypes of progenitors and differentiated cells produced. Using the wrong subtype may have unforeseen consequences. For example, the endothelium in medium-large arteries is continuous, whereas in the liver and kidney, the endothelium may be fenestrated or sinusoidal owing to their respective functions. Developing iECs that form a continuous monolayer is crucial for arterial function. Similarly, SMCs display significant heterogeneity in normal anatomy and are characteristically different depending on their embryonic origin (e.g., neural crest, mesodermal) and local factors (biochemical, extracellular matrix and physical) [94,95]. Thus, they may experience differing production of growth factors, morphology and resistance to vascular disease [96]. For example, cell proliferation and DNA synthesis are increased by TGF-β1 in neural crest-SMCs but decreased in mesoderm-SMCs [97]. In addition, venous SMCs tend to be less differentiated, hold greater proliferative capabilities and have higher tendencies to develop atherosclerosis in venous grafts compared to the arterial phenotype [98]. Consequently, each artery possesses a distinct mosaic pattern of SMCs geared towards their specific regional function. Thus, research should understand whether transplanting venous or lymphatic phenotypes of iSMCs and iECs can achieve the desired therapeutic benefits in arterial structures. Identifying markers may help elucidate specific sub-phenotypes better and help identify molecules that direct cells to a specific phenotype.

Furthermore, quantifying arterial and venous markers and morphological changes to categorise heterogeneity is inadequate; protocols to screen for heterogeneous functional responses are needed. For example, using lentiviral integration of GCaMP6f, a sensitive Ca^{2+} sensor, into SMCs, Cuenca et al. could observe and quantify dynamic intracellular calcium changes and analyse vasoactive heterogeneity [99]. Similar technologies that can assess variations in acetylated-LDL uptake and NO excretion in an EC population will give more holistic interpretations of functional heterogeneity.

Cellular tracking technology may also be helpful for differentiating reprogrammed cells from the starting population, and tracking cellular changes and cell division during reprogramming. Several methods have already been reported [100,101]. ScarTrace is a single-cell sequencing approach which adds fluorescent tags to cells enabling tracking in multiple locations on an organism [102]. Similarly, the Cre-loxP recombinase system activates GFP (green fluorescent protein) expression in specific cells and permits tracking their off-spring in vivo, which may prove to be useful in tracking the differentiation of induced-vascular progenitor cells [100].

7.3. Factor Delivery Systems & Viral Integration

With the end goal of clinical translation, factor and molecule delivery systems should be safe and target the desired cell type and, where appropriate, the specific genes. Lentiviral delivery systems are commonplace in reprogramming strategies for their long-term gene expression and packaging capacity for large transgene sequences [27]. However, they possess several safety risks due to unpredictable transgene insertion, which can cause insertional mutagenesis and inadvertently activate proto-oncogenes or silence tumour suppressors, increasing cancer risk [103]. Other non-integrating viruses like adenoviruses and Sendai viruses are reportedly safer as they harbour low immunogenic properties and rarely insert into host DNA. However, adenoviruses possess smaller insert sizes and are difficult to direct to specific cells [27]. Thus, given the advantages and disadvantages of different viral vectors, research should continue exploring both lentiviral and non-lentiviral vectors and analyse what developments and modifications can improve their safety profiles, gene expression stability and packaging capacities. For example, monitoring technology for replication-competent viruses and protocols for reducing infectious virus particles can lower the risk of mutations [104].

CRISPR/Cas9 is a growing gene activation approach that has garnered considerable attention in reprogramming various somatic cells into iPSCs [105–107]. Recently, CRISPR was used to endogenously activate gene promoters of the GNT gene, which successfully reprogrammed fibroblasts into cardiovascular progenitors [108]. This gene editing approach is a safe alternative as it avoids integrating viral vectors and exogenous expression of pluripotency factors. Other transgene-free vectors, such as small molecules (DKK3, TLR3 agonists and miRs) and plasmids, are viable alternatives. However, their use is rare, and their efficiency and safety profiles are still under review [12].

7.4. Tumorigenicity Risk

No study observed the development of tumours after transplanting the induced cells into mice over the respective observation periods. While this suggests a low tumorigenicity risk, the risk is not absent. For example, one study did find that one OSKM-derived induced-vascular progenitor clone did form a tumour [20]. Hence, sorting and purification methods must be robust and reliable. Although differentiated cells may pose a lower tumour risk to progenitors, we must not forget that reprogrammed cells undergo gross metabolic, chromosomal, genetic and epigenetic alterations. Thus, even if differentiated cells are transplanted, there is an unknown risk that they harbour abnormalities which adversely affect cell function or have a preponderance towards cancer formation. For example, a dysfunctional cell could induce pathogenic changes, such as promoting thrombosis through overexpression of adhesion molecules. Studies should look for multi-year assessments of tumorigenicity in longer-living animals (e.g., pigs and sheep) and closer animal relatives to humans. Furthermore, there is an inherent tumour risk with using pluripotency factors, notably c-MYC (in OSKM), which has been shown as both dispensable and a facilitator of lineage-conversion [109,110]. As further research occurs, we may find efficient non-oncogenic factors for direct cellular reprogramming.

7.5. Recapitulating Disease

There is growing evidence that direct reprogrammed cells and iPSC-derivates may retain disease signatures [111,112]. This finding has both positive and negative consequences. On the one hand, reprogrammed cells could allow in vitro modelling of patient-specific diseases and drug responses. Strategies could be employed to identify how different reprogramming approaches may respond depending on the underlying disease process. On the other hand, we are presented with another significant challenge: how to erase disease signatures successfully. This research topic will become much more pronounced as direct reprogramming advances towards clinical application.

7.6. 2D In Vitro Analysis vs. 3D Microenvironments

The studies analysed thus far are based mainly on 2D analysis, which fails to recapitulate in vivo 3D microenvironments. Several direct reprogramming studies have highlighted improved conversion efficacies and cellular maturation with mimics of 3D microenvironments compared to 2D in vitro studies which may underestimate reprogramming efficiencies [82,113–115]. We have already discussed how TEVGs can act as both investigative and therapeutic platforms by emulating specific 3D microenvironments to facilitate reprogramming and cell viability. Another avenue is vascular-on-chip models (VoCs), which provide a dynamic platform to test physiological and pathogenic processes on reprogrammed cells. For example, one could assess the endothelial barrier function of iECs or the haemostatic, immune and inflammatory responses to endothelial injury, as already exemplified by several groups [116,117]. Cuenca et al. used a VoC model and observed how the co-culturing of iSMCs and iECs with each other, adding mural cells and mechanical forces propagated further cell maturation, network self-assembly and stabilisation [99]. VoCs also hold potential for experimental analysis on the effect of drugs on diseased arterial cells and reprogrammed cells in a highly individualised manner whereby cells and blood can be taken directly from a patient. Such approaches provide holistic interpretations of reprogrammed cell activity under in vivo settings and are the next logical step towards clinical application and personalised medicine.

8. Conclusions

Direct reprogramming has resulted in a paradigm shift away from traditional stem cell reprogramming approaches. By introducing pluripotency factors, lineage-specific transcription factors, and small molecules, alongside curated culture conditions and mediums, somatic cells have successfully converted into iECs, iSMCs, and iVPCs without a pluripotent intermediate step. Emerging evidence has collectively shown huge therapeutic potentials for these vascular cells directly reprogrammed from other somatic cells (mainly fibroblasts) (Figure 5). Data from the preclinical studies discussed in this review have confirmed the potential for significant therapeutic benefits from increased angiogenesis and reduced ischaemia by directly incorporating reprogrammed cells with endogenous vasculature and paracrine modulation. Importantly, the advent of computational processing for reprogramming factor identification, TEVGs for disease modelling and therapeutics, 3D platforms such as VoCs, and more sophisticated gene/factor delivery systems will provide a giant leap in direct reprogramming research. However, several challenges still need to be addressed, notably the low conversion efficacies, heterogeneity of reprogrammed populations and the risk of integrating gene delivery systems such as lentiviruses. Further work should aim to identify effective reprogramming factors and comprehensively elucidate their underlying molecular pathways. As research continues and the standardisation of materials, cultures, and protocols becomes widespread, translation to human studies and clinical application for CVD treatment may appear sooner rather than later.

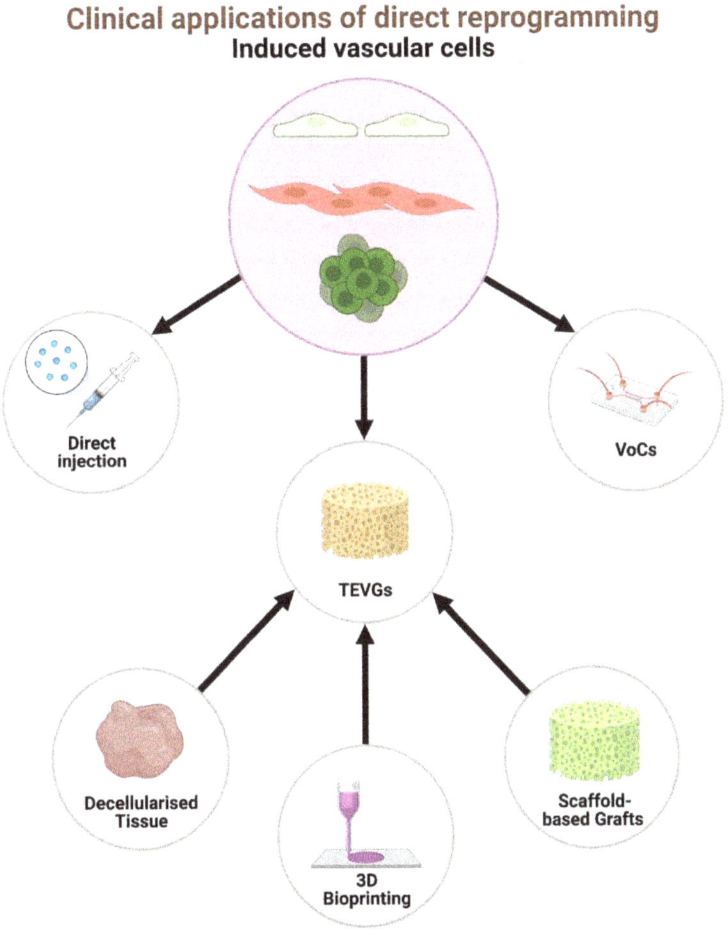

Figure 5. Spider diagram summarising the potential clinical applications of vascular cells generated through cellular reprogramming, such as cell therapy for vascular regeneration, tissue-engineered vascular grafts (TEVGs) for vascular grafting, and vascular on chips (VoCs) for disease modelling and high-throughput drug screening.

Author Contributions: Conceptualization, S.G.S. and Q.X.; methodology, S.G.S. and J.Y.L.; formal analysis and investigation, S.G.S. and J.Y.L.; writing—original draft preparation, S.G.S. and J.Y.L.; writing—review and editing, S.G.S. and Q.X.; supervision, Q.X.; project administration, Q.X.; funding acquisition, Q.X. All authors have read and agreed to the published version of the manuscript.

Funding: This work was partially supported by the British Heart Foundation (PG/15/11/31279, PG/15/86/31723, PG/16/1/31892, and PG/20/10458). This work forms part of the research portfolio for the National Institute for Health Research Biomedical Research Centre at Barts.

Conflicts of Interest: The authors declare no conflict of interest, and the funders had no role in the design of the study; in the collection, analyses, or interpretation of data; in the writing of the manuscript, or in the decision to publish the results.

References

1. Roth, G.A.; Mensah, G.A.; Johnson, C.O.; Addolorato, G.; Ammirati, E.; Baddour, L.M.; Barengo, N.C.; Beaton, A.Z.; Benjamin, E.J.; Benziger, C.P.; et al. Global Burden of Cardiovascular Diseases and Risk Factors, 1990–2019: Update From the GBD 2019 Study. *J. Am. Coll. Cardiol.* **2020**, *76*, 2982–3021. [CrossRef] [PubMed]
2. McKavanagh, P.; Yanagawa, B.; Zawadowski, G.; Cheema, A. Management and Prevention of Saphenous Vein Graft Failure: A Review. *Cardiol. Ther.* **2017**, *6*, 203–223. [CrossRef] [PubMed]
3. SabikIII, J.F. Understanding Saphenous Vein Graft Patency. *Circulation* **2011**, *124*, 273–275. [CrossRef] [PubMed]
4. Pashneh-Tala, S.; MacNeil, S.; Claeyssens, F. The tissue-engineered vascular graft—Past, present, and future. *Tissue Eng. Part B Rev.* **2016**, *22*, 68–100. [CrossRef] [PubMed]
5. Wang, H.; Yang, Y.; Liu, J.; Qian, L. Direct cell reprogramming: Approaches, mechanisms and progress. *Nat. Rev. Mol. Cell Biol.* **2021**, *22*, 410–424. [CrossRef]
6. Takahashi, K.; Yamanaka, S. Induction of Pluripotent Stem Cells from Mouse Embryonic and Adult Fibroblast Cultures by Defined Factors. *Cell* **2006**, *126*, 663–676. [CrossRef]
7. Lee, S.; Kim, J.E.; AL Johnson, B.; Andukuri, A.; Yoon, Y.-S. Direct reprogramming into endothelial cells: A new source for vascular regeneration. *Regen. Med.* **2017**, *12*, 317–320. [CrossRef]
8. Kurian, L.; Sancho-Martinez, I.; Nivet, E.; Aguirre, A.; Moon, K.; Pendaries, C.; Volle-Challier, C.; Bono, F.; Herbert, J.-M.; Pulecio, J.; et al. Conversion of human fibroblasts to angioblast-like progenitor cells. *Nat. Methods* **2013**, *10*, 77–83. [CrossRef]
9. Sayed, N.; Wong, W.T.; Ospino, F.; Meng, S.; Lee, J.; Jha, A.; Dexheimer, P.; Aronow, B.J.; Cooke, J.P. Transdifferentiation of human fibroblasts to endothelial cells role of innate immunity. *Circulation* **2015**, *131*, 300–309. [CrossRef]
10. Margariti, A.; Winkler, B.; Karamariti, E.; Zampetaki, A.; Tsai, T.-N.; Baban, D.; Ragoussis, J.; Huang, Y.; Han, J.-D.J.; Zeng, L.; et al. Direct reprogramming of fibroblasts into endothelial cells capable of angiogenesis and reendothelization in tissue-engineered vessels. *Proc. Natl. Acad. Sci. USA* **2012**, *109*, 13793–13798. [CrossRef]
11. Li, J.; Huang, N.F.; Zou, J.; Laurent, T.J.; Lee, J.C.; Okogbaa, J.; Cooke, J.P.; Ding, S. Conversion of Human Fibroblasts to Functional Endothelial Cells by Defined Factors. *Arter. Thromb. Vasc. Biol.* **2013**, *33*, 1366–1375. [CrossRef]
12. Hong, X.; Margariti, A.; Le Bras, A.; Jacquet, L.; Kong, W.; Hu, Y.; Xu, Q. Transdifferentiated Human Vascular Smooth Muscle Cells are a New Potential Cell Source for Endothelial Regeneration. *Sci. Rep.* **2017**, *7*, 5590. [CrossRef] [PubMed]
13. Chen, T.; Karamariti, E.; Hong, X.; Deng, J.; Wu, Y.; Gu, W.; Simpson, R.; Wong, M.M.; Yu, B.; Hu, Y.; et al. DKK3 (Dikkopf-3) Transdifferentiates Fibroblasts Into Functional Endothelial Cells—Brief Report. *Arter. Thromb. Vasc. Biol.* **2019**, *39*, 765–773. [CrossRef] [PubMed]
14. Ginsberg, M.; James, D.; Ding, B.-S.; Nolan, D.; Geng, F.; Butler, J.M.; Schachterle, W.; Pulijaal, V.R.; Mathew, S.; Chasen, S.T.; et al. Efficient Direct Reprogramming of Mature Amniotic Cells into Endothelial Cells by ETS Factors and TGFβ Suppression. *Cell* **2012**, *151*, 559–575. [CrossRef]
15. Wong, W.T.J.; Cooke, J.P. Therapeutic transdifferentiation of human fibroblasts into endothelial cells using forced expression of lineage-specific transcription factors. *J. Tissue Eng.* **2016**, *7*, 2041731416628329. [CrossRef] [PubMed]
16. Kim, J.-J.; Kim, D.-H.; Lee, J.Y.; Lee, B.-C.; Kang, I.; Kook, M.G.; Kong, D.; Choi, S.W.; Woo, H.-M.; Kim, D.-I.; et al. cAMP/EPAC Signaling Enables ETV2 to Induce Endothelial Cells with High Angiogenesis Potential. *Mol. Ther.* **2020**, *28*, 466–478. [CrossRef] [PubMed]
17. Morita, R.; Suzuki, M.; Kasahara, H.; Shimizu, N.; Shichita, T.; Sekiya, T.; Kimura, A.; Sasaki, K.-I.; Yasukawa, H.; Yoshimura, A. ETS transcription factor ETV2 directly converts human fibroblasts into functional endothelial cells. *Proc. Natl. Acad. Sci. USA* **2015**, *112*, 160–165. [CrossRef]
18. Lee, S.; Park, C.; Han, J.W.; Kim, J.Y.; Cho, K.; Kim, E.J.; Kim, S.; Lee, S.-J.; Oh, S.Y.; Tanaka, Y.; et al. Direct Reprogramming of Human Dermal Fibroblasts Into Endothelial Cells Using ER71/ETV2. *Circ. Res.* **2017**, *120*, 848–861. [CrossRef]
19. McCoy, M.G.; Pérez-Cremades, D.; Belkin, N.; Peng, W.; Zhang, B.; Chen, J.; Sachan, M.; Wara, A.K.M.K.; Zhuang, R.; Cheng, H.S.; et al. A miRNA cassette reprograms smooth muscle cells into endothelial cells. *FASEB J.* **2022**, *36*, e22239. [CrossRef]
20. Yin, L.; Ohanyan, V.; Pung, Y.F.; DeLucia, A.; Bailey, E.; Enrick, M.; Stevanov, K.; Kolz, C.L.; Guarini, G.; Chilian, W.M. Induction of Vascular Progenitor Cells From Endothelial Cells Stimulates Coronary Collateral Growth. *Circ. Res.* **2012**, *110*, 241–252. [CrossRef]

21. Wasteson, P.; Johansson, B.R.; Jukkola, T.; Breuer, S.; Akyürek, L.M.; Partanen, J.; Lindahl, P. Developmental origin of smooth muscle cells in the descending aorta in mice. *Development* **2008**, *135*, 1823–1832. [CrossRef] [PubMed]
22. Bray, S.J. Notch signalling: A simple pathway becomes complex. *Nat. Rev. Mol. Cell Biol.* **2006**, *7*, 678–689. [CrossRef] [PubMed]
23. Cheng, Y.; Gu, W.; Zhang, G.; Guo, X. Notch1 activation of Jagged1 contributes to differentiation of mesenchymal stem cells into endothelial cells under cigarette smoke extract exposure. *BMC Pulm. Med.* **2022**, *22*, 139. [CrossRef] [PubMed]
24. Manderfield, L.J.; High, F.A.; Engleka, K.A.; Liu, F.; Li, L.; Rentschler, S.; Epstein, J.A. Notch Activation of Jagged1 Contributes to the Assembly of the Arterial Wall. *Circulation* **2012**, *125*, 314–323. [CrossRef] [PubMed]
25. Akil, A.; Gutiérrez-García, A.K.; Guenter, R.; Rose, J.B.; Beck, A.W.; Chen, H.; Ren, B. Notch Signaling in Vascular Endothelial Cells, Angiogenesis, and Tumor Progression: An Update and Prospective. *Front. Cell Dev. Biol.* **2021**, *9*, 177. [CrossRef]
26. Zouein, F.A.; Booz, G.W.; Altara, R. STAT3 and Endothelial Cell—Cardiomyocyte Dialog in Cardiac Remodeling. *Front. Cardiovasc. Med.* **2019**, *6*, 50. [CrossRef]
27. Zheng, C.-X.; Wang, S.-M.; Bai, Y.-H.; Luo, T.-T.; Wang, J.-Q.; Dai, C.-Q.; Guo, B.-L.; Luo, S.-C.; Wang, D.-H.; Yang, Y.-L.; et al. Lentiviral Vectors and Adeno-Associated Virus Vectors: Useful Tools for Gene Transfer in Pain Research. *Anat. Rec.* **2018**, *301*, 825–836. [CrossRef]
28. Yu, B.; Kiechl, S.; Qi, D.; Wang, X.; Song, Y.; Weger, S.; Mayr, A.; Le Bras, A.; Karamariti, E.; Zhang, Z.; et al. A Cytokine-Like Protein Dickkopf-Related Protein 3 Is Atheroprotective. *Circulation* **2017**, *136*, 1022–1036. [CrossRef]
29. Gong, W.; Das, S.; Sierra-Pagan, J.E.; Skie, E.; Dsouza, N.; Larson, T.A.; Garry, M.G.; Luzete-Monteiro, E.; Zaret, K.S.; Garry, D.J. ETV2 functions as a pioneer factor to regulate and reprogram the endothelial lineage. *Nat. Cell Biol.* **2022**, *24*, 672–684. [CrossRef]
30. Park, C.; Lee, T.-J.; Bhang, S.H.; Liu, F.; Nakamura, R.; Oladipupo, S.S.; Pitha-Rowe, I.; Capoccia, B.; Choi, H.S.; Kim, T.M.; et al. Injury-Mediated Vascular Regeneration Requires Endothelial ER71/ETV2. *Arter. Thromb. Vasc. Biol.* **2016**, *36*, 86–96. [CrossRef]
31. Hayashi, M.; Pluchinotta, M.; Momiyama, A.; Tanaka, Y.; Nishikawa, S.-I.; Kataoka, H. Endothelialization and altered hematopoiesis by persistent Etv2 expression in mice. *Exp. Hematol.* **2012**, *40*, 738–750.e11. [CrossRef] [PubMed]
32. Koyano-Nakagawa, N.; Garry, D.J. Etv2 as an essential regulator of mesodermal lineage development. *Cardiovasc. Res.* **2017**, *113*, 1294–1306. [CrossRef] [PubMed]
33. Schachterle, W.; Badwe, C.R.; Palikuqi, B.; Kunar, B.; Ginsberg, M.; Lis, R.; Yokoyama, M.; Elemento, O.; Scandura, J.M.; Rafii, S. Sox17 drives functional engraftment of endothelium converted from non-vascular cells. *Nat. Commun.* **2017**, *8*, 13963. [CrossRef]
34. Han, J.-K.; Chang, S.-H.; Cho, H.-J.; Choi, S.-B.; Ahn, H.-S.; Lee, J.; Jeong, H.; Youn, S.-W.; Lee, H.-J.; Kwon, Y.-W.; et al. Direct Conversion of Adult Skin Fibroblasts to Endothelial Cells by Defined Factors. *Circulation* **2014**, *130*, 1168–1178. [CrossRef] [PubMed]
35. Hywood, J.D.; Sadeghipour, S.; Clayton, Z.E.; Yuan, J.; Stubbs, C.; Wong, J.W.T.; Cooke, J.P.; Patel, S. Induced endothelial cells from peripheral arterial disease patients and neonatal fibroblasts have comparable angiogenic properties. *PLoS ONE* **2021**, *16*, e0255075. [CrossRef]
36. Ju, H.; Zhang, C.; Lu, W. Progress in heterologous biosynthesis of forskolin. *J. Ind. Microbiol. Biotechnol.* **2021**, *48*, 9. [CrossRef] [PubMed]
37. Salehi, B.; Staniak, M.; Czopek, K.; Stępień, A.; Dua, K.; Wadhwa, R.; Chellappan, D.K.; Sytar, O.; Brestic, M.; Bhat, N.G.; et al. The Therapeutic Potential of the Labdane Diterpenoid Forskolin. *Appl. Sci.* **2019**, *9*, 4089. [CrossRef]
38. Palikuqi, B.; Nguyen, D.-H.T.; Li, G.; Schreiner, R.; Pellegata, A.F.; Liu, Y.; Redmond, D.; Geng, F.; Lin, Y.; Gómez-Salinero, J.M.; et al. Adaptable haemodynamic endothelial cells for organogenesis and tumorigenesis. *Nature* **2020**, *585*, 426–432. [CrossRef]
39. Lai, L.; Reineke, E.; Hamilton, D.; Cooke, J.P. Glycolytic Switch Is Required for Transdifferentiation to Endothelial Lineage. *Circulation* **2019**, *139*, 119–133. [CrossRef]
40. Meng, S.; Zhou, G.; Gu, Q.; Chanda, P.K.; Ospino, F.; Cooke, J.P. Transdifferentiation Requires iNOS Activation. *Circ. Res.* **2016**, *119*, e129–e138. [CrossRef]
41. Galagali, H.; Kim, J.K. The multifaceted roles of microRNAs in differentiation. *Curr. Opin. Cell Biol.* **2020**, *67*, 118–140. [CrossRef] [PubMed]
42. Lawson, C.; Wolf, S. ICAM-1 signaling in endothelial cells. *Pharmacol. Rep.* **2009**, *61*, 22–32. [CrossRef]
43. Kong, D.-H.; Kim, Y.K.; Kim, M.R.; Jang, J.H.; Lee, S. Emerging Roles of Vascular Cell Adhesion Molecule-1 (VCAM-1) in Immunological Disorders and Cancer. *Int. J. Mol. Sci.* **2018**, *19*, 1057. [CrossRef] [PubMed]
44. Fontaine, M.; Herkenne, S.; Ek, O.; Paquot, A.; Boeckx, A.; Paques, C.; Nivelles, O.; Thiry, M.; Struman, I. Extracellular Vesicles Mediate Communication between Endothelial and Vascular Smooth Muscle Cells. *Int. J. Mol. Sci.* **2021**, *23*, 331. [CrossRef] [PubMed]
45. Karamariti, E.; Margariti, A.; Winkler, B.; Wang, X.; Hong, X.; Baban, D.; Ragoussis, J.; Huang, Y.; Han, J.-D.J.; Wong, M.M.; et al. Smooth Muscle Cells Differentiated From Reprogrammed Embryonic Lung Fibroblasts Through DKK3 Signaling Are Potent for Tissue Engineering of Vascular Grafts. *Circ. Res.* **2013**, *112*, 1433–1443. [CrossRef]
46. Rensen, S.S.; Doevendans, P.A.; van Eys, G.J. Regulation and characteristics of vascular smooth muscle cell phenotypic diversity. *Neth. Heart J.* **2007**, *15*, 100–108. [CrossRef] [PubMed]
47. Karamariti, E.; Zhai, C.; Yu, B.; Qiao, L.; Wang, Z.; Potter, C.M.; Wong, M.M.; Simpson, R.M.; Zhang, Z.; Wang, X.; et al. DKK3 (Dickkopf 3) Alters Atherosclerotic Plaque Phenotype Involving Vascular Progenitor and Fibroblast Differentiation Into Smooth Muscle Cells. *Arter. Thromb. Vasc. Biol.* **2018**, *38*, 425–437. [CrossRef]

48. Hirai, H.; Yang, B.; Garcia-Barrio, M.T.; Rom, O.; Ma, P.X.; Zhang, J.; Chen, Y.E. Direct Reprogramming of Fibroblasts Into Smooth Muscle-Like Cells With Defined Transcription Factors—Brief Report. *Arter. Thromb. Vasc. Biol.* **2018**, *38*, 2191–2197. [CrossRef]
49. Turner, E.C.; Huang, C.-L.; Govindarajan, K.; Caplice, N.M. Identification of a Klf4-dependent upstream repressor region mediating transcriptional regulation of the myocardin gene in human smooth muscle cells. *Biochim. et Biophys. Acta Gene Regul. Mech.* **2013**, *1829*, 1191–1201. [CrossRef]
50. Zhou, B.; Zeng, S.; Li, N.; Yu, L.; Yang, G.; Yang, Y.; Zhang, X.; Fang, M.; Xia, J.; Xu, Y. Angiogenic Factor with G Patch and FHA Domains 1 Is a Novel Regulator of Vascular Injury. *Arter. Thromb. Vasc. Biol.* **2017**, *37*, 675–684. [CrossRef]
51. Psaltis, P.J.; Simari, R.D. Vascular Wall Progenitor Cells in Health and Disease. *Circ. Res.* **2015**, *116*, 1392–1412. [CrossRef] [PubMed]
52. Zhang, L.; Jambusaria, A.; Hong, Z.; Marsboom, G.; Toth, P.T.; Herbert, B.-S.; Malik, A.B.; Rehman, J. SOX17 Regulates Conversion of Human Fibroblasts Into Endothelial Cells and Erythroblasts by Dedifferentiation Into CD34$^+$ Progenitor Cells. *Circulation* **2017**, *135*, 2505–2523. [CrossRef] [PubMed]
53. Zhang, Y.; Cao, N.; Huang, Y.; Spencer, C.I.; Fu, J.-D.; Yu, C.; Liu, K.; Nie, B.; Xu, T.; Li, K.; et al. Expandable Cardiovascular Progenitor Cells Reprogrammed from Fibroblasts. *Cell Stem Cell* **2016**, *18*, 368–381. [CrossRef] [PubMed]
54. Van Pham, P.; Vu, N.B.; Nguyen, H.T.; Huynh, O.T.; Truong, M.T.-H. Significant improvement of direct reprogramming efficacy of fibroblasts into progenitor endothelial cells by ETV2 and hypoxia. *Stem Cell Res. Ther.* **2016**, *7*, 104. [CrossRef] [PubMed]
55. Park, S.Y.; Lee, H.; Kwon, Y.W.; Park, M.R.; Kim, J.H.; Kim, J.B. *Etv2*- and *Fli1*-Induced Vascular Progenitor Cells Enhance Functional Recovery in Ischemic Vascular Disease Model—Brief Report. *Arter. Thromb. Vasc. Biol.* **2020**, *40*, e105–e113. [CrossRef]
56. Sidney, L.E.; Branch, M.J.; Dunphy, S.E.; Dua, H.S.; Hopkinson, A. Concise Review: Evidence for CD34 as a Common Marker for Diverse Progenitors. *Stem Cells* **2014**, *32*, 1380–1389. [CrossRef]
57. Van Pham, P.; Vu, N.B.; Dao, T.T.-T.; Le, H.T.-N.; Phi, L.T.; Phan, N.K. Production of endothelial progenitor cells from skin fibroblasts by direct reprogramming for clinical usages. *Vitr. Cell. Dev. Biol. Anim.* **2017**, *53*, 207–216. [CrossRef]
58. Jochems, C.E.A.; van der Valk, J.B.; Stafleu, F.R.; Baumans, V. The Use of Fetal Bovine Serum: Ethical or Scientific Problem? *Altern. Lab. Anim.* **2002**, *30*, 219–227. [CrossRef]
59. Blum, K.M.; Zbinden, J.C.; Ramachandra, A.B.; Lindsey, S.E.; Szafron, J.M.; Reinhardt, J.W.; Heitkemper, M.; Best, C.A.; Mirhaidari, G.J.M.; Chang, Y.-C.; et al. Tissue engineered vascular grafts transform into autologous neovessels capable of native function and growth. *Commun. Med.* **2022**, *2*, 3. [CrossRef]
60. Mallis, P.; Kostakis, A.; Stavropoulos-Giokas, C.; Michalopoulos, E. Future Perspectives in Small-Diameter Vascular Graft Engineering. *Bioengineering* **2020**, *7*, 160. [CrossRef]
61. Matsuzaki, Y.; John, K.; Shoji, T.; Shinoka, T. The Evolution of Tissue Engineered Vascular Graft Technologies: From Preclinical Trials to Advancing Patient Care. *Appl. Sci.* **2019**, *9*, 1274. [CrossRef] [PubMed]
62. Generali, M.; Casanova, E.A.; Kehl, D.; Wanner, D.; Hoerstrup, S.P.; Cinelli, P.; Weber, B. Autologous endothelialized small-caliber vascular grafts engineered from blood-derived induced pluripotent stem cells. *Acta Biomater.* **2019**, *97*, 333–343. [CrossRef] [PubMed]
63. Saito, J.; Kaneko, M.; Ishikawa, Y.; Yokoyama, U. Challenges and Possibilities of Cell-Based Tissue-Engineered Vascular Grafts. *Cyborg Bionic Syst.* **2021**, *2021*, 1532103. [CrossRef] [PubMed]
64. Hu, K.; Li, Y.; Ke, Z.; Yang, H.; Lu, C.; Li, Y.; Guo, Y.; Wang, W. History, progress and future challenges of artificial blood vessels: A narrative review. *Biomater. Transl.* **2022**, *3*, 81. [CrossRef]
65. Moore, M.J.; Tan, R.P.; Yang, N.; Rnjak-Kovacina, J.; Wise, S.G. Bioengineering artificial blood vessels from natural materials. *Trends Biotechnol.* **2022**, *40*, 693–707. [CrossRef]
66. Obiweluozor, F.O.; Emechebe, G.A.; Kim, D.-W.; Cho, H.-J.; Park, C.H.; Kim, C.S.; Jeong, I.S. Considerations in the Development of Small-Diameter Vascular Graft as an Alternative for Bypass and Reconstructive Surgeries: A Review. *Cardiovasc. Eng. Technol.* **2020**, *11*, 495–521. [CrossRef]
67. Liao, J.; Xu, B.; Zhang, R.; Fan, Y.; Xie, H.; Li, X. Applications of decellularized materials in tissue engineering: Advantages, drawbacks and current improvements, and future perspectives. *J. Mater. Chem. B* **2020**, *8*, 10023–10049. [CrossRef]
68. Jin, Y.; Cho, S.-W. Bioengineering platforms for cell therapeutics derived from pluripotent and direct reprogramming. *APL Bioeng.* **2021**, *5*, 031501. [CrossRef]
69. Jiang, Y.; Li, R.; Han, C.; Huang, L. Extracellular matrix grafts: From preparation to application (Review). *Int. J. Mol. Med.* **2021**, *47*, 463–474. [CrossRef]
70. Wang, X.; Chan, V.; Corridon, P.R. Decellularized blood vessel development: Current state-of-the-art and future directions. *Front. Bioeng. Biotechnol.* **2022**, *10*, 951644. [CrossRef]
71. Khanna, A.; Zamani, M.; Huang, N.F. Extracellular Matrix-Based Biomaterials for Cardiovascular Tissue Engineering. *J. Cardiovasc. Dev. Dis.* **2021**, *8*, 137. [CrossRef] [PubMed]
72. Ji, H.; Atchison, L.; Chen, Z.; Chakraborty, S.; Jung, Y.; Truskey, G.A.; Christoforou, N.; Leong, K.W. Transdifferentiation of human endothelial progenitors into smooth muscle cells. *Biomaterials* **2016**, *85*, 180–194. [CrossRef] [PubMed]
73. Jin, Y.; Lee, J.S.; Kim, J.; Min, S.; Wi, S.; Yu, J.H.; Chang, G.-E.; Cho, A.-N.; Choi, Y.; Ahn, D.-H.; et al. Three-dimensional brain-like microenvironments facilitate the direct reprogramming of fibroblasts into therapeutic neurons. *Nat. Biomed. Eng.* **2018**, *2*, 522–539. [CrossRef] [PubMed]

4. Jin, Y.; Kim, J.; Lee, J.S.; Min, S.; Kim, S.; Ahn, D.-H.; Kim, Y.-G.; Cho, S.-W. Vascularized Liver Organoids Generated Using Induced Hepatic Tissue and Dynamic Liver-Specific Microenvironment as a Drug Testing Platform. *Adv. Funct. Mater.* **2018**, *28*, 1801954. [CrossRef]
5. Chen, S.-G.; Ugwu, F.; Li, W.-C.; Caplice, N.M.; Petcu, E.B.; Yip, S.P.; Huang, C.-L. Vascular Tissue Engineering: Advanced Techniques and Gene Editing in Stem Cells for Graft Generation. *Tissue Eng. Part B Rev.* **2021**, *27*, 14–28. [CrossRef]
6. Liguori, G.; Sinkunas, V.; Liguori, T.; Moretto, E.; Sharma, P.; Harmsen, M.; Moreira, L. Decellularized Arterial Extracellular Matrix-Based Hydrogel Supports 3D Bioprinting of the Media Layer of Small-Caliber Blood Vessels I Circulation. *Circ. Cell Tissue Eng.* **2019**, *140*, A14119.
7. Ho, L.; Hsu, S.-H. Cell reprogramming by 3D bioprinting of human fibroblasts in polyurethane hydrogel for fabrication of neural-like constructs. *Acta Biomater.* **2018**, *70*, 57–70. [CrossRef]
8. Lui, C.; Chin, A.F.; Park, S.; Yeung, E.; Kwon, C.; Tomaselli, G.; Chen, Y.; Hibino, N. Mechanical stimulation enhances development of scaffold-free, 3D-printed, engineered heart tissue grafts. *J. Tissue Eng. Regen. Med.* **2021**, *15*, 503–512. [CrossRef]
9. Sato, Y.; Yamamoto, K.; Horiguchi, S.; Tahara, Y.; Nakai, K.; Kotani, S.-I.; Oseko, F.; Pezzotti, G.; Yamamoto, T.; Kishida, T.; et al. Nanogel tectonic porous 3D scaffold for direct reprogramming fibroblasts into osteoblasts and bone regeneration. *Sci. Rep.* **2018**, *8*, 15824. [CrossRef]
10. Wang, Y.; Hu, J.; Jiao, J.; Liu, Z.; Zhou, Z.; Zhao, C.; Chang, L.-J.; Chen, Y.E.; Ma, P.X.; Yang, B. Engineering vascular tissue with functional smooth muscle cells derived from human iPS cells and nanofibrous scaffolds. *Biomaterials* **2014**, *35*, 8960–8969. [CrossRef] [PubMed]
11. Carlson, A.L.; Bennett, N.; Francis, N.; Halikere, A.; Clarke, S.; Moore, J.C.; Hart, R.; Paradiso, K.; Wernig, M.; Kohn, J.; et al. Generation and transplantation of reprogrammed human neurons in the brain using 3D microtopographic scaffolds. *Nat. Commun.* **2016**, *7*, 10862. [CrossRef] [PubMed]
12. Li, N.; Xue, F.; Zhang, H.; Sanyour, H.J.; Rickel, A.P.; Uttecht, A.; Fanta, B.; Hu, J.; Hong, Z. Fabrication and Characterization of Pectin Hydrogel Nanofiber Scaffolds for Differentiation of Mesenchymal Stem Cells into Vascular Cells. *ACS Biomater. Sci. Eng.* **2019**, *5*, 6511–6519. [CrossRef] [PubMed]
13. Liu, C.; Huang, Y.; Pang, M.; Yang, Y.; Linshan, L.; Liu, L.; Shu, T.; Zhou, W.; Wang, X.; Rong, L.; et al. Tissue-Engineered Regeneration of Completely Transected Spinal Cord Using Induced Neural Stem Cells and Gelatin-Electrospun Poly (Lactide-Co-Glycolide)/Polyethylene Glycol Scaffolds. *PLoS ONE* **2015**, *10*, e0117709. [CrossRef]
14. Haddad, T.; Noel, S.; Liberelle, B.; El Ayoubi, R.; Ajji, A.; De Crescenzo, G. Fabrication and surface modification of poly lactic acid (PLA) scaffolds with epidermal growth factor for neural tissue engineering. *Biomatter* **2016**, *6*, e1231276. [CrossRef] [PubMed]
15. Lo, B.; Parham, L. Ethical Issues in Stem Cell Research. *Endocr. Rev.* **2009**, *30*, 204–213. [CrossRef] [PubMed]
16. Byrnes, W.M.; Center, T.N.C.B. Direct Reprogramming and Ethics in Stem Cell Research. *Natl. Cathol. Bioeth. Q.* **2008**, *8*, 277–290. [CrossRef]
17. Dotson, G.A.; Ryan, C.W.; Chen, C.; Muir, L.; Rajapakse, I. Cellular reprogramming: Mathematics meets medicine. *WIREs Mech. Dis.* **2020**, *13*, e1515. [CrossRef]
18. Tran, A.; Yang, P.; Yang, J.Y.H.; Ormerod, J. Computational approaches for direct cell reprogramming: From the bulk omics era to the single cell era. *Brief. Funct. Genom.* **2022**, *21*, 270–279. [CrossRef]
19. Eguchi, R.; Hamano, M.; Iwata, M.; Nakamura, T.; Oki, S.; Yamanishi, Y. TRANSDIRE: Data-driven direct reprogramming by a pioneer factor-guided trans-omics approach. *Bioinformatics* **2022**, *38*, 2839–2846. [CrossRef]
20. Rackham, O.; The FANTOM Consortium; Firas, J.; Fang, H.; Oates, M.E.; Holmes, M.; Knaupp, A.; Suzuki, H.; Nefzger, C.; Daub, C.O.; et al. A predictive computational framework for direct reprogramming between human cell types. *Nat. Genet.* **2016**, *48*, 331–335. [CrossRef]
21. Ronquist, S.; Patterson, G.; Muir, L.A.; Lindsly, S.; Chen, H.; Brown, M.; Wicha, M.S.; Bloch, A.; Brockett, R.; Rajapakse, I. Algorithm for cellular reprogramming. *Proc. Natl. Acad. Sci. USA* **2017**, *114*, 11832–11837. [CrossRef] [PubMed]
22. Gam, R.; Sung, M.; Pandurangan, A.P. Experimental and Computational Approaches to Direct Cell Reprogramming: Recent Advancement and Future Challenges. *Cells* **2019**, *8*, 1189. [CrossRef] [PubMed]
23. Guerrero-Ramirez, G.; Valdez-Cordoba, C.; Islas-Cisneros, J.; Trevino, V. Computational approaches for predicting key transcription factors in targeted cell reprogramming (Review). *Mol. Med. Rep.* **2018**, *18*, 1225–1237. [CrossRef] [PubMed]
24. Majesky, M.W. Developmental Basis of Vascular Smooth Muscle Diversity. *Arter. Thromb. Vasc. Biol.* **2007**, *27*, 1248–1258. [CrossRef] [PubMed]
25. Wang, G.; Jacquet, L.; Karamariti, E.; Xu, Q. Origin and differentiation of vascular smooth muscle cells. *J. Physiol.* **2015**, *593*, 3013–3030. [CrossRef]
26. Shen, M.; Quertermous, T.; Fischbein, M.P.; Wu, J.C. Generation of Vascular Smooth Muscle Cells From Induced Pluripotent Stem Cells. *Circ. Res.* **2021**, *128*, 670–686. [CrossRef]
27. Topouzis, S.; Majesky, M.W. Smooth Muscle Lineage Diversity in the Chick Embryo. Two types of aortic smooth muscle cell differ in growth and receptor-mediated transcriptional responses to transforming growth factor-beta. *Dev. Biol.* **1996**, *178*, 430–445. [CrossRef]
28. Wong, A.P.; Nili, N.; Strauss, B.H. In vitro differences between venous and arterial-derived smooth muscle cells: Potential modulatory role of decorin. *Cardiovasc. Res.* **2005**, *65*, 702–710. [CrossRef]

99. Cuenca, M.V.; Cochrane, A.; Hil, F.E.V.D.; de Vries, A.A.; Oberstein, S.A.L.; Mummery, C.L.; Orlova, V.V. Engineered 3D vessel-on-chip using hiPSC-derived endothelial- and vascular smooth muscle cells. *Stem Cell Rep.* **2021**, *16*, 2159–2168. [CrossRef]
100. Yao, M.; Ren, T.; Pan, Y.; Xue, X.; Li, R.; Zhang, L.; Li, Y.; Huang, K. A New Generation of Lineage Tracing Dynamically Records Cell Fate Choices. *Int. J. Mol. Sci.* **2022**, *23*, 5021. [CrossRef]
101. Zhang, Y.; Zeng, F.; Han, X.; Weng, J.; Gao, Y. Lineage tracing: Technology tool for exploring the development, regeneration, and disease of the digestive system. *Stem Cell Res. Ther.* **2020**, *11*, 438. [CrossRef] [PubMed]
102. Alemany, A.; Florescu, M.; Baron, C.S.; Peterson-Maduro, J.; van Oudenaarden, A. Whole-organism clone tracing using single-cell sequencing. *Nature* **2018**, *556*, 108–112. [CrossRef] [PubMed]
103. Schlimgen, R.; Howard, J.; Wooley, D.; Thompson, M.; Baden, L.R.; Yang, O.O.; Christiani, D.C.; Mostoslavsky, G.; Diamond, D.V.; Duane, E.G.; et al. Risks Associated with Lentiviral Vector Exposures and Prevention Strategies. *J. Occup. Environ. Med.* **2016**, *58*, 1159–1166. [CrossRef] [PubMed]
104. Connolly, J.B. Lentiviruses in gene therapy clinical research. *Gene Ther.* **2002**, *9*, 1730–1734. [CrossRef]
105. Rubio, A.; Luoni, M.; Giannelli, S.G.; Radice, I.; Iannielli, A.; Cancellieri, C.; Di Berardino, C.; Regalia, G.; Lazzari, G.; Menegon, A.; et al. Rapid and efficient CRISPR/Cas9 gene inactivation in human neurons during human pluripotent stem cell differentiation and direct reprogramming. *Sci. Rep.* **2016**, *6*, 37540. [CrossRef] [PubMed]
106. Sokka, J.; Yoshihara, M.; Kvist, J.; Laiho, L.; Warren, A.; Stadelmann, C.; Jouhilahti, E.-M.; Kilpinen, H.; Balboa, D.; Katayama, S.; et al. CRISPR activation enables high-fidelity reprogramming into human pluripotent stem cells. *Stem Cell Rep.* **2022**, *17*, 413–426. [CrossRef] [PubMed]
107. Weltner, J.; Balboa, D.; Katayama, S.; Bespalov, M.; Krjutškov, K.; Jouhilahti, E.-M.; Trokovic, R.; Kere, J.; Otonkoski, T. Human pluripotent reprogramming with CRISPR activators. *Nat. Commun.* **2018**, *9*, 2643. [CrossRef] [PubMed]
108. Jiang, L.; Liang, J.; Huang, W.; Ma, J.; Park, K.H.; Wu, Z.; Chen, P.; Zhu, H.; Ma, J.-J.; Cai, W.; et al. CRISPR activation of endogenous genes reprograms fibroblasts into cardiovascular progenitor cells for myocardial infarction therapy. *Mol. Ther.* **2022**, *30*, 54–74. [CrossRef]
109. Deng, J.; Luo, K.; Xu, P.; Jiang, Q.; Wang, Y.; Yao, Y.; Chen, X.; Cheng, F.; Xie, D.; Deng, H. High-efficiency c-Myc-mediated induction of functional hepatoblasts from the human umbilical cord mesenchymal stem cells. *Stem Cell Res. Ther.* **2021**, *12*, 375. [CrossRef]
110. Wernig, M.; Meissner, A.; Cassady, J.P.; Jaenisch, R. c-Myc Is Dispensable for Direct Reprogramming of Mouse Fibroblasts. *Cell Stem Cell* **2008**, *2*, 10–12. [CrossRef]
111. Bersini, S.; Schulte, R.; Huang, L.; Tsai, H.; Hetzer, M.W. Direct reprogramming of human smooth muscle and vascular endothelial cells reveals defects associated with aging and Hutchinson-Gilford progeria syndrome. *Elife* **2020**, *9*, e54383. [CrossRef] [PubMed]
112. Maffioletti, S.M.; Sarcar, S.; Henderson, A.B.; Mannhardt, I.; Pinton, L.; Moyle, L.A.; Steele-Stallard, H.; Cappellari, O.; Wells, K.E.; Ferrari, G.; et al. Three-Dimensional Human iPSC-Derived Artificial Skeletal Muscles Model Muscular Dystrophies and Enable Multilineage Tissue Engineering. *Cell Rep.* **2018**, *23*, 899–908. [CrossRef] [PubMed]
113. Li, Y.; Dal-Pra, S.; Mirotsou, M.; Jayawardena, T.M.; Hodgkinson, C.P.; Bursac, N.; Dzau, V.J. Tissue-engineered 3-dimensional (3D) microenvironment enhances the direct reprogramming of fibroblasts into cardiomyocytes by microRNAs. *Sci. Rep.* **2016**, *6*, 38815. [CrossRef] [PubMed]
114. Smith, A.W.; Hoyne, J.D.; Nguyen, P.K.; McCreedy, D.A.; Aly, H.; Efimov, I.R.; Rentschler, S.; Elbert, D.L. Direct reprogramming of mouse fibroblasts to cardiomyocyte-like cells using Yamanaka factors on engineered poly(ethylene glycol) (PEG) hydrogels. *Biomaterials* **2013**, *34*, 6559–6571. [CrossRef] [PubMed]
115. Paoletti, C.; Divieto, C.; Chiono, V. Impact of Biomaterials on Differentiation and Reprogramming Approaches for the Generation of Functional Cardiomyocytes. *Cells* **2018**, *7*, 114. [CrossRef] [PubMed]
116. Moses, S.R.; Adorno, J.J.; Palmer, A.F.; Song, J.W. Vessel-on-a-chip models for studying microvascular physiology, transport, and function in vitro. *Am. J. Physiol. Physiol.* **2021**, *320*, C92–C105. [CrossRef]
117. Kim, S.; Kim, W.; Lim, S.; Jeon, J.S. Vasculature-On-A-Chip for In Vitro Disease Models. *Bioengineering* **2017**, *4*, 8. [CrossRef]

Disclaimer/Publisher's Note: The statements, opinions and data contained in all publications are solely those of the individual author(s) and contributor(s) and not of MDPI and/or the editor(s). MDPI and/or the editor(s) disclaim responsibility for any injury to people or property resulting from any ideas, methods, instructions or products referred to in the content.

Article

Finite Element Analysis of the Non-Uniform Degradation of Biodegradable Vascular Stents

Hanbing Zhang, Tianming Du, Shiliang Chen, Yang Liu, Yujia Yang, Qianwen Hou and Aike Qiao *

Faculty of Environment and Life, Beijing University of Technology, Beijing 100124, China
* Correspondence: qak@bjut.edu.cn

Abstract: Most of the studies on the finite element analysis (FEA) of biodegradable vascular stents (BVSs) during the degradation process have limited the accuracy of the simulation results due to the application of the uniform degradation model. This paper aims to establish an FEA model for the non-uniform degradation of BVSs by considering factors such as the dynamic changes of the corrosion properties and material properties of the element, as well as the pitting corrosion and stress corrosion. The results revealed that adjusting the corrosion rate according to the number of exposed surfaces of the element and reducing the stress threshold according to the corrosion status accelerates the degradation time of BVSs by 26% and 25%, respectively, compared with the uniform degradation model. The addition of the pitting model reduces the service life of the BVSs by up to 12%. The effective support of the stent to the vessel could reach at least 60% of the treatment effect before the vessel collapsed. These data indicate that the proposed non-uniform degradation model of BVSs with multiple factors produces different phenomena compared with the commonly used models and make the numerical simulation results more consistent with the real degradation scenario.

Keywords: biodegradable vascular stents; continuum damage mechanics; finite element method; non-uniform degradation

Citation: Zhang, H.; Du, T.; Chen, S.; Liu, Y.; Yang, Y.; Hou, Q.; Qiao, A. Finite Element Analysis of the Non-Uniform Degradation of Biodegradable Vascular Stents. *J. Funct. Biomater.* **2022**, *13*, 152. https://doi.org/10.3390/jfb13030152

Academic Editors: Hae-Won Kim and Håvard J. Haugen

Received: 2 August 2022
Accepted: 9 September 2022
Published: 14 September 2022

Publisher's Note: MDPI stays neutral with regard to jurisdictional claims in published maps and institutional affiliations.

Copyright: © 2022 by the authors. Licensee MDPI, Basel, Switzerland. This article is an open access article distributed under the terms and conditions of the Creative Commons Attribution (CC BY) license (https://creativecommons.org/licenses/by/4.0/).

1. Introduction

In-stent restenosis (ISR) is a major problem limiting the treatment effect of percutaneous coronary intervention [1]. The biodegradable vascular stent (BVS) is the fourth-generation endovascular stent capable of promoting effective vascular remodeling and restoring the elasticity and diastolic–systolic functions of vessels [2]. The BVSs have a greater potential for avoiding ISR than permanent ones [3–5].

The BVSs function immediately after they are implanted. The degradation of BVSs involves multiple disciplines such as electrochemistry, metal physics, thermodynamics, materials science, and mechanics [6,7]. The corrosion mechanism of BVSs includes but is not limited to uniform corrosion, pitting corrosion, intergranular corrosion, stress corrosion, and fatigue corrosion [8]. Compared with in vivo and in vitro research, in silico research is widely used in the preliminary study of medical implant developments because it can fully capture the degradation behavior of bioabsorbable materials and the advantages of low cost and time saving [9]. Continuum damage mechanics (CDM) is a common phenomenological approach used to simulate material corrosion for the above-mentioned mechanisms in the corrosive environment under the finite element analysis (FEA) framework.

The uniform degradation of BVSs from the global corrosion that occurs at the entire free surface exposed to a corrosive environment. Uniform corrosion is the most common global corrosion which was mostly used in early studies to simulate stent degradation. Uniform corrosion is the overall thinning caused by chemical and electrochemical reactions on the surface, with an even corrosion rate. The uniform corrosion was simulated using the FEA by uniformly removing elements from BVSs' outer surfaces [10]. Further, modifications were made to the uniform corrosion model by using arbitrary Lagrangian–Eulerian adaptive meshing [11].

The non-uniform degradation of BVSs from the local corrosion that occurs at specific spots on the surface seriously affects the therapeutic effect [12]. BVSs' mechanical strength and geometry change during the degradation [13]. The rapid structural weakening caused by local corrosion may lead to adverse events such as fast restenosis due to vessel collapse [14,15].

Stress corrosion, the corrosion under the combined action of tensile stress and blood, is one of the key causes of non-uniform degradation and early failure of BVSs [16]. The stent may crack perpendicular to the stress direction under the safe load. The residual stress after the stent implantation and the working stress during service are the driving stresses of stress corrosion [17]. The CDM method analyzes the stress, strain, and induced damage process of the stent material from the perspective of continuum mechanics [12]. A CDM-based degradation model coupled the stress corrosion model with the uniform corrosion adapted from the work of da Costa-Mattos et al. was established [18,19]. The initial results of the simulations were considered phenomenologically consistent with experimental observations. This coupled formulation was further studied and completed in vitro corrosion tests on two BVSs designs and compared results against simulated predictions for the same designs [20]. The stress corrosion threshold changed dynamically with the material properties and mechanical properties during degradation. However, this issue was not considered in previous studies.

Pitting corrosion is local corrosion caused by the surface defects during stent manufacturing and the stress concentration during angioplasty [8]. A model by expanding on the CDM uniform corrosion element removal model through the inclusion of the assignment and transfer of pitting parameters to capture the effects of pitting corrosion, and results showed agreement with the tests [10]. A physically based pitting corrosion model was established to consider the multi-physical process of magnesium stent degradation, including the transport of magnesium ions. The gaps between phenomenological and physical corrosion models of absorbable mental stents could be addressed [21]. However, the influencing factors of pitting corrosion were studied independently in most previous research, and there is a lack of studies that combine pitting corrosion with uniform corrosion and stress corrosion.

Mass transfer and flow shear stress due to blood circulation can accelerate corrosion (including local and global corrosion etc.). Flow shear stress increased the average uniform corrosion rate, the coverage and depth of local corrosion, and the removal rate of corrosion products from pits [22]. The effect of blood flow should be considered for BVSs degradation.

The BVSs' edges and corners are susceptible to corrosion, and one example is that the corners would be rounded [23,24]. The damaged edges and corners would change the cross-sectional shape and mechanical properties of the BVSs, which in turn affects the degradation. In the FEA framework, the elements at the edges and corners have more surfaces exposed to the corrosive environment. A uniform and stress corrosion model considering multi-dimensional effects on up to three exposed surfaces was developed and was applied in degradation simulations of biodegradable magnesium alloy stents [25]. As the number of exposed surfaces changes dynamically during the degradation, more exposed surfaces should be considered. A multi-dimensional pitting corrosion model that takes the number of exposed surfaces of the element and the interaction of adjacent elements into account was established and the proportional link between the corrosion rate and the number of exposed surfaces was revealed in the results [26]. However, stress corrosion and the interaction between the stent and the vessel were not considered.

Iron (Fe), magnesium (Mg) and their alloys have been extensively studied as BVS metals. Both Mg and Fe are essential trace elements in the human body. Mg exhibits excellent biocompatibility, low thrombogenicity, and is critical in many cellular functions [27]. However, the inherently rapid corrosion of Mg in physiological environments is a major concern [28]. Fe and its alloys possess high radial strength, allowing the use of stents with substantially thinner struts and thereby reducing the restenosis rate [29]. The degradation of iron takes too long to be fully degraded over its expected service period and it produces a relatively large volume of iron oxide products which might not be safely metabolized in the body [30,31]. Despite much research that has been completed to tailor the properties of Mg and Fe through alloying, advanced processing, and manufacturing routes, further improvements are still needed to identify an ideal stent material [32–34]. In 2013, zinc (Zn) was introduced as an alternative to Mg and Fe, mainly due to its moderate corrosion rate in simulated body fluid [29,35]. Potential toxicity and biocompatibility aspects are important for medical implants [36]. Zinc alloys are biocompatible with acceptable or zero cytotoxicity [37]. As one of the necessary trace elements for humans, zinc can reduce the inflammatory reaction after stent implantation and promote the regeneration of the vascular intima [38]. Therefore, zinc alloys could be used to manufacture BVSs [39].

This paper aims to establish an FEA model for the degradation of zinc alloy stents by simultaneously considering the combined effects of such factors as the number of exposed surfaces in the blood flow, the dynamic changes of material properties, and the addition of pitting corrosion, which may lead to non-uniform degradation. The degradation process and service performance are explained by the phenomenological description of the damage to the BVSs. The corrosion feature of this model will be revealed by comparing the influences of the different factors herein.

2. Materials and Methods

The relationship between the number of exposed surfaces and the corrosion rate was considered, and the stress corrosion threshold of an element was dynamically adjusted with the degradation process. Pitting corrosion was combined with uniform corrosion and stress corrosion. The non-uniform degradation of the BVSs under the combined action of multiple corrosion factors was simulated by setting the material properties and corrosion properties for individual elements under the FEA framework. The service performance and support time of the degradable stent were investigated.

2.1. Stent Degradation Model

Based on the CDM principle, the macroscopic damage and strength weakening effect of the stent were simulated by establishing a scalar field composed of the damage variable D of each element. The damage variable D was a gradually cumulative and monotonically increasing function. The effective stress during the damage process can be calculated with Equation (1).

$$\sigma = \bar{\sigma}(1 - D) \tag{1}$$

where σ is the effective stress, $\bar{\sigma}$ is the undamaged stress, and D is the damage variable which increases monotonously from 0 to 1. Linear attenuation adjustment was made to the material properties during degradation through D [18]. When D equals 0, the material is undamaged, while when D approximately equals 1, the material is completely damaged, and the element is removed from the model.

The occurrence of uniform corrosion is only judged by whether the element is at the corrosion surface regardless of the stress state of the stent. An electrochemical reaction of zinc and other metals in the stent with electrolytes in the blood occurs on the initial corrosion surface. The reaction between corrosion products and the blood further exposes the zinc alloy to the blood environment, thus updating the corrosion surface. For the exposed element located on the corrosion surface, the evolution equation of the uniform corrosion damage parameter D_U can be described with Equation (2) [18].

$$\dot{D}_U = dD_U = \frac{\delta_U}{L_e} k_U dt \tag{2}$$

where k_U is a kinetic parameter related to the uniform corrosion process, δ_U is a characteristic dimension of the uniform corrosion process, and L_e is the characteristic length of a finite element. Based on the immersion test in our previous research, the average degradation rate from 0 to 42 days was 0.188 mm/y, which is very close to the degradation rates of zinc alloys with different compositions in other literature [40]. The detailed kinetic parameters k_U and δ_U of uniform corrosion are obtained from references and listed in Table 1 [17,18].

Table 1. Parameters for the stent degradation model [17–19].

Parameters	δ_U	k_U	δ_{SC}	S	R	σ_{th}	β
Value	0.1 mm	0.05/h	0.07 mm	0.005 mm²√h /N	2	66 MPa	0.8

For the elements exposed to the corrosive environment during the degradation, k_U was dynamically adjusted in proportion to the number of surfaces exposed to the corrosive environment [25], so as to speed up the degradation of the element at the edges and corners of the stent. Considering the convective–diffusion effect of wall shear stress generated by blood flow on stent corrosion products, only the corrosion effect of blood on stent was considered in the current model.

Stress corrosion is relevant to the stress state during the degradation. The stress corrosion threshold is the stress level at which the material can resist crack extension in the corrosive environment. The occurrence of stress corrosion was judged by comparing the equivalent stress σ_{eq}^* with the stress corrosion threshold σ_{th}. The maximum principal stress of the element was chosen as the σ_{eq}^* in this study. The evolution equation of the element's stress corrosion damage parameter D_{SC} is described with Equations (3) and (4) [18].

If $\sigma_{eq}^* < \sigma_{th}$:

$$\dot{D}_{SC} = 0, \tag{3}$$

if $\sigma_{eq}^* \geq \sigma_{th} > 0$:

$$\dot{D}_{SC} = dD_{SC} = \frac{L_e}{\delta_{SC}} \left(\frac{S\sigma_{eq}^*}{1 - D_{SC}} \right)^R dt, \tag{4}$$

where δ_{SC} is a characteristic dimension of the stress corrosion process, S and R are the functions of the corrosive environment related to the kinetics of stress corrosion. Based on the da Costa-Mattos' research [19], S and R are constants due to the constant pH in human's corrosive environment. σ_{th} is closely related to the combination of material composition, metallurgical conditions, and corrosive environment, and usually ranges from 30% of the yield stress to 90% of the ultimate tensile stress [41]. The stress corrosion threshold was set to 30% of the yield stress σ_s of zinc alloy to ensure safety due to a lack of research on such materials at present. The details of these parameters are listed in Table 1.

The material properties including Young's modulus and yield stress of the BVSs change with the corrosion [6]. Considering the complexity of corrosion products and uncertainty in stress corrosion experiments, the σ_s was dynamically adjusted by D during degradation to objectively describe the constitutive relationship of the material based on the work of our previous study [17].

As random local corrosion, a pitting corrosion parameter λ_e was randomly assigned to the element. The evolution equation of the pitting damage parameter D_P was described with Equation (5) [10]. The pitting corrosion parameter was transferred to its adjacent elements when an element was removed from the model as shown in Equation (6).

$$\dot{D}_P = dD_P = \frac{\delta_U}{L_e} \lambda_e k_U dt, \tag{5}$$

$$\lambda_e = \beta \lambda_n, \tag{6}$$

where λ_n is the pitting parameter of the removed element and β is a dimensionless parameter that controls the acceleration of pit growth [10]. Then the growing of pitting pits was described. The details of these parameters are listed in Table 1.

The above three types of corrosion occur independently. A linear superposition of different scalar fields from different degradation processes was assumed [18]. The total damage variable D was the linear algebraic sum of each corrosion damage parameter as shown in Equation (7).

$$D = D_U + D_{SC} + D_P. \tag{7}$$

Further, a dynamic degradation model considering uniform corrosion, stress corrosion, and pitting corrosion was established (Figure 1). The characteristic length L_e of the elements was introduced in the evolution process of the above three corrosion mechanisms so that the damage variable D could reflect the damage "per unit element". The influence of the mesh size on the accuracy of the stent degradation model could be avoided to ensure the spatial synchronization between the model and the actual degradation.

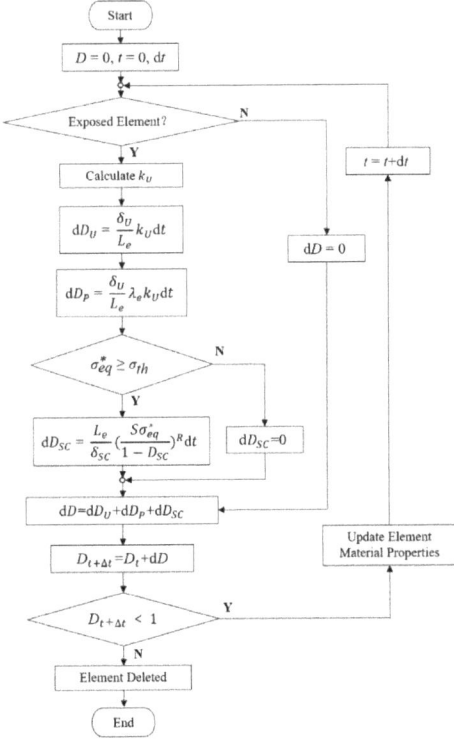

Figure 1. Calculation of damage variables and update of material properties for individual elements.

2.2. Geometry Model

The FEA model was composed of an artery, two sets of the eight-corolla stents, and a balloon (Figure 2a). The stent has a thickness of 0.25 mm with its structs, and the artery has an inner diameter of 4.6 mm and a thickness of 1 mm [17]. The artery length was chosen according to three ring lengths to keep a similar artery/ring length ratio of about 1.5 [17]. The balloon has a length of 12 mm and a diameter of 3mm. No stenosis was added to the vessel as this research only focused on stent degradation. Using the Bottom-Up meshing technique, the quadrilateral mesh was divided on the inner surface of the stent, and properly densified the mesh in the corolla; then the stent thickness was offset externally (Figure 2b). The stent meshed with 68,016 nodes and 47,360 C3D8R elements with a characteristic length of 50 μm based on the mesh convergence study.

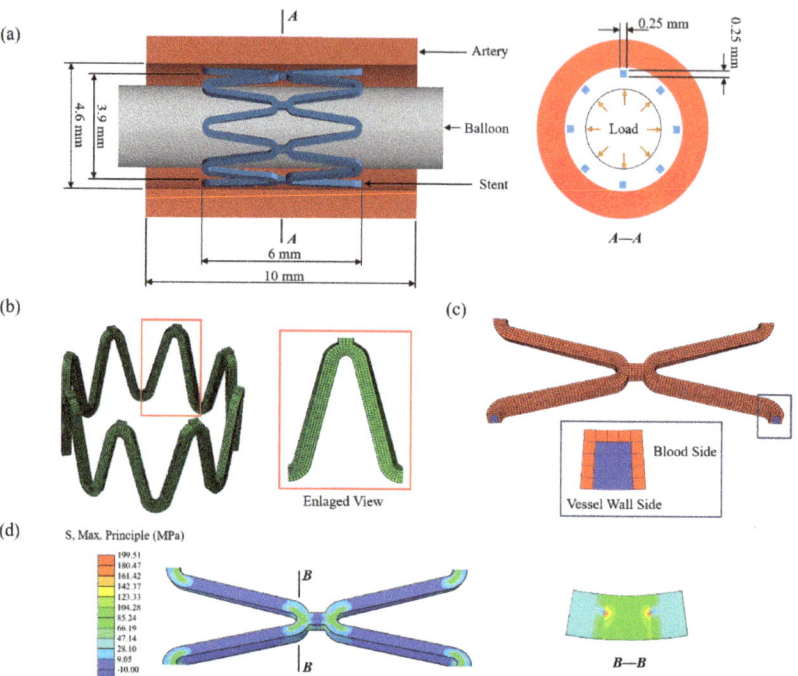

Figure 2. (a) The finite element model of the artery, stent, and balloon; (b) the mesh of the stent with an enlarged view of the stent strut; (c) the initial corrosion surface of the stent; (d) the residual stress of the stent after deployment.

The stent was modeled as a homogeneous, isotropic, and elastoplastic material. The material properties of the zinc alloy used in this study were measured by a tensile test [18]. The value of 74,300 MPa as Young's modulus of the undamaged zinc alloy was obtained by tests and the value of 300 MPa was assigned for the completely damaged zinc alloy and linearly decayed over this range by D during degradation (Table 2). The balloon was modeled as a cylindrical membrane with a thickness of 0.02 mm. The vessel was modeled as an incompressible material described by a third-order Ogden isotropic hyperelastic material model to simulate its highly nonlinear mechanical [42].

Table 2. Material properties [17].

Part	Density (g/cm³)	Young's Modulus (MPa)	Poisson's Ratio	Yield Stress (MPa)
Stent	8.5	74,300 (Max)~300 (Min)	0.3	220
Artery	1.12	-	-	-
Balloon	1.256	920	0.4	-

2.3. Transfer of Corrosive Properties

All three types of the above-mentioned corrosion only occur on the stent surface in contact with the corrosive environment for alloy-based BVSs. The corrosion surface is continuously updated with the degradation. The element on the corrosion surface has its own corrosion properties, including the exposure property of whether it is in direct contact with the corrosive environment, the number of exposed surfaces, and the pitting corrosion parameter. The corrosion properties will be transferred to the adjacent elements after a complete corrosion element is removed.

The definite location between elements facilitates the identification of corrosion surface and the update of exposure-related corrosion properties of the elements during degradation.

The removal of an exposed element only changes the corrosion properties of its face-adjacent elements. The transfer of corrosion properties includes updating the exposure property and increasing the number of exposed surfaces for the face-adjacent elements of the removed element and passing the pitting parameters to them.

The serial number of the element and corresponding node information was obtained after meshing, and the adjacent relationship between elements was obtained by calculating the number of shared nodes through the program. For the C3D8R element used in this research, the element located inside ha six face-adjacent elements, otherwise, it is on the surface. The initial corrosion surface of the stent was set as the elements in contact with blood by identifying the contact pairs between the stent and the vessel after implantation and excluding the surface elements only in contact with the vessel wall (Figure 2c). Then the number of exposed surfaces for the elements located on the initial corrosion surface was counted, and pitting corrosion parameters were assigned by using the Weibull random-number generator. Initial corrosion properties of the elements and adjacent relationships between elements were saved in external files.

2.4. Finite Element Analysis

The finite element analysis was divided into two parts: stent deployment and stent degradation. Firstly, the stent deployment was simulated by expanding the stent to 1.1 times the diameter of the vessel by applying displacement to the inner surface of the balloon (Figure 2a). The process was achieved by finite element analysis of the structural mechanics of the stent-vessel model with the Abaqus 6.14/Explicit solver (DS-SIMULIA, Providence, RI, USA) due to its superior contact enforcement method. The balloon was removed from the artery after stent deployment. The residual stress of the stent after deployment acted as the initial stress driving stent degradation (Figure 2d). The corrosion at this stage could be ignored as the stent deployment is short compared to the degeneration process. The general contact was set between each part, the coefficient of friction was assumed to be 0.2 for the tangential contract, and the normal contact was set as the default "hard" contact [10]. The constraint at axial and circumferential was applied at both ends of the artery to simulate the pull of the surrounding artery. The average ratio of kinetic energy to internal energy was maintained below 5% to minimize the effects of dynamics and ensure that the analysis process is quasi-static.

The continuous damage evolution process of zinc alloys was simulated by using Abaqus/Explicit solver and two user-defined subroutines of VEXTERNALDB and VUSDFLD. VEXTERNALDB was used for reading and writing external files in Explicit analysis. The damage variable functions at each integration point of the element in the stent model were defined in the degradation process, and these functions were calculated by VUSDFLD. Corrosion properties of the removed elements in the increment were transferred according to the adjacent relationship of the elements while returned at the end of each increment (Figure 3).

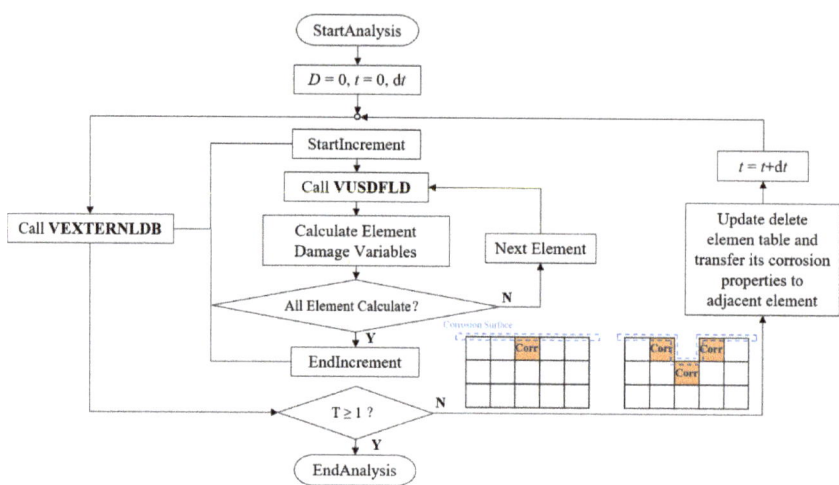

Figure 3. The stent's degradation was achieved through VUSDFLD and VEXTERNALDB subroutines in ABAQUS/Explicit.

The remained volume of the stent decreased as the element was gradually removed during the degradation. The corresponding time and the volume of the removed elements were the outputs in this process. The volume of the removed element was recorded as the cube of the characteristic length L_e. The mass loss during stent degradation is shown in Equation (8).

$$\text{mass loss} = \frac{\sum \rho \cdot L_e^3}{\text{initial mass of the stent}} \times 100\%, \tag{8}$$

The supporting performance is an important indicator for evaluating the service performance of degradable stents. It expresses effective support time and loss of mechanical strength of the stent. The vessel recoil during degradation is described in Equation (9).

$$\text{vessel recoil} = \frac{\text{vessel diameter during degradation} - \text{vessel diameter before stent development}}{\text{vessel diameter after stent development} - \text{vessel diameter before stent development}} \times 100\%, \tag{9}$$

The following five circumstances were simulated: (I) Case U: uniform corrosion, with constant corrosion rate; (II) Case U_{surf}: uniform corrosion, with the corrosion rate adjusted according to the number of exposed surfaces; (III) Case $U_{surf}SC$: add stress corrosion based on Case U_{surf}, with constant stress corrosion threshold; (IV) Case $U_{surf}SC_{th}$: add stress corrosion based on Case U_{surf}, with dynamic change of stress corrosion threshold according to the corroded status; (V) Case $U_{surf}SC_{th}P$: add pitting corrosion based on Case $U_{surf}SC_{th}$. Six simulations on different pitting corrosion distribution locations were conducted to eliminate the randomness of pitting corrosion.

3. Results

3.1. Effects of the Number of Exposed Surfaces on Stent Degradation

In Case U (Figure 4a), the damage variables were nearly identical for all exposed elements, and thus the stent struts became thinner due to the simultaneous removal of the elements on the corrosion surface of the stents. In Case U_{surf} (Figure 4b), the elements at the edges of the stent degraded first as their damage variables were greater than others due to the two surfaces being exposed to the corrosive environment. Subsequently, the number of exposed surfaces of its face-adjacent element increased, and the degradation accelerated with the increase in the rate of damage variable accumulation. The originally square cross-section first became round and further became thinner.

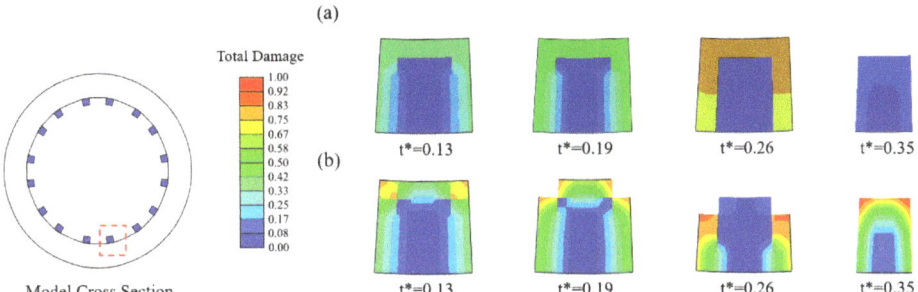

Figure 4. Changes in the cross-section of struts during degradation. Each view is shown for the normalized time unit. (**a**) Case U; (**b**) Case U_{surf}.

It can be observed in the mass loss curve that the the dynamic change of the corrosion rate with the number of exposed surfaces accelerated the time to complete degradation of the stent by 25% (Figure 5). The "stair" caused by the simultaneous removal of elements on the corrosion surface can be observed in the mass loss curve of Case U. Elements with approximately the characteristic length accumulated the same damage and were removed from the model simultaneously due to the consistent corrosion rate. The corrosion surface area of the stent became smaller with the degradation, and the height of the "stair" in the curve became lower. The mass loss curve of Case U_{surf} showed a gradual increase without the obvious mutation in Case U.

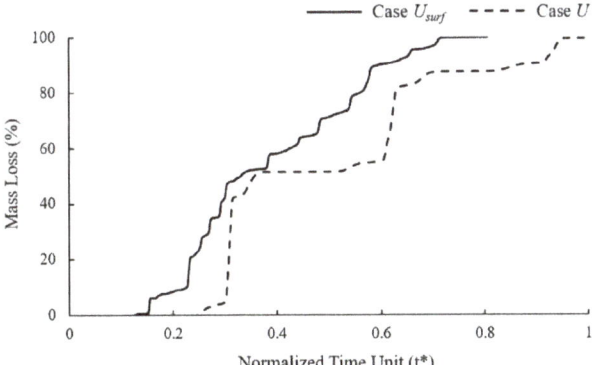

Figure 5. Mass loss of Case U and Case U_{surf} during stent degradation.

3.2. Effects of Dynamic Changes of Stress Corrosion Threshold on Stent Degradation

As local corrosion, stress corrosion occurs in a small area but can greatly impact the structure. Comparing the simulation results of Case $U_{surf}SC_{th}$ and $U_{surf}SC$, it can be found that Case $U_{surf}SC_{th}$ mainly showed the fracture at the corolla of the stent (Figure 6a), and Case $U_{surf}SC$ mainly showed the uniform thinning of the struts (Figure 6b). The degradation of stents in both cases first occurred in the high tensile stress area inside the corolla, and the struts gradually became thinner under the dominant effect of uniform corrosion in the subsequent degradation. The struts of Case $U_{surf}SC$ were thinner than those of Case $U_{surf}SC_{th}$ when the stent fractured.

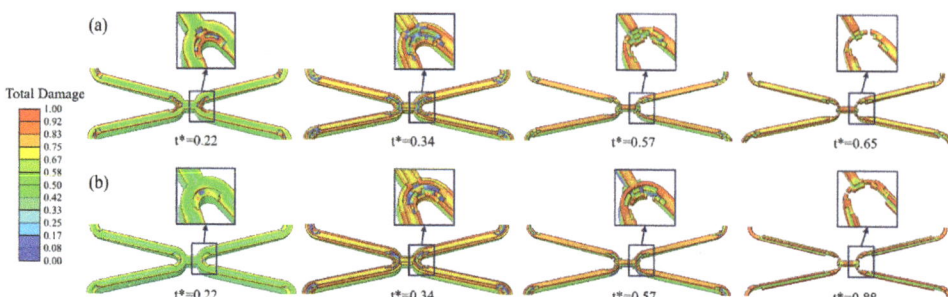

Figure 6. The degradation form of the stent until fracture. Each view is shown for the normalized time unit. (**a**) Case $U_{surf}SC_{th}$; (**b**) Case $U_{surf}SC$.

In Case $U_{surf}SC_{th}$ and $U_{surf}SC$, the removed elements under the dominant action of stress corrosion were mainly distributed in the high tensile stress area at the corolla of the stent, while those under the dominant action of uniform corrosion were in various areas of the stent (Figure 7). Figure 7a shows that more elements were removed under the dominant effect of stress corrosion as the dynamic reduction in the stress corrosion threshold made the element subject to stress corrosion due to the tensile stress reaching the threshold. The dynamic change of stress corrosion threshold accelerated the degradation of the high tensile stress area and thus sped up the stent fracture.

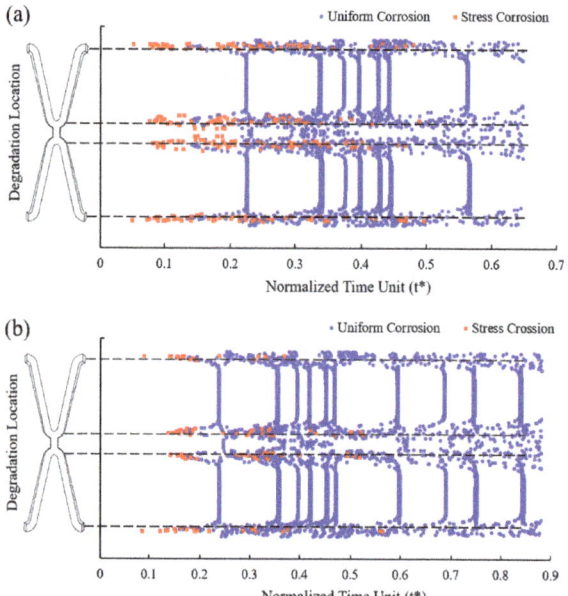

Figure 7. The location of the removed elements and the dominant factors leading to degradation. (**a**) Case $U_{surf}SC_{th}$; (**b**) Case $U_{surf}SC$.

In mass loss curves (Figure 8), the tendency of Case $U_{surf}SC_{th}$ and $U_{surf}SC$ were roughly the same, while the stent fracture in Case $U_{surf}SC_{th}$ was earlier than that in Case $U_{surf}SC$. The mass loss of the stent increased slowly in the early stage of degradation, and the change was more obvious in the later stage. The mass loss curve showed noticeable changes when the element on the edges or other elements on the corrosion surface were removed simultaneously. The mass loss of Case $U_{surf}SC_{th}$ and $U_{surf}SC$ were about 65% and 83%, respectively, when the stent fractured. The dynamic reduction in the stress threshold could accelerate the time of stent degradation by 26% and reduce mass loss by 22% when the stent fractured.

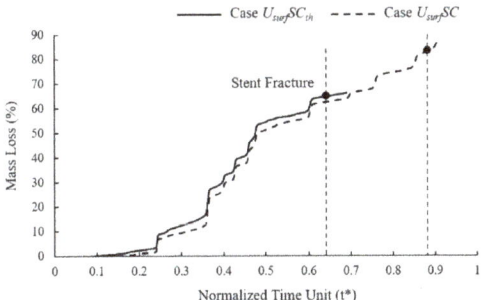

Figure 8. Mass loss of Case $U_{surf}SC_{th}$ and Case $U_{surf}SC$ during stent degradation.

3.3. Effects of the Addition of Pitting Corrosion on Stent Degradation

Pits caused by pitting corrosion can be seen during the degradation and would develop further over time. The addition of the pitting corrosion model accelerated the degradation. Especially when the pitting corrosion occurs at critical locations of the corolla, the local fracture would be obviously accelerated (Figure 9).

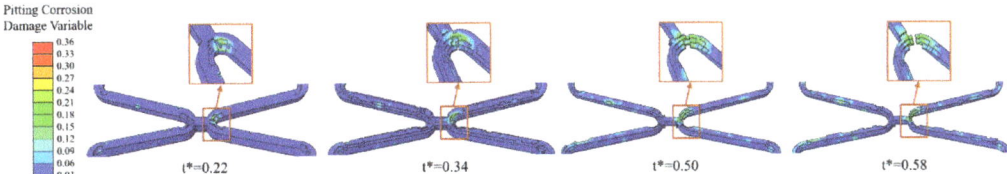

Figure 9. The degradation form of the stent until fracture of Case $U_{surf}SC_{th}P$ when the pitting corrosion occurs at critical locations of the corolla. Each view is shown for the normalized time unit.

In six different sets of simulations, the mass loss curves remained almost consistent despite some differences in the degradation due to the pitting corrosion parameters assigned to the surface elements being the same with different locations (Figure 10a). Compared to the model without pitting corrosion, the degradation was accelerated to a certain extent regardless of the locations of pitting corrosion, and the time of stent fracture was in $t^* = 0.58{\sim}0.64$.

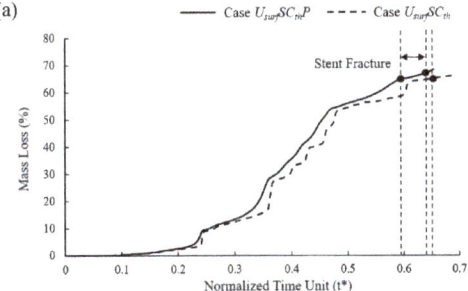

Figure 10. *Cont.*

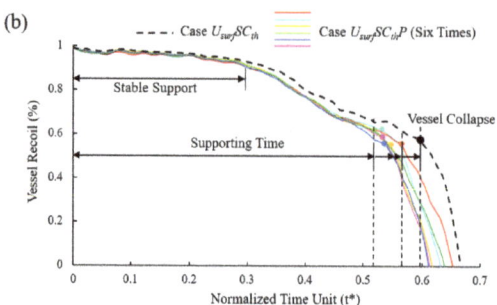

Figure 10. (a) Mass loss of Case $U_{surf}SC_{th}P$ and Case $U_{surf}SC_{th}$ during degradation; (b) vessel recoil due to loss of mechanical strength of stents during degradation.

In the initial stage of stent degradation, there was a stable effective support time when the vessel diameter hardly changed (Figure 10b). After $t^* = 0.3$, the diameter of the vessel began to change significantly from the increase in stent mass loss after the struts become thinner. It is still considered that the stent could provide effective support for the vessel in this stage. There would be a sudden loss of support at a certain moment around $t^* = 0.52 \sim 0.57$, and the vessel would collapse before the stent fracture. No matter how the pitting corrosion is distributed, the supporting performance and support time were reduced compared to Case $U_{surf}SC_{th}$. The supporting time was reduced by up to 12% when pitting corrosion was located in the critical location of the stent. Before the vessel collapse, the stent could provide effective support for the vessel with the diameter difference reaching more than 60% before and after stent implantation.

4. Discussion

In this study, an FEA model based on the CDM principle was established to describe the degradation process and service performance of the BVSs by the phenomenological description of the damage. We made several improvements to the degradation model by considering the combined effect of multiple corrosion factors to explore a more realistic non-uniform degradation process.

The corrosion rate of the element was dynamically adjusted according to the number of surfaces of the element exposed to the corrosive environment during degradation. The results indicated that the elements with more exposed surfaces at the edges of the stent degraded first, and the cross-section of the stent gradually became round from square (Figure 4). It also smoothened the edges and corners that emerged during the degradation. This phenomenon is consistent with the work of Grogan et al. [43]. The proposed model can successfully capture the accelerated degradation of elements at the edges and corners of the stent under multi-dimensional attack [25]. The relationship between multi-dimension corrosions and mechanical property is reflected in this model. The multi-dimension corrosions increased the corrosion rates, and then reduced mechanical strength. During degradation, the number of exposed surfaces changed dynamically, especially the elements in corroded areas with poor mechanical integrity have more exposed surfaces and are more vulnerable to corrosion attack. Adjusting a higher corrosion rate for elements with more exposed surfaces, i.e., those at the edges and corners, helps to make the simulated degradation more realistic [44,45].

Stress corrosion plays a vital role in the degradation process [12]. The structure may fail under the safe load due to the mutual promotion of electrochemical corrosion in the corrosive environment and the mechanical damage under stress. Stress corrosion is the dominant factor affecting element removal at the beginning of the degradation. The elements subject to stress corrosion are mainly concentrated in the high tensile stress area at the corolla (Figure 7). These areas are critical in affecting the supporting performance, despite its small area on the stent. As mentioned in work by Li et al. [46], the stress corrosion threshold affected the corrosion rate, and thus stress threshold was an essential parameter for the degradation process. It is necessary to consider the stress threshold in more detail, such as the dynamic changes with the degradation progress. The dynamic adjustment of the stress threshold changes the degradation morphology and accelerates the degradation of the stent. This work might be helpful for studying the problem of non-uniform degradation and early failure after stent implantation [16].

In this research, the ratio of pitting corrosion to uniform corrosion was controlled by limiting the range of pitting corrosion parameters [13]. As a kind of local corrosion, pitting corrosion is mainly affected by the defect location during stent production and implantation. The above two factors can be controlled in the current production process; thus, pitting corrosion will not become the dominant factor for the degradation. This work confirmed that the pitting corrosion occurs independently during degradation and is unaffected by either the stress state or the material composition [43]. However, pitting corrosion plays a significant role in accelerating the local fracture of the stent compared with critical locations without pitting corrosion (Figure 9). Pitting corrosion in high-stress regions changes the shape of the stent, and the resulting stress concentration accelerates the development of stress corrosion. Therefore, it is necessary to consider the pitting corrosion effect in the degradation of the BVSs.

Stent degradation is a complex process in which many factors interact with each other. As consistent with previous work [44], the material properties of the stents are continuously weakened, and the geometry of the struts is gradually changed with the degradation. The mechanical strength of the stent decreases gradually with the decrease in the remaining volume (Figure 10b). Most of the mass loss of the stent comes from uniform corrosion, while the local stress corrosion and pitting corrosion plays a more critical role in the service life of the stent. As can be seen, the stent degrades rapidly after a stable support time [25]. The stents would suddenly lose support to the vessel and lead the vessel to collapse at one moment in the degradation process. However, we found that this moment precedes when the stent loses its mechanical integrity. Therefore, the effective support time of the BVSs should adapt to the vascular remodeling time to prevent adverse events such as the sudden collapse of the vessel.

Some limitations might exist in this study. Only blood was considered as the corrosive environment in this paper, while the vessel wall is also one. The corrosion rate of the elements only in contact with the vessel wall was temporarily set to be 0 due to the lack of the corrosion rate in the corrosive environment of the vessel wall, which is lower in the absence of blood flow [22]. Both corrosive environments will be considered in future work, and the corrosion rate needs to be differentiated. Especially for locations with stent malapposition, the stent is separated from the vessel wall and suspended in the blood. The outer surface of the struts is completely exposed to the blood, which might be more likely to fracture than the case where the stent is in contact with the vessel wall. This method might be helpful for exploring the degradation and fracture of the BVSs with malapposition. We will calibrate the model by experiments in future work to obtain the degradation morphology in more detail and predict the service time of the stent more accurately.

5. Conclusions

In this paper, a BVS non-uniform degradation model was established considering the combined action of multiple factors. The results demonstrated that both the increase in the number of exposed surfaces of the element and the change in the stress corrosion threshold expedited BVSs' degradation process compared with the uniform degradation model. In particular, the addition of the pitting corrosion model significantly affected the service life of BVSs. The analysis proved that the stent could effectively support for the vessel by reaching more than 60% treatment effect during service. The phenomena of the numerical simulation results with the proposed model were more consistent with the real degradation scenario than the commonly used models. The corrosion rate was adjusted based on hemodynamic results that will be considered in future work to optimize the present model, especially in the case of stent malapposition. This work might provide a numerical simulation method and scientific basis for performance evaluation, the optimization of structural design, and the development of other alloys for BVSs.

Author Contributions: Conceptualization, H.Z. and A.Q.; methodology, T.D.; software, H.Z. and Y.L.; validation, Y.Y. and Q.H.; data curation, S.C.; writing—original draft preparation, H.Z. and Y.L.; writing—review and editing, H.Z. and A.Q.; supervision, A.Q. All authors have read and agreed to the published version of the manuscript.

Funding: This research was funded by the National Natural Science Foundation of China (12172018), and the Joint Program of Beijing Municipal—Beijing Natural Science Foundation (KZ202110005004).

Institutional Review Board Statement: Not applicable.

Informed Consent Statement: Not applicable.

Data Availability Statement: Not applicable.

Conflicts of Interest: The authors declare no conflict of interest. The funders had no role in the design of the study; in the collection, analyses, or interpretation of data; in the writing of the manuscript; or in the decision to publish the results.

References

1. Canfield, J.; Totary-Jain, H. 40 years of percutaneous coronary intervention: History and future directions. *J. Pers. Med.* **2018**, *8*, 33. [CrossRef] [PubMed]
2. Morlacchi, S.; Pennati, G.; Petrini, L.; Dubini, G.; Migliavacca, F. Influence of plaque calcifications on coronary stent fracture: A numerical fatigue life analysis including cardiac wall movement. *J. Biomech.* **2014**, *47*, 899–907. [CrossRef] [PubMed]
3. Bourantas, C.V.; Zhang, Y.; Farooq, V.; Garcia-Garcia, H.M.; Onuma, Y.; Serruys, P.W. Bioresorbable scaffolds: Current evidence and ongoing clinical trials. *Curr. Cardiol. Rep.* **2012**, *14*, 626–634. [CrossRef] [PubMed]
4. Iqbal, J.; Onuma, Y.; Ormiston, J.; Abizaid, A.; Waksman, R.; Serruys, P. Bioresorbable scaffolds: Rationale, current status, challenges, and future. *Eur. Heart J.* **2014**, *35*, 765–776. [CrossRef] [PubMed]
5. Sotomi, Y.; Onuma, Y.; Collet, C.; Tenekecioglu, E.; Virmani, R.; Kleiman, N.S.; Serruys, P.W. Bioresorbable scaffold: The emerging reality and future directions. *Circ. Res.* **2017**, *120*, 1341–1352. [CrossRef]
6. Li, J.; Zheng, F.; Qiu, X.; Wan, P.; Tan, L.; Yang, K. Finite element analyses for optimization design of biodegradable magnesium alloy stent. *Mater. Sci. Eng. C* **2014**, *42*, 705–714. [CrossRef]
7. Shen, Z.; Zhao, M.; Zhou, X.; Yang, H.; Liu, J.; Guo, H.; Zheng, Y.; Yang, J. A numerical corrosion-fatigue model for biodegradable mg alloy stents. *Acta Biomater.* **2019**, *97*, 671–680. [CrossRef]
8. Boland, E.L.; Shine, C.J.; Kelly, N.; Sweeney, C.A.; Mchugh, P.E. A review of material degradation modelling for the analysis and design of bioabsorbable stents. *Ann. Biomed. Eng.* **2016**, *44*, 341–356. [CrossRef]
9. Jamari, J.; Ammarullah, M.I.; Santoso, G.; Sugiharto, S.; Supriyono, T.; van der Heide, E. In silico contact pressure of metal-on-metal total hip implant with different materials subjected to gait loading. *Metals* **2022**, *12*, 1241. [CrossRef]
10. Grogan, J.A.; O'Brien, B.J.; Leen, S.B.; Mchugh, P.E. A corrosion model for bioabsorbable metallic stents. *Acta Biomater.* **2011**, *7*, 3523–3533. [CrossRef]
11. Grogan, J.A.; Leen, S.B.; Mchugh, P.E. Optimizing the design of a bioabsorbable metal stent using computer simulation methods. *Biomaterials* **2013**, *34*, 8049–8060. [CrossRef] [PubMed]
12. Wu, W.; Gastaldi, D.; Yang, K.; Tan, L.; Petrini, L.; Migliavacca, F. Finite element analyses for design evaluation of biodegradable magnesium alloy stents in arterial vessels. *Mater. Sci. Eng. B* **2011**, *176*, 1733–1740. [CrossRef]
13. Debusschere, N.; Segers, P.; Dubruel, P.; Verhegghe, B.; De Beule, M. A computational framework to model degradation of biocorrodible metal stents using an implicit finite element solver. *Ann. Biomed. Eng.* **2016**, *44*, 382–390. [CrossRef] [PubMed]
14. Pinto Slottow, T.L.; Pakala, R.; Waksman, R. Serial imaging and histology illustrating the degradation of a bioabsorbable magnesium stent in a porcine coronary artery. *Eur. Heart J.* **2008**, *29*, 314. [CrossRef] [PubMed]
15. Barlis, P.; Virmani, R.; Sheppard, M.N.; Tanigawa, J.; Di Mario, C. Angiographic and histological assessment of successfully treated late acute stent thrombosis secondary to a sirolimus-eluting stent. *Eur. Heart J.* **2007**, *28*, 1675. [CrossRef]
16. Wang, P.; Ferralis, N.; Conway, C.; Grossman, J.C.; Edelman, E.R. Strain-induced accelerated asymmetric spatial degradation of polymeric vascular scaffolds. *Proc. Natl. Acad. Sci. USA* **2018**, *115*, 2640–2645. [CrossRef]
17. Cui, X.; Peng, K.; Liu, S.; Ren, Q.; Li, G.; Gu, Z.; Qiao, A. A computational modelling of the mechanical performance of a bioabsorbable stent undergoing cyclic loading. *Procedia Struct. Integr.* **2019**, *15*, 67–74. [CrossRef]
18. Gastaldi, D.; Sassi, V.; Petrini, L.; Vedani, M.; Trasatti, S.; Migliavacca, F. Continuum damage model for bioresorbable magnesium alloy devices—Application to coronary stents. *J. Mech. Behav. Biomed.* **2011**, *4*, 352–365. [CrossRef]

9. da Costa-Mattos, H.S.; Bastos, I.N.; Gomes, J.A.C.P. A simple model for slow strain rate and constant load corrosion tests of austenitic stainless steel in acid aqueous solution containing sodium chloride. *Corros. Sci.* **2008**, *50*, 2858–2866. [CrossRef]
10. Wu, W.; Chen, S.; Gastaldi, D.; Petrini, L.; Mantovani, D.; Yang, K.; Tan, L.; Migliavacca, F. Experimental data confirm numerical modeling of the degradation process of magnesium alloys stents. *Acta Biomater.* **2013**, *9*, 8730–8739. [CrossRef]
11. Grogan, J.A.; Leen, S.B.; Mchugh, P.E. A physical corrosion model for bioabsorbable metal stents. *Acta Biomater.* **2014**, *10*, 2313–2322. [CrossRef] [PubMed]
12. Wang, J.; Giridharan, V.; Shanov, V.; Xu, Z.; Collins, B.; White, L.; Jang, Y.; Sankar, J.; Huang, N.; Yun, Y. Flow-induced corrosion behavior of absorbable magnesium-based stents. *Acta Biomater.* **2014**, *10*, 5213–5223. [CrossRef] [PubMed]
13. Han, P.; Tan, M.; Zhang, S.; Ji, W.; Li, J.; Zhang, X.; Zhao, C.; Zheng, Y.; Chai, Y. Shape and site dependent in vivo degradation of Mg-Zn pins in rabbit femoral condyle. *Int. J. Mol. Sci.* **2014**, *15*, 2959–2970. [CrossRef]
14. von de Hoh, N.; Bormann, D.; Lucas, A.; Thorey, F.; Meyer-Lindenberg, A. Comparison of the in vivo degradation progress of solid magnesium alloy cylinders and screw-shaped magnesium alloy cylinders in a rabbit model. *Mater. Sci. Forum* **2010**, *638–642*, 742–747. [CrossRef]
15. Gao, Y.; Wang, L.; Gu, X.; Chu, Z.; Guo, M.; Fan, Y. A quantitative study on magnesium alloy stent biodegradation. *J. Biomech.* **2018**, *74*, 98–105. [CrossRef] [PubMed]
16. Shi, W.; Li, H.; Mitchell, K.; Zhang, C.; Zhu, T.; Jin, Y.; Zhao, D. A multi-dimensional non-uniform corrosion model for bioabsorbable metallic vascular stents. *Acta Biomater.* **2021**, *131*, 572–580. [CrossRef]
17. Saris, N.L.; Mervaala, E.; Karppanen, H.; Khawaja, J.A.; Lewenstam, A. Magnesium: An update on physiological, clinical and analytical aspects. *Clin. Chim. Acta* **2000**, *294*, 1–26. [CrossRef]
18. Tzion-Mottye, L.B.; Ron, T.; Eliezer, D.; Aghion, E. The effect of Mn on the mechanical properties and in vitro behavior of biodegradable Zn-2%Fe alloy. *Metals* **2022**, *12*, 1291. [CrossRef]
19. Mostaed, E.; Sikora-Jasinskabc, M.; Drelicha, J.W.; Vedanib, M. Zinc-based alloys for degradable vascular stent applications. *Acta Biomater.* **2018**, *71*, 1–23. [CrossRef]
20. Peuster, M.; Hesse, C.; Schloo, T.; Fink, C.; Beerbaum, P.; von Schnakenburg, C. Long-term biocompatibility of a corrodible peripheral iron stent in the porcine descending aorta. *Biomaterials* **2006**, *27*, 4955–4962. [CrossRef]
21. Pierson, D.; Edick, J.; Tauscher, A.; Pokorney, E.; Bowen, P.; Gelbaugh, J.; Stinson, J.; Getty, H.; Lee, C.H.J.; Drelich, J.; et al. A simplified in vivo approach for evaluating the bioabsorbable behavior of candidate stent materials. *J. Biomed. Mater. Res. B Appl. Biomater.* **2012**, *100B*, 58–67. [CrossRef] [PubMed]
22. Zheng, Y.; Gu, X.; Witte, F. Biodegradable metals. *Mater. Sci. Eng. R Rep.* **2014**, *77*, 1–34. [CrossRef]
23. Bikora-Jasinska, M.; Paternoster, C.; Mostaed, E.; Tolouei, R.; Casati, R.; Vedani, M.; Mantovani, D. Synthesis, mechanical properties and corrosion behavior of powder metallurgy processed Fe/Mg2Si composites for biodegradable implant applications. *Mater. Sci. Eng. C* **2017**, *81*, 511–521. [CrossRef]
24. Hiromoto, S.; Yamamoto, A. Control of degradation rate of bioabsorbable magnesium by anodization and steam treatment. *Mater. Sci. Eng. C* **2010**, *30*, 1085–1093. [CrossRef]
25. Bowen, P.K.; Drelich, J.; Goldman, J. Zinc exhibits ideal physiological corrosion behavior for bioabsorbable stentst *Adv. Mater.* **2013**, *25*, 2577–2582.
26. Jamari, J.; Ammarullah, M.I.; Santoso, G.; Sugiharto, S.; Supriyono, T.; Prakoso, A.T.; Basri, H.; van der Heide, E. Computational contact pressure prediction of CoCrMo, SS 316L and Ti6Al4V femoral head against UHMWPE acetabular cup under gait cycle. *J. Funct. Biomater.* **2022**, *13*, 64. [CrossRef]
27. Prasadh, S.; Raguraman, S.; Wong, R.; Gupta, M. Current status and outlook of temporary implants (Magnesium/Zinc) in cardiovascular applications. *Metals* **2022**, *12*, 999. [CrossRef]
28. Su, Y.; Cockerill, I.; Wang, Y.; Qin, Y.; Chang, L.; Zheng, Y.; Zhu, D. Zinc-based biomaterials for regeneration and therapy. *Trends Biotechnol.* **2019**, *37*, 428–441. [CrossRef]
29. Hernández-Escobar, D.; Champagne, S.; Yilmazer, H.; Dikici, B.; Boehlert, C.J.; Hermawan, H. Current status and perspectives of zinc-based absorbable alloys for biomedical applications. *Acta Biomater.* **2019**, *97*, 1–22. [CrossRef]
30. Li, H.; Xie, X.; Zheng, Y.; Cong, Y.; Zhou, F.; Qiu, K.; Wang, X.; Chen, S.; Huang, L.; Tian, L.; et al. Development of biodegradable Zn-1X binary alloys with nutrient alloying elements Mg, Ca and Sr. *Sci. Rep.* **2015**, *5*, 10719. [CrossRef]
31. Winzer, N.; Atrens, A.; Song, G.; Ghali, E.; Dietzel, W.; Kainer, K.U.; Hort, N.; Blawert, C. A critical review of the stress corrosion cracking (SCC) of magnesium alloys. *Adv. Eng. Mater.* **2005**, *7*, 659–693. [CrossRef]
32. Martin, D.; Boyle, F. Finite element analysis of balloon-expandable coronary stent deployment: Influence of angioplasty balloon configuration. *Int. J. Numer. Methods Biomed. Eng.* **2013**, *29*, 1161–1175. [CrossRef] [PubMed]
33. Grogan, J.A.; Leen, S.B.; Mchugh, P.E. A Phenomenological Model of Corrosion in Biodegradable Metallic Stents. In Proceedings of the ASME 2010 Summer Bioengineering Conference, Naples, FL, USA, 16–19 June 2010.
34. Du, X.; Jin, L. Meso-scale numerical investigation on cracking of cover concrete induced by corrosion of reinforcing steel. *Eng. Fail. Anal.* **2014**, *39*, 21–33. [CrossRef]
35. Jin, L.; Zhang, R.; Du, X.; Li, Y. Investigation on the cracking behavior of concrete cover induced by corner located rebar corrosion. *Eng. Fail. Anal.* **2015**, *52*, 129–143. [CrossRef]
36. Li, G.; Zhu, S.; Nie, J.; Zheng, Y.; Sun, Z. Investigating the stress corrosion cracking of a biodegradable Zn-0.8 wt%Li alloy in simulated body fluid. *Bioact. Mater.* **2021**, *6*, 1468–1478. [CrossRef]

Journal of Functional Biomaterials

Article

A Finite Element Investigation on Material and Design Parameters of Ventricular Septal Defect Occluder Devices

Zhuo Zhang [1], Yan Xiong [1,*], Jinpeng Hu [2], Xuying Guo [2], Xianchun Xu [2], Juan Chen [2], Yunbing Wang [3,*] and Yu Chen [4]

[1] School of Mechanical Engineering, Sichuan University, Chengdu 610000, China
[2] Shanghai Shape Memory Alloy Co., Ltd., Shanghai 200000, China
[3] College of Biomedical Engineering, Sichuan University, Chengdu 610000, China
[4] Department of Applied Mechanics, Sichuan University, Chengdu 610000, China
* Correspondence: xy@scu.edu.cn (Y.X.); yunbing.wang@scu.edu.cn (Y.W.)

Abstract: Background and Objective: Ventricular septal defects (VSDs) are the most common form of congenital heart defects. The incidence of VSD accounts for 40% of all congenital heart defects (CHDs). With the development of interventional therapy technology, transcatheter VSD closure was introduced as an alternative to open heart surgery. Clinical trials of VSD occluders have yielded promising results, and with the development of new material technologies, biodegradable materials have been introduced into the application of occluders. At present, the research on the mechanical properties of occluders is focused on experimental and clinical trials, and numerical simulation is still a considerable challenge due to the braided nature of the VSD occluder. Finite element analysis (FEA) has proven to be a valid and efficient method to virtually investigate and optimize the mechanical behavior of minimally invasive devices. The objective of this study is to explore the axial resistive performance through experimental and computational testing, and to present the systematic evaluation of the effect of various material and braid parameters by FEA. Methods: In this study, an experimental test was used to investigate the axial resistive force (ARF) of VSD Nitinol occluders under axial displacement loading (ADL), then the corresponding numerical simulation was developed and compared with the experimental results to verify the effectiveness. Based on the above validation, numerical simulations of VSD occluders with different materials (polydioxanone (PDO) and Nitinol with different austenite moduli) and braid parameters (wire density, wire diameter, and angle between left and right discs) provided a clear presentation of mechanical behaviors that included the maximal axial resistive force (MARF), maximal axial displacement (MAD) and initial axial stiffness (IAS), the stress distribution and the maximum principal strain distribution of the device under ADL. Results: The results showed that: (1) In the experimental testing, the axial resistive force (ARF) of the tested occluder, caused by axial displacement loading (ADL), was recorded and it increased linearly from 0 to 4.91 N before reducing. Subsequent computational testing showed that a similar performance in the ARF was experienced, albeit that the peak value of ARF was smaller. (2) The investigated design parameters of wire density, wire diameter and the angle between the left and right discs demonstrated an effective improvement (7.59%, 9.48%, 1.28%, respectively, for MARF, and 1.28%, 1.80%, 3.07%, respectively, for IAS) for the mechanical performance for Nitinol occluders. (3) The most influencing factor was the material; the performance rose by 30% as the Nitinol austenite modulus (EA) increased by 10,000 MPa. The performance of Nitinol was better than that of PDO for certain wire diameters, and the performance improved more obviously (1.80% for Nitinol and 0.64% for PDO in IAS, 9.48% for Nitinol and 2.00% for PDO in MARF) with the increase in wire diameter. (4) For all of the models, the maximum stresses under ADL were distributed at the edge of the disc on the loaded side of the occluders. Conclusions: The experimental testing presented in the study showed that the mechanical performance of the Nitinol occluder and the MARF prove that it has sufficient ability to resist falling out from its intended placement. This study also represents the first experimentally validated computational model of braided occluders, and provides a perception of the influence of geometrical and material parameters in these systems. The results could further

Citation: Zhang, Z.; Xiong, Y.; Hu, J.; Guo, X.; Xu, X.; Chen, J.; Wang, Y.; Chen, Y. A Finite Element Investigation on Material and Design Parameters of Ventricular Septal Defect Occluder Devices. *J. Funct. Biomater.* **2022**, *13*, 182. https://doi.org/10.3390/jfb13040182

Academic Editor: Pedro Morouço

Received: 31 August 2022
Accepted: 1 October 2022
Published: 9 October 2022

Publisher's Note: MDPI stays neutral with regard to jurisdictional claims in published maps and institutional affiliations.

Copyright: © 2022 by the authors. Licensee MDPI, Basel, Switzerland. This article is an open access article distributed under the terms and conditions of the Creative Commons Attribution (CC BY) license (https://creativecommons.org/licenses/by/4.0/).

provide meaningful suggestions for the design of biodegradable VSD closure devices and to realize a series of applications for biodegradable materials in VSD.

Keywords: nitinol; polydioxanone; self-expanding occluder; braided wire occluder; finite element analysis (FEA); mechanical performance

1. Introduction

Congenital heart defects (CHD) refer to the general structural abnormality of the heart or thoracic great vessels at birth, with an incidence of 4–50 per 1000 live births [1]. Ventricular septal defect (VSD) is the most common CHD and accounts for about 40% of all CHDs [2]. It occurs in isolation or in combination with other structural defects. Nowadays, with the development of interventional therapy for CHD, the implantation of occlusion devices has become a widely accepted and highly-effective treatment for occluding abnormal blood/thrombus flow within the heart. It is less invasive, avoids extracorporeal circulatory support and surgical scars, and offers a faster recovery [3].

Due to their great applicability, a large variety of closure device designs exist on the market employing different types of materials, different geometries, and deployment mechanisms. Currently, the most commonly used clinical interventions are nickel–titanium alloy occluders. Although they have shown good near- to mid-term efficacy, the occurrence of complications is still not considered to be negligible, and there are potential disadvantages and safety risks for metal blockers to remain in the body permanently. The ideal occluder should be biodegradable, and the degradation products should be non-toxic, harmless, and fully absorbed. Polydioxanone (PDO) is a biodegradable polymer which can be completely absorbed in about 6 months, with the degradation products mainly being excreted through urine, and the rest discharged by digestion or as carbon dioxide [4]. PDO has demonstrated in potential applications such as intragastric stents and drug delivery systems [5,6].

The existing work usually evaluates the effectiveness of the mechanical properties of the closure devices through experiments and clinical experience, resulting in a long research and development cycle, high cost, and lack of predictability. Based on experimental rest, Thepphithak et al. [7] evaluated the transformation and mechanical behavior of the nitinol atrial septal defect (ASD) occluder through differential scanning calorimetry (DSC) measurements and pull tests; Wu et al. [8] performed the axial compress in vitro tests of the fully absorbable occluders. And through animal experiments, Huang et al. [9] performed percutaneous transcatheter closure of interventional created VSDs in 16 dogs with the occluders, and obtained gross pathology and histopathology at 6, 12, and 24 weeks of follow-up. Some scholars have validated the application of VSD through clinical experience, such as Tzikas et al. [10], who reported on 19 patients who underwent transcatheter perimembranous ventricular septal defect (PMVSD) closure using a second-generation occluder device, and suggested that the procedure was feasible, safe, and effective. Lee et al. [11] put forward that transcatheter closure of PMVSD with the Amplatzer ductal occlude (ADO) was a safe and promising treatment option, but long-term follow-up in a large number of patients would be warranted.

Numerical simulation represents a powerful support tool providing engineers and clinicians with more detailed information about the effects on the overall biomechanical properties of the material and design parameters of VSD closure devices. Nevertheless, few works focusing on numerical modeling of VSD occlusion devices have been presented; only Li et al. [12] has so far carried out radial compression and axial bending finite element analysis of ventricular septal defect occluders. Nonetheless, only the waist of the device is modeled in a braided structure, while the two discs were simplified into an entity. This work is intended to establish a complete braided numerical model of VSD occluders, and based on the experimental verification of the manufacturer's product, to provide a reference for subsequent numerical research.

The stability test is required for occluders to ensure that they remain in place without defect, in accordance with the pharmaceutical industry standards. Also, due to the insufficient mechanical properties of the degradable material, there is a possible problem of insufficient clamping force when the material is changed to a biodegradable material. In this study, the VSD closure device developed by Shanghai Shape Memory Alloy Company was taken as a prototype, which consists of interlaced wires that slide and rotate during deformation. The stability tests were performed as required to investigate the mechanical properties of a commercially available occluder under axial displacement loading (ADL). An accurate finite element model was developed and used to perform computational tests to systematically compare and evaluate the effects on stability of the wire material properties and braid parameters, including the degradability of materials, wire density, wire diameter, and the angle between the left and right discs.

This computational modeling would help to determine the relevant material and design parameters of wire-braided VSD occlusion devices by simulating a wide range of occlusion device configurations. It would further provide references to optimize the VSD occlusion device design. This method could also be further extended to numerical studies of biodegradable occlusion devices and atrial septal occluders.

2. Materials and Methods

2.1. Geometry of Commercial VSD Occluder and Modeling

A commercially available occluder (Shanghai Shape Memory Alloy Company, Shanghai, China) for the closure of ventricular septal defects (VSD) was selected as the basic design as shown in Figure 1a, which consists of interlaced wires that slide and rotate as they deform. It is a self-expanding and self-centering occlusion device, prepared by weaving and thermoforming technology using 72 nitinol monofilaments with a diameter of 0.11 mm [13]. The most important part of the occluder device is the double-disc structure, the left and right circular discs, which are linked together by a short connecting waist, as shown in Figure 1b [14].

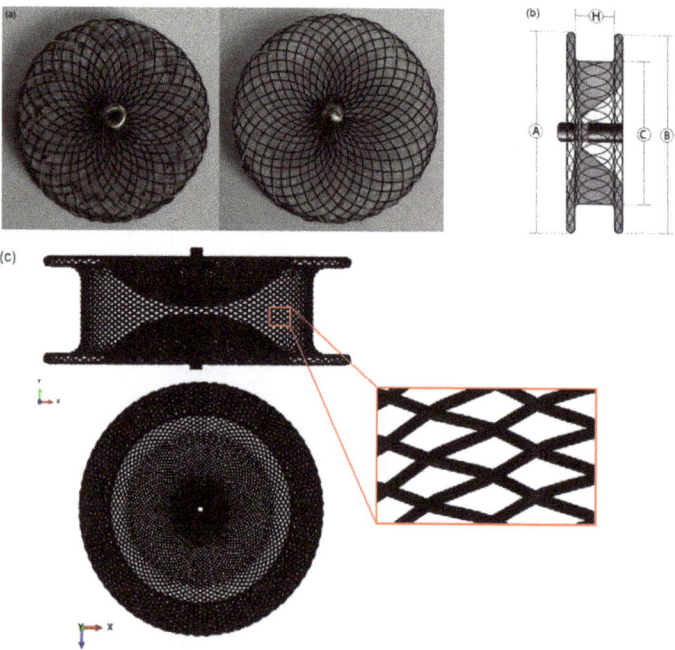

Figure 1. (**a**) VSD occluder product, (**b**) device specification, and (**c**) Numerical 3D model of VSD occluder.

The 3D geometric model was established and imported into ABAQUS2021 software. The complete numerical 3D geometry of the braided occluder was established as shown in Figure 1c. General contact was applied between the wires, with a friction coefficient of 0.25, so that wire-to-wire contact and cross-slip are implemented [15].

2.2. Experimental Testing

A ventricular septal defect is a birth defect of the heart in which there is a hole in the wall (septum) that separates the two lower chambers (ventricles) of the heart. The occluders are used to plug a hole that is not meant to be present. The device is delivered percutaneously to the hole (the defected site of the heart) in the sheath, and it is then allowed to self-expand to its original double-disc shape when the sheath is removed [16]. After the device is placed properly, the discs of the occluder clamp the VSD, while the waist of the occluder supports the VSD hole. Ideally, the structure of the device will ensure its effective self-placement without falling out, which ensures the stability of the device. However, it is necessary to convincingly prove the effectiveness of the device clamping feature, since fall-out of the occluder would be life-threatening for the patient. Therefore, we designed an experimental test: the occluder was put into the silica model (Figure 2a,b), which was used to simulate the defect, and was then fixed to the test machine, as shown in Figure 2c. After finishing the assembly setup, the loading rod was displaced to make sure it pushed down slowly until the occluder fell out from the silica model (Figure 2d). The value of the axial displacement (AD) and axial resistive force (ARF) of the control rod was recorded.

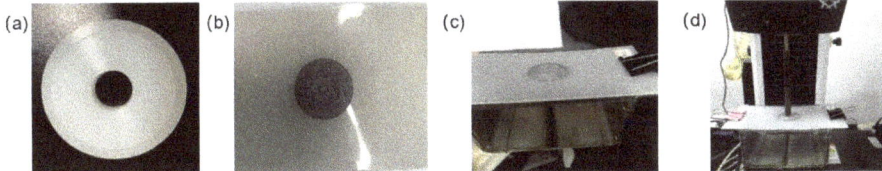

Figure 2. (a) Silica defect model, (b) silica model blocked by occluder, (c) clamp to fix silica model and (d) the loading rod to apply axial loading.

2.3. Computational Testing

FEA is a lower-cost method to not only replicate experiments but to also carry out repeated tests, finding the most vulnerable point in the model, and optimizing it [17]. A computational framework based on the experimental testing was developed to predict the mechanical performance of wire braid occluders and to explore the role of geometric and material properties on device mechanics. The computational framework was compared and validated by experimental axial testing. A systematic evaluation of the effect of wire braid parameters (e.g., wire density, wire diameter and the angle between left and right discs) and wire materials (Nitinol material properties and biodegradable PDO material) on axial loading response of these devices is presented.

2.3.1. Material Properties

The device is constructed from self-expanding Nitinol mesh, due to its remarkable super-elasticity and shape memory, as well its good biocompatibility and corrosion resistance [18]. The Nitinol occluder is easily crimped at low temperatures and reverts back to its original profile after being released from the sheath at the body temperature [19]. In this study, the Nitinol material was modeled in ABAQUS with a user material subroutine interface (UMAT) [20]. The material properties of this model are shown in Table 1.

Table 1. Nitinol material properties.

Symbol	Parameter	Value
E_A	Austenite elasticity (MPa)	46,728
v_A	Austenite Poisson's ratio	0.33
E_M	Martensite elasticity	25,199
v_M	Martemosite Poisson's radio	0.33
ε^L	Transformation strain	0.0426
$(\delta\sigma/\delta T)_L$	$(\delta\sigma/\delta T)$ loading	4.5
σ^S_L	Start of transformation loading	358.2
σ^E_L	End of transformation loading	437.8
T_0	Reference temperature	0
$(\delta\sigma/\delta T)_U$	$(\delta\sigma/\delta T)$ unloading	4.5
σ^S_U	Start of transformation unloading	124.5
σ^E_U	End of transformation unloading	17.75
σ^S_{CL}	Start of transformation stress during loading in compression, as a positive value	537.3
ε^L_V	Volumetric transformation strain	0.0426

The PDO used in the study had a linear elastic behavior in tension, with a Young's modulus of 1200 MPa and a tensile strength of 326 MPa in the longitudinal direction. For the simulation, the Poisson ratio was set to 0.35.

2.3.2. Boundary and Loading Conditions

All of the FEA processes were undertaken in ABAQUS. The assembly is shown in Figure 3. Besides the occluder, the geometry of the defect and the loading rod were also modeled, which were simplified as rigid bodies controlled by reference points (RP). The translations and rotations of the defective part were limited in all directions. The same limitation was imposed on the loading rod part except for in the Y-axis direction, which applied a displacement as shown in Figure 3a. All the parts had meshed as presented in Figure 3b. The braided occluders were meshed with the 2-nodes linear Timoshenko beam elements (B31), and the defect, as well as loading rod, were both meshed with the 4-nodes 3-D bilinear rigid quadrilateral elements (R3D4). There were 21,886 elements in the occluders, 1513 elements in the defect and 175 elements in the loading rod. Because of the high nonlinearity, a quasi-static analysis was performed by a dynamic explicit algorithm. The quasi-static is guaranteed by making sure that the kinetic energy was less than 5% of the internal strain energy [21].

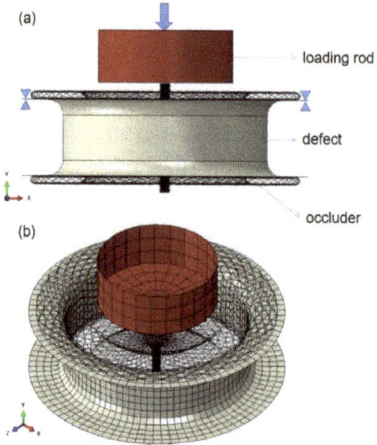

Figure 3. (**a**) Boundary condition and loading, and (**b**) assembly mesh.

3. Model Validation

As shown in Figure 4a, both the computational testing and experimental testing predicted similar axial resistive force (ARF) trends for braided occluders under axial displacement loading (ADL). Both test results showed ARF increased firstly (from 0 to 4.91 N in experimental testing and from 0 to 4.21 N in computational testing) and decreased later when applying displacement loading. Although the maximum value (14.02% lower) and inflection point (15.6% lower) of the computational testing data are underpredicted, which is a common feature in such models [22], the trend of ARF showed a good agreement with experimental testing, especially in the increasing and stable stage (AD is from 1 to 6 mm of computational testing), and these two stages were the stages we cared most about in this study.

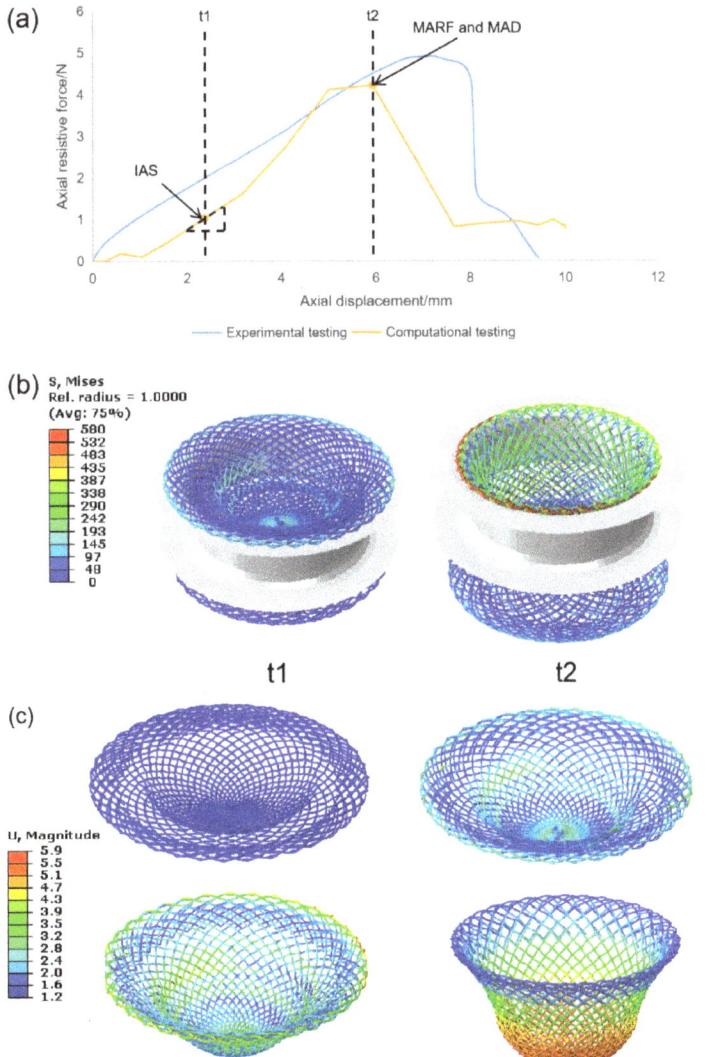

Figure 4. (a) Comparison of experimental and computational testing data for devices under ADL. (b) Device deformation at t1 and t2. (c) The deformation process on the loading discs of the occluder in different displacement loading.

For the comparison of stability characteristics of the device, the initial axial stiffness (IAS) of the occluders at t1 in the increasing stage, as well as the maximal axial resistive force (MARF) and the maximal axial displacement (MAD) at t2 in the stable stage is as shown in Figure 4a. The value of IAS, MARF, and MAD in experimental testing were 0.66 N/mm, 4.91 N, and 7.03 mm, respectively, whereas in the computational testing, they were 0.74 N/mm (overpredicted 12%), 4.21 N (underpredicted 14%) and 5.93 mm (underpredicted 15%), respectively. This error can be partially explained by the complicated deformations of the braided device, where there are substantial self-contacts. Nonetheless, the computational testing method is acceptable to predict the axial response for devices with different parameters.

Figure 4b showed that in moment t1, the occluder clamped the margin of VSD to resist axial loading by deformation until it came to the threshold at moment t2, where the device could barely clamp thereafter. The deformation process is presented in Figure 4c. As the axial loading was applied, the loading disc of the device underwent visible bending and stretching, resulting in transforming the device into concave shape. Moreover, the waist part was elongated and accompanied by a change in the pitch angles.

4. Parameter Study

4.1. Occluder System Parameter

A systematic evaluation of the effect of geometric and material properties on the functional performance of occluders under the axial loading conditions was carried out by the computational testing method. The commercial wire braid occluder was chosen to be the baseline device geometry with 72 wires of 0.11 mm diameter and an angle of 0° between the left and right disc (two discs parallel to each other). The influence of the Nitinol austenite modulus, wire density, wire diameter, and the angle between the left and right discs were evaluated in wire braid occluders, according to the values shown in Table 2. The model of the occluders with various wire densities (n) and various angles between the left and right discs (α) are presented in Figure 5.

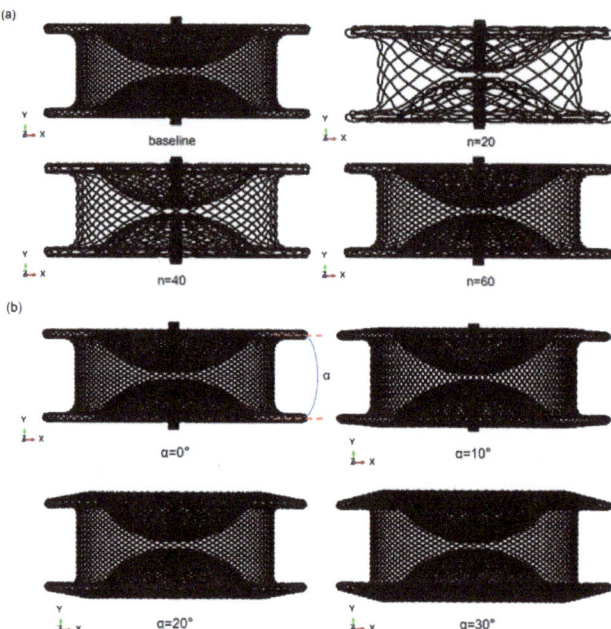

Figure 5. The model with various wire densities (**a**) and various angles between left and right discs (**b**).

Table 2. The baseline parameter properties for a braided occluder and parameter variations.

Wire Parameter	Baseline	Variations	
Nitinol austenite modulus (E_A)	46,728	30,000; 40,000; 60,000	MPa
Wire density (n)	72	20; 40; 60	wires
Wire diameter (d)	0.11	0.07; 0.09; 0.13	mm
Angle between left and right discs (α)	0	10; 20; 30	°

4.2. The Results

The effect of key parameters on the axial performance of wire braid occluders is shown in Figure 6.

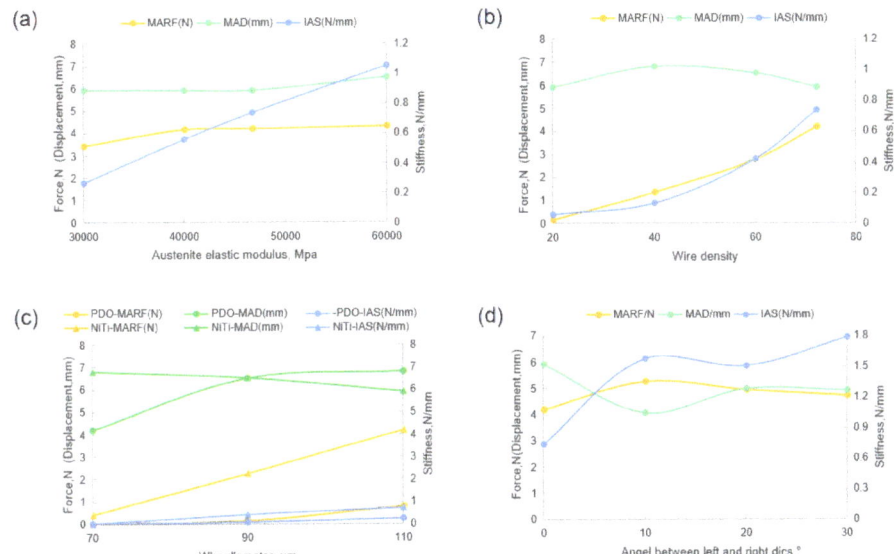

Figure 6. Plots demonstrate the effects of (**a**) Nitinol austenite elastic modulus, (**b**) wire density, (**c**) wire diameter with PDO and Nitinol, and (**d**) angle between left and right discs on MARF, MAD and IAS.

The results showed that increasing the austenite elastic modulus of Nitinol causes an obvious increase in IAS, MARF, and MAD of the device (increased by 30%, 30%, and 20%, respectively, with every 10,000 MPa increasement), as shown in Figure 6a. It can be seen in Figure 6b that the effects of wire density on the performance cannot be disregarded, as for MARF with a 7.59% linear increase, for IAS with a nonlinear increase of about 1.28% more or less.

The most significant parameter was wire diameter. Figure 6c shows the effects of different wire diameters with PDO and Nitinol materials on the MARF, MAD, and IAS. With the increase in the diameter of the braided wires, the mechanical properties of the occluders made up of both Nitinol and PDO materials had similar trends. The IAS and MARF increased linearly by 1.80% (Nitinol), 0.64% (PDO) and 9.48% (Nitinol), 2.00% (PDO), respectively. The increasing trend in mechanical properties for the PDO material was obviously weaker than that for Nitinol. The performance of the PDO occluders was over 60% lower in IAS and 70% lower in MARF than that of the Nitinol occluders. For a given wire diameter occluder, the mechanical properties of Nitinol were also weaker compared to PDO. The wire diameter of the occluder was 0.07 mm, the IAS of the Nitinol occluder was 0.019 N/mm and that of the PDO was 0.013 N/mm (reduced by 31%). When the wire diameter was 0.09 and 0.11 mm, the difference in IAS was more obvious, which decreased by more than 60%. The MARF of the PDO was also significantly lower than that of Nitinol for these three different wire diameters. In addition, the MAD of the occluder under MARF loading was also different for various braid materials and braiding wire diameters. With the increase in wire diameter, the MAD of the Nitinol occluder decreased slowly, while for the PDO, it was just the reverse, which still needs to be verified and analyzed in subsequent tests.

As shown in Figure 6d, there was also a difference when the angle between the left and right discs appropriately increased, such that, if ignoring the result of the angle at 10°, the IAS and MARF increased by 3.07% and 1.28%, respectively, while the MAD decreased by 2.02% (increased angle resulted in a decrease in disc diameter). The effect of the wire diameter on the performance of braided products was also investigated in terms of both Nitinol and PDO materials.

The deformations of all the devices with MARF after axial loading are shown in Figure 7a–d. The models are divided into 4 groups according to various design parameters from sp1 to sp16 as follows: (a) sp1 to sp4: wire density, (b) sp5 to sp8: Nitinol austenite elastic modulus, (c) sp9 to sp12: wire diameter, and (d) sp13 to sp16: angle between left and right discs of the occluder. In each group, the parameters of the corresponding design variables were changed to explore the different effects of the parameters on the mechanical response under axial displacement load.

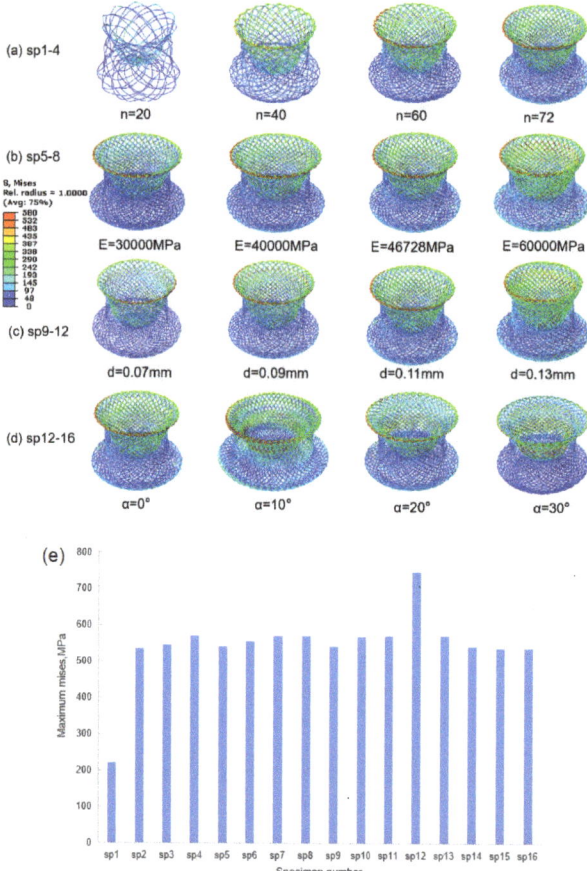

Figure 7. Contour plots showing the effect of, (**a**) wire density, (**b**) Nitinol austenite elastic modulus, (**c**) wire diameter, and (**d**) angle between left and right discs on the stress distribution of the Nitinol device under ADL; (**e**) maximum stress on devices.

The results showed that, except for sp1, all of the models had similar deformations. The margin of the discs, where the maximal stress fields of the device are located, were deformed to resist the loading. All values for maximal stress were over σ^E_L (437.8 MPa, end of the transformation loading for the simulated Nitinol material from austenite to martensite) except sp1, as shown in Figure 7e, which means that the transformations were completely finished in these high-stress fields.

Figure 8 shows the stress-colored map of the maximum axial resistance of the PDO occluder, in which the stress distribution at the edge of the disc on the loaded side of the occluder was significantly larger than that on the rest of the parts. The maximum stress value distributed at the edge of the disc for different braided wire diameters of 0.07, 0.09, and 0.11 mm under the maximum axial load was 21, 26, and 47MPa, respectively. The stress distribution for the NiTi alloy occluder under maximum axial load is shown by the stress-colored map, and presents a similar pattern to that of the PDO occluder, where all of the stress distribution is at the edge of the disc on the loaded side, but the maximum stresses are significantly larger than that of the PDO occluders, which are 542.1 MPa, 568 MPa, 570.3 MPa, respectively.

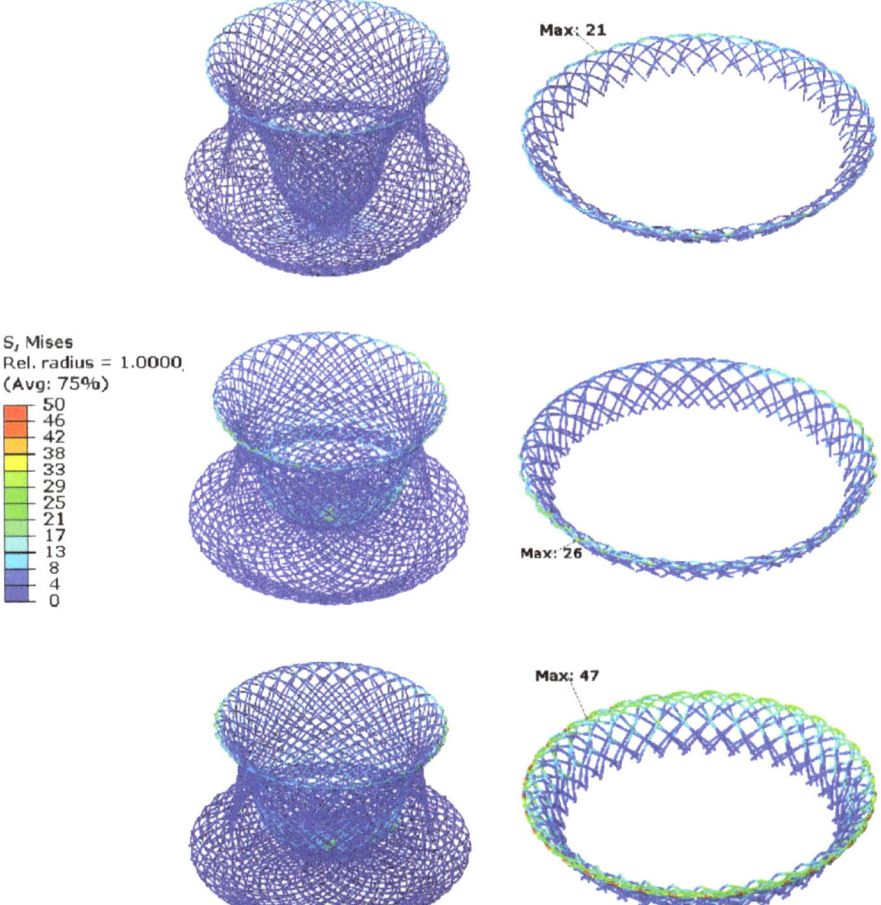

Figure 8. Contour plots showing the effect of wire diameter on the stress distribution of the PDO device under ADL.

5. Discussion

Nowadays, the implantation of occlusion devices has become widely accepted as a highly effective way to treat abnormal blood/thrombus flow within the heart. Studies of occlusion devices have previously focused on the materials and configurations through experiments and clinical experience, leading to a longer research cycle, high cost and lack of predictability. In this study, not only were the mechanics of the self-expanding wire-braided VSD occluders evaluated through experimental testing, but a systematic evaluation was also developed to investigate the effects of various material and braid parameters using computational simulation.

Based on analytical and experimental testing results, it was demonstrated that axial resistance force would increase to the maximum followed by a decrease, as displacement loading was slowly implemented. The computational method also identified that the stiffness of the device could be effectively improved in at least three aspects, such as the material of the braided wire and the method of weaving, as well as the shape of the hot-pressed double-discs.

Finite element analysis (FEA) has been widely implemented as a productivity tool for design engineers to reduce both development time and cost. It provides a clear presentation of the mechanical response that will help practitioners to avoid mistakes and conduct design improvement for implanted medical devices [23]. However, few studies have applied FEA to the research of occlusion devices. An accurate finite element model would be required to enable the correct estimation of the mechanical behavior of braided closure devices. To the authors' knowledge, this study represents the first computational model for the braided VSD occluders, which were experimentally validated under ADL and they have great potential in the future design of these devices.

It has been proven that for braided stents, increasing the austenite elastic modulus of Nitinol causes a linear increase in initial radial stiffness [22]. In this study, the FEA testing results showed that for braided occluders, the austenite elastic modulus of Nitinol has the same influence on initial axial stiffness. Nevertheless, the wire density and wire diameter have different effects on initial axial stiffness and initial radical stiffness. For initial radial stiffness, increasing wire density and wire diameter resulted in a linear and non-linear increase, respectively [16,24]. While for initial axial stiffness, this study showed that they cause a non-linear and linear increase separately. The difference might be explained by different deformations under their corresponding loading conditions. During the radial crimping, the straight cylindrical braided stent undergoes a change in the pitch angles, resulting in a diameter reduction as well as a longitudinal elongation of the whole device [25,26]. Applying axial loading, the double-disc braided occluder undergoes obvious bending of the loading disc, besides elongation and change in the pitch angles. The loading disc undergoes concave deformation, the change in the curvature contributes to the increase in bending stress, and eventually a decreased diameter results from being unable to suffer larger loading.

After implanting occlusion devices, the defect will be endothelialized and covered by newly formed autologous tissues. Studies have shown that it takes no longer than 6 months. Therefore, the occluders play a role as a temporary bridge for the self-repair of the heart and should be biodegradable [27]. On the other hand, the pressure difference between the left and right ventricles is about 6.65—14.67 kPa (50—110 mmHg). For the VSD with a diameter of 12 mm, the largest force it would experience from blood would be no more than 1.65 N. It could be noticed that the devices can tolerate a force as large as 9 N and 4.2 N from the experimental and computational testing results, respectively. That means that the Nitinol occluders can effectively seal the defect with a satisfactory mechanical property to avoid falling out. However, it cannot be denied that metal occluders exist in vivo permanently. The weaving material is expected to evolve from non-degradable Nitinol to biodegradable polylatide (PLA), polydioxanone (PDO), polycaprolactone (PCL), etc. These biodegradable materials are non-metallic materials with relatively poor mechanical properties in some aspects, such as initial axial stiffness, compared to Nitinol (lower 30%

with same wire diameter). Therefore, new methods for braiding and shaping structures need to be developed to improve the mechanical properties of the biodegradable materials. What has been brought up in this study, through the investigations into the mechanical properties with the FEA method, can be further used for biodegradable materials in the future.

Several limitations and hypotheses of this study need to be mentioned. In computational testing, the defect was assumed to suffer the loading with little deformation, so as to be regarded as a rigid body. The frictional coefficient of all the contact was also assumed to be identical to 0.2. These assumptions were beneficial to FEA numerical computation, but at the risk of losing accuracy. Additionally, the computational testing of parameter studies should also be conducted by in vitro experimental testing to validate the results. When using the biodegradable materials instead of Nitinol to test, the results would be more clinically meaningful. Besides, more testing under different loading conditions should be carried out to fully validate the model and the differences between experimental and FEA testing.

6. Conclusions

In summary, this study used experimental tests to investigate the mechanical properties of VSD Nitinol occluders under axial loading and MARF conditions, to prove that they have the sufficient ability to resist falling out from their correct placements. The first experimentally validated computational models of braided occluders were established, and they provided a clear presentation by simulating the process with FEA. Furthermore, the mechanical behavior of different occluder materials and design parameters were also presented by means of FEA. Our results demonstrated that material is the key governing parameter that affects the axial resistive performance of wire-braided occluders. The axial performance of PDO devices is relatively weaker than Nitinol devices; using larger wire diameters and a greater wire density could bring about significant improvement. It is expected that the outcomes of this study could be used as a reference for the future design of biodegradable braided occluders.

Author Contributions: Conceptualization, Z.Z. and Y.X.; methodology, Z.Z. and J.H.; validation, Z.Z., X.G. and X.X.; formal analysis, Z.Z. and J.H.; writing—original draft preparation, Z.Z.; writing—review and editing, Z.Z. and Y.X.; project administration, J.C., Y.W. and Y.C.; funding acquisition, Y.C. All authors have read and agreed to the published version of the manuscript.

Funding: This study was funded by the National Key R&D Program of China (grant number 12172239).

Data Availability Statement: Not applicable.

Conflicts of Interest: The authors declare no conflict of interest.

References

1. Hoffman, J.I.E.; Kaplan, S. The incidence of congenital heart disease. *J. Am. Coll. Cardiol.* **2002**, *39*, 1890–1900. [CrossRef]
2. Zhao, L.J.; Han, B.; Zhang, J.J.; Yi, Y.C.; Jiang, D.D.; Lyu, J.L. Transcatheter closure of congenital perimembranous ventricular septal defect using the Amplatzer duct occluder 2. *Cardiol. Young* **2018**, *28*, 447–453. [CrossRef]
3. Lock, J.E.; Block, P.C.; McKay, R.G.; Baim, D.S.; Keane, J.F. Transcatheter closure of ventricular septal defects. *Circulation* **1988**, *78*, 361–368. [CrossRef]
4. Goonoo, N.; Jeetah, R.; Bhaw-Luximon, A.; Jhurry, D. Polydioxanone-based bio-materials for tissue engineering and drug/gene delivery applications. *Eur. J. Pharm. Biopharm.* **2015**, *97*, 371–391. [CrossRef] [PubMed]
5. Wang, C.; Zhang, P. Design and characterization of PDO biodegradable intravascular stents. *Text. Res. J.* **2017**, *87*, 1968–1976. [CrossRef]
6. Xie, X.; Zheng, X.; Han, Z.; Chen, Y.; Zheng, Z.; Zheng, B.; He, X.; Wang, Y.; Kaplan, D.L.; Li, Y.; et al. A Biodegradable Stent with Surface Functionalization of Combined-Therapy Drugs for Colorectal Cancer. *Adv. Healthc. Mater.* **2018**, *7*, 1801213. [CrossRef]
7. Rueangmontree, T.; Khantachawana, A. Fabrication and Mechanical Property Evaluation of ASD Occluders Made from Shape Memory Alloys. In *Materials Science Forum*; Trans Tech Publications Ltd.: Wollerau, Switzerland, 2019; pp. 151–155.
8. Wu, S.W.; Li, C.J.; Wang, F.J.; Wang, L.; Li, Y.M.; Sun, K. Preparation and mechanical performance evaluation of twocomponent fully absorbable occluder. *J. Donghua Univ. (Natl. Sci.)* **2021**, *47*, 7. (In Chinese)

9. Huang, X.M.; Zhu, Y.F.; Cao, J.; Hu, J.Q.; Bai, Y.; Jiang, H.B.; Li, Z.F.; Chen, Y.; Wang, W.; Qin, Y.W. Development and preclinical evaluation of a biodegradable ventricular septal defect occluder. *Catheter. Cardiovasc. Interv.* **2013**, *81*, 324–330. [CrossRef] [PubMed]
10. Tzikas, A.; Ibrahim, R.; Velasco-Sanchez, D.; Freixa, X.; Alburquenque, M.; Khairy, P.; Bass, J.L.; Ramirez, J.; Aguirre, D.; Miro, J. Transcatheter closure of perimembranous ventricular septal defect with the Amplatzer membranous VSD occluder 2: Initial world experience and one-year follow-up. *Catheter. Cardiovasc. Interv.* **2014**, *83*, 571–580. [CrossRef] [PubMed]
11. Lee, S.M.; Song, J.Y.; Choi, J.Y.; Paik, J.S.; Chang, S.I.; Shim, W.S.; Kim, S.H. Transcatheter closure of perimembranous ventricular septal defect using Amplatzer ductal occluder. *Catheter. Cardiovasc. Interv.* **2013**, *82*, 1141–1146. [CrossRef] [PubMed]
12. Li, Y.; Sun, K.; Song, C. Finite Element Modeling and Analysis of Ventricular Septal Defect Occluders. *J. Med. Biomech.* **2018**, *33*, 18–23.
13. Lin, C.; Liu, L.; Liu, Y.; Leng, J. Recent developments in next-generation occlusion devices. *Acta Biomater.* **2021**, *128* (Suppl. S2), 100–119. [CrossRef] [PubMed]
14. Thanopoulos, B.V.D.; Laskari, C.V.; Tsaousis, G.S.; Zarayelyan, A.; Vekiou, A.; Papadopoulos, G.S. Closure of atrial septal defects with the Amplatzer occlusion device: Preliminary results. *J. Am. Coll. Cardiol.* **1998**, *31*, 1110–1116. [CrossRef]
15. Kelly, N.; McGrath, D.J.; Sweeney, C.A.; Kurtenbach, K.; Grogan, J.A.; Jockenhoevel, S.; O'Brien, B.J.; Bruzzi, M.; McHugh, P.E. Comparison of computational modelling techniques for braided stent analysis. *Comput. Methods Biomech. Biomed. Eng.* **2019**, *22*, 1334–1344. [CrossRef] [PubMed]
16. Venkatraman, S.; Yingying, H.; Wong, Y.S. Bio-absorbable cardiovascular implants: Status and prognosis. *JOM* **2020**, *72*, 1833–1844. [CrossRef]
17. Liu, Z.; Wu, L.; Yang, J.; Cui, F.; Ho, P.; Wang, L.; Dong, J.; Chen, G. Thoracic aorta stent grafts design in terms of biomechanical investigations into flexibility. *Math. Biosci. Eng.* **2021**, *18*, 800–816. [CrossRef] [PubMed]
18. Stoeckel, D.; Pelton, A.; Duerig, T. Self-expanding Nitinol stents: Material and design considerations. *Eur. Radiol.* **2004**, *14*, 292–301. [CrossRef] [PubMed]
19. Kleinstreuer, C.; Li, Z.; Basciano, C.; Seelecke, S.; Farber, M. Computational mechanics of Nitinol stent grafts. *J. Biomech.* **2008**, *41*, 2370–2378. [CrossRef]
20. Kumar, G.P.; Cui, F.; Danpinid, A.; Su, B.; Hon, J.K.F.; Leo, H.L. Design and finite element-based fatigue prediction of a new self-expandable percutaneous mitral valve stent. *Comput. -Aided Des.* **2013**, *45*, 1153–1158. [CrossRef]
21. Choi, H.H.; Hwang, S.M.; Kang, Y.H.; Kim, J.; Kang, B.S. Comparison of implicit and explicit finite-element methods for the hydroforming process of an automobile lower arm. *Int. J. Adv. Manuf. Technol.* **2002**, *20*, 407–413. [CrossRef]
22. Mckenna, C.G.; Vaughan, T.J. A finite element investigation on design parameters of bare and polymer-covered self-expanding wire braided stents. *J. Mech. Behav. Biomed. Mater.* **2021**, *115*, 104305. [CrossRef] [PubMed]
23. Kurowski, P. Finite Element Analysis for Design Engineers. In *Finite Element Analysis for Design Engineers*; SAE: Warrendale, PA, USA, 2017; pp. i–xiii.
24. Ni, X.Y.; Pan, C.W.; Gangadhara Prusty, B. Numerical investigations of the mechanical properties of a braided non-vascular stent design using finite element method. *Comput. Methods Biomech. Biomed. Eng.* **2015**, *18*, 1117–1125. [CrossRef]
25. Zaccaria, A.; Pennati, G.; Petrini, L. Analytical methods for braided stents design and comparison with FEA. *J. Mech. Behav. Biomed. Mater.* **2021**, *119*, 104560. [CrossRef] [PubMed]
26. Shanahan, C.; Tiernan, P.; Tofail SA, M. Looped ends versus open ends braided stent: A comparison of the mechanical behaviour using analytical and numerical methods. *J. Mech. Behav. Biomed. Mater.* **2017**, *75*, 581–591. [CrossRef]
27. Liu, S.J.; Peng, K.M.; Hsiao, C.Y.; Liu, K.S.; Chung, H.T.; Chen, J.K. Novel biodegradable polycaprolactone occlusion device combining nanofibrous PLGA/collagen membrane for closure of atrial septal defect (ASD). *Ann. Biomed. Eng.* **2011**, *39*, 2759–2766. [CrossRef]

Article

Physical and Chemical Characterization of Biomineralized Collagen with Different Microstructures

Tianming Du [1], Yumiao Niu [1], Youjun Liu [1], Haisheng Yang [1], Aike Qiao [1,*] and Xufeng Niu [2,*]

[1] Beijing International Science and Technology Cooperation Base for Intelligent Physiological Measurement and Clinical Transformation, Department of Biomedical Engineering, Faculty of Environment and Life, Beijing University of Technology, Beijing 100124, China; dutianming@bjut.edu.cn (T.D.); yumiaoniu@163.com (Y.N.); lyjlma@bjut.edu.cn (Y.L.); haisheng.yang@bjut.edu.cn (H.Y.)
[2] Key Laboratory of Biomechanics and Mechanobiology (Beihang University), Ministry of Education, Beijing Advanced Innovation Center for Biomedical Engineering, School of Biological Science and Medical Engineering, Beihang University, Beijing 100083, China
* Correspondence: qak@bjut.edu.cn (A.Q.); nxf@buaa.edu.cn (X.N.)

Abstract: Mineralized collagen is the basic unit in hierarchically organized natural bone with different structures. Polyacrylic acid (PAA) and periodic fluid shear stress (FSS) are the most common chemical and physical means to induce intrafibrillar mineralization. In the present study, non-mineralized collagen, extrafibrillar mineralized (EM) collagen, intrafibrillar mineralized (IM) collagen, and hierarchical intrafibrillar mineralized (HIM) collagen induced by PAA and FSS were prepared, respectively. The physical and chemical properties of these mineralized collagens with different microstructures were systematically investigated afterwards. Transmission electron microscopy (TEM) and scanning electron microscopy (SEM) showed that mineralized collagen with different microstructures was prepared successfully. The pore density of the mineralized collagen scaffold is higher under the action of periodic FSS. Fourier transform infrared spectroscopy (FTIR) analysis showed the formation of the hydroxyapatite (HA) crystal. A significant improvement in the pore density, hydrophilicity, enzymatic stability, and thermal stability of the mineralized collagen indicated that the IM collagen under the action of periodic FSS was beneficial for maintaining collagen activity. HIM collagen fibers, which are prepared under the co-action of periodic FSS and sodium tripolyphosphate (TPP), may pave the way for new bone substitute material applications.

Keywords: mineralized collagen; microstructure; physical characterization; chemical characterization

1. Introduction

Bone, as a mechanically adaptive organ, can adapt to a variety of external forces by adjusting its complex hierarchical structure [1]. The basic unit of the complex hierarchical structure of bone is mineralized collagen, which consists of collagen fibers and apatite [2,3]. Collagen fibers are the most resourceful proteins in animals and have good mechanical sensitivity [1,4,5]. Hydroxyapatite (HA) crystals are typical representatives of apatite, which is transformed from amorphous calcium phosphate (ACP). ACP and HA crystals have different calcium (Ca): phosphorus (P) ratio ratios and crystallinity [6]. Therefore, it is an effective mechanical factor due to its higher stiffness and crystallinity [7–10]. HA can effectively promote the growth of osteoblast with high crystallinity. In the process of mineralization, collagen can induce the nucleation and crystallization of ACP, and then the ACP grow along the long axis of the aligned collagen fibers, eventually forming mineralized collagen with different microstructures [11–15]. Mineralized collagen is a response for the second level for the highly anisotropic hierarchical structure of bone. The highly anisotropic hierarchical structure is critical to the good mechanical and biological activity properties of bone [16].

Currently, mineralized collagen mainly includes extrafibrillar mineralized (EM) collagen and intrafibrillar mineralized (IM) collagen. Since the structure of the IM collagen is closer to the natural mineralized collagen structure, the study of the IM collagen has gained more attention. The IM collagen is divided into the normal IM collagen and the hierarchical intrafibrillar mineralized (HIM) collagen. According to the polymer-induced liquid phase precursor mineralization process, the polyanionic compounds, such as polyacrylic acid (PAA), can stabilize the ACP precursors as sequestration analogues to induce normal IM collagen [17,18]. Meanwhile, polyphosphate compounds, such as sodium tripolyphosphate (TPP), can regulate HIM collagen as template analogues with the co-presence of PAA [19–22]. However, the crystal transformation is slower and the mineralized collagen fibers are arranged poorly under the action of PAA [8,23,24]. Since the crystal size and crystallinity will affect the mechanical and biological activity properties of the materials, in order to prepare a better bone substitute material with the arrangement microstructure and the high crystal conversion degree, it is necessary to prepare an arranged HIM collagen under the action of external force.

Among the external force, the periodic fluid shear stress (FSS) is regarded as the most important mechanical stimulation mode for the bone matrix [25,26]. Correspondingly, the collagen mineralization can be guided by the external mechanical environment [27–29]. According to our previous study, the periodic FSS can replace the PAA to induce a highly arranged IM collagen and a highly arranged HIM collagen with the co-presence of TPP [30,31]. HIM collagen could better promote the biological responses [32]. Nevertheless, in the bone, these mineralized collagens with different microstructures may exist at the same time and result in the formation of the complex structure of the bone [33]. The physical and chemical properties of the materials are the key to the biocompatibility. The basic physical and chemical properties of these mineralized collagens with different microstructures have not been systematically analyzed.

Hence, we prepared mineralized collagen fibers with different microstructures under the action of PAA, periodic FSS, and periodic FSS-TPP, respectively. The morphology of the mineralized collagen was studied by scanning electron microscopy (SEM) and transmission electron microscopy (TEM). The molecular structure and the crystal structure of the mineralized collagen fibers was investigated by Fourier transform infrared spectroscopy (FTIR). The hydrophilicity, the stability, and the salinity of the mineralized collagen fibers were examined by analyzing the contact angles, the enzymatic times, and the thermogravimetry (TG) [34]. Our aim was to analyze the physical and chemical properties of the mineralized collagen with different mineralized microstructures, provide basic data support for the further exploration of the osteogenesis properties of those mineralized collagen fibers with different microstructures, and finally prepare advanced bone substitute materials.

2. Materials and Methods

2.1. Preparation of Mineralized Collagen with Different Microstructures

The collagen source solution was obtained from BD Biocoat (No. 354236, Corning, New York, NY, USA). The collagen was rat tail collagen (type I). The concentration of collagen was around 3–5 mg/mL. The collagen source solution was mixed with 0.1 M $CaCl_2$ and 0.1 M $(NH_4)_2HPO_4$ according to the molar ratio of Ca: P in HA. In the mineralization system, the molar ratio of calcium-to-phosphorus was kept to 1.67. The pH of the mineralization system was adjusted to 7.4 by 1 M NaOH and 0.1 M ammonium hydroxide. Then, the collagen concentration was adjusted to 1 mg/mL by distilled water, and Ca concentration was regulated to 5 mM by distilled water in the mineralization reaction system. In addition, 1 mg/mL PAA was added into the mineralization system, and its final concentration was 0 and 30 µg/mL for the preparation of EM collagen and normal IM collagen, respectively.

Meanwhile, the mineralization experiments were repeated with the same collagen and Ca concentration but adding the periodic FSS adjustment step during the mineralization process. A cone-and-plate viscometer (Brookfield R/S-CPS+ rheometer, Rheo3000,

Brookfield, CT, USA) was used to provide 1.0 Pa of periodic FSS (every period is two hours, which involves work for 1 h and rest for 1 h) for the arranged IM collagen.

In addition, the collagen phosphorylated under the action of 3% TPP ($Na_5P_3O_{10}$, No. T5508, Sigma-Aldrich, Saint Louis, MO, USA) for 5 h was mineralized using the same procedure to prepare the arranged HIM collagen under the co-action of periodic FSS [35].

All the mineralization reactions were kept at room temperature (RT) for 24 h. After being mineralized for 24 h, these samples were washed three times and excess water was removed by centrifugation. The samples were centrifuged at 5000 rpm for 10 min (Universal 320R, Hettich, Vlotho, Germany) and then, further lyophilized (Christ Alpha 1-4 LD, Christ, Osterode am Harz, Germany) for the following analysis.

2.2. Transmission Electron Microscopy (TEM) Observations

The internal morphologies of the mineralized collagen with different microstructures were analyzed by TEM (JEM-2100F, JEOL, Tokyo, Japan). The lyophilized mineralized collagen samples with different microstructures were embedded in spurr resin at 65 °C for 24 h without staining. The resin-embedded samples were cut into thin slices with a thickness of less than 100 nm by an ultraslicer (Leica UC7, speed 2.0 mm/s, Wetzlar, Germany). Ultrathin sections were transferred to copper mesh. These ultrathin samples were observed by TEM, and an analysis of the crystallization of the mineralized collagen was carried out by a selected area electron diffraction (SAED) at 200 kV.

2.3. Scanning Electron Microscopy (SEM) Observations

SEM (LEO1530VP, Munich, Germany) was used to observe the morphology of the mineralized collagen with different microstructures. All the lyophilized mineralized collagen samples were cut into small patches (5 × 5 mm) and fixed on the special sample stage for SEM with the conductive tapes. Then, the sample stage with the mineralized specimens was sputter-coated with a layer of gold using an ion sputter coater (SBC-12, Beijing Zhongke Instrument Co., Ltd., Beijing, China) at 0.1 Torr, 15 mA for 90 s total. Finally, the gold-sprayed sample was placed in the sample chamber of the SEM and observed under the condition of an accelerating voltage of 15 kV.

2.4. Fourier Transform Infrared Spectra (FTIR) Measurements

The molecular structure and the crystallinity of the mineralized collagen were analyzed by FTIR (NICOLET380, Boston, MA, USA). The FTIR measurements were performed using the potassium bromide pellet pressing method. In the tablets, the mass ratio of mineralized collagen and potassium bromide was controlled between 1:150 and 1:250. All spectra were recorded in the range of 400~4000 cm^{-1} at a resolution of 4 cm^{-1} intervals, and spectra plots represented 32 scans [36].

2.5. Hydrophilicity of Mineralized Collagen

The hydrophilicity of the mineralized collagen sponges can be analyzed by measuring the contact angle of the lyophilized collagen sponge surface using a contact-angle-measuring instrument (JC2000DM, Beijing Zhongyi Kexin Technology Co., Ltd., Beijing, China) [37]. The sessile drop contact angle measurements are carried out by 1 µL distilled water drops, gently delivered from a capillary tip onto the mineralized collagen surfaces. Due to the better hydrophilicity of the collagen material, we analyzed the dynamic contact angle of the different materials by taking photographs every 0.5 s for 30 s. The dynamic contact angles of the different sponges were measured three times.

2.6. Enzyme Measurements

A certain amount (1 mg) of the lyophilized mineralized collagen was placed in a 357 units/mL collagenase solution. The collagenase solution was prepared with Tris-HCl buffer at pH 7.4, 0.05 M. In order to simulate the enzymatic hydrolysis process in vivo, these samples with collagenase solution were dissolved in a water bath at 37 °C. The time of the dissolution of these samples can be seen as the time of enzymolysis.

2.7. Thermogravimetry (TG) Measurements

Weight loss can be used to analyze the mineralization degree and structural stability. The weight losses of the mineralized collagen with different microstructures (3~5 mg) were measured using a TG analyzer (LABSYS evo, Setsys, Caluire, France). According to the degradation temperature of the collagen and HA, the conditions of the TG test are as follows: the heating temperature range is from 50 °C to 800 °C, the heating rate is 20 °C min^{-1}, and the atmosphere is Ar.

2.8. Statistical Analysis

A statistical analysis was performed using one-way analysis of variance (ANOVA) on SPSS 19.0. Corresponding *p*-values of less than 0.05 were considered significant.

3. Results and Discussion

During the collagen mineralization, ACP nucleates, grows, and crystallizes along collagen fibrils to form different internal mineralized structures. The internal microstructures of mineralized collagen under different conditions were observed by unstained TEM. Nonmineralized collagen is transparent under the TEM (Figure 1A). As a typical polyanionic compound, PAA is used to stabilize ACP precursors and induce the formation of nano-sized ACPs. Without the action of PAA and periodic FSS, ACP was too large to permeate into collagen fibers, large-sized ACPs attached only on the surface of collagen and formed EM collagen (Figure 1B, triangles). Meanwhile, with the action of PAA, the nano-sized ACP was formed; the nano-sized ACPs were smaller than the collagen gap regains, so they was able to penetrate into the collagen fibers through the gap regains and form normal IM collagen (Figure 1C). Arranged IM collagen formed under the regulation of periodic FSS (Figure 1D), and arranged HIM collagen formed under the co-action of periodic FSS and TPP (Figure 1E, yellow lines). The selected area electron diffraction (SAED) of the mineralized collagen produced ring patterns, proving that the ACP particles have transformed from a continuous amorphous mineral phase into a crystal phase after being mineralized for 24 h (Figure 1F). The SAED of the mineralized collagen produced crystalline patterns containing specific diffusive rings corresponding to (002), (211), and (004) reflections (Figure 1F). The specific diffusive ring patterns were the typical characteristics of highly aligned HA nanoplatelets. Under the action of PAA, normal IM collagen was randomly oriented (Figure 1C, red lines). The arranged IM collagen and HIM collagen both oriented well under the action of periodic FSS (Figure 1D,E, red lines). This proves that periodic FSS is beneficial to the neat arrangement of mineralized collagen.

SEM was used to observe the micro-spatial network structure of the mineralized collagen with different microstructures (Figure 2). All mineralized collagen sponges possess a three-dimensional interconnected porous structure, which has no significant difference before and after mineralization (Figure 2A). However, the high magnifications of these mineralized collagens (Figure 2B–F) showed the differences between these mineralized collagens. Before mineralization, the surface of collagen was very smooth. After mineralization, the surface of the EM collagen was rough, although the large ACP particles and the IM collagen were smoother than the EM collagen. Among them, we found that more nanosized spherical ACP particles were observed on the surface of the normal IM collagen induced by PAA (Figure 2D, triangle). The reason is that PAA induces intrafibrillar mineralization by stabilizing the ACPs. Furthermore, periodic FSS can reduce the size of the ACPs by its own perturbation, inducing the formation of intrafibrillar mineralization [8]. More specifically, periodic FSS can arrange the collagen fibers neatly (Figure 2E,F, arrows) and induce the mineralized phosphorylated collagen with a stripe-like structure (Figure 2F, triangles), which is the oriented HIM collagen. In addition, the oriented IM and HIM collagen under the action of periodic FSS has high pore density. The results confirmed that we have made mineralized collagen with different microstructures successfully, and periodic FSS is favorable for making the oriented HIM collagen.

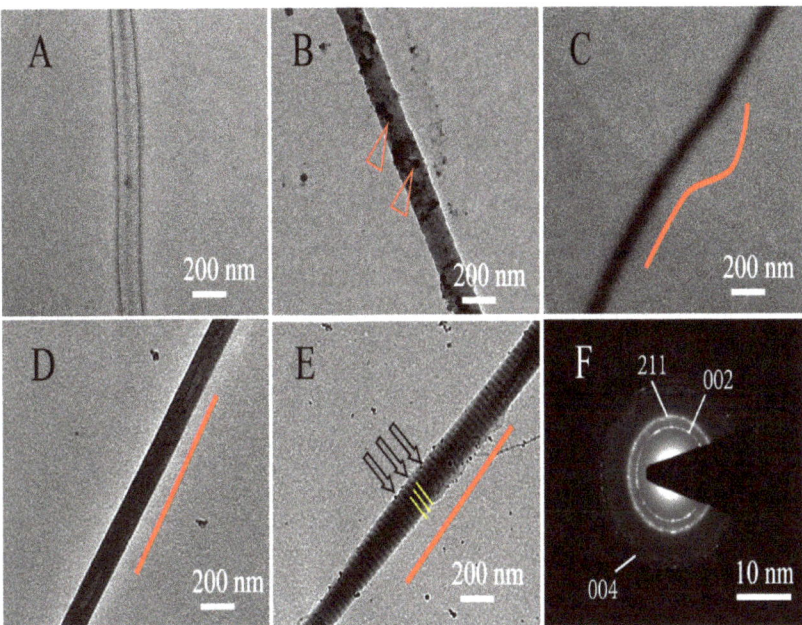

Figure 1. The internal morphological characteristics of different mineralized collagens observed by TEM. (**A**): The non-mineralized (NM) collagen; (**B**): the EM collagen; (**C**): the PAA-induced normal IM collagen; (**D**): the periodic FSS-induced arranged IM collagen; and (**E**): the periodic FSS and TPP-co-induced arranged HIM collagen. The NM collagen is transparent (**A**). There are some large-sized ACPs attached to the surface of the EM collagen (**B**, triangles). The IM and HIM collagen oriented under the action of the periodic FSS (**D,E**, red line). (**F**): The SAED of the mineralized collagen produced ring patterns. The ring patterns along (002), (211), and (004) were characteristic of highly crystalline, anisotropic HA (**F**).

FTIR spectroscopy can directly reflect the conformational structure of a compound molecule through the absorption peak of the characteristic groups. Therefore, FTIR spectroscopy was applied to measure the changes of the collagen molecule structure before and after mineralization. In the FTIR spectroscopy, amide A, amide B, amide I, amide II, and amide III are the absorption peaks of the classical structure of collagen (Figure 3a). The presence of these absorption peaks of the classical structure of the collagen triple helix structure means that the collagen triple helix structure was kept well after mineralization. During the process of mineralization, the Ca and P ions first aggregated to form ACPs; then, the ACPs transformed calcium-deficient apatite and HA crystals. ACP, calcium deficient apatite, and HA crystals have different Ca:P ratio ratios and crystallinity. The transformation of ACP to HA can be differentiated by the appearance of absorption peaks of phosphate groups around 1030 cm^{-1} and the single peak around 580 cm^{-1} gradually splitting into two peaks around 600 and 560 cm^{-1} (Figure 3b, dotted lines in yellow boxes). The higher the crystallinity, the sharper the peak shape. The lower the crystallinity, the smaller the half width of the peak shape. Therefore, by observing the change of the phosphate group absorption peaks of the mineralized collagen under different conditions, we found that the characteristic peaks at 562, 600, and 1029 cm^{-1} of the collagen mineralized with the action of the periodic FSS were the highest and sharpest (Figure 3b-D,b-E arrows), indicating that the periodic FSS could promote the transformation of HA crystals and improve their crystallization rate. Meanwhile, the PAA-induced normal IM collagen has the lower absorption of characteristic peaks about the formation of HA around 560, 600, and 1030 cm^{-1} than other groups (Figure 3b-C); the reason is that PAA stabilizes the size of ACP by blocking the conversion of ACP, which in turn leads to intrafibrillar mineralization.

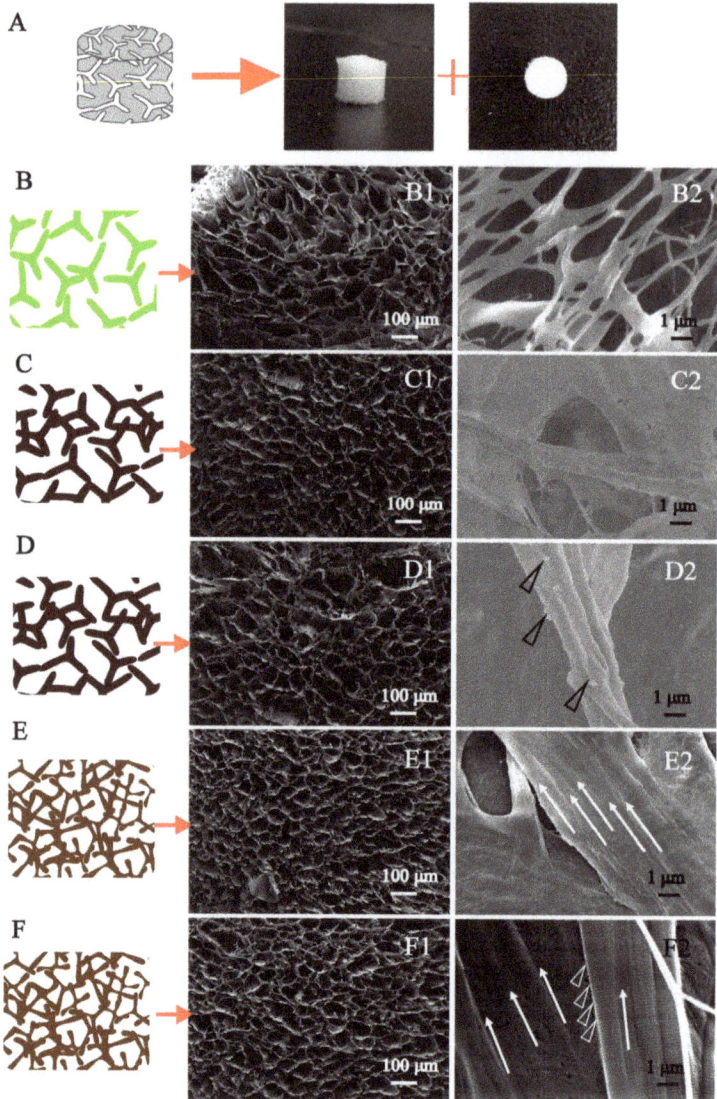

Figure 2. The SEM images of different mineralized collagens. (**A**): The three-dimensional structure of the collagen scaffolds; (**B**): the NM collagen; (**C**): the EM collagen; (**D**): the PAA-induced normal IM collagen; (**E**): the periodic FSS-induced oriented IM collagen; and (**F**): the periodic FSS and TPP-co-induced oriented HIM collagen. In the collagen scaffolds, the pore density has increased after mineralization, especially under the action of periodic FSS. The high magnification images of these mineralized collagens with different microstructures (**B1–F2**) show that the surface of the collagen is very smooth before mineralization. After mineralization, the EM collagen was rough and the IM collagen was smoother. The spherical particles were formed on the surface of the normal IM collagen induced by PAA (**D**, triangle), and collagen fibers were arranged neatly and induced by periodic FSS (**E**,**F**, arrows). The oriented HIM collagen has stripe-like structures (**F**, triangles).

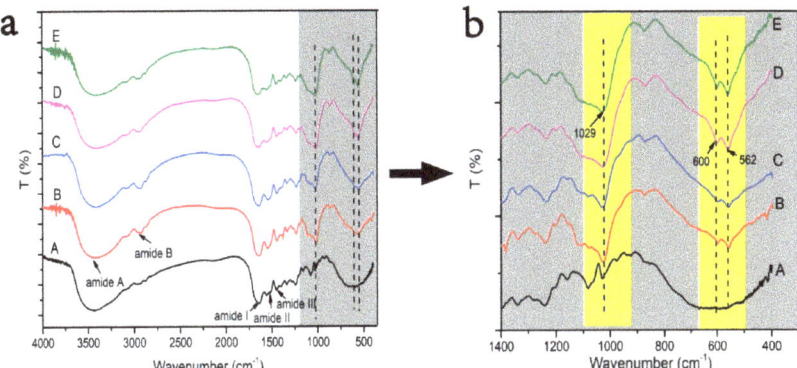

Figure 3. The FTIR spectra of different mineralized collagens. In (**a,b**), A: the NM collagen; B: the EM collagen; C: the PAA-induced normal IM collagen; D: the periodic FSS-induced oriented IM collagen; and E: the periodic FSS and TPP-co-induced oriented HIM collagen. (**b**) is an enlargement of the gray box in (**a**). The characteristic peaks of HA at 562, 600, and 1029 cm^{-1} were the highest and sharpest under the condition of periodic FSS (D and E).

Furthermore, collagen, as a common biological material, has good hydrophilicity. The hydrophilic properties of the mineralized collagen are related to the network structure and the available hydrophilic groups of collagen sponges. We can analyze its hydrophilic properties after mineralization, through the dynamic contact angle. According to the changes of contact angles with increasing time, we drew dynamic contact angle curves of the mineralized collagen sponges (Figure 4). From the curves, we observed that the hydrophilic properties of the collagen sponges were good whether mineralized or not (Figure 4). The best contact angle was only maintained for 4 s; after mineralization, the contact angle can reach 10 s (Figure 4). Non-mineralized collagen has the best hydrophilic properties, and the EM collagen has the worst hydrophilic properties, indicating that this property mainly correlated to the structure of the collagen. HA or ACP particles attached on the surface of collagen can cover the hydrophilic groups, which is in accordance with the SEM analysis. Meanwhile, this study showed that the hydrophilic properties after mineralization with the action of periodic FSS were slightly less numerous than in the other groups, and the contact angles was 80°, which could be due to the arrangement of the collagen. It is worth noting that after 11 s these are all superhydrophilic, and for the proposed application, the differences in the first 11s regarding the contact angle values need to be further tested. Consistent with our TEM and SEM analysis, extrafibrillar and intrafibrillar mineralization had different levels of influence on the microstructure of the collagen sponges, which means that the contact angle values are in line with the TEM and SEM data.

Collagenases can specifically hydrolyze collagen by breaking the helical structure of the collagen and the links between the molecular chains under physiological pH and temperature conditions [38]. The enzymatic time of the collagen is reduced after mineralization. EM collagen has the lowest enzymatic time (Figure 5). The reason is that on the one hand, there are many hydrogen bonds between the helical structure of the collagen, and the hydrogen bonds can maintain the structural stability of the collagen. During the process of the HA crystal growth and deposition, the hydrogen bonding between the collagen is destroyed, resulting in the decrease of the structural stability of the collagen, which can take less time to be dissolved by collagenase. On the other hand, the larger size of the ACP, the easier it is to deposit on the surface of the collagen fibers, when mineralized collagen with the same mass, the proportion, and level of contact with the surface of collagen in the EM collagen is smaller, resulting in the enzymatic time become shorter. In addition, the PAA and the FSS all can reduce the size of the ACP. Because there is little difference between the PAA-induced normal IM collagen and the periodic FSS-induced oriented IM collagen and HIM collagen, we inferred that the conversion of the ACP has an influence on the collagen structure and enzymatic stability.

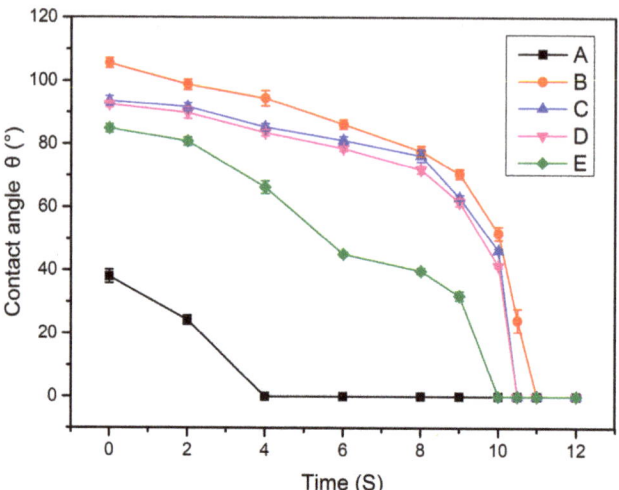

Figure 4. The dynamic contact angles of the different mineralized collagens. A: the NM collagen; B: the EM collagen; C: the PAA-induced normal IM collagen; D: the periodic FSS-induced oriented IM collagen; and E: the periodic FSS and TPP-co-induced oriented HIM collagen. The change of contact angles with the time was quick from 40° to 0°. The EM collagen has poorer hydrophilicity than IM collagen. The periodic FSS and TPP-co-induced oriented HIM collagen have the best hydrophilicity after mineralization, and their contact angles are 80°.

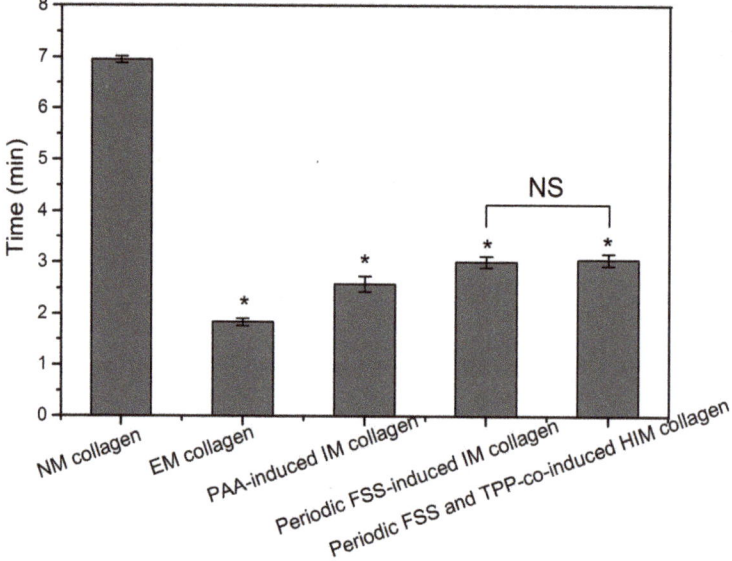

Figure 5. The enzymatic time of the different mineralized collagens. The NM collagen has the longest enzymatic time, and the EM collagen has the shortest enzymatic time. The enzymatic time was longer than other mineralization conditions under the action of periodic FSS. * $p < 0.05$ vs. non-mineralized collagen group, NS: no significant difference.

Collagen, as the organic matter in the mineralized collagen, can be gradually degraded with the increase of temperature, and the weight is obviously lost. HA, as an inorganic matter in the mineralized collagen, does not degrade with the increase of temperature, and there is no significant weight loss. Therefore, the ratio of inorganic to organic in the mineralized collagen can be determined by analyzing their weight losses. With the increase of temperature, two weight loss steps can be distinguished in the TG curves of the collagen. As the temperature increases, the binding water is lost first in the range of 30–150 °C (Figure 6 green box 1). In the range of 250–600 °C, collagen chains were broken and the structure were degraded (Figure 6 yellow box 2). The degradation of the chains means the destruction of the secondary conformation structure. Most of weight losses occur in the degradation process. Due to the fact HA can only lose the weight of the binding water, the weight loss is very slight, and the weight losses of the mineralized collagen, regardless of the mineralized microstructure, is lower than that of the non-mineralized collagen (Figure 6). Compared with the different mineralized collagens, we found that the EM collagen has the least weight losses. The oriented IM and HIM collagens have greater weight loss (Figure 6D,E). However, we found that the second stage of the weightlessness of the EM collagen appears significantly earlier, indicating that the extrafibrillar mineralization has destroyed the collagen secondary conformation structure (Figure 6B, arrow in yellow box 2). The difference in weight losses between the different mineralized collagens corresponded well with the dehydration of HA. The lower the weight losses of the mineralized collagen, the higher the content of HA in the collagen fibers. The results are consistent with the enzymatic time of the different mineralized collagens. The reason for this is that when the ACP particle sizes become smaller, their specific surface area increases, and the specific surface area of the collagen fibers is limited, resulting in the reduction of the deposition of the ACPs. Furthermore, although the perturbation of the periodic FSS can promote the formation of intrafibrillar mineralization, there is a certain obstacle to the ACP entering into the collagen fibers.

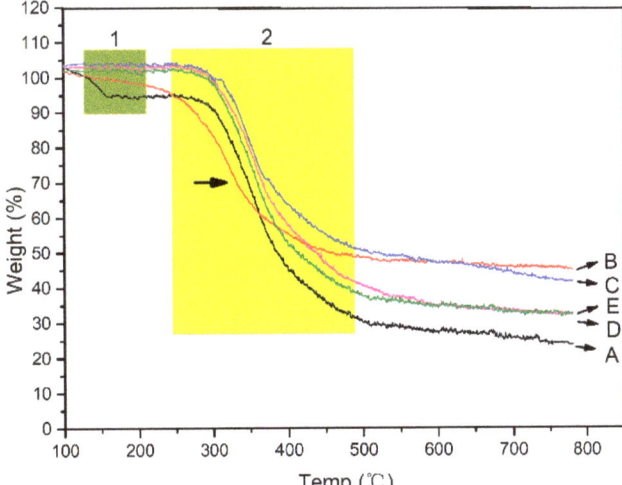

Figure 6. The TG curves of the different mineralized collagens. The curves from A to E were the NM collagen, the EM collagen, the PAA-induced normal IM collagen, the periodic FSS-induced oriented IM collagen, and the periodic FSS and TPP-co-induced oriented HIM collagen. The NM collagen has the largest weight losses, and the EM collagen has the smallest weight losses. The EM collagen first begins the second stage of the weight loss.

4. Conclusions

Since the mineralized collagen is the basic bone matrix of the hierarchically organized structures of the natural bone, there are many different mineralized microstructures of the mineralized collagen in the bone. Therefore, this study prepared the mineralized collagen with different microstructures induced under different conditions, such as the PAA and the periodic FSS. Then, we systematically analyzed their physical and chemical properties. Multiple sources of evidence were provided to prove that the addition of the PAA and the periodic FSS to the collagen mineralization system can induce intrafibrillar mineralization. The periodic FSS has improved the hydrophilicity, the enzymatic stability, and the crystal conversion of the mineralized collagen. Good hydrophilicity, enzyme stability, and crystal conversion rate are the basic conditions for the cell growth and biocompatibility of biomaterials. These experiments have laid the foundation for subsequent animal experiments and cell experiment evaluations of these mineralized collagens with different microstructures as bone substitute materials. The periodic FSS and the TPP co-induced oriented HIM collagen can be better matrix materials for the design of bone substitute materials with better bone repair abilities and may pave the way for tissue engineering applications.

Author Contributions: T.D. performed the mineralization experiments, analyzed the data, and wrote the manuscript. Y.N. performed TEM examination. Y.L. and H.Y. provided suggestions on the experimental design and manuscript writing. A.Q. and X.N. supervised the project and wrote the manuscript. All authors discussed the results and revised the manuscript. All authors have read and agreed to the published version of the manuscript.

Funding: This research was funded by [China Postdoctoral Science Foundation funded project] grant number [2021TQ0020], [the National Natural Science Foundation of China] grant numbers [11872097, 12172018 and 11832003], [the National Key R&D Program of China] grant numbers [2020YFC0122204], [the Beijing Natural Science Foundation] grant numbers [7202003], [t the Joint Program of Beijing Municipal-Beijing Natural Science Foundation] grant numbers [KZ202110005004]. And the APC was funded by [KZ202110005004].

Acknowledgments: Thanks are extended to the Large-Scale Instruments and Equipments Sharing Platform of the Beijing University of Technology for the TEM and TG measurements.

Conflicts of Interest: There is no conflict to declare.

References

1. Fratzl, P.; Gupta, H.; Paschalis, E.; Roschger, P. Structure and mechanical quality of the collagen–mineral nano-composite in bone. *J. Mater. Chem.* **2004**, *14*, 2115–2123. [CrossRef]
2. Rollo, J.M.D.A.; Boffa, R.S.; Cesar, R.; Schwab, D.C.; Leivas, T.P. Assessment of Trabecular Bones Microarchitectures and Crystal Structure of Hydroxyapatite in Bone Osteoporosis with Application of the Rietveld Method. *Procedia Eng.* **2015**, *110*, 8–14. [CrossRef]
3. He, C.; Xiao, G.; Jin, X.; Sun, C.; Ma, P.X. Electrodeposition on Nanofibrous Polymer Scaffolds: Rapid Mineralization, Tunable Calcium Phosphate Composition and Topography. *Adv. Funct. Mater.* **2010**, *20*, 3568–3576. [CrossRef] [PubMed]
4. Rho, J.; Kuhn-Spearing, L.; Zioupos, P. Mechanical properties and the hierarchical structure of bone. *Med. Eng. Phys.* **1998**, *20*, 92–102. [CrossRef]
5. Li, Z.; Du, T.; Ruan, C.; Niu, X. Bioinspired mineralized collagen scaffolds for bone tissue engineering. *Bioact. Mater.* **2021**, *6*, 1491–1511. [CrossRef]
6. Jie, Z.; Yu, L.; Sun, W.; Yang, X. First detection, characterization, and application of amorphous calcium phosphate in dentistry. *J. Dent. Sci.* **2012**, *7*, 316–323.
7. Ayatollahi, M.R.; Yahya, M.Y.; Shirazi, H.A.; Hassan, S.A. Mechanical and tribological properties of hydroxyapatite nanoparticles extracted from natural bovine bone and the bone cement developed by nano-sized bovine hydroxyapatite filler. *Ceram. Int.* **2015**, *41*, 10818–10827. [CrossRef]
8. Du, T.; Niu, X.; Hou, S.; Xu, M.; Li, Z.; Li, P.; Fan, Y. Highly aligned hierarchical intrafibrillar mineralization of collagen induced by periodic fluid shear stress. *J. Mater. Chem. B* **2020**, *8*, 2562–2572. [CrossRef]
9. Du, T.; Niu, Y.; Jia, Z.; Liu, Y.; Qiao, A.; Yang, H.; Niu, X. Orthophosphate and alkaline phosphatase induced the formation of apatite with different multilayered structures and mineralization balance. *Nanoscale* **2022**, *14*, 1814–1825. [CrossRef]
10. Hassan, M.; Sulaiman, M.; Yuvaraju, P.D.; Galiwango, E.; Rehman, I.u.; Al-Marzouqi, A.H.; Khaleel, A.; Mohsin, S. Biomimetic PLGA/Strontium-Zinc Nano Hydroxyapatite Composite Scaffolds for Bone Regeneration. *J. Func. Biomater.* **2022**, *13*, 13. [CrossRef]

1. Nudelman, F.; Pieterse, K.; George, A.; Bomans, P.H.; Friedrich, H.; Brylka, L.J.; Hilbers, P.A.; de With, G.; Sommerdijk, N.A. The role of collagen in bone apatite formation in the presence of hydroxyapatite nucleation inhibitors. *Nat. Mater.* **2010**, *9*, 1004–1009. [CrossRef] [PubMed]
2. Meng, Q.; An, S.; Damion, R.A.; Jin, Z.; Wilcox, R.; Fisher, J.; Jones, A. The effect of collagen fibril orientation on the biphasic mechanics of articular cartilage. *J. Mech. Behav. Biomed. Mater.* **2017**, *65*, 439–453. [CrossRef] [PubMed]
3. Hu, C.; Zhang, L.; Wei, M. Development of Biomimetic Scaffolds with Both Intrafibrillar and Extrafibrillar Mineralization. *ACS Biomater. Sci. Eng.* **2015**, *1*, 669–676. [CrossRef] [PubMed]
4. Hartgerink, J.D.; Beniash, E.; Stupp, S.I. Self-Assembly and Mineralization of Peptide-Amphiphile Nanofibers. *Science* **2010**, *294*, 1684–1688. [CrossRef] [PubMed]
5. Liu, Y.; Li, N.; Qi, Y.P.; Dai, L.; Bryan, T.E.; Mao, J.; Pashley, D.H.; Tay, F.R. Intrafibrillar collagen mineralization produced by biomimetic hierarchical nanoapatite assembly. *Adv. Mater.* **2011**, *23*, 975–980. [CrossRef]
6. Wegst, U.G.; Bai, H.; Saiz, E.; Tomsia, A.P.; Ritchie, R.O. Bioinspired structural materials. *Nat. Mater.* **2015**, *14*, 23. [CrossRef]
7. Niu, L.; Jee, S.E.; Jiao, K.; Tonggu, L.; Li, M.; Wang, L.; Yang, Y.; Bian, J.; Breschi, L.; Jang, S.S. Collagen intrafibrillar mineralisation as a result of the balance between osmotic equilibrium and electroneutrality. *Nat. Mater.* **2017**, *16*, 370. [CrossRef]
8. Du, T.; Niu, X.; Hou, S.; Li, Z.; Li, P.; Fan, Y. Apatite minerals derived from collagen phosphorylation modification induce the hierarchical intrafibrillar mineralization of collagen fibers. *J. Biomed. Mater. Res. A* **2019**, *107*, 2403–2413. [CrossRef]
9. Niu, X.; Feng, Q.; Wang, M.; Guo, X.; Zheng, Q. Porous nano-HA/collagen/PLLA scaffold containing chitosan microspheres for controlled delivery of synthetic peptide derived from BMP-2. *J. Control. Release* **2009**, *134*, 111–117. [CrossRef]
10. Nudelman, F.; Lausch, A.J.; Sommerdijk, N.A.J.M.; Sone, E.D. In Vitro models of collagen biomineralization. *J. Struct. Biol.* **2013**, *183*, 258–269. [CrossRef]
11. Matlahov, I.; Ilinevul, T.; Abayev, M.; Lee, E.M.Y.; Nadavtsubery, M.; Keinanadamsky, K.; Gray, J.J.; Goobes, G. Interfacial Mineral-Peptide Properties of A Mineral Binding Peptide from Osteonectin and Bone-like Apatite. *Chem. Mater.* **2015**, *27*, 5562–5569. [CrossRef]
12. Du, T.; Yang, H.; Niu, X. Phosphorus-containing compounds regulate mineralization. *Mater. Today Chem.* **2021**, *22*, 100579. [CrossRef]
13. Wang, Y.; Azaïs, T.; Robin, M.; Vallée, A.; Catania, C.; Legriel, P.; Pehau-Arnaudet, G.; Babonneau, F.; Giraud-Guille, M.-M.; Nassif, N. The predominant role of collagen in the nucleation, growth, structure and orientation of bone apatite. *Nat. Mater.* **2012**, *11*, 724–733. [CrossRef] [PubMed]
14. Jiao, K.; Niu, L.N.; Ma, C.F.; Huang, X.Q.; Pei, D.D.; Luo, T.; Huang, Q.; Chen, J.H.; Tay, F.R. Complementarity and uncertainty in intrafibrillar mineralization of collagen. *Adv. Funct. Mater.* **2016**, *26*, 6858–6875. [CrossRef]
15. Weiner, S.; Traub, W.; Wagner, H.D. Lamellar bone: Structure-function relations. *J. Struct. Biol.* **1999**, *126*, 241–255. [CrossRef]
16. Burger, E.H.; Klein-Nulend, J. Mechanotransduction in bone—Role of the lacuno-canalicular network. *FASEB J.* **1999**, *13*, S101–S112. [CrossRef]
17. Ponik, S.M.; Triplett, J.W.; Pavalko, F.M. Osteoblasts and osteocytes respond differently to oscillatory and unidirectional fluid flow profiles. *J. Cell. Biochem.* **2007**, *100*, 794–807. [CrossRef]
18. Yeatts, A.B.; Fisher, J.P. Bone tissue engineering bioreactors: Dynamic culture and the influence of shear stress. *Bone* **2011**, *48*, 171–181. [CrossRef]
19. Huo, B.; Lu, X.L.; Hung, C.T.; Costa, K.D.; Xu, Q.; Whitesides, G.M.; Guo, X.E. Fluid flow induced calcium response in bone cell network. *Cell. Mol. Bioeng.* **2008**, *1*, 58–66. [CrossRef]
20. Niu, X.; Wang, L.; Tian, F.; Wang, L.; Li, P.; Feng, Q.; Fan, Y. Shear-mediated crystallization from amorphous calcium phosphate to bone apatite. *J. Mech. Behav. Biomed. Mater.* **2016**, *54*, 131–140. [CrossRef]
21. Niu, X.; Fan, R.; Guo, X.; Du, T.; Yang, Z.; Feng, Q.; Fan, Y. Shear-mediated orientational mineralization of bone apatite on collagen fibrils. *J. Mater. Chem. B* **2017**, *5*, 9141–9147. [CrossRef] [PubMed]
22. Liu, Y.; Liu, S.; Luo, D.; Xue, Z.; Yang, X.; Gu, L.; Zhou, Y.; Wang, T. Hierarchically Staggered Nanostructure of Mineralized Collagen as a Bone-Grafting Scaffold. *Adv. Mater.* **2016**, *28*, 8740–8748. [CrossRef] [PubMed]
23. Allo, B.; Costa, D.; Dixon, S.; Mequanint, K.; Rizkalla, A. Bioactive and biodegradable nanocomposites and hybrid biomaterials for bone regeneration. *J. Func. Biomater.* **2012**, *3*, 432–463. [CrossRef] [PubMed]
24. Riedel, S.; Ward, D.; Kudláčková, R.; Mazur, K.; Bačáková, L.; Kerns, J.G.; Allinson, S.L.; Ashton, L.; Koniezcny, R.; Mayr, S.G. Electron Beam-Treated Enzymatically Mineralized Gelatin Hydrogels for Bone Tissue Engineering. *J. Func. Biomater.* **2021**, *12*, 57. [CrossRef]
25. Wang, H.; Du, T.; Li, R.; Main, R.P.; Yang, H. Interactive effects of various loading parameters on the fluid dynamics within the lacunar-canalicular system for a single osteocyte. *Bone* **2022**, *158*, 116367. [CrossRef]
26. Grandfield, K.; Herber, R.; Chen, L.; Djomehri, S.; Tam, C.; Lee, J.-H.; Brown, E.; Woolwine, W.R.; Curtis, D.; Ryder, M. Strain-guided mineralization in the bone-PDL-cementum complex of a rat periodontium. *Bone Rep.* **2015**, *3*, 20–31. [CrossRef]
27. Elliott, J.; Woodward, J.; Umarji, A.; Mei, Y.; Tona, A. The effect of surface chemistry on the formation of thin films of native fibrillar collagen. *Biomaterials* **2007**, *28*, 576–585. [CrossRef]
28. Prakasam, M.; Locs, J.; Salma-Ancane, K.; Loca, D.; Largeteau, A.; Berzina-Cimdina, L. Biodegradable materials and metallic implants—A review. *J. Func. Biomater.* **2017**, *8*, 44. [CrossRef]

Article

Anti-Inflammatory and Mineralization Effects of an ASP/PLGA-ASP/ACP/PLLA-PLGA Composite Membrane as a Dental Pulp Capping Agent

Wenjuan Yan [1,2,†], Fenghe Yang [3,†], Zhongning Liu [4], Quan Wen [2], Yike Gao [1], Xufeng Niu [3,*] and Yuming Zhao [1,*]

1. Department of Pediatric Dentistry, Peking University School and Hospital of Stomatology & National Center of Stomatology & National Clinical Research Center for Oral Diseases & National Engineering Research Center of Oral Biomaterials and Digital Medical Devices & Beijing Key Laboratory of Digital Stomatology & Research Center of Engineering and Technology for Computerized Dentistry Ministry of Health & NMPK Key Laboratory for Dental Materials, Beijing 100081, China; yanwenjuan85@163.com (W.Y.); gaoyike@pku.edu.cn (Y.G.)
2. Department of First Clinical Division, Peking University School and Hospital of Stomatology & National Center of Stomatology & National Clinical Research Center for Oral Diseases & National Engineering Research Center of Oral Biomaterials and Digital Medical Devices & Beijing Key Laboratory of Digital Stomatology & Research Center of Engineering and Technology for Computerized Dentistry Ministry of Health & NMPK Key Laboratory for Dental Materials, Beijing 100081, China; wincky.ls@163.com
3. Key Laboratory of Biomechanics and Mechanobiology, Ministry of Education, Beijing Advanced Innovation Center for Biomedical Engineering, School of Biological Science and Medical Engineering, Beihang University, Beijing 100083, China; yfh1993@buaa.edu.cn
4. Department of Prosthodontics, Peking University School and Hospital of Stomatology & National Center of Stomatology & National Clinical Research Center for Oral Diseases & National Engineering Research Center of Oral Biomaterials and Digital Medical Devices & Beijing Key Laboratory of Digital Stomatology & Research Center of Engineering and Technology for Computerized Dentistry Ministry of Health& NMPK Key Laboratory for Dental Materials, Beijing 100081, China; lzn45@163.com

* Correspondence: nxf@buaa.edu.cn (X.N.); yuming_zhao@hotmail.com (Y.Z.)
† These authors contributed equally to this work.

Abstract: Dental pulp is essential for the development and long-term preservation of teeth. Dental trauma and caries often lead to pulp inflammation. Vital pulp therapy using dental pulp-capping materials is an approach to preserving the vitality of injured dental pulp. Most pulp-capping materials used in clinics have good biocompatibility to promote mineralization, but their anti-inflammatory effect is weak. Therefore, the failure rate will increase when dental pulp inflammation is severe. The present study developed an amorphous calcium phosphate/poly (L-lactic acid)-poly (lactic-co-glycolic acid) membrane compounded with aspirin (hereafter known as ASP/PLGA-ASP/ACP/PLLA-PLGA). The composite membrane, used as a pulp-capping material, effectively achieved the rapid release of high concentrations of the anti-inflammatory drug aspirin during the early stages as well as the long-term release of low concentrations of aspirin and calcium/phosphorus ions during the later stages, which could repair inflamed dental pulp and promote mineralization. Meanwhile, the composite membrane promoted the proliferation of inflamed dental pulp stem cells, downregulated the expression of inflammatory markers, upregulated the expression of mineralization-related markers, and induced the formation of stronger reparative dentin in the rat pulpitis model. These findings indicate that this material may be suitable for use as a pulp-capping material in clinical applications.

Keywords: aspirin; anti-inflammation; amorphous calcium phosphate composite scaffold; dental pulp-capping material; dental pulp stem cell

Citation: Yan, W.; Yang, F.; Liu, Z.; Wen, Q.; Gao, Y.; Niu, X.; Zhao, Y. Anti-Inflammatory and Mineralization Effects of an ASP/PLGA-ASP/ACP/PLLA-PLGA Composite Membrane as a Dental Pulp Capping Agent. *J. Funct. Biomater.* **2022**, *13*, 106. https://doi.org/10.3390/jfb13030106

Academic Editor: James Kit-hon Tsoi

Received: 7 July 2022
Accepted: 26 July 2022
Published: 29 July 2022

Publisher's Note: MDPI stays neutral with regard to jurisdictional claims in published maps and institutional affiliations.

Copyright: © 2022 by the authors. Licensee MDPI, Basel, Switzerland. This article is an open access article distributed under the terms and conditions of the Creative Commons Attribution (CC BY) license (https://creativecommons.org/licenses/by/4.0/).

1. Introduction

The preservation of vital dental pulp is essential for maintaining tooth sensation, nutrition, immunity, and dentin formation; these aspects contribute to long-term prognoses [1,2].

Therefore, vital dental pulp preservation is an important goal of modern minimally invasive dentistry [3]. Currently, the most common clinical pulp-capping agent material is mineral trioxide aggregate (MTA), regarded as the clinical standard for pulp capping [4]. MTA has an excellent effect on the promotion of the formation of reparative dentin; however, its anti-inflammation effect was relatively weak [5]. Therefore, the success rate of vital pulp preservation is relatively low in the presence of inflammation [6,7]. Investigations of novel pulp-capping materials with anti-inflammatory effects could improve the success rate of vital pulp preservation and expand its indications, which have important clinical significance and practical value.

Ideal pulp-capping material for early-stage pulpitis should rapidly inhibit inflammation to create a beneficial microenvironment for dental pulp repair, after which it should promote long-term mineralization to form a hard tissue barrier [8–10]. In order to fulfill these requirements, there is a need for multilayer biomaterials with sequential release properties. Aspirin (ASP) is a nonsteroidal anti-inflammatory drug widely used in clinics. Its main functions include analgesia, antipyretic and anti-inflammatory; in addition, it has properties that affect the immune response and exert anti-infection and anti-biofilm activity [11]. Aspirin can reduce the expression levels of inflammatory factors in dental pulp stem cells [12], which indicates that aspirin could play an effective role in controlling pulp inflammation. Amorphous calcium phosphate (ACP) has been demonstrated to have good biodegradability and bioactivity and no cytotoxicity; ACP scaffolds have been shown to continuously release calcium and phosphorus ions. It has been widely applied in the biomedical field due to its excellent bioactivity, high cell adhesion, and adjustable biodegradation rate [13–16]. The biodegradable materials poly (L-lactic acid) (PLLA) and poly (lactic-co-glycolic acid) (PLGA) can achieve the controlled release or sustained release of drugs by adjusting the degradation rate [17,18].

The dental pulp-capping material was designed as a three-layer composite membrane. The lower layer was an ASP/PLGA membrane that contained high concentrations of ASP; the middle layer was an ASP/ACP/PLLA membrane that contained low concentrations of ASP and ACP, and the outer layer was a PLGA membrane. The lower ASP/PLGA membrane can rapidly release aspirin to control pulp inflammation; the middle ASP/ACP/PLLA membrane can perform the long-term release of calcium/phosphorus ions and aspirin to promote the formation of reparative dentin; and the outer PLGA membrane layer improves the superficial hydrophilicity of the composite membrane, which is beneficial to improving the sealing property of the pulp-capping material.

Pulpitis is inflammation caused by bacteria or toxins invading the dental pulp in the center of the tooth. Gram-negative bacteria are common microorganisms that cause pulp inflammation and necrosis. Lipopolysaccharide (LPS) is the main molecular component of these bacterial cell walls [19]. In vivo, it can activate monocyte macrophages, endothelial cells, and epithelial cells, through the cell signal transduction system, cause the synthesis and release of various cytokines and inflammatory mediators, and then initiate a series of reactions in the body. It is often used to induce inflammatory models [20]. This study aimed to evaluate the effect of the composite membrane on the proliferation, migration, and differentiation of dental pulp cells under an LPS-induced inflammation condition in vitro, as well as its potential role in the formation of reparative dentin in a rat pulpitis model in vivo, so as to provide support for its future clinical application.

2. Materials and Methods

2.1. Preparation of the ASP/ACP/PLLA Electrospun Membrane

The membrane was fabricated by using a multilayering electrospinning method. Briefly, a PLLA polymer solution (50% w/v) with aspirin at 0.12 wt% PLLA and ACP powders at 10 wt% PLLA were prepared in a 3:1 DCM: DMF solvent mixture. The solution was stirred for 1 h to obtain a uniform suspension, after which this suspension was loaded into a 20 mL syringe and electrospun at 20 kV under a steady flow rate of 1.2 mL/h. The distance between the collection foil and the syringe needle was set as 10 cm. After electrospinning for 2.5 h, the resulting ASP/ACP/PLLA membrane was separated from the collection foil and dried in a desiccator overnight. The dried ASP/ACP/PLLA membrane was cut into small pieces of equal weight and put on the collection foil for further electrospinning.

2.2. Preparation of the ASP/PLGA-ASP/ACP/PLLA-PLGA Electrospun Composite Membrane

A polymer solution containing 30% w/v PLGA with aspirin at 1.2 wt% PLGA was prepared in a 3:1 DCM: DMF solvent mixture. The solution was electrospun on one side of the small pieces of the ASP/ACP/PLLA membrane we prepared before at 20 kV under a steady flow rate of 1.6 mL/h with a distance of 10 cm for 0.5 h to obtain the ASP/PLGA-ASP/ACP/PLLA electrospun membrane. The other side of the ASP/ACP/PLLA membrane was electrospun for 0.5 h using a pure PLGA solution under the same conditions to obtain the final ASP/PLGA-ASP/ACP/PLLA-PLGA membrane. The resulting multilayer membranes were collected and dried in a desiccator overnight.

2.3. Characterization of the Composite Membrane

The membranes were mounted on metal stubs and coated with platinum for microscopical observation. Both the surface and cross-section morphologies were observed using a scanning electron microscope (SEM, FEI Quanta 250 FEG, Hillsboro, OR, USA). The diameter distribution of the nanofibers was analyzed using the image analysis program ImageJ. An energy-dispersive spectrometer (EDS) was used to determine the chemical composition of the electrospun membrane. The contact angle of the ASP/ACP/PLLA and ASP/PLGA-ASP/ACP/PLLA-PLGA membranes was measured by a contact angle measurement system (JC2000FM, POWER EACH, Shanghai, China) with a water droplet size of 0.5 µL. Moreover, transmission electron microscope (TEM) and X-ray diffraction (XRD) were employed here to identify that the ACP we loaded is in amorphous phases. For the TEM (JEOLJEM-2100, JEOL Ltd., Akishima-shi, Japan) observation, the samples were ultrasonicated in alcohol for 30 min and picked up with copper net films to observe their surface morphologies. Meanwhile, the XRD patterns were obtained from a Rigaku D/Max X-ray diffractometer (Rigaku Corporation, Tokyo, Japan) with Cu Kα radiation at 40 kV, 120 mA, and a scanning rate of 10°/min from 3° to 70°.

2.4. In Vitro Degradation Experiments

In vitro degradation experiments were conducted by monitoring the media pH and mass loss of the electrospun membrane during the course of 21 days. Briefly, 20 mg of the ASP/PLGA-ASP/ACP/PLLA-PLGA membrane were immersed in 3 mL of phosphate buffered solution (PBS, 0.1 M, pH 7.4) and kept at 37 °C incubator shaker. The PBS media were changed every three days with fresh PBS. The replaced PBS was collected, and the pH value of this degradation media was measured by using a pH meter (Leici PHS-3C, Shangai INESA Scientific Instrument Co., Ltd., Shanghai, China). The mass loss of the composite membrane was examined on 3, 7, 14 and 21 days using an electronic balance, and the membrane was lyophilized for 24 h before measuring.

2.5. In Vitro Release Experiments

In vitro release experiments were carried out to analyze the ion release (Ca^{2+}, PO_4^{3-}) and aspirin release from the membrane over a period of 21 days. Before the release experiments, an acceleration study was firstly carried out according to our previous work to calculate the aspirin loading efficiency in the membrane. In order to study the release in vitro, 20 mg of the ASP/PLGA-ASP/ACP/PLLA-PLGA membrane were placed in test tubes with 3 mL of physiological saline (for Ca^{2+}, PO_4^{3-} measurement) or PBS (for aspirin measurement) and kept in an incubator shaker that is maintained at 37 °C. At predetermined time intervals, the release media in test tubes were collected and replaced with an equal amount of fresh media. The Ca^{2+}, PO_4^{3-}, and aspirin concentrations were tested by a calcium colorimetric assay kit (Sigma-Aldrich, Saint Louis, MO, USA) and a phosphate ion colorimetric assay kit (Sigma-Aldrich, Saint Louis, MO, USA), as well as a multimode microplate reader (Varioskan Flash, Thermo Fisher Scientific, Waltham, MA, USA) at 278 nm/298 nm, respectively.

2.6. Cell Culture

Dental pulp stem cells (DPCs) were provided by the Oral Stem Cell Bank (Beijing, China). The DPCs were cultured in an α-modified minimum essential medium (α-MEM; Gibco, St. Louis, MO, USA) with 10% fetal bovine serum (Gibco, St. Louis, MO, USA) and 100 U/mL penicillin-streptomycin (Sigma-Aldrich, St. Louis, MO, USA) at 37 °C in a 5% CO_2 incubator. The medium was changed every 3 days. Cells were used in passages 3–5 for all of the experiments.

2.7. Cell Proliferation Assay

The effect of the composite membrane on DPCs proliferation was assessed by using a Cell Counting Kit-8 (CCK-8) assay (Dojindo, Kumamoto, Japan). DPCs were seeded at a density of 3×10^3 cells/well in 96-well plates (n = 3 wells per group), cultured, and treated with the culture medium containing ASP/PLGA-ASP/ACP/PLLA-PLGA composite membrane extract, while the α-MEM was used for the control group. At 1, 3, 5, and 7 days, 10 µL of the CCK-8 solution was added to each well. The absorbance was measured at 450 nm by using an Absorbance Microplate Reader (BioTek Instruments, Winooski, VT, USA). The DPCs were treated with 1 µg/mL of lipopolysaccharide (LPS) to establish a model of inflamed dental pulp stem cells (iDPCs). The cells were treated separately with ASP, ACP/PLLA-PLGA scaffold extract, PLGA-PLLA-PLGA scaffold extract, or ASP/PLGA-ASP/ACP/PLLA-PLGA composite membrane extract; the α-MEM was used for the control group. At 1, 3, 5, and 7 days, CCK-8 assays were conducted as described previously.

2.8. Cell Adhesion and Migration

Cell adhesion was observed by using phalloidin/4,6-diamino-2-phenylindole (DAPI; Invitrogen, Carlsbad, CA, USA) staining. DPCs were seeded at a density of 1×10^4 cells/well in a 24-well plate with a composite membrane or a glass slide at the bottom and cultured in the α-MEM. After 24 h, the cells were washed and fixed with paraformaldehyde. Phalloidin staining was performed for 30 min, and the nuclei were stained with DAPI for 10 min. After the cells had been washed with phosphate-buffered saline, cell skeletons were observed by using a confocal microscope (Zeiss LSM 7 Duo, Jena, Germany). The cell scratch test was used to examine the impact of each membrane treatment on DPCs migration. DPCs were seeded at 1×10^5 cells/well in a 6-well plate. When the cells reached 90% confluence, a clear and straight scratch was drawn by using a 1 mL pipette tip on the bottom of the plate; the scratch was measured by using a ruler. Exfoliated cells were washed away using phosphate-buffered saline, and the remaining DPCs were treated with the membrane extract liquid and α-MEM. The initial position and width of the scratch were recorded by photography using an inverted microscope. After culturing at 37 °C with 5% CO_2 in a humidified incubator for 12 h, photographs were collected by using the inverted microscope at the same observation point to record the distance of cell migration; the distance was measured and analyzed by using ImageJ software.

2.9. Alizarin Red Staining (ARS)

ARS was performed to evaluate intracellular mineral deposition. DPCs were seeded at 6×10^4 cells/well in a 12-well plate and were then treated with LPS. Subsequently, they were cultured with ASP, ACP/PLLA-PLGA membrane extract, or ASP/PLGA-ASP/ACP/PLLA-PLGA composite membrane extract, α-MEM only was used for the control group. When the cells reached approximately 80% confluence, the medium was replaced with osteogenic differentiation media that contained 100 nM dexamethasone, 10 mM sodium β-glycerophosphate, and 10 nM L-ascorbic acid (Sigma-Aldrich, Saint Louis, MO, USA). After being cultured for 14 days, the cells were then fixed in 4% paraformaldehyde, washed with phosphate-buffered saline to remove fixative residues, and stained using a 2% ARS solution (Sigma-Aldrich, Saint Louis, MO, USA). The samples were subjected to a microscopy analysis for the qualitative evaluation of the differentiation. For the quantitative evaluation, ARS was dissolved in 5% perchloric acid; cellular absorbance was then measured at 490 nm using a Varioskan Flash (Thermo Fisher Scientific, Waltham, MA, USA) microplate spectrophotometer.

2.10. Quantitative Polymerase Chain Reaction (qPCR)

DPCs were seeded at 1×10^5 cells/well in a 6-well plate and were then treated with LPS Cells were cultured for 7 days in osteogenic media with the membrane extract. The relative expression levels of osteogenic/odontogenic markers alkaline phosphatase (ALP), dentin sialophosphoprotein (DSPP), dentin matrix protein1 (DMP1), Bone Sialoprotein (BSP) and inflammatory cytokines tumor necrosis factor-α (TNF-α), interleukin-1β (IL-1β), interleukin-6 (IL-6), interleukin-8 (IL-8) were evaluated by using qPCR. Ribonucleic acid (RNA) was extracted from each group of DPCs by using TRIzol (Invitrogen); the total RNA was then reverse-transcribed to cDNA by using the Moloney murine leukemia virus reverse transcriptase (Promega, Madison, WI, USA). qPCR was performed by using SYBR Green Master Mix (Roche, Indianapolis, IN, USA) with 0.5 μL cDNA and 200 nM specific primers (listed in Table 1); the housekeeping gene GAPDH served as an internal control. The reactions were performed by using an ABI PRISM 7500 Sequence Detection System (Applied Biosystems, Foster City, CA, USA). The results were analyzed by using Prism 6 software.

Table 1. Primers used for quantitative PCR.

Target Gene		Sequence
GAPDH	forward:	GAGAAGGCTGGGGCTCATTT
	reverse:	TAAGCAGTTGGTGGTGCAGG
ALP	forward:	ATCTTCCTGGGCGATGGGAT
	reverse:	CCACATATGGGAAGCGGTCC
DMP1	forward:	TTGTGAACTACGGAGGGTAGAGG
	reverse:	CTGCTCTCCAAGGGTGGTG
DSPP	forward:	CATGGGCCATTCCAGTTCCTC
	reverse:	TCATGCACCAGGACACCACT
BSP	forward:	CGATTTCCAGTTCAGGGCAGT
	reverse:	TCCATAGCCCAGTGTTGTAGC
TNF-α	forward:	CACTTTGGAGTGATCGGCCC
	reverse:	CAGCTTGAGGGTTTGCTACAAC
IL-1β	forward:	TTCGAGGCACAAGGCACAA
	reverse:	TGGCTGCTTCAGACACTTGAG
IL-6	forward:	CATCCTCGACGGCATCTCAG
	reverse:	TCACCAGGCAAGTCTCCTCA
IL-8	forward:	AGTTTTTGAAGAGGGCTGAGA
	reverse:	TGCTTGAAGTTTCACTGGCATC

2.11. Enzyme-Linked Immunosorbent Assay (ELISA)

DPCs were seeded at a density of 1×10^5 cells/well in a 6-well plate and were then treated with LPS. Cells treated with the membrane extract were cultured for 7 days. Cell supernatants were collected; the levels of inflammatory cytokines TNF-α, IL-1β, and IL-6 were measured using enzyme-linked immunosorbent assay kits (MM-0181H2, MM-0049H2, and MM-0122H2; Enzyme Immunobiology). The absorbance was measured at 450 nm by using a microplate reader. The concentrations of the cytokines in the cells were analyzed by using Prism 6.0 software.

2.12. Western Blot

DPCs were cultured for 14 days in osteogenic media with the membrane extract for Western blot assays. The cells were harvested with a protein lysis buffer (Applygen, Beijing, China). The protein concentration was determined by a BCA protein analysis kit

(Pierce, Rockford, IL, USA). Equal aliquots of 40 µg of total protein were separated by sodium dodecyl sulfate-polyacrylamide gel electrophoresis (SDS–PAGE) and transferred to polyvinylidene difluoride membranes (PVDF) (Millipore, Bedford, MA, USA). After blocking in 5% nonfat dry milk for 1 h, the proteins of interest were probed with primary antibodies overnight at 4 °C: DSPP, DMP1, and osteocalcin (OCN) (Santa Cruz Biotechnology, Santa Cruz, CA, USA). After being incubated with secondary antibodies for 1 h, the immunoblots were detected by the Western enhanced chemiluminescence blotting kit (ECL, Applygen, Beijing, China).

2.13. Pulp Capping in the Rat Model

Healthy upper first molars from Sprague Dawley rats were used to establish the experimental pulpitis model for this investigation. Rats were anesthetized by using intraperitoneal injections of 10% chloral hydrate (Hushi, Shanghai, China). Cavity preparation and pulp chamber opening were performed by using high-speed turbine tooth drill burs (BR49, ISO001008) with a terminal diameter of 0.5 mm. Coronal pulpal tissue was damaged by using sterile #40 K-files, treated for 30 min with sterile cotton balls that had been soaked in 1 µg/mL of LPS, and then rinse with saline; bleeding was controlled with a small cotton pellet with gentle pressure. The ASP/PLGA-ASP/ACP/PLLA-PLGA composite membrane was placed in the cavity over the exposed pulp. Mineral trioxide aggregate (MTA; Dentsply, York, PA, USA) was used to treat the contralateral cavity as a control. The cavity was then sealed with glass-ionomer cement (GIC; Fuji IX, Shizuoka, Japan). After 8 weeks in a specific pathogen-free laboratory, the rats were sacrificed by CO_2 inhalation. The specimens were then fixed, decalcified, and sectioned (4 µm). Hematoxylin–eosin staining was performed, and DMP1 expression was detected via immunohistochemistry.

2.14. Statistical Analysis

Statistical analyses were performed by using SPSS Statistics 23.0 software (IBM Corp., Armonk, NY, USA). The groups were compared by using a one-way analysis of variance. p-values < 0.05 were considered statistically significant.

3. Results

3.1. Characterization of Electrospinning Composite Membranes

Figure 1A shows the morphology of ASP/PLGA-ASP/ACP/PLLA-PLGA membranes under a camera. Membranes were observed with white and smooth surface features. SEM was further used to determine the morphological details of the membranes. It can be seen in Figure 1E–H that the electrospun membranes were composed of many nanofibers, the diameters of which mainly ranged from 600 nm to 1400 nm. Although the nanofibers of the ASP/PLGA-ASP/ACP/PLLA-PLGA and ASP/ACP/PLLA membranes exhibited no difference at low magnifications, there are some protuberances presenting on the PLLA membrane at higher magnifications. EDS mapping (Figure 2A) was employed here to analyze the chemical composition of these protuberances. The results revealed strong Ca and P signals at the protuberance site, indicating these protuberances were the agglomerated ACP powders that we added in. To determine that the ACP powders were in amorphous phases, TEM and XRD were carried out here. Figure 2B reveals the TEM morphology of ACP samples. The samples showed a curvilinear appearance with a typical diffraction pattern of an amorphous halo ring. Furthermore, the XRD pattern (Figure 2C) showed no discernable peaks of crystalline calcium phosphate but a characteristic hump of an amorphous phase at around 30°, which was consistent with the TEM results. Apart from the difference in the nanofibers, there were also some differences in the cross-section structure between the ASP/ACP/PLLA and ASP/PLGA-ASP/ACP/PLLA-PLGA membranes. The cross-section images of the ASP/PLGA-ASP/ACP/PLLA-PLGA membrane (Figure 1C) clearly exhibited a sandwich structure composed of three layers of membranes, while the ASP/ACP/PLLA membrane was only observed with a single layer of the membrane (Figure 1D). The contact angle of ASP/ACP/PLLA (inner layer) and PLGA (outer layer) was tested to find out the hydrophilic change in the membrane before and after PLGA coverage. Figure 1B shows that the ASP/PLGA-ASP/ACP/PLLA-PLGA membrane possessed a smaller contact angle (87.1 ± 3.4°) as compared to that of the ASP/ACP/PLLA membrane (126.8 ± 5.1°), showing the improvement effect of PLGA coating on the hydrophilicity of the ASP/ACP/PLLA membrane.

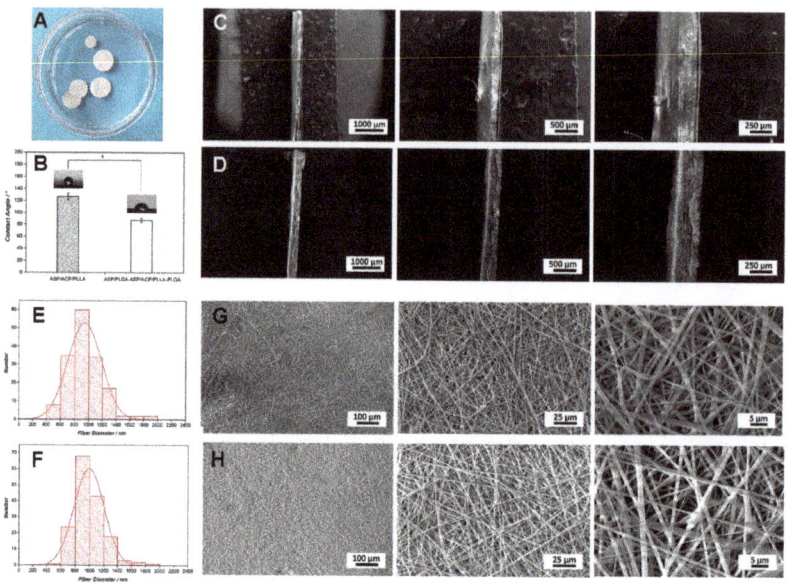

Figure 1. The surface and cross-section morphology, diameter distribution, and hydrophilicity of ASP/PLGA-ASP/ACP/PLLA-PLGA membranes. (**A**) The morphology of ASP/PLGA-ASP/ACP/PLLA-PLGA membranes under the camera. (**B**) The contact angle of the ASP/PLGA-ASP/ACP/PLLA-PLGA membrane was significantly lower than that of the ASP/ACP/PLLA membrane. (**C–H**) Diameter distribution and surface as well as cross-section microscopic images of electrospun membranes. (**C,E,G**) ASP/PLGA-ASP/ACP/PLLA-PLGA membrane; (**D,F,H**) ASP/ACP/PLLA membrane. * $p < 0.05$.

3.2. In Vitro Degradation and Release Experiments

The in vitro degradation experiment was conducted by measuring the mass loss and media pH variation over a period of 21 days. The mass loss of the electrospun membrane stayed at a slow but constant speed during the whole in vitro experiment. It can be seen from Figure 3A that the remaining mass of the membrane was 85% of the original mass at the end of the study. Figure 3B shows that the pH value of the degradation media was lower in the first nine days while being higher the rest of the time. During the whole degradation experiment, the pH value varied in a relatively small range: from 6.8 to 7.2. As for the in vitro release experiments, both the aspirin and ion release profiles were tested. Figure 2D shows that almost all of the aspirin was successfully loaded in the ASP/ACP/PLLA and ASP/PLGA-ASP/ACP/PLLA-PLGA membranes. The aspirin release profile of the ASP/PLGA-ASP/ACP/PLLA-PLGA membrane is shown in Figure 3C. It can be seen that the aspirin release rate was fast in the first week; more than half of the loading amount was released during this time. After day seven, a decreased release rate was observed, and almost 80% of the aspirin was released from the membrane at the end of the experiment. Figure 3D exhibits the release curves of Ca^{2+} and PO_4^{3-}. Compared with aspirin, the ion release rates were somewhat faster, with the observation of an obvious initial burst release. Approximately 75% of the Ca^{2+} and PO_4^{3-} were released from the membrane on the first day. After the burst release, the Ca^{2+} and PO_4^{3-} were released in a slower but constant pattern until day 21.

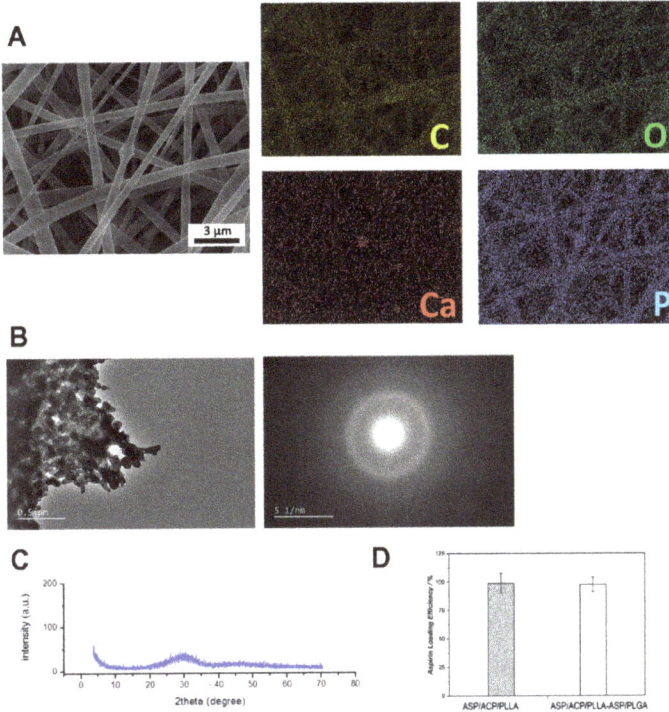

Figure 2. The element composition of ASP/PLGA-ASP/ACP/PLLA-PLGA membranes and the crystalline phases of ACP. (**A**) C, O, Ca, and P element distribution of the ASP/ACP/PLLA membrane. (**B**) TEM images and (**C**) XRD pattern of the ACP sample. (**D**) The aspirin loading efficiency of the ASP/ACP/PLLA and ASP/PLGA-ASP/ACP/PLLA-PLGA membranes.

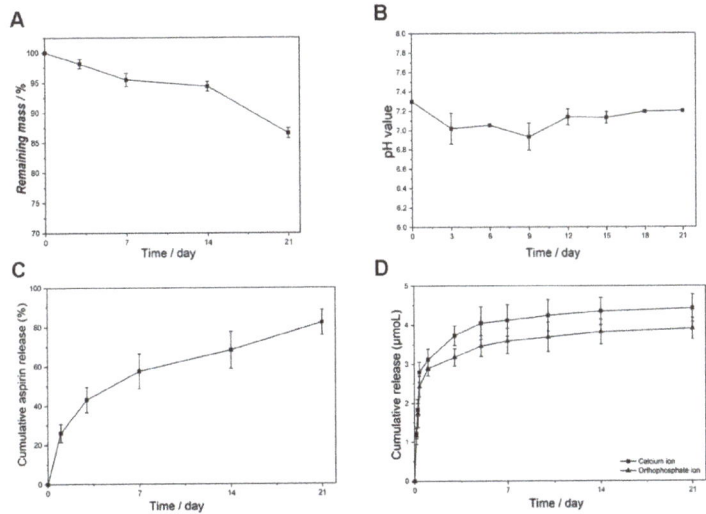

Figure 3. The degradation and release properties of ASP/PLGA-ASP/ACP/PLLA-PLGA membranes. (**A**) The remaining mass, (**B**) media pH value, (**C**) aspirin release profile, and (**D**) ion release profiles over a period of 21 days.

3.3. Effects of the Composite Membrane on DPCs Proliferation, Adhesion and Migration

The CCK-8 analysis showed that the DPCs treated with the composite membrane extract grew stably, with proliferation rates higher than the control group on day seven (Figure 4A). DPCs proliferation rate significantly decreased after LPS treatment. The proliferation rate increased when ASP was added; it further increased after ASP/PLGA-ASP/ACP/PLLA-PLGA treatment. There were significant differences among groups in terms of the numbers of viable cells on 1, 5, and 7 days. (p < 0.05) (Figure 4B). DPCs were inoculated on the composite membrane surface and stained with phalloidin/DAPI. Figure 4C indicates that DPCs could colonize and grow on the composite membrane. The cell scratch test was used to examine cell migration abilities; the DPCs migration rate was significantly higher after treatment with ASP and the composite membrane compared with other groups (Figure 4D,E).

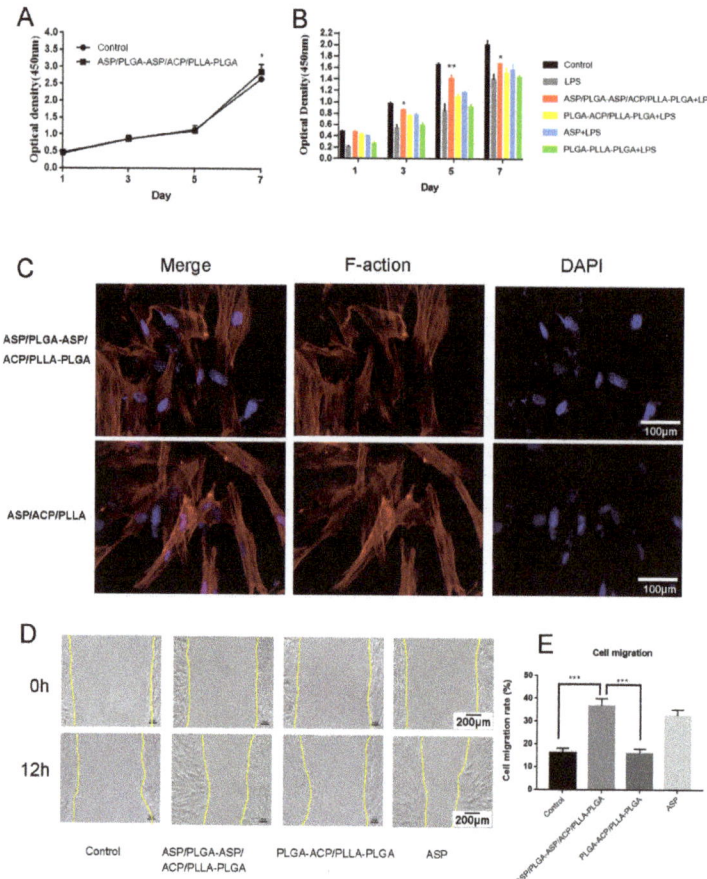

Figure 4. Effects of the ASP/PLGA-ASP/ACP/PLLA-PLGA composite membrane on DPCs proliferation, adhesion, and migration. (**A**) A CCK-8 assay showed stable DPCs growth. There was a significant difference in the cell proliferation rate between the ASP/PLGA-ASP/ACP/PLLA-PLGA and the control group on day 7 (p < 0.05). (**B**) A CCK8 assay showed the effects of the ASP/PLGA-ASP/ACP/PLLA-PLGA composite membrane on DPCs proliferation after LPS treatment. (**C**) Adhered DAPI-labeled DPCs on the ASP/PLGA-ASP/ACP/PLLA-PLGA and ASP/ACP/PLLA membranes were visualized by using a confocal microscope at 24 h. (**D**,**E**) Cell migration rates in the ASP/PLGA-ASP/ACP/PLLA-PLGA and other groups. (* $p < 0.05$, ** $p < 0.01$, and *** $p < 0.001$).

3.4. Effects of Composite Membrane on the Mineralization and Odontogenic Differentiation of iDPCs

The mineralization-inducing ability of the composite membrane was assessed by using ARS. After 21 days of culturing in osteogenic media, mineralized nodules were stained red with ARS. Mineralized nodule formation significantly decreased after LPS treatment, while it increased after the addition of ASP and PLGA-ACP/PLLA-PLGA. The ASP/PLGA-ASP/ACP/PLLA-PLGA group demonstrated significantly more mineralization according to the ARS findings (Figure 5A). The odontogenic differentiation ability was investigated via qPCR and Western blot through the detection of differentiation-related biomarkers. Both RNA expression levels of ALP, DMP1, DSPP, BSP and protein expression levels of DMP1, DSPP, OCN were reduced in dental pulp cells under inflammatory conditions. they recovered or increased after ASP or PLGA-ACP/PLLA-PLGA treatment and significantly increased after ASP/PLGA-ASP/ACP/PLLA-PLGA membrane treatment. (Figure 5B,C).

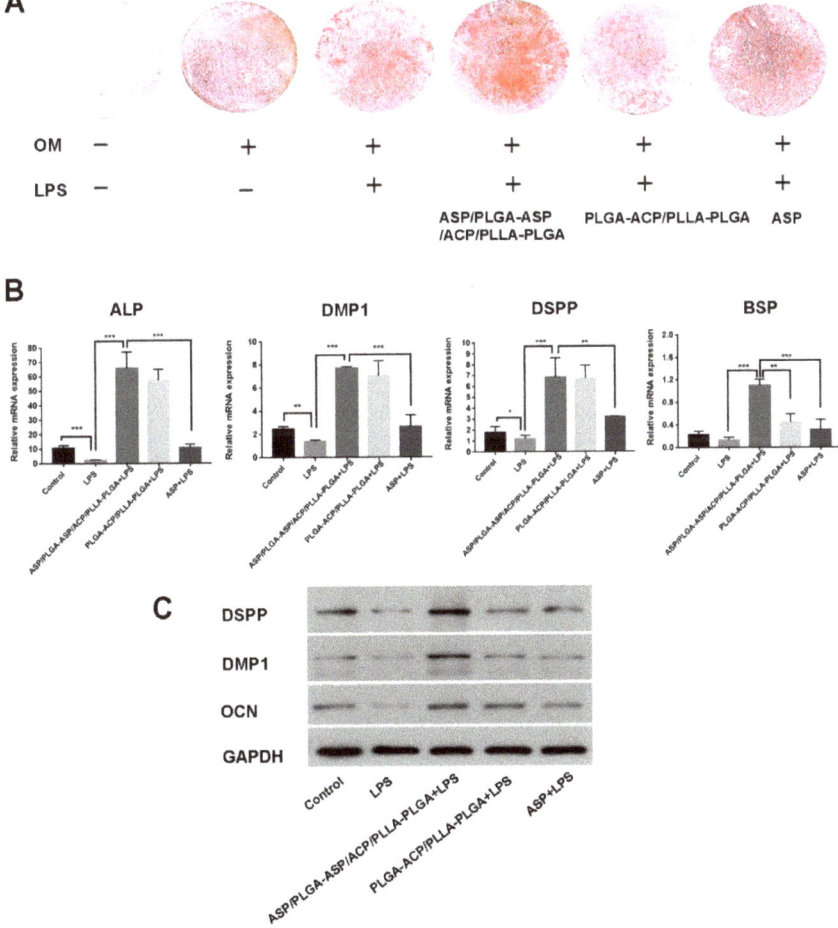

Figure 5. Effects of the ASP/PLGA-ASP/ACP/PLLA-PLGA composite membrane on mineralization and odontogenesis expression in DPCs. (**A**) Mineralized nodule formation was detected by using alizarin red staining on day 21. (**B**) mRNA expression levels of ALP, DMP1, DSPP, and BSP were detected by using qPCR on day 14. (**C**) The protein expression levels of DSPP, DMP1, and OCN by using Western blot on day 14. (* $p < 0.05$, ** $p < 0.01$, and *** $p < 0.001$).

3.5. Effects of the Composite Membrane on Inflammatory Cytokine Expression in iDPCs

In order to assess the anti-inflammatory effects of the composite membrane, the levels of inflammatory biomarkers were measured by using qPCR. The mRNA expression levels of the inflammatory cytokines IL-1β, IL-6, TNF-α, and IL-8 were higher in iDPCs than in DPCs. After treatment with ASP or the ASP/PLGA-ASP/ACP/PLLA-PLGA composite membrane, the mRNA expression levels of inflammatory cytokines significantly decreased in iDPCs (Figure 6A). Enzyme-linked immunosorbent assays also showed that ASP and composite membrane treatments significantly downregulated the expression levels of TNF-α, IL-1β, and IL-6 (Figure 6B).

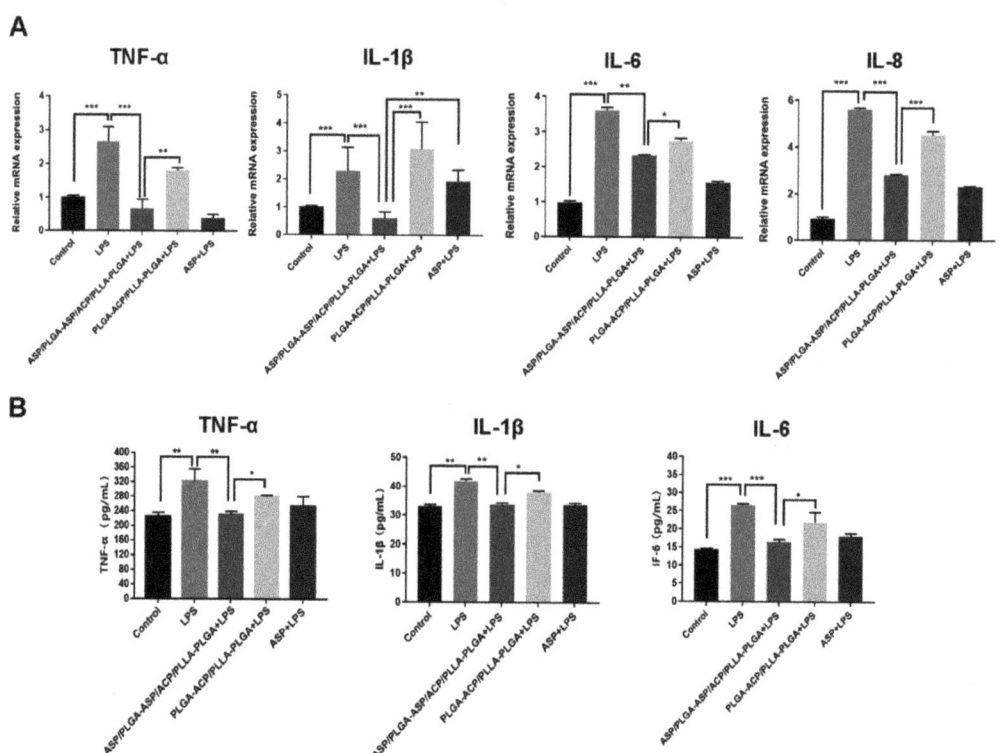

Figure 6. Effects of the ASP/PLGA-ASP/ACP/PLLA-PLGA composite membrane on cytokine expression in DPCs. (**A**) The mRNA expression levels of IL-1β, IL-6, IL-8, and TNF-α were detected by a qPCR analysis on day 7. (**B**) The IL-1β, IL-6, and TNF-α expression levels in the cell supernatant were determined by using enzyme-linked immunosorbent assays. (* $p < 0.05$, ** $p < 0.01$, and *** $p < 0.001$).

3.6. Effect of the ASP/PLGA-ASP/ACP/PLLA-PLGA Composite Membrane for Dental-Pulp-Capping In Vivo

The in vivo effects of the ASP/PLGA-ASP/ACP/PLLA-PLGA composite membrane as a pulp-capping material was investigated in a rat dental pulpitis model. Pulp tissues treated with LPS became necrotic at eight weeks (Figure 7B,C). Newly formed dentin-like tissue was observed in both the ASP/PLGA-ASP/ACP/PLLA-PLGA (Figure 7D) and MTA groups (Figure 7E) at eight weeks. The immunohistochemical staining results showed higher DMP1 expression in the ASP/PLGA-ASP/ACP/PLLA-PLGA group than in the MTA group (Figure 7I,J).

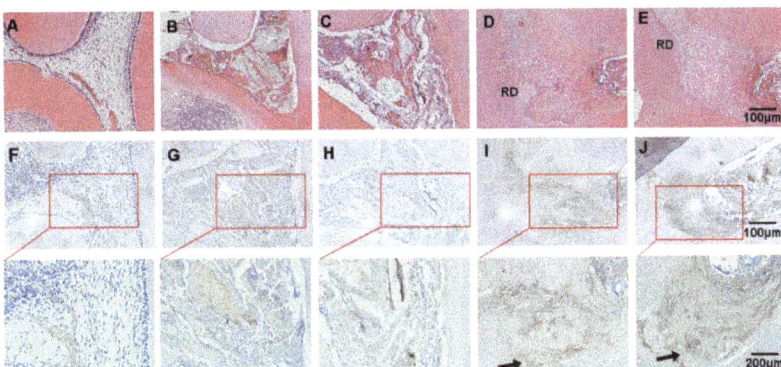

Figure 7. Hematoxylin–eosin staining (**A–E**) and immunohistochemical staining (**F–J**) showed newly formed tissues. (**A,F**) Normal group, (**B,G**) LPS group, (**C,H**) LPS + GIC group, (**D,I**) MTA + GIC group, and (**E,J**) ASP/PLGA-ASP/ACP/PLLA-PLGA + GIC group. RD: reparative dentin-like tissue. The arrows indicate DMP1-positive cells.

4. Discussion

Dental caries and trauma often lead to dental pulp exposure. The dysplasia of enamel and dentin caused by some genetic diseases may also lead to dental pulp tissue being vulnerable to damage [21–23]. Vital pulp therapy is important for clinical treatment because vital pulp tissue promotes root development in young permanent teeth [24,25]; it also provides resistance to biting forces, thus reducing the risk of root fractures [26,27]. The long-term prognosis of teeth treated by root canal therapy is worse than vital pulp teeth [28]. The biological basis for vital pulp therapy lies in the repair and regeneration capacity of the dental pulp. The dental pulp contains many undifferentiated mesenchymal stem cells with high proliferation and multi-differentiation potentials [29]. When the pulp is infected, odontoblasts become seriously damaged, and various inflammatory cytokines are produced [30]. Under these circumstances, prompt infection control allows DPCs to migrate to the injury site and repair the defective dentin, thus forming a hard tissue barrier and enabling the re-establishment of tissue homeostasis and health [31,32]. Conversely, persistent inflammation induces an immune response that can lead to tissue necrosis. Therefore, the timely control of early pulpitis is crucial for dental pulp repair and regeneration [33–35].

The anti-inflammatory drug, ASP, used in this study exerts favorable anti-inflammatory effects by inhibiting cell cyclooxygenase, blocking the conversion of prostaglandins into arachidonic acid, and reducing the production of inflammatory mediators. It can also inhibit the expression of nuclear-factor-κB-mediated inflammatory cytokines (e.g., TNF-α), which has important anti-inflammatory effects [36–39]. Aspirin is widely used in bone regeneration [40]. Studies have confirmed that aspirin can promote the osteogenic differentiation of mesenchymal stem cells and inhibits the expression of inflammatory factors to activate osteoblasts and inhibit osteoclasts [41]. The effect of aspirin on bone regeneration is dose-dependent. The intervention of low-dose aspirin on mesenchymal cells may inhibit the differentiation and formation of RANKL-induced osteoclasts as well as significantly reduce the number of trap-positive osteoclasts [42]. The dose of aspirin is an important factor affecting cell proliferation, differentiation, and migration. High doses of aspirin will inhibit cell proliferation [43]. In recent years, many studies have loaded aspirin on different materials, and the composite application of these materials is conducive to controlling the released concentration of aspirin. Similarly, the effect of aspirin on dental pulp cells is dose-dependent [44]. Low concentrations of ASP can promote DPCs proliferation and reduce the expression levels of inflammatory cytokines, including TNF-α and IL-6. Low concentrations of ASP can also promote odontoblast differentiation of DPCs through the

Wnt/β-catenin pathway [44–46]. However, there is some evidence that high concentrations of ASP can inhibit DPCs proliferation and lead to cell death [47]. Therefore, an optimal ASP concentration is necessary for effective inflammation control and the avoidance of cell death during pulp capping. In this study, the lower layer of the composite membrane quickly released high concentrations of ASP to inhibit the early stages of inflammation, the middle layer allowed for the long-term slow release of low concentrations of ASP during the later stages of inflammation, thus promoting DPCs proliferation and differentiation. When studying the effect of anti-inflammatory drugs on cells in an inflammatory state, in vitro inflammatory models are usually used. LPS is often used to establish an inflammatory microenvironment in vitro [48–50]. According to previous studies, treating human dental pulp cells (HDPCs) with LPS with a concentration of 1 µg/mL is the appropriate method to simulate an inflammatory microenvironment in vitro [51]. Therefore, this study used 1 µg/mL LPS to establish an in vitro inflammatory microenvironment model of HDPCs for in vitro research. Our results showed that the expression levels of the inflammatory cytokines IL-1β, IL-6, IL-8, and TNF-α were significantly increased in LPS-induced inflammatory-reacted dental pulp cells, but they decreased after treatment with the composite membrane. Thus, ASP has a positive effect on the repair of the inflammatory dental pulp.

Ideally, pulp capping should be able to repair the damaged pulp and the material can be replaced by reparative dentin. Therefore, the pulp-capping material should be biodegradable. Meanwhile, some physical and chemical properties of pulp-capping materials, such as the pH value and calcium ion release, are also crucial because they will have a direct impact on the repair process of dental pulp tissue [52,53]. ACP is a hydroxyapatite precursor with good biocompatibility and biodegradability [54,55]. It can alter cell function and tissue differentiation by releasing calcium ions to promote cell osteogenic differentiation [56]. ACP/PLLA scaffolds can continuously release calcium and phosphorus ions. The biological activity of a pulp-capping agent is related to its ability to release calcium ions. Studies have confirmed that increasing the concentration of calcium ions can increase the expression of osteocalcin (OPN) and bone morphogenetic protein-2 (BMP-2) as well as stimulate the release of proteoglycans and growth factors of a mineralized dentin matrix. Then, the undifferentiated pulp stem cells migrate to the injury site, proliferate, and differentiate to form an extracellular matrix and promote mineralization [52,57]. The ACP-mediated directional differentiation of mesenchymal stem cells is affected by the concentration of free calcium ions; if the concentration decreases in the surrounding environment, osteogenic differentiation is reduced [58]. As a drug delivery carrier and scaffold material, PLLA has some problems, such as its weak stability of hydrophilic components and the burst release phenomenon, which need to be improved [59]. For the slow and long-term release of calcium ions, our ACP/PLLA scaffolds were coated with the PLGA membrane, which is biocompatible and has a sustained release capacity. This membrane can regulate the degradation rate, ensure the long-term maintenance of the drug concentration, and avoid "peak and valley" phenomena [60,61]. In this study, the release of calcium and phosphorus ions from the composite membrane was steady, suggesting that the structure of the composite membrane could achieve the controlled release of calcium and phosphorus ions. The metabolites of PLLA in the body are lactic acid, and the metabolites of PLGA in the body are lactic acid and glycolic acid. These monomers are metabolized in the body through the tricarboxylic acid cycle, and the final metabolites, carbon dioxide and water, have no toxic effect on cells [62–65]. PLGA and PLLA can be used as suitable microcarriers to achieve sustained drug release [66,67]. As a scaffold material, PLGA is conducive to the colonization of cells (such as SHED, DPSCs, and dental pulp fibroblasts), and on PLGA scaffolds, stem cells can differentiate into odontoblast-like cells and endothelial cells, producing tissues similar to dental pulp and dentin [68]. The outer PLGA membrane layer improves the superficial hydrophilicity of the composite membrane [69,70]. This hydrophilic property ameliorates the biocompatibility of the ASP/PLGA-ASP/ACP/PLLA-PLGA membrane, as well as increasing its favorability for cell adhesion [71,72]. The rapid adhesion of cells will initiate the differentiation process in the early stage, resulting in a satisfactory odontogenic differentiation at the end [73–75]. Moreover, an acidic environment not only suppresses the viability of bone-related cells but also inhibits their mineralization process [76,77]. Previous studies have confirmed that an increase in alkaline pH will enhance the expression of BMP-2 mRNA and ALP activity in dental pulp cells, but that exorbitant pH will lead to excessive calcium ion release, induce dystrophic calcification in exposed dental pulp areas, reduce the volume of the reparative dental pulp, and may hinder any future dental pulp treatment [57,78]. In this work, the addition of ACP, an alkaline agent [79], in the membrane neutralizes the acidic degradation byproducts of PLLA and PLGA in addition to maintaining a neutral environment, which we believe is also conducive to odontogenic differentiation.

In this study, the effect of the composite membrane on iDPCs proliferation and odontoblastic differentiation after LPS treatment was more than the effects of ASP or ACP/PLLA/PLGA scaffolds alone. This finding is potentially the case because the early release of ASP controlled inflammation and created a suitable environment for cell proliferation and differentiation, while the subsequent controlled release of ASP and ACP promoted a microenvironment that aided in pulp tissue repair.

The purpose of this paper was to solve the problem of preserving vital pulp in clinical practice. The rat pulp-capping model was used for eight weeks of observations. In future studies, the observation time should be extended, and large animals should be used to further observe the pulp-capping effect of the material. In addition, for the release of anti-inflammatory drugs, such as aspirin, in pulp-capping materials, inflammatory responsive controlled release materials can be further designed for different inflammatory states so as to more accurately regulate the pulp inflammation state. This paper provides a new idea and theoretical basis for the design and application of vital dental pulp preservation materials.

5. Conclusions

In this study, the composite application of aspirin, amorphous calcium phosphate, and PLLA/PLGA scaffold materials was used to develop a pulp-capping material with multiple effects for the repair of the injured dental pulp. The ASP/PLGA-ASP/ACP/PLLA-PLGA composite membrane developed in this study allowed for the sequential release of the anti-inflammatory drug aspirin as well as the calcium/phosphorus ions required for mineralization; The mass loss of the membrane stayed at a slow but constant speed, with a relatively stable pH value during degradation; it also promoted the proliferation and odontogenic differentiation of iDPCs, while inhibiting inflammatory cytokine release. It had the dual effects of reducing inflammation and promoting mineralized barrier formation. The ASP/PLGA-ASP/ACP/PLLA-PLGA composite membrane represents a promising vital pulp conservation material with the potential for clinical applications.

The application of composite materials for pulp therapy can combine the advantages of multiple materials, but it also faces great challenges because the mechanism of action between composites is complex and needs to be further explored. We will further evaluate the biological properties of the material and conduct long-term in vivo studies to assess the effectiveness of ASP/PLGA-ASP/ACP/PLLA-PLGA composite membrane as a vital pulp-capping agent in clinical applications.

Author Contributions: Conceptualization, Y.Z. and X.N.; Methodology, W.Y., F.Y. and Z.L.; Software, F.Y. and Z.L.; Formal Analysis, W.Y. and F.Y.; Investigation, W.Y., F.Y. and Z.L.; Resources, Y.G. and Q.W.; Data Curation, W.Y. and F.Y.; Writing—Original Draft Preparation, W.Y. and F.Y.; Writing—Review and Editing, Y.Z., X.N., W.Y., F.Y. and Q.W.; Project Administration, Y.Z. and X.N.; Funding Acquisition, Y.Z., Z.L. and X.N. All authors have read and agreed to the published version of the manuscript.

Funding: This research was funded by Beijing Natural Science Foundation (7212135), National Natural Science Foundation of China (11872097) and Beijing Nova Program of Science and Technology (Z191100001119096).

Institutional Review Board Statement: The animal study protocol was approved by Experimental Animal Care and Ethical Committee of Peking University (protocol code: LA2020171).

Informed Consent Statement: Not applicable.

Data Availability Statement: The data presented in this study are available on request from the corresponding author.

Conflicts of Interest: The authors declare no conflict of interest.

Abbreviations

BSP	Bone Sialoprotein
DCM	Dichloromethane
DMF	Dimethyl Formamide
DMP	Dentin matrix protein
DPCs	Dental pulp stem cells
DSPP	Dentin sialophosphoprotein
EDS	Energy-dispersive spectrometer
GIC	Glass-ionomer cement
HDPCs	Human dental pulp cells
iDPCs	Inflamed dental pulp stem cells
IL	Interleukin
LPS	Lipopolysaccharide
MTA	Mineral trioxide aggregate
OCN	Osteocalcin
OPN	Osteopontin
PBS	Phosphate buffered solution
PLGA	Poly (lactic-co-glycolic acid)
PLLA	Poly (L-lactic acid)
qPCR	Quantitative Polymerase Chain Reaction
RNA	Ribonucleic Acid
SEM	Scanning electron microscope
SHED	Stem cells from human exfoliated deciduous teeth
TEM	Transmission electron microscope
TNF	Tumor necrosis factor
XRD	X-ray diffraction
α-MEM	Minimum Essential Medium-Alpha

References

1. Huang, G.T. Dental pulp and dentin tissue engineering and regeneration: Advancement and challenge. *Front. Biosci.* **2011**, *3*, 788–800. [CrossRef] [PubMed]
2. Aguilar, P.; Linsuwanont, P. Vital pulp therapy in vital permanent teeth with cariously exposed pulp: A systematic review. *J. Endod.* **2011**, *37*, 581–587. [CrossRef] [PubMed]
3. Hanna, S.N.; Perez Alfayate, R.; Prichard, J. Vital Pulp Therapy an Insight Over the Available Literature and Future Expectations. *Eur. Endod. J.* **2020**, *5*, 46–53. [PubMed]
4. Kunert, M.; Lukomska-Szymanska, M. Bio-Inductive Materials in Direct and Indirect Pulp Capping-A Review Article. *Materials* **2020**, *13*, 1204. [CrossRef] [PubMed]
5. Kim, D.H.; Jang, J.H.; Lee, B.N. Bio-Inductive Materials in Direct and Indirect Pulp s of Human and Rat Dental Pulps In Vitro and In Vivo. *J. Endod.* **2018**, *44*, 1534–1541. [CrossRef] [PubMed]
6. Zanini, M.; Meyer, E.; Simon, S. Pulp Inflammation Diagnosis from Clinical to Inflammatory Mediators: A Systematic Review. *J. Endod.* **2017**, *43*, 1033–1051. [CrossRef]
7. Mejàre, I.A.; Axelsson, S.; Davidson, T.; Frisk, F.; Hakeberg, M.; Kvist, T.; Norlund, A.; Petersson, A.; Portenier, I.; Sandberg, H.; et al. Diagnosis of the condition of the dental pulp: A systematic review. *Int. Endod. J.* **2012**, *45*, 597–613. [CrossRef]
8. Parirokh, M.; Torabinejad, M.; Dummer, P. Mineral trioxide aggregate and other bioactive endodontic cements: An updated overview—Part I: Vital pulp therapy. *Int. Endod. J.* **2017**, *51*, 177–205. [CrossRef]
9. da Rosa, W.L.; Cocco, A.R.; Silva, T.M.; Mesquita, L.C.; Galarca, A.D.; Silva, A.F.; Piva, E. Current trends and future perspectives of dental pulp capping materials: A systematic review. *J. Biomed. Mater. Res. B Appl. Biomater.* **2018**, *106*, 1358–1368. [CrossRef]
10. Tziafas, D.; Smith, A.; Lesot, H. Designing new treatment strategies in vital pulp therapy. *J. Dent.* **1999**, *28*, 77–92. [CrossRef]
11. Di Bella, S.; Luzzati, R.; Principe, L. Aspirin and Infection: A Narrative Review. *Biomedicines* **2022**, *10*, 263. [CrossRef]
12. Liu, Y.; Chen, C.; Liu, S.; Liu, D.; Xu, X.; Chen, X.; Shi, S. Acetylsalicylic Acid Treatment Improves Differentiation and Immunomodulation of SHED. *J. Dent. Res.* **2014**, *94*, 209–218. [CrossRef]

3. Bienek, D.R.; Skrtic, D. Utility of Amorphous Calcium Phosphate-Based Scaffolds in Dental/Biomedical Applications. *Biointerface Res. Appl. Chem.* **2017**, *7*, 1989–1994.
4. Niu, X.; Liu, Z.; Tian, F.; Chen, S.; Lei, L.; Jiang, T.; Feng, Q.; Fan, Y. Sustained delivery of calcium and orthophosphate ions from amorphous calcium phosphate and poly(L-lactic acid)-based electrospinning nanofibrous scaffold. *Sci. Rep.* **2017**, *7*, srep45655. [CrossRef]
5. Anderson, J.M.; Shive, M.S. Biodegradation and biocompatibility of PLA and PLGA microspheres. *Adv. Drug Deliv. Rev.* **1997**, *28*, 5–24. [CrossRef]
6. Su, Y.; Zhang, B.; Sun, R.; Liu, W.; Zhu, Q.; Zhang, X.; Wang, R.; Chen, C. PLGA-based biodegradable microspheres in drug delivery: Recent advances in research and application. *Drug Deliv.* **2021**, *28*, 1397–1418. [CrossRef]
7. Vlachopoulos, A.; Karlioti, G.; Balla, E.; Daniilidis, V.; Kalamas, T.; Stefanidou, M.; Bikiaris, N.D.; Christodoulou, E.; Koumentakou, I.; Karavas, E.; et al. Poly(Lactic Acid)-Based Microparticles for Drug Delivery Applications: An Overview of Recent Advances. *Pharmaceutics* **2022**, *14*, 359. [CrossRef]
8. Ding, D.; Zhu, Q. Recent advances of PLGA micro/nanoparticles for the delivery of biomacromolecular therapeutics. *Mater. Sci. Eng. C* **2018**, *92*, 1041–1060. [CrossRef]
9. Martin, F.E.; Nadkarni, M.A.; Jacques, N.A.; Hunter, N. Quantitative Microbiological Study of Human Carious Dentine by Culture and Real-Time PCR: Association of Anaerobes with Histopathological Changes in Chronic Pulpitis. *J. Clin. Microbiol.* **2002**, *40*, 1698–1704. [CrossRef]
10. Lee, S.-I.; Min, K.-S.; Bae, W.-J.; Lee, Y.-M.; Lee, S.-Y.; Lee, E.-S.; Kim, E.-C. Role of SIRT1 in Heat Stress- and Lipopolysaccharide-induced Immune and Defense Gene Expression in Human Dental Pulp Cells. *J. Endod.* **2011**, *37*, 1525–1530. [CrossRef]
11. Barron, M.J.; McDonnell, S.T.; MacKie, I.; Dixon, M.J. Hereditary dentine disorders: Dentinogenesis imperfecta and dentine dysplasia. *Orphanet J. Rare Dis.* **2008**, *3*, 31. [CrossRef] [PubMed]
12. Crawford, P.J.; Aldred, M.; Bloch-Zupan, A. Amelogenesis imperfecta. *Orphanet J. Rare Dis.* **2007**, *2*, 17. [CrossRef] [PubMed]
13. Minervini, G.; Romano, A.; Petruzzi, M.; Maio, C.; Serpico, R.; Lucchese, A.; Candotto, V.; Di Stasio, D. Telescopic overdenture on natural teeth: Prosthetic rehabilitation on (OFD) syndromic patient and a review on available literature. *J. Biol. Regul. Homeost. Agents* **2018**, *32*, 131–134. [PubMed]
14. Kratunova, E.; Silva, D. Pulp therapy for primary and immature permanent teeth: An overview. *Gen. Dent.* **2018**, *66*, 30–38. [PubMed]
15. Yong, D.; Cathro, P. Conservative pulp therapy in the management of reversible and irreversible pulpitis. *Aust. Dent. J.* **2021**, *66* (Suppl. 1), S1–S14. [CrossRef]
16. Dammaschke, T.; Leidinger, J.; Schäfer, E. Long-term evaluation of direct pulp capping–treatment outcomes over an average period of 6.1 years. *Clin. Oral. Investig.* **2010**, *14*, 559–567. [CrossRef]
17. Torabinejad, M.; Corr, R.; Handysides, R.; Shabahang, S. Outcomes of nonsurgical retreatment and endodontic surgery: A systematic review. *J. Endod.* **2009**, *35*, 930–937. [CrossRef]
18. Caplan, D.J.; Cai, J.; Yin, G.; White, B.A. Root Canal Filled Versus Non-Root Canal Filled Teeth: A Retrospective Comparison of Survival Times. *J. Public Health Dent.* **2005**, *65*, 90–96. [CrossRef]
19. d'Aquino, R.; Graziano, A.; Sampaolesi, M.; Laino, G.; Pirozzi, G.; De Rosa, A.; Papaccio, G. Human postnatal dental pulp cells co-differentiate into osteoblasts and endotheliocytes: A pivotal synergy leading to adult bone tissue formation. *Cell Death Differ.* **2007**, *14*, 1162–1171. [CrossRef]
30. Gronthos, S.; Mankani, M.; Brahim, J.; Robey, P.G.; Shi, S. Postnatal human dental pulp stem cells (DPSCs) in vitro and in vivo. *Proc. Natl. Acad. Sci. USA* **2000**, *97*, 13625–13630. [CrossRef]
31. Piva, E.; Tarlé, S.A.; Nör, J.E.; Zou, D.; Hatfield, E.; Guinn, T.; Eubanks, E.J.; Kaigler, D. Dental Pulp Tissue Regeneration Using Dental Pulp Stem Cells Isolated and Expanded in Human Serum. *J. Endod.* **2017**, *43*, 568–574. [CrossRef]
32. Farges, J.-C.; Alliot-Licht, B.; Renard, E.; Ducret, M.; Gaudin, A.; Smith, A.J.; Cooper, P.R. Dental Pulp Defence and Repair Mechanisms in Dental Caries. *Mediat. Inflamm.* **2015**, *2015*, 230251. [CrossRef]
33. Huang, H.; Zhao, N.; Xu, X.; Xu, Y.; Li, S.; Zhang, J.; Yang, P. Dose-specific effects of tumor necrosis factor alpha on osteogenic differentiation of mesenchymal stem cells. *Cell Prolif.* **2011**, *44*, 420–427. [CrossRef]
34. Colombo, J.S.; Moore, A.N.; Hartgerink, J.D.; D'Souza, R.N. Scaffolds to Control Inflammation and Facilitate Dental Pulp Regeneration. *J. Endod.* **2014**, *40*, S6–S12. [CrossRef]
35. Kearney, M.; Cooper, P.R.; Smith, A.J.; Duncan, H.F. Epigenetic Approaches to the Treatment of Dental Pulp Inflammation and Repair: Opportunities and Obstacles. *Front. Genet.* **2018**, *9*, 311. [CrossRef]
36. Fujita, T.; Kutsumi, H.; Sanuki, T.; Hayakumo, T.; Azuma, T. Adherence to the preventive strategies for nonsteroidal anti-inflammatory drug- or low-dose aspirin-induced gastrointestinal injuries. *J. Gastroenterol.* **2013**, *48*, 559–573. [CrossRef]
37. Trabert, B.; Ness, R.B.; Lo-Ciganic, W.-H.; Murphy, M.A.; Goode, E.L.; Poole, E.M.; Brinton, L.A.; Webb, P.M.; Nagle, C.M.; Jordan, S.J.; et al. Aspirin, Nonaspirin Nonsteroidal Anti-inflammatory Drug, and Acetaminophen Use and Risk of Invasive Epithelial Ovarian Cancer: A Pooled Analysis in the Ovarian Cancer Association Consortium. *JNCI J. Natl. Cancer Inst.* **2014**, *106*, djt431. [CrossRef]
38. Chokshi, R.; Bennett, O.; Zhelay, T.; Kozak, J.A. NSAIDs Naproxen, Ibuprofen, Salicylate, and Aspirin Inhibit TRPM7 Channels by Cytosolic Acidification. *Front. Physiol.* **2021**, *12*, 727549. [CrossRef]

39. Brox, R.; Hackstein, H. Physiologically relevant aspirin concentrations trigger immunostimulatory cytokine production by human leukocytes. *PLoS ONE* **2021**, *16*, e0254606. [CrossRef]
40. Fattahi, R.; Mohebichamkhorami, F.; Khani, M.M.; Soleimani, M.; Hosseinzadeh, S. Aspirin effect on bone remodeling and skeletal regeneration: Review article. *Tissue Cell* **2022**, *76*, 101753. [CrossRef]
41. Ren, L.; Pan, S.; Li, H.; Li, Y.; He, L.; Zhang, S.; Che, J.; Niu, Y. Effects of aspirin-loaded graphene oxide coating of a titanium surface on proliferation and osteogenic differentiation of MC3T3-E1 cells. *Sci. Rep.* **2018**, *8*, 1–13. [CrossRef] [PubMed]
42. Shi, S.; Yamaza, T.; Akiyama, K. Is aspirin treatment an appropriate intervention to osteoporosis? *Future Rheumatol.* **2008**, *3*, 499–502. [PubMed]
43. Müller, M.; Raabe, O.; Addicks, K.; Wenisch, S.; Arnhold, S. Effects of non-steroidal anti-inflammatory drugs on proliferation, differentiation and migration in equine mesenchymal stem cells. *Cell Biol. Int.* **2011**, *35*, 235–248. [CrossRef] [PubMed]
44. Yuan, M.; Zhan, Y.; Hu, W.; Li, Y.; Xie, X.; Miao, N.; Jin, H.; Zhang, B. Aspirin promotes osteogenic differentiation of human dental pulp stem cells. *Int. J. Mol. Med.* **2018**, *42*, 1967–1976. [CrossRef]
45. Rankin, R.; Lundy, F.T.; Schock, B.C. A connectivity mapping approach predicted acetylsalicylic acid (aspirin) to induce osteo/odontogenic differentiation of dental pulp cells. *Int. Endod. J.* **2020**, *53*, 834–845. [CrossRef]
46. Chang, M.C.; Tsai, Y.L.; Chang, H.H. IL-1β-induced MCP-1 expression and secretion of human dental pulp cells is related to TAK1, MEK/ERK, and PI3K/Akt signaling pathways. *Arch. Oral. Biol.* **2016**, *61*, 16–22. [CrossRef]
47. Li, J.Y.; Wang, S.N.; Dong, Y.M. Anti-inflammatory and repaired effects of non-steroidal anti-inflammatory drugs on human dental pulp cells. *Beijing Da Xue Xue Bao Yi Xue Ban* **2020**, *52*, 24–29.
48. Marconi, G.D.; Fonticoli, L.; Guarnieri, S.; Cavalcanti, M.F.X.B.; Franchi, S.; Gatta, V.; Trubiani, O.; Pizzicannella, J.; Diomede, F. Ascorbic Acid: A New Player of Epigenetic Regulation in LPS-gingivalis Treated Human Periodontal Ligament Stem Cells. *Oxid. Med. Cell. Longev.* **2021**, *2021*, 6679708. [CrossRef]
49. Zhu, N.; Chatzistavrou, X.; Ge, L.; Qin, M.; Papagerakis, P.; Wang, Y. Biological properties of modified bioactive glass on dental pulp cells. *J. Dent.* **2019**, *83*, 18–26. [CrossRef]
50. Saga, R.; Matsuya, Y.; Takahashi, R.; Hasegawa, K.; Date, H.; Hosokawa, Y. 4-Methylumbelliferone administration enhances radiosensitivity of human fibrosarcoma by intercellular communication. *Sci. Rep.* **2021**, *11*, 8258. [CrossRef]
51. Bindal, P.; Ramasamy, T.S.; Abu Kasim, N.H.; Gnanasegaran, N.; Chai, W.L. Immune responses of human dental pulp stem cells in lipopolysaccharide-induced microenvironment. *Cell Biol. Int.* **2018**, *42*, 832–840. [CrossRef]
52. Rashid, F.; Shiba, H.; Mizuno, N.; Mouri, Y.; Fujita, T.; Shinohara, H.; Ogawa, T.; Kawaguchi, H.; Kurihara, H. The Effect of Extracellular Calcium Ion on Gene Expression of Bone-related Proteins in Human Pulp Cells. *J. Endod.* **2003**, *29*, 104–107. [CrossRef]
53. Gandolfi, M.G.; Siboni, F.; Primus, C.M.; Prati, C. Ion Release, Porosity, Solubility, and Bioactivity of MTA Plus Tricalcium Silicate. *J. Endod.* **2014**, *40*, 1632–1637. [CrossRef]
54. Niu, X.; Chen, S.; Tian, F.; Wang, L.; Feng, Q.; Fan, Y. Hydrolytic conversion of amorphous calcium phosphate into apatite accompanied by sustained calcium and orthophosphate ions release. *Mater. Sci. Eng. C* **2017**, *70*, 1120–1124. [CrossRef]
55. Bose, S.; Tarafder, S. Calcium phosphate ceramic systems in growth factor and drug delivery for bone tissue engineering: A review. *Acta Biomater.* **2012**, *8*, 1401–1421. [CrossRef]
56. Chen, X.; Wang, J.; Chen, Y. Roles of calcium phosphate-mediated integrin expression and MAPK signaling pathways in the osteoblastic differentiation of mesenchymal stem cells. *J. Mater. Chem. B* **2016**, *4*, 2280–2289. [CrossRef]
57. Petta, T.M.; Pedroni, A.C.F.; Saavedra, D.F.; Faial, K.D.C.F.; Marques, M.M.; Couto, R.S.D. The effect of three different pulp capping cements on mineralization of dental pulp stem cells. *Dent. Mater. J.* **2020**, *39*, 222–228. [CrossRef]
58. Gröninger, O.; Hess, S.; Mohn, D. Directing Stem Cell Commitment by Amorphous Calcium Phosphate Nanoparticles Incorporated in PLGA: Relevance of the Free Calcium Ion Concentration. *Int. J. Mol. Sci.* **2020**, *21*, 2627. [CrossRef]
59. Im, S.H.; Im, D.H.; Park, S.J.; Chung, J.J.; Jung, Y.; Kim, S.H. Stereocomplex Polylactide for Drug Delivery and Biomedical Applications: A Review. *Molecules* **2021**, *26*, 2846. [CrossRef]
60. Zhang, Z.; Hu, J.; Ma, P.X. Nanofiber-based delivery of bioactive agents and stem cells to bone sites. *Adv. Drug Deliv. Rev.* **2012**, *64*, 1129–1141. [CrossRef]
61. Mathieu, S.; Jeanneau, C.; Sheibat-Othman, N.; Kalaji, N.; Fessi, H.; About, I. Usefulness of controlled release of growth factors in investigating the early events of dentin-pulp regeneration. *J. Endod.* **2013**, *39*, 228–235. [CrossRef]
62. Capuana, E.; Lopresti, F.; Ceraulo, M.; La Carrubba, V. Poly-l-Lactic Acid (PLLA)-Based Biomaterials for Regenerative Medicine: A Review on Processing and Applications. *Polymers* **2022**, *14*, 1153. [CrossRef]
63. Farah, S.; Anderson, D.G.; Langer, R. Physical and mechanical properties of PLA, and their functions in widespread applications—A comprehensive review. *Adv. Drug Deliv. Rev.* **2016**, *107*, 367–392. [CrossRef]
64. Sadat Tabatabaei Mirakabad, F.; Nejati-Koshki, K.; Akbarzadeh, A. PLGA-based nanoparticles as cancer drug delivery systems. *Asian Pac. J. Cancer Prev.* **2014**, *15*, 517–535. [CrossRef]
65. Panyam, J.; Labhasetwar, V. Biodegradable nanoparticles for drug and gene delivery to cells and tissue. *Adv. Drug Deliv. Rev.* **2003**, *55*, 329–347. [CrossRef]
66. Jain, R.A. The manufacturing techniques of various drug loaded biodegradable poly(lactide-co-glycolide) (PLGA) devices. *Biomaterials* **2000**, *21*, 2475–2490. [CrossRef]

67. Parhizkar, A.; Asgary, S. Local Drug Delivery Systems for Vital Pulp Therapy: A New Hope. *Int. J. Biomater.* **2021**, *2021*, 5584268. [CrossRef]
68. Moussa, D.G.; Aparicio, C. Present and future of tissue engineering scaffolds for dentin-pulp complex regeneration. *J. Tissue Eng. Regen. Med.* **2019**, *13*, 58–75. [CrossRef]
69. Martins, C.; Sousa, F.; Araújo, F.; Sarmento, B. Functionalizing PLGA and PLGA Derivatives for Drug Delivery and Tissue Regeneration Applications. *Adv. Healthc. Mater.* **2018**, *7*, 10. [CrossRef]
70. Oh, S.H.; Kim, J.H.; Song, K.S. Peripheral nerve regeneration within an asymmetrically porous PLGA/Pluronic F127 nerve guide conduit. *Biomaterials* **2008**, *29*, 1601–1609. [CrossRef]
71. Kim, K.; Yu, M.; Zong, X. Control of degradation rate and hydrophilicity in electrospun non-woven poly (D, L-lactide) nanofiber scaffolds for biomedical applications. *Biomaterials* **2003**, *24*, 4977–4985. [CrossRef]
72. Alves, C.M.; Yang, Y.; Marton, D.; Carnes, D.L.; Ong, J.L.; Sylvia, V.L.; Dean, D.D.; Reis, R.L.; Agrawal, C.M. Plasma surface modification of poly(D,L-lactic acid) as a tool to enhance protein adsorption and the attachment of different cell types. *J. Biomed. Mater. Res. Part B Appl. Biomater.* **2008**, *87B*, 59–66. [CrossRef] [PubMed]
73. Le Saux, G.; Wu, M.C.; Toledo, E. Cell-Cell Adhesion-Driven Contact Guidance and Its Effect on Human Mesenchymal Stem Cell Differentiation. *ACS Appl. Mater. Interfaces* **2020**, *12*, 22399–22409. [CrossRef] [PubMed]
74. Kang, J.; Rajangam, T.; Rhie, J.; Kim, S. Characterization of cell signaling, morphology, and differentiation potential of human mesenchymal stem cells based on cell adhesion mechanism. *J. Cell Physiol.* **2020**, *235*, 6915–6928. [CrossRef]
75. Rocha, C.V.; Gonçalves, V.; da Silva, M.C.; Bañobre-López, M.; Gallo, J. PLGA-Based Composites for Various Biomedical Applications. *Int. J. Mol. Sci.* **2022**, *23*, 2034. [CrossRef]
76. Sui, G.; Yang, X.; Mei, F. Poly-L-lactic acid/hydroxyapatite hybrid membrane for bone tissue regeneration. *J. Biomed. Mater. Res. A* **2007**, *82*, 445–454. [CrossRef]
77. Sittinger, M.; Reitzel, D.; Dauner, M. Resorbable polyesters in cartilage engineering: Affinity and biocompatibility of polymer fiber structures to chondrocytes. *J. Biomed. Mater. Res.* **1996**, *33*, 57–63. [CrossRef]
78. Okabe, T.; Sakamoto, M.; Takeuchi, H.; Matsushima, K. Effects of pH on mineralization ability of human dental pulp cells. *J. Endod.* **2006**, *32*, 198–201. [CrossRef]
79. Cross, K.J.; Huq, N.L.; Reynolds, E.C. Casein phosphopeptides in oral health—Chemistry and clinical applications. *Curr. Pharm. Des.* **2007**, *13*, 793–800. [CrossRef]

Article

A Porous Hydrogel with High Mechanical Strength and Biocompatibility for Bone Tissue Engineering

Changxin Xiang [1], Xinyan Zhang [1], Jianan Zhang [1], Weiyi Chen [1], Xiaona Li [1,*], Xiaochun Wei [2,*] and Pengcui Li [2]

1 College of Biomedical Engineering, Taiyuan University of Technology, Taiyuan 030024, China
2 Shanxi Key Laboratory of Bone and Soft Tissue Injury Repair, Department of Orthopedics, The Second Hospital of Shanxi Medical University, Taiyuan 030001, China
* Correspondence: lixiaona@tyut.edu.cn (X.L.); sdeygksys@163.com (X.W.)

Abstract: Polyvinyl alcohol (PVA) hydrogels are considered to be ideal materials for tissue engineering due to their high water content, low frictional behavior, and good biocompatibility. However, their limited mechanical properties restrict them from being applied when repairing load-bearing tissue. Inspired by the composition of mussels, we fabricated polyvinyl alcohol/hydroxyapatite/tannic acid (PVA/HA/TA) hydrogels through a facile freeze–thawing method. The resulting composite hydrogels exhibited high moisture content, porous structures, and good mechanical properties. The compressive strength and tensile strength of PVA hydrogels were improved from 0.77 ± 0.11 MPa and 0.08 ± 0.01 MPa to approximately 3.69 ± 0.41 MPa and 0.43 ± 0.01 MPa, respectively, for the PVA/HA/1.5TA hydrogel. The toughness and the compressive elastic modulus of PVA/HA/1.5TA hydrogel also attained 0.86 ± 0.02 MJm^{-3} and 0.11 ± 0.02 MPa, which was approximately 11 times and 5 times higher than the PVA hydrogel, respectively. The PVA/HA/1.5TA hydrogel also exhibited fatigue resistance abilities. The mechanical properties of the composite hydrogels were improved through the introduction of TA. Furthermore, in vitro PVA/HA/1.5TA hydrogel showed excellent cytocompatibility by promoting cell proliferation in vitro. Scanning electron microscopy analysis indicated that PVA/HA/1.5TA hydrogels provided favorable circumstances for cell adhesion. The aforementioned results also indicate that the composite hydrogels had potential applications in bone tissue engineering, and this study provides a facile method to improve the mechanical properties of PVA hydrogel.

Keywords: bone regeneration; hydrogels; mechanical properties; polyvinyl alcohol; tannic acid

1. Introduction

Bone is the main weight-bearing tissue of the human body, and plays an essential role in supporting, protecting, transporting, and producing blood cells [1]. Specifically, tens of millions of people suffer from bone defects caused by trauma, developmental deformities, or tumor resection, which can influence their quality of life [2–4]. Natural bone can self-heal, to a certain extent, but large bone regeneration remains a major challenge, especially for load-bearing bone defects [5,6]. Although traditional autologous allografts are used clinically to repair large segments of bone defects, they still show various disadvantages, including a shortage of sources, additional trauma, need for a second surgery, and the high risk of disease transmission [6–8]. These complications motivate researchers to find suitable methods to repair defects.

Tissue engineering is an alternative strategy that uses a combination of seed cells, biological factors, and scaffolds to repair or replace damaged tissues [9,10]. The ideal bone scaffolds should require suitable mechanical properties, inter-connectivity, porous structure, and excellent biocompatibility [11,12]. Among the optional material candidates, hydrogel plays a vital role in biomedical and biopharmaceutical treatments due to its resemblance to the extracellular matrix of natural tissues [13–15].

Polyvinyl alcohol (PVA) hydrogels have been widely used in tissue engineering due to their high water content, non-toxicity, and good biocompatibility [16]. However, the application of PVA hydrogels in bone tissue regeneration is restricted by their limited mechanical strength [17]. PVA hydrogels can be synthesized by physical and chemical methods. PVA hydrogels prepared by physical freeze–thawing methods can form crystal nuclei and hydrogen bonds to increase the crosslinked density of hydrogel networks, which can improve the mechanical properties [18]. Due to hydroxyapatite (HA) being one of the main components of bone tissue, it is often added in PVA hydrogels to mimic the structure and composition of natural bone tissue [19]. Recently, there have been significant efforts to design PVA/HA hydrogels for bone tissue engineering. HA was deposited in situ to fabricate the PVA/HA hydrogels with porous structures, but the compressive strength only attained 0.8 MPa [20]. A porous polyvinyl alcohol/sodium alginate/hydroxyapatite composite hydrogel with tunable structure and mechanical properties was fabricated by a dual-crosslinking method [16]. However, the pore size was about 1μm and the optimized compressive modulus was about 50 kPa. Hydroxypropyl guar was used to improve mechanical properties of PVA/HA hydrogels and the compressive strength the hydrogels attained was approximately 7 MPa [21], but the addition of HPG decreased the cell viability of MC3T3-E1 on HPG/PVA/n-HA. By utilizing in situ HA synthesis, the biocompatibility of PVA/HA hydrogels was enhanced [22]. However, there were no pores on the surface of the composite hydrogels, which hindered the nutrient and waste transportation of cells. Therefore, it is still a challenge to synthesize porous PVA/HA hydrogels with excellent mechanical properties and good biocompatibility.

Inspired by the structure and composition of mussels, catechol-based molecules, peptides, and polymers have been widely applied in tissue engineering [23]. For example, some researchers use the catechol groups in polydopamine (PDA) to form covalent/non-covalent bonds with different materials to modify the hydrogels [24]. However, their expensive price and dark color limit the practical applications of PDA. Tannic acid (TA) is moderate in color and cheaper than PDA. TA is a natural polyphenol, which is composed of pyrogallol and catechol groups [24]. TA can be crosslinked with other macromolecules through hydrogen bonds, ionic bonds, coordination bonds, and hydrophobic interactions [25–27]. The introduction of TA into PVA hydrogels was reported to improve the mechanical properties of hydrogels through hydrogen bonds [25–27].

In this study, PVA was used as the raw material to fabricate the composite hydrogels with the introduction of TA and HA into the PVA hydrogel using a facile physical method. The chemical and physical properties, microstructures, mechanical properties, and biocompatibility of PVA/HA/TA hydrogels were examined. Our study demonstrated that the introduction of TA and HA into the PVA hydrogel could improve mechanical strength and provide a suitable microenvironment for cell adhesion, proliferation, and composite hydrogels. These results showed potential applications in bone tissue engineering.

2. Materials and Methods

2.1. Materials

PVA-117 (Mw = 145,000) and TA (Aladdin Reagent Co., Ltd., Shanghai, China). Hydroxyapatite (Macklin Reagent Co., Ltd., Shanghai, China). Mouse pre-osteoblasts (MC3T3-E1) cells (Fenghui Biotechnology Co., Changsha, China), Fetal bovine serum (FBS), alpha minimum essential medium (α-MEM), and penicillin/streptomycin (Gibco, Invitrogen, Carlsbad, CA, USA). CCK8 kit and Live/Dead assay kit (Beyotime Biological Co. Ltd., Shanghai, China), Deionized water (UPR-IV-10 T, China).

2.2. Preparation of PVA/HA/TA Hydrogels

PVA (10 g, 10 wt%) and TA (0, 0.5 g, 1 g and 1.5 g, 0.5 wt%, 1 wt% and 1.5 wt%) were dissolved in 100 mL deionized water with the assistance of vigorous stirring for 3 h at 90 °C. Then, HA (5 g, 5 wt%) powders were added into the mixed solution and stirred to disperse homogeneously for 1 h at 90 °C. The final mixed solution was held for 1 h at room temperature to remove air bubbles. Next, the solution was poured into molds. Finally, the solution was frozen at −20 °C for 20 h and thawed at room temperature for 4 h. After five cycles of the freeze–thawing process, PVA/HA/TA hydrogels were prepared. These processes were the same as the synthesis of PVA hydrogel. The hydrogels with different contents of TA were named PVA, PVA/HA, PVA/HA/0.5TA, PVA/HA/1.0TA, and PVA/HA/1.5TA, respectively.

2.3. Fourier Transform Infrared Spectroscopy

Fourier transform infrared spectroscopy (FTIR) spectra of HA, PVA, TA, PVA/HA, and PVA/HA/0.5TA were obtained using an FTIR Spectrometer (Bruker, Alpha, Germany) in the transmittance mode within the range of 4000–450 cm^{-1} at a resolution of 4 cm^{-1} intervals, and the spectra plots represented 32 scans. Samples were mixed with KBr pellets to prepare the specimens for FTIR.

2.4. Scanning Electron Microscope

The microstructures of HA, PVA, PVA/HA, PVA/HA/0.5TA, PVA/HA/1TA, and PVA/HA/1.5TA hydrogels were observed by a scanning electron microscope (SEM, JEOL JSM-7100F, Japan) with an accelerating voltage of 5 kV. Before observation, the samples were freeze-dried and all samples were sputtered with platinum (Pt) prior to testing. The drying process consisted of putting the composite hydrogels into the liquid nitrogen for 5 min to pre-freeze before the pre-frozen hydrogels were moved into a freeze-dryer for 2 days of lyophilization.

2.5. Porosity

The porosity of the composite hydrogels was determined by a previous solvent immersion method, with slight modifications [28]. The freeze-dried hydrogels were cut into the same shape, weighed, and submerged into alcohol for 3 min. After taking out the samples, the extra alcohol was removed and the weight of the samples was measured. The porosity was calculated as follows:

$$\text{Porosity}(\%) = \frac{W_2 - W_1}{V\rho} \times 100\%$$

where W_2 and W_1 are the weight of the sample before and after immersion, respectively, V was the volume of the samples before immersion, and ρ was the density of alcohol. All tests were repeated 3 times.

2.6. Water Content

The water content of the composite hydrogels was calculated through the weight change of hydrogels before and after removing water. Filter paper was used to remove water from the surface of composite hydrogels before they were weighed. The moisture content was calculated using the followed equation:

$$\text{Water content}(\%) = \frac{W_4 - W_3}{W_4} \times 100\%$$

W_4 and W_3 were the weight of the hydrogels before and after freeze-drying, respectively. All tests were repeated three times.

2.7. Mechanical Tests

The uniaxial tensile test was conducted on an electronic universal testing machine (Instron 5544, Boston, MA, USA) at room temperature. The samples for the compression tests were prepared in a cylinder shape (diameter of 15 mm, thickness of 10 mm) with a 2000 N sensor. The samples for tensile tests were prepared in a dumbbell shape (length of 30 mm, width of 4 mm, and thickness of 1 mm) with a 50 N sensor. The tensile strength and elongation of the hydrogels were measured at a strain rate of 50 mm·min^{-1} until breakage occurred. For the compression tests, the hydrogels were placed between the self-leveling plates and compressed at a rate of 5 mm·min^{-1} until the maximum strain reached 95%. The compressive elastic modulus was calculated at a slope range of 5–20% strain on the stress–strain curves. Toughness was obtained from the tensile stress–strain as the following equation:

$$\text{Toughness} = \int_0^{\varepsilon_{max}} \sigma d\varepsilon$$

where σ was the stress (MPa) and ε was the strain. For each sample, tests were measured at least 3 times and the average was reported.

Cyclic loading and unloading tests were also conducted on an electronic universal testing machine. The samples for the tests were prepared in the same way as the aforementioned samples. The rates of the tensile and compression tests were at 50 mm·min^{-1} and 5 mm·min^{-1}, respectively. The cyclic loading–unloading test was performed to investigate the energy dissipation mechanism of the hydrogels. The dissipated energy was calculated as the following equation:

$$U_{hys} = \int_0^{\varepsilon_x} (\sigma_{load} - \sigma_{unload}) d\varepsilon$$

here, σ_{load} and σ_{unload} were the loading and unloading stress. For each sample, tests were measured at least three times.

2.8. Cell Culture

The MC3T3-E1 cells were cultured in the alpha minimum essential medium (α-MEM, Gibco, Carlsbad, CA, USA), which contains 10% fetal bovine serum (FBS, Gibco, Carlsbad, CA, USA) and 1% penicillin-streptomycin (Gibco, Carlsbad, CA, USA) solution for the 5% CO_2 incubator at 37 °C. The culture medium was refreshed every other day.

2.9. Cell Viability

All hydrogels were sterilized by autoclaving for one day and were washed by PBS three times before cell seeding. The as-prepared hydrogels were placed in a 24-well plate and were immersed in an α-MEM medium for 1 day. MC3T3-E1 was seeded onto the disk, PVA, PVA/HA, and PVA/HA/1.5TA at a density of 2×10^4 cells per well. Then, cells were incubated for 1 or 5 days. A live/dead assay (Beyotime, Nantong, China) kit was used to determine the cytotoxicity of PVA/HA/1.5TA hydrogels. The medium was removed and the samples were washed with PBS twice, and the live/dead working solution was then added to each well. The live cells were stained with green, and dead cells were stained with red.

2.10. Cell Proliferation

A cell counting kit-8 (CCK-8, Beyotime, Nantong, China) was used to determine cell proliferation. The medium was removed and the samples were placed in a new 24-well culture plate after being washed with PBS twice. The 10% CCK-8 fresh culture medium was added to every well and incubated for 4 h in the dark at 37 °C. The optical density (OD) was measured by using a microplate reader (Biorad iMark, Hercules, CA, USA) at a wavelength of 450 nm.

2.11. Cell Morphology

The cell morphology on the hydrogels was observed by SEM. Cells were seeded on the 24 well-plates. After cultivating for 1 day, samples were rinsed with PBS three times and fixed with 2.5% glutaraldehyde. The samples were then dehydrated in sequence using a gradient series of ethanol (50%, 60%, 70%, 80%, 90% and 100%) for 10 min. The acquired samples were observed by SEM with platinum (Pt).

2.12. Statistical Analysis

All data were presented as mean ± standard error values, and analyzed using SPSS 14.0. A Student's *t*-test or one-way ANOVA was used to evaluate the difference in means between different groups; $p < 0.05$ was considered statistically significant.

3. Results and Discussion

3.1. Preparation of PVA/HA/TA Hydrogels

The preparation process of the composite hydrogels is briefly shown in Figure 1a. Figure 1b shows the chemical structures of the PVA and TA molecules. The TA molecule possessed a five-polyphenol-arm structure with 25 hydroxyl groups, which could form hydrogen bonds with hydroxyl groups on PVA and HA. As the contents of TA increased, the color of the hydrogels turned dark and opaque, as shown at Figure 1a. The hydroxyl groups between the PVA and TA molecules could form hydrogen bonds to build a robust physical crosslinked network. During the cyclic freeze–thawing processes, crystalline of PVA chains was also associated with multiple hydrogen bonds among HA, TA, and PVA molecules to densify the molecule network, as shown at Figure 1b. HA also served to mimic the natural bone structure and composition in PVA/HA/TA hydrogels.

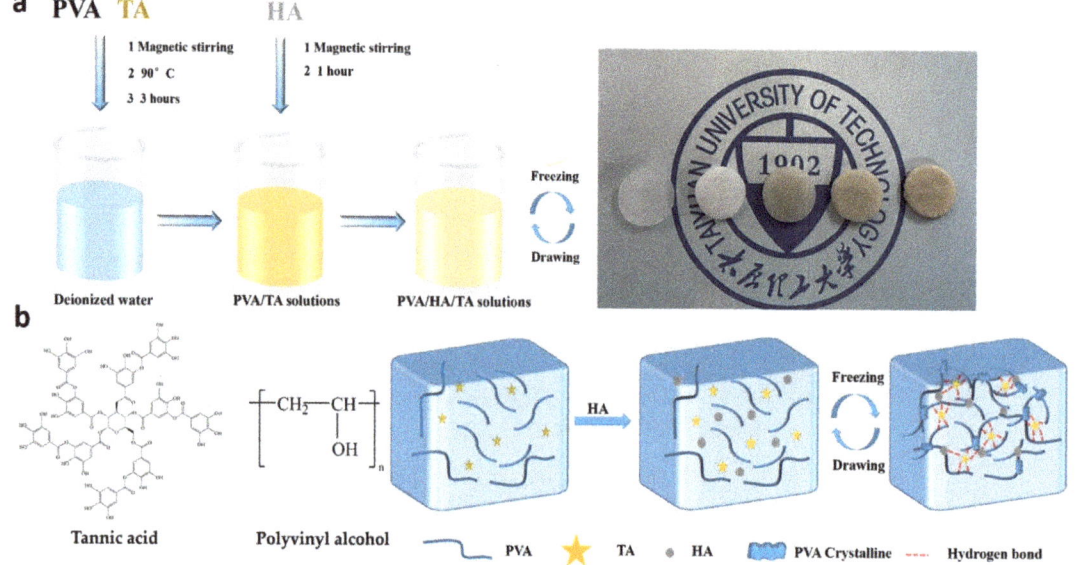

Figure 1. Schematic illustration of the preparation process (**a**) for the PVA/HA/TA hydrogel (from left to right, the hydrogels are PVA, PVA/HA, PVA/HA/0.5TA, PVA/HA/1.0TA, and PVA/HA/1.5TA) and the chemical structures (**b**) of the PVA/HA/TA hydrogel.

3.2. Characterization of PVA/HA/TA Hydrogels

3.2.1. Fourier Transform Infrared Spectroscopy Analysis of Hydrogels

To analyze the compositions and structural changes of the composite hydrogels, the pure PVA, HA, TA, PVA/HA, and PVA/HA/TA hydrogels were investigated by Fourier transform infrared spectroscopy (FTIR) analysis, as shown in Figure 2. The PVA exhibited characteristic absorption bands of -OH stretching vibrations at 3277 cm^{-1}, -CH$_2$ symmetric stretching vibrations at 2943 cm^{-1}, and C-O stretching vibration at 1086 cm^{-1}. The HA exhibited the characteristic absorption bands of PO$_4^{3-}$ at 1044 cm^{-1} and 962 cm^{-1}. After the introduction of TA into the PVA/HA hydrogels, the absorption bands of the -OH stretching vibration shifted to lower wavenumbers at 3271 cm^{-1}. According to previous studies, the formation of hydrogen bonds intra- and inter-molecule reduced the force constants of the -OH groups, which resulted in the shift of vibrational frequencies to lower wavenumbers [25–27]. The significant shifts of the absorption bands toward lower wavenumbers were due to the formation of hydrogen bonds when adding TA into the PVA/HA hydrogels. In addition, PVA/HA/TA hydrogels exhibited characteristic vibration peaks of TA at 1712 cm^{-1} (C=O), 1536 cm^{-1}, and 1443 cm^{-1} (aromatic C-C), which implied the successful synthesis of PVA/HA/TA hydrogels via a facile and reliable method.

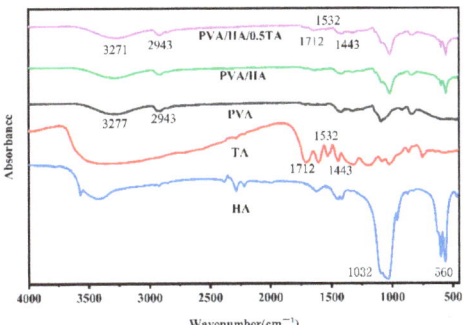

Figure 2. The FTIR spectra of PVA, HA, TA, PVA/HA, and PVA/HA/TA hydrogels.

3.2.2. Microstructure of Hydrogels

The microstructure of the hydrogels was characterized by scanning electron microscopy (SEM). As illustrated in Figure 3, the microstructures of the PVA, PVA/HA, PVA/HA/0.5TA, PVA/HA/1.0TA, and PVA/HA/1.5TA hydrogels presented interconnected porous structures. The interconnected microstructures of the composite hydrogels were beneficial for cell attachment and proliferation, which could promote new bone formation. Meanwhile, the numerous pore structures provide abundant space for drug diffusion and release [29]. After adding the HA, the microstructure of PVA hydrogels remained unchanged and the distribution of the pores was still uniform. Compared with PVA and PVA/HA hydrogels, the pore size of PVA/HA/TA hydrogels became smaller due to the dense crosslinked network. When the content of TA attained 1.5%, continuous pores still existed, but the diameter of the pores was reduced to less than 10μm. The TA molecules might serve as new crosslinked points to form a dense crosslinked network [30].

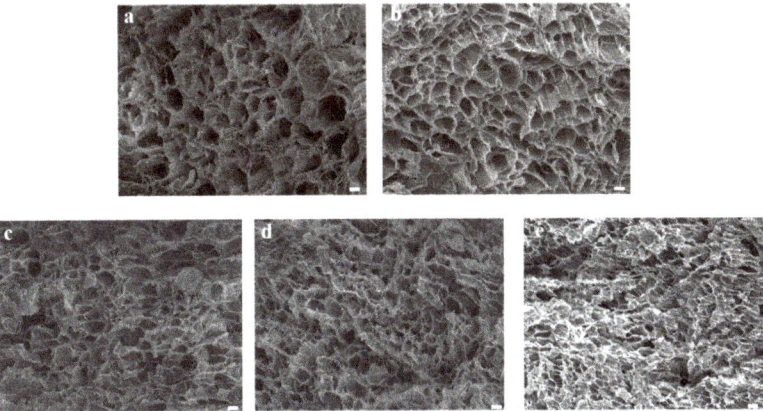

Figure 3. The SEM image of PVA (**a**), PVA/HA (**b**), PVA/HA/0.5TA (**c**), PVA/HA/1.0TA (**d**), and PVA/HA/1.5TA (**e**) hydrogels. Scale bars: 10 μm.

3.2.3. Porosity and Water Content of Hydrogels

The porosities of the PVA, PVA/HA, PVA/HA/0.5TA, PVA/HA/1.0TA, and PVA/HA/1.5TA hydrogels were 88.4 ± 0.5%, 84.1 ± 1.2%, 80.7 ± 0.6%, 78.0 ± 1.2%, and 72.6 ± 0.7%, respectively (Figure 4a). The newly formed hydrogen bonds resulted in a significant decrease in the porosity ratio of the composite hydrogels ($p < 0.05$). However, appropriate porosity was beneficial for cell proliferation and nutrient transportation [29], and the high porosity provided abundant space for drug diffusion and controlled release [30]. Previous studies have demonstrated that suitable porosity (70%) of hydrogel scaffolds is required to create new bone tissue [31].

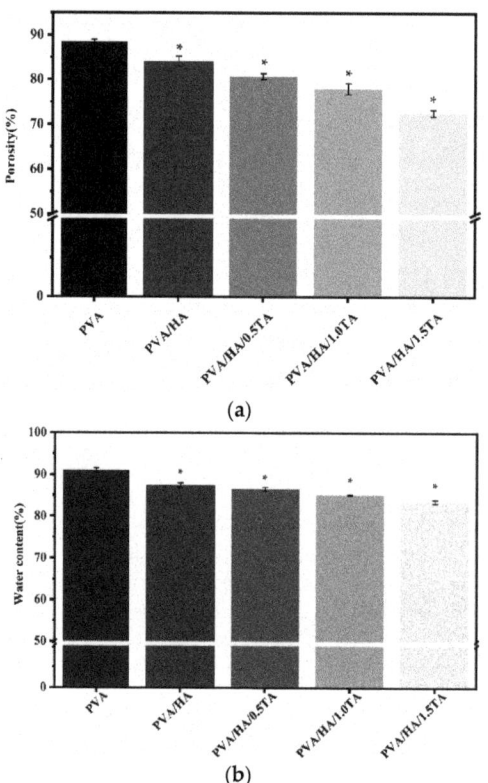

Figure 4. The porosity (a) and water content (b) of PVA, PVA/HA, PVA/HA/0.5TA, PVA/HA/1.0TA, and PVA/HA/1.5TA hydrogels. * $p < 0.05$ compared to PVA group.

The water content was an important characteristic for hydrogels as tissue regenerative materials, which directly influenced the nutrient transport [32]. As shown in Figure 4b, the water contents of the PVA, PVA/HA, PVA/HA/0.5TA, PVA/HA/1.0TA, and PVA/HA/1.5TA hydrogels were 91.1 ± 0.4%, 87.4 ± 0.5%, 86.5 ± 0.5%, 85.1 ± 0.1%, and 83.5 ± 0.4%, respectively. The water content of the composite hydrogels decreased significantly with the increase in TA content ($p < 0.05$). PVA contained many hydroxyl groups and it was easy to form hydrogen bonds with TA, which resulted in a lower water content, though the water content of the PVA/HA/1.5TA hydrogels still attained 83.51%.

3.2.4. Mechanical Properties of Hydrogels

An ideal hydrogel for tissue engineering should satisfy the mechanical requirements of different tissues. Excellent mechanical performance is a necessary requirement for hydrogels applied in bone-tissue engineering. Therefore, a series of mechanical experiments were carried out with composite hydrogels. The PVA/HA/1.5TA hydrogel was able to bear bending and knotting and could withstand compression to a large strain without the occurrence of damage (Figure 5a). The PVA/HA/1.5TA hydrogel could load with approximately 2.0 L water (Figure 5b). The tensile and compressive stress–strain curves of the composite hydrogels are presented in Figure 5c,d. The mechanical performances of composite hydrogels were improved after the introduction of TA. The compressive strengths of the PVA, PVA/HA, PVA/HA/0.5TA, PVA/HA/1.0TA, and PVA/HA/1.5TA hydrogels were 0.77 ± 0.11 MPa, 1.20 ± 0.05 MPa, 1.60 ± 0.23 MPa, 2.05 ± 0.16 MPa, and 3.69 ± 0.41 MPa, respectively. The compressive elastic modulus of the PVA/HA/1.5TA hydrogels was 0.111 ± 0.17 MPa, which was five times higher than the PVA hydrogels ($p < 0.05$). The tensile strength of the PVA hydrogel was only 0.08 ± 0.01 MPa, but the tensile strength of PVA/HA/1.5TA attained 0.42 ± 0.01 MPa after the addition of TA. The fracture toughness also significantly increased from 0.07 ± 0.01 MJm^{-3} for the PVA hydrogel to 0.86 ± 0.02 MJm^{-3} for PVA/HA/1.5TA ($p < 0.05$). In a previous study, a biomimetic porous hydrogel was fabricated using interactions between amino hydroxyapatite and gelatin/gellan gum, the compressive stress of this hydrogel at 80% strain only attained approximately 0.7 MPa [33]. When the PVA was solely combined with HA, the mechanical properties of the PVA/HA hydrogel were only slightly improved. Therefore, the introduction of TA was favorable for the improvement of mechanical strength and toughness. The improvements of the tensile and compressive performance of the composite hydrogels were due to the numerous hydrogen bonds that formed between the abundant hydroxyl groups of PVA, HA, and TA. In comparison to previous studies, a nanomaterial was introduced into the PVA hydrogels to improve their mechanical properties, and the compressive strength was only 600 kPa [34]. By adding a magnesium oxide nanoparticle and black phosphorus nanosheet into polyvinyl alcohol/chitosan hydrogel, a multifunctional hydrogel was fabricated to repair the bone defects. However, the compressive strength and elastic modulus of the hydrogels only attained 70 kPa and 3 kPa, respectively [35]. The PVA/HA/TA hydrogels in our study not only exhibited porous structures and high water contents, but also exhibited excellent mechanical properties. The strain of PVA/HA/0.5TA was $248.6 \pm 0.1\%$ and the strain of PVA/HA/1.5TA was $221.5 \pm 2.0\%$ (Figure 5g).

The energy dissipation ability of the hydrogels influenced their practical applications [36]. Hydrogen bonds were able to store and dissipate the energy effectively to improve the mechanical properties of composite hydrogels [37]. In this study, the PVA/HA hydrogel and PVA/HA/1.5TA hydrogel were chosen to confirm the dissipation ability of hydrogels at varying strains. As shown in Figure 6, the PVA/HA/1.5TA hydrogel presented pronounced hysteresis loops during the loading–unloading processes. The PVA/HA/1.5TA hydrogel showed more dissipated energy than the PVA/HA hydrogel did. The dissipated energy of the PVA/HA/1.5TA hydrogel increased two times and four times, respectively, in comparison to the PVA/HA hydrogel at 80% compressive strain and at 200% tensile strain; this may be due to richer hydrogen bonds in the PVA/HA/1.5TA hydrogel. Previous studies have demonstrated that the strong hydrogen bonds could serve as permanent crosslink points, while weak hydrogen bonds could break and reconstruct to dissipate the energy in PVA/TA hydrogels during the loading–unloading cycles [26].

For application as bone-tissue materials, hydrogels were required to bear repetitive force for a short period of time. In this study, ten cyclic loading–unloading compressive tests were performed on the PVA/HA/1.5TA hydrogel at fixed strains to evaluate its stability. As shown in Figure 7, the maximum force decreased slightly after every cycle in the force–time curve and the residual strains remained small in the stress–strain curve. The stress–strain of PVA/HA/1.5TA hydrogel nearly coincide from the third to tenth loading cycles. After ten successive cycles, the maximum stress still remained 74.1% of the initial stress. The PVA/HA/1.5TA hydrogels exhibited excellent energy dissipation abilities and good mechanical stability. The physical crosslinked network may dynamically break and reconstruct during the loading–unloading processes.

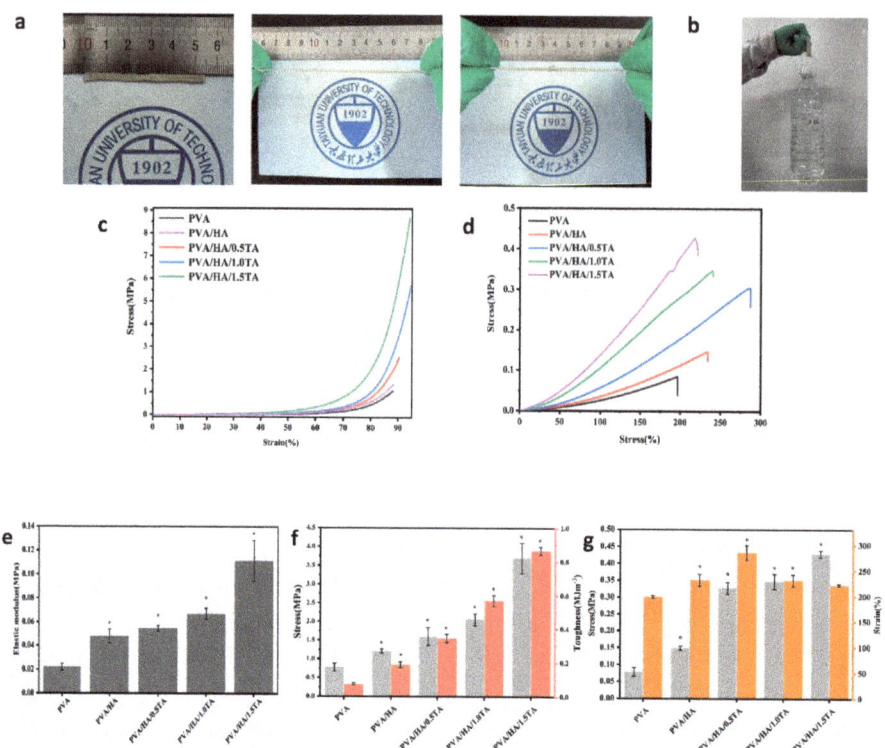

Figure 5. Photographs of PVA/HA/TA hydrogel stretching (**a**) while withstanding a weight of 2 L water. (**b**,**c**) Compressive curves of the PVA, PVA/HA, PVA/HA/0.5TA, PVA/HA/1.0TA, and PVA/HA/1.5TA hydrogels. (**d**) Tensile curves of the PVA, PVA/HA, PVA/HA/0.5TA, PVA/HA/1.0TA, and PVA/HA/1.5TA hydrogels. (**e**) Elastic modulus of PVA, PVA/HA, PVA/HA/0.5TA, PVA/HA/1.0TA, and PVA/HA/1.5TA hydrogels. (**f**) Compressive stress and toughness of PVA, PVA/HA, PVA/HA/0.5TA, PVA/HA/1.0TA, and PVA/HA/1.5TA hydrogels. (**g**) Tensile strength and breakage elongation of PVA, PVA/HA, PVA/HA/0.5TA, PVA/HA/1.0TA, and PVA/HA/1.5TA hydrogels. * $p < 0.05$ compared to the PVA group.

3.3. Biological Properties of Hydrogels

3.3.1. Cell Viability and Proliferation

The PVA/HA/1.5TA hydrogel exhibited excellent mechanical properties with high water content and porosity, so it was chosen for cell experiments. Cell viability was evaluated by using live/dead staining assays at 1 and 5 days. (Figure 8a). All hydrogels were shown to be nontoxic to cells with few dead cells during the culture period. The density of the cells on the hydrogels increased with culture time. Compared to the PVA hydrogel, more cells were observed on PVA/HA/1.5TA hydrogel, which indicates that the TA and HA enhanced cell viability. Cell proliferation on the different hydrogels was assessed by CCK-8 proliferation assays at 1, 3, and 5 days. According to Figure 8b, the CCK-8 absorption of cells on each hydrogel increased with culture time, which indicates that the PVA/HA/1.5TA hydrogel did not affect the MC3T3-E1 proliferation. The cell proliferation of the control group was higher than the PVA group at 1 day ($p < 0.05$). During the experimental groups, the PVA/HA/1.5TA was shown to have higher proliferation than the PVA group at 3 days ($p < 0.05$). After 5 days, the absorbance values of the PVA/HA/1.5TA composite hydrogel were higher than the PVA and PVA/HA hydrogels ($p < 0.05$). PVA/HA/1.5TA hydrogel could promote cell proliferation better than PVA hydrogel. The aforementioned results proved that the biocompatibility of the PVA/HA/1.5TA hydrogel is good.

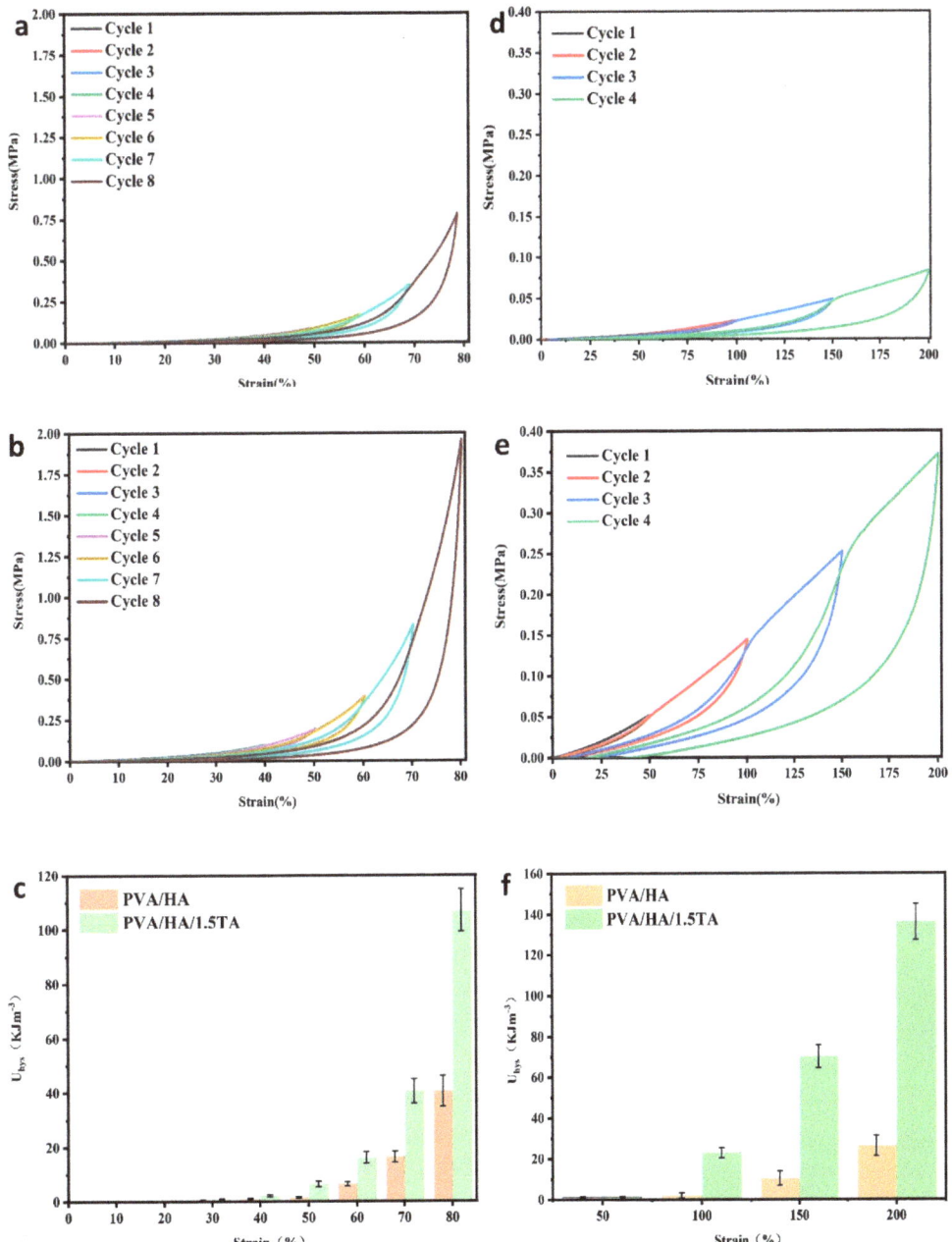

Figure 6. (a) Loading–unloading compressive test and tensile test of PVA/HA (**a,d**) and PVA/HA/1.5TA (**b,e**) under different strain values and the calculated dissipated energy (curve area) (**c,f**) of the PVA/HA and PVA/HA/1.5TA hydrogels during the loading–unloading cycles with different strains.

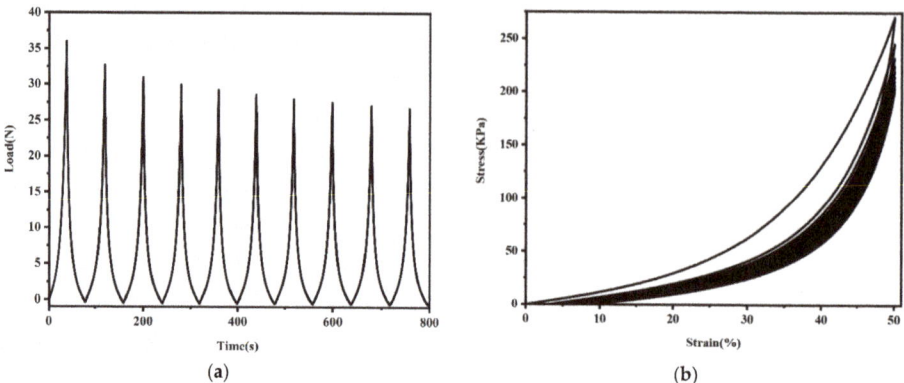

Figure 7. The load–time curve (**a**) and stress–strain curve (**b**) of ten successive compression–relaxation cycles of the PVA/HA/1.5TA hydrogel.

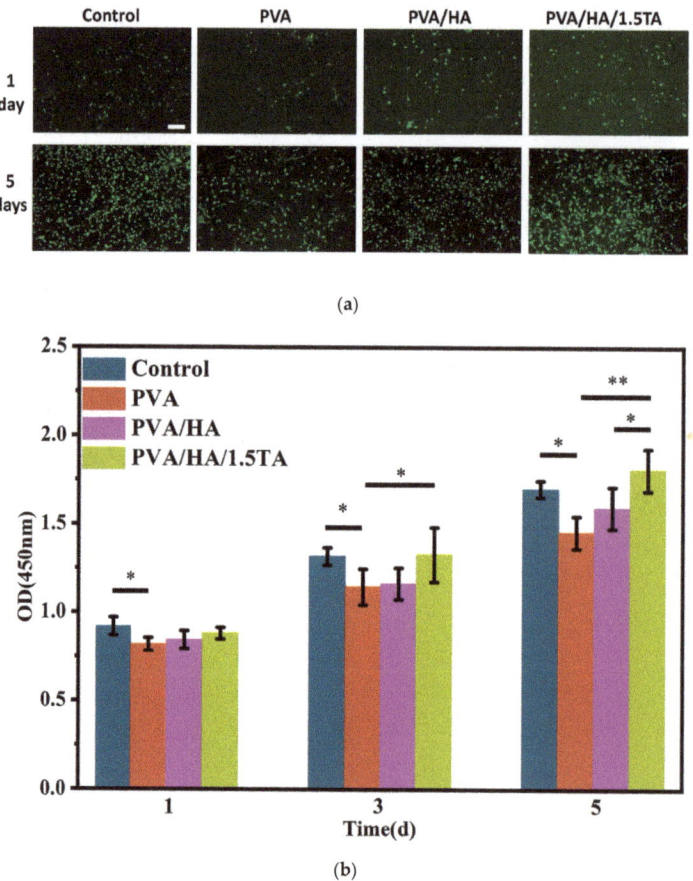

Figure 8. Cell viability and proliferation of MC3T3-E1 cells. (**a**) Live/dead assay for the viability of MC3T3-E1 cells on different hydrogels for 1 day and 5 days. (**b**) CCK8 assays for the proliferation of MC3T3-E1 cells on different hydrogels for 1, 3, and 5 day(s). (* $p < 0.05$, ** $p < 0.01$). Scale bar: 200 um.

3.3.2. Cell Morphology

SEM images presented how the different hydrogels affected the spread of MC3T3-E1 cells (Figure 9). The cells spread on the surface of the PVA/HA/1.5TA hydrogels with obvious pseudopodia for 1 day. In contrast, cells on the PVA and PVA/HA hydrogels were not fully extended, which indicates that the PVA/HA/1.5TA hydrogels could provide a better environment for MC3T3-E1 cell adhesion than PVA and PVA/HA hydrogels. The PVA exhibited a low adhesion ability for cells due to the hydrophilic materials [38,39]. On the PVA/HA hydrogel, we observed that the cells were attached and had begun to spread. After the introduction of TA, it could be seen that there was more cell adhesion. Previous studies showed that bone-cells tended to grow on rigid substrates [20]. Therefore, the PVA/HA/1.5TA hydrogels might provide a suitable microenvironment for MC3T3-E1 cell growth due to their excellent mechanical properties and proper porosity.

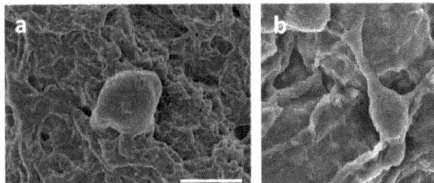
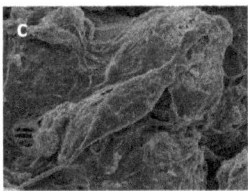

Figure 9. Cell morphologies of MC3T3-E1 on the PVA (**a**), PVA/HA (**b**), and PVA/HA/1.5TA (**c**) hydrogels. Scale bar: 10 um.

4. Conclusions

In this study, a new composite hydrogel with high water content, porous structures, excellent mechanical properties, and good biocompatibility were fabricated via a facile physical method. The high porosity and porous structure of the PVA/HA/TA hydrogels are beneficial for cell nutrient and waste transport. The microstructures and mechanical properties of the composite hydrogels could be flexibly adjusted by adjusting the content of TA. In addition, the PVA/HA/TA hydrogels were shown to be more favorable for cell proliferation, spreading, and adhesion. This study shows the potential applications of composite hydrogels in bone tissue engineering. However, the mechanical properties of these composite hydrogels were still lower than natural bone, and the biofunctions of composite hydrogels in vivo also need to be evaluated in further studies.

Author Contributions: Conceptualization, X.L. and X.W.; data curation, C.X.; formal analysis J.Z.; funding acquisition, W.C. and P.L.; investigation, C.X.; methodology X.Z.; project administration, C.X.; resources, P.L.; supervision, X.L.; writing—original draft, C.X.; writing—review and editing, C.X. All authors have read and agreed to the published version of the manuscript.

Funding: This work was supported by the National Natural Science Foundation of China, grant numbers 11632013 and 82172503.

Institutional Review Board Statement: Not applicable.

Informed Consent Statement: Not applicable.

Data Availability Statement: The data generated from the study is clearly presented and discussed in the manuscript.

Conflicts of Interest: The authors declare no conflict of interest.

References

1. Man, K.; Alcala, C.; Mekhileri, N.V.; Lim, K.S.; Jiang, L.H.; Woodfield, T.B.F.; Yang, X.B.B. GelMA Hydrogel Reinforced with 3D Printed PEGT/PBT Scaffolds for Supporting Epigenetically-Activated Human Bone Marrow Stromal Cells for Bone Repair. *J. Funct. Biomater.* **2022**, *13*, 41. [CrossRef] [PubMed]
2. Yu, L.Y.; Xia, K.; Zhou, J.; Hu, Z.A.; Yin, X.; Zhou, C.C.; Zou, S.J.; Liu, J. circ_0003204 regulates the osteogenic differentiation of human adipose-derived stem cells via miR-370-3p/HDAC4 axis. *Int. J. Oral Sci.* **2022**, *14*, 30. [CrossRef]
3. Vallet-Regi, M.; Ruiz-Hernandez, E. Bioceramics: From Bone Regeneration to Cancer Nanomedicine. *Adv. Mater.* **2011**, *23*, 5177–5218. [CrossRef]
4. Salhotra, A.; Shah, H.N.; Levi, B.; Longaker, M.T. Mechanisms of bone development and repair. *Nat. Rev. Mol. Cell Biol.* **2020**, *21*, 696–711. [CrossRef]
5. Jamari, J.; Ammarullah, M.I.; Santoso, G.; Sugiharto, S.; Supriyono, T.; Prakoso, A.T.; Basri, H.; van der Heide, E. Computational Contact Pressure Prediction of CoCrMo, SS 316L and Ti6Al4V Femoral Head against UHMWPE Acetabular Cup under Gait Cycle. *J. Funct. Biomater.* **2022**, *13*, 64. [CrossRef]
6. Wang, W.H.; Yeung, K.W.K. Bone grafts and biomaterials substitutes for bone defect repair: A review. *Bioact. Mater.* **2017**, *2*, 224–247. [CrossRef] [PubMed]
7. Zhang, M.; Matinlinna, J.P.; Tsoi, J.K.H.; Liu, W.L.; Cui, X.; Lu, W.W.; Pan, H.B. Recent developments in biomaterials for long-bone segmental defect reconstruction: A narrative overview. *J. Orthop. Transl.* **2020**, *22*, 26–33. [CrossRef] [PubMed]
8. Magalhaes, L.; Andrade, D.B.; Bezerra, R.D.S.; Morais, A.I.S.; Oliveira, F.C.; Rizzo, M.S.; Silva-Filho, E.C.; Lobo, A.O. Nanocomposite Hydrogel Produced from PEGDA and Laponite for Bone Regeneration. *J. Funct. Biomater.* **2022**, *13*, 53. [CrossRef] [PubMed]
9. Yousefi, A.M.; Hoque, M.E.; Prasad, R.; Uth, N. Current strategies in multiphasic scaffold design for osteochondral tissue engineering: A review. *J. Biomed. Mater. Res. Part A* **2015**, *103*, 2460–2481. [CrossRef] [PubMed]
10. Langer, R.; Vacanti, J.P. Tissue engineering. *Science* **1993**, *260*, 920–926. [CrossRef]
11. Dimitriou, R.; Jones, E.; McGonagle, D.; Giannoudis, P.V. Bone regeneration: Current concepts and future directions. *BMC Med.* **2011**, *9*, 66.
12. Haugen, H.J.; Lyngstadaas, S.P.; Rossi, F.; Perale, G. Bone grafts: Which is the ideal biomaterial? *J. Clin. Periodontol.* **2019**, *46*, 92–102. [CrossRef] [PubMed]
13. Xue, X.; Hu, Y.; Wang, S.C.; Chen, X.; Jiang, Y.Y.; Su, J.C. Fabrication of physical and chemical crosslinked hydrogels for bone tissue engineering. *Bioact. Mater.* **2022**, *12*, 327–339. [PubMed]
14. Rial-Hermida, M.I.; Rey-Rico, A.; Blanco-Fernandez, B.; Carballo-Pedrares, N.; Byrne, E.M.; Mano, J.F. Recent Progress on Polysaccharide-Based Hydrogels for Controlled Delivery of Therapeutic Biomolecules. *ACS Biomater. Sci. Eng.* **2021**, *7*, 4102–4127. [CrossRef]
15. Chandel, A.K.S.; Kannan, D.; Nutan, B.; Singh, S.; Jewrajka, S.K. Dually crosslinked injectable hydrogels of poly(ethylene glycol) and poly (2-dimethylamino)ethyl methacrylate -b-poly(N-isopropyl acrylamide) as a wound healing promoter. *J. Mat. Chem. B* **2017**, *5*, 4955–4965. [CrossRef] [PubMed]
16. Xu, M.J.; Qin, M.; Zhang, X.M.; Zhang, X.Y.; Li, J.X.; Hu, Y.C.; Chen, W.Y.; Huang, D. Porous PVA/SA/HA hydrogels fabricated by dual-crosslinking method for bone tissue engineering. *J. Biomater. Sci. Polym. Ed.* **2020**, *31*, 816–831. [CrossRef]
17. Kumar, A.; Han, S.S. Enhanced mechanical, biomineralization, and cellular response of nanocomposite hydrogels by bioactive glass and halloysite nanotubes for bone tissue regeneration. *Mater. Sci. Eng. C* **2021**, *128*, 112236. [CrossRef]
18. Wang, Y.Q.; Xue, Y.A.; Wang, J.H.; Zhu, Y.P.; Zhu, Y.; Zhang, X.H.; Liao, J.W.; Li, X.N.; Wu, X.G.; Qin, Y.X.; et al. A Composite Hydrogel with High Mechanical Strength, Fluorescence, and Degradable Behavior for Bone Tissue Engineering. *Polymers* **2019**, *11*, 1112. [CrossRef]
19. Kumar, A.; Negi, Y.S.; Choudhary, V.; Bhardwaj, N.K. Fabrication of poly (vinyl alcohol)/ovalbumin/cellulose nanocrystals/nanohydroxyapatite based biocomposite scaffolds. *Int. J. Polym. Mater. Polym. Biomat.* **2016**, *65*, 191–201. [CrossRef]
20. Li, W.X.; Wang, D.; Yang, W.; Song, Y. Compressive mechanical properties and microstructure of PVA-HA hydrogels for cartilage repair. *RSC Adv.* **2016**, *6*, 20166–20172. [CrossRef]
21. Parameswaran-Thankam, A.; Al-Anbaky, Q.; Al-karakooly, Z.; RanguMagar, A.B.; Chhetri, B.P.; Ali, N.; Ghosh, A. Fabrication and characterization of hydroxypropyl guar-poly (vinyl alcohol)-nano hydroxyapatite composite hydrogels for bone tissue engineering. *J. Biomater. Sci. Polym. Ed.* **2018**, *29*, 2083–2105. [CrossRef] [PubMed]
22. Chocholata, P.; Kulda, V.; Dvorakova, J.; Supova, M.; Zaloudkova, M.; Babuska, V. In Situ Hydroxyapatite Synthesis Enhances Biocompatibility of PVA/HA Hydrogels. *Int. J. Mol. Sci.* **2021**, *22*, 9335. [CrossRef] [PubMed]
23. Han, H.; Lee, K. Systematic Approach to Mimic Phenolic Natural Polymers for Biofabrication. *Polymers* **2022**, *14*, 1282. [CrossRef]
24. Liu, B.C.; Wang, Y.; Miao, Y.; Zhang, X.Y.; Fan, Z.X.; Singh, G.; Zhang, X.Y.; Xu, K.G.; Li, B.Y.; Hu, Z.Q.; et al. Hydrogen bonds autonomously powered gelatin methacrylate hydrogels with super-elasticity, self-heal and underwater self-adhesion for sutureless skin and stomach surgery and E-skin. *Biomaterials* **2018**, *171*, 83–96. [CrossRef] [PubMed]
25. Azadikhah, F.; Karimi, A.R. Injectable photosensitizing supramolecular hydrogels: A robust physically cross-linked system based on polyvinyl alcohol/chitosan/tannic acid with self-healing and antioxidant properties. *React. Funct. Polym.* **2022**, *173*, 105212. [CrossRef]

26. Chen, Y.N.; Peng, L.F.; Liu, T.Q.; Wang, Y.X.; Shi, S.J.; Wang, H.L. Poly(vinyl alcohol)-Tannic Acid Hydrogels with Excellent Mechanical Properties and Shape Memory Behaviors. *ACS Appl. Mater. Interfaces* **2016**, *8*, 27199–27206. [CrossRef]
27. Zhan, Y.; Xing, Y.; Ji, Q.; Ma, X.; Xia, Y. Strain-sensitive alginate/polyvinyl alcohol composite hydrogels with Janus hierarchy and conductivity mediated by tannic acid. *Int. J. Biol. Macromol.* **2022**, *212*, 202–210. [CrossRef] [PubMed]
28. Singh Chandel, A.K.; Ohta, S.; Taniguchi, M.; Yoshida, H.; Tanaka, D.; Omichi, K.; Shimizu, A.; Isaji, M.; Hasegawa, K.; Ito, T. Balance of antiperitoneal adhesion, hemostasis, and operability of compressed bilayer ultrapure alginate sponges. *Biomater. Adv.* **2022**, *137*, 212825. [CrossRef]
29. Tohamy, K.M.; Soliman, I.E.; Mabrouk, M.; ElShebiney, S.; Beherei, H.H.; Aboelnasr, M.A.; Das, D.B. Novel polysaccharide hybrid scaffold loaded with hydroxyapatite: Fabrication, bioactivity, and in vivo study. *Mater. Sci. Eng. C* **2018**, *93*, 1–11. [CrossRef]
30. Li, S.Y.; Deng, R.L.; Forouzanfar, T.; Wu, G.; Quan, D.P.; Zhou, M. Decellularized Periosteum-Derived Hydrogels Promote the Proliferation, Migration and Osteogenic Differentiation of Human Umbilical Cord Mesenchymal Stem Cells. *Gels* **2022**, *8*, 294. [CrossRef]
31. Tohamy, K.M.; Mabrouk, M.; Soliman, I.E.; Beherei, H.H.; Aboelnasr, M.A. Novel alginate/hydroxyethyl cellulose/hydroxyapatite composite scaffold for bone regeneration: In vitro cell viability and proliferation of human mesenchymal stem cells. *Int. J. Biol. Macromol.* **2018**, *112*, 448–460. [PubMed]
32. Jena, S.R.; Dalei, G.; Das, S.; Nayak, J.; Pradhan, M.; Samanta, L. Harnessing the potential of dialdehyde alginate-xanthan gum hydrogels as niche bioscaffolds for tissue engineering. *Int. J. Biol. Macromol.* **2022**, *207*, 493–506. [CrossRef] [PubMed]
33. Wang, L.Y.; Li, M.Y.; Li, X.M.; Liu, J.; Mao, Y.J.; Tang, K.Y. A Biomimetic Hybrid Hydrogel Based on the Interactions between Amino Hydroxyapatite and Gelatin/Gellan Gum. *Macromol. Mater. Eng.* **2020**, *305*, 2000188. [CrossRef]
34. Subramanian, S.A.; Oh, S.; Mariadoss, A.V.A.; Chae, S.; Dhandapani, S.; Parasuraman, P.S.; Song, S.Y.; Woo, C.; Dong, X.; Choi, J.Y.; et al. Tunable mechanical properties of Mo3Se3-poly vinyl alcohol-based/silk fibroin-based nanowire ensure the regeneration mechanism in tenocytes derived from human bone marrow stem cells. *Int. J. Biol. Macromol.* **2022**, *210*, 196–207. [CrossRef] [PubMed]
35. Qing, Y.A.; Wang, H.; Lou, Y.; Fang, X.; Li, S.H.; Wang, X.Y.; Gao, X.; Qin, Y.G. Chemotactic ion-releasing hydrogel for synergistic antibacterial and bone regeneration. *Mater. Today Chem.* **2022**, *24*, 100894. [CrossRef]
36. Gan, S.C.; Lin, W.N.; Zou, Y.L.; Xu, B.; Zhang, X.; Zhao, J.H.; Rong, J.H. Nano-hydroxyapatite enhanced double network hydrogels with excellent mechanical properties for potential application in cartilage repair. *Carbohydr. Polym.* **2020**, *229*, 115523. [CrossRef]
37. Jiang, P.; Lin, P.; Yang, C.; Qin, H.L.; Wang, X.L.; Zhou, F. 3D Printing of Dual-Physical Cross-linking Hydrogel with Ultrahigh Strength and Toughness. *Chem. Mater.* **2020**, *32*, 9983–9995. [CrossRef]
38. Bhowmick, S.; Koul, V. Assessment of PVA/silver nanocomposite hydrogel patch as antimicrobial dressing scaffold: Synthesis, characterization and biological evaluation. *Mater. Sci. Eng. C* **2016**, *59*, 109–119. [CrossRef]
39. Wan, W.K.; Bannerman, A.D.; Yang, L.F.; Mak, H. Poly(Vinyl Alcohol) Cryogels for Biomedical Applications. In *Polymeric Cryogels: Macroporous Gels with Remarkable Properties*; Okay, O., Ed.; Springer: Berlin, Germany, 2014; Volume 263, pp. 283–321.

Journal of
Functional
Biomaterials

MDPI

Review

Biomechanical Characteristics and Analysis Approaches of Bone and Bone Substitute Materials

Yumiao Niu [†], Tianming Du [*,†] and Youjun Liu [*]

Faculty of Environment and Life, Beijing University of Technology, Beijing 100124, China
* Correspondence: dutianming@bjut.edu.cn (T.D.); lyjlma@bjut.edu.cn (Y.L.)
† These authors contributed equally to this work.

Abstract: Bone has a special structure that is both stiff and elastic, and the composition of bone confers it with an exceptional mechanical property. However, bone substitute materials that are made of the same hydroxyapatite (HA) and collagen do not offer the same mechanical properties. It is important for bionic bone preparation to understand the structure of bone and the mineralization process and factors. In this paper, the research on the mineralization of collagen is reviewed in terms of the mechanical properties in recent years. Firstly, the structure and mechanical properties of bone are analyzed, and the differences of bone in different parts are described. Then, different scaffolds for bone repair are suggested considering bone repair sites. Mineralized collagen seems to be a better option for new composite scaffolds. Last, the paper introduces the most common method to prepare mineralized collagen and summarizes the factors influencing collagen mineralization and methods to analyze its mechanical properties. In conclusion, mineralized collagen is thought to be an ideal bone substitute material because it promotes faster development. Among the factors that promote collagen mineralization, more attention should be given to the mechanical loading factors of bone.

Keywords: bone; biomaterial; collagen; mineralization; biomechanics

Citation: Niu, Y.; Du, T.; Liu, Y. Biomechanical Characteristics and Analysis Approaches of Bone and Bone Substitute Materials. *J. Funct. Biomater.* **2023**, *14*, 212. https://doi.org/10.3390/jfb14040212

Academic Editor: Elisa Boanini

Received: 1 March 2023
Revised: 24 March 2023
Accepted: 4 April 2023
Published: 11 April 2023

Copyright: © 2023 by the authors. Licensee MDPI, Basel, Switzerland. This article is an open access article distributed under the terms and conditions of the Creative Commons Attribution (CC BY) license (https://creativecommons.org/licenses/by/4.0/).

1. Introduction

Bone is a stiff, strong, and tough organ that serves as a vital supporting organ in the body. It is composed of a hierarchically structured and naturally optimized bone matrix. Furthermore, it has an abundance of blood vessels and nerves that can constantly carry out metabolism, growth, and development, as well as reconstruct, repair, and regenerate [1]. The human body contains a total of 206 bones. Except for the six auditory ossicles that belong to the receptors, bones are classified according to their location as the skull, vertebrae, and limb bones [2]. Bone is classified into four types based on its morphological characteristics: flat bone (such as the spine), long bone (such as the humerus, femur, etc.), short bone (such as the carpal bone), and irregular bone (such as plate scapula). The long bones provide structural support for the motor system and support body movement, whereas the flat, short, and irregular bones can fill and protect the body (such as the skull) and help the body complete life activities more flexibly and efficiently (such as sesamoid bone) [2].

1.1. Composition and Structure of Natural Bone

Bone tissue is a complex structure composed of inorganic and organic matter, making it one of the most complex compounds in nature. It is primarily composed of inorganic (65%) and organic phases (30%) [3]. The perfect combination of organic and inorganic materials gives the bones good stiffness and toughness. The inorganic phase is composed of calcium phosphate, primarily HA (HA, $Ca_{10}(PO_4)_6(OH)_2$), and is stiff and strong, making it an ideal carrier for mineralized collagen [4]. Type I collagen is the main organic component of the bone matrix. Osteoblasts are the cells that produce collagen. The general process is as follows: first, the collagen polypeptide helical chain is synthesized, and then the peptide chain is modified by amino acid (proline and lysine) hydroxylation to self-assemble the triple helix to form collagen fibrils [5]. Individual collagen fibrils are approximately 1.6 nm in diameter and 300 nm in length [6]. Collagen fiber mineralization begins when the body synthesizes collagen fibrils; that is, the collagen fibers are forming in a periodic structure. While collagen fibers assemble, the inside and outside gaps are filled with HA nanocrystals, and mineralized collagen is formed [7]. These mineralized collagen fibers serve as the foundation for cortical and cancellous bone [8].

The structure of natural bone tissue has a multiscale and multilevel range from micro to macro. Weiner and Wagner first proposed a seven-level hierarchical structure of bone tissue by studying the femur [9]. From macro to micro, the sequence is as follows: the whole bone tissue, cancellous and dense bone, cylindrical Haversian canal (bone unit), the parallel or staggered arrangement of mineralized collagen fiber bundles, mineralized collagen fiber bundles, micron-scale mineralized collagen fibers, and nanoscale HA and collagen molecules [9]. Procollagen microfibers are the smallest unit of bone tissue composition, and it is assembled in an orderly hierarchical manner to form a macroscopic bone in general. Reznikov et al. further classified bone tissue into a nine-level hierarchical structure based on this [10]. The previously proposed hierarchical structure theory was expanded to include structures at the histological hierarchy (100 nm) and lamellar bone structures (10 μm). They then revealed the crystal morphology and orientation patterns by extracting slices of the lower femoral neck, using scanning transmission electron microscopy and three-dimensional (3D) reconstruction and electron diffraction, and combing them with crystallographic data to establish the corresponding model and broaden the structure of the bone to 12 layers (Figure 1) [11].

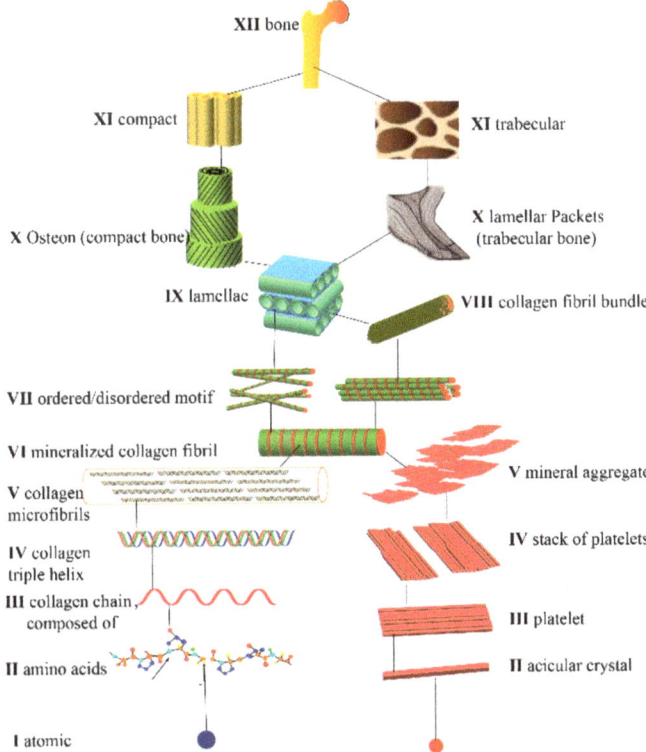

Figure 1. Twelve-level hierarchical structure of bone.

1.2. Mechanical Properties of Natural Bone

Bone possesses the exceptional properties of both collagen and HA, namely rigidity and toughness [12], making it an ideal structural material for the human body that is light but strong. Numerous studies have revealed that bone strength is affected not only by its composition but also by bone mass, geometry, and microstructure. The anisotropic behavior of bone materials and the magnitude of stress intensity vary slightly across the bone [13]. On the microscopic level, the needle-shaped inorganic salt crystals are primarily arranged longitudinally along the collagen fibrils, whose primary function is to connect and constrain the longitudinal fibers so that they are not unstable under compressive and bending loads [14]. Collagen, on the other hand, binds to inorganic salt crystals, and collagen is a common biopolymer that can provide toughness to biologically hard tissue materials [15]. The hollow beam structure of bone can greatly improve the bending strength without increasing the weight [16,17] (Figure 2A). Furthermore, the internal organization of the bone demonstrates that it is a reasonable load-bearing structure. According to the comprehensive stress analysis, the area that bears high stress also has high strength. The arrangement of femoral trabeculae, for example, is very similar to the stress distribution. To withstand greater stress, materials with higher density and strength are arranged in the internal structure of bone in the high-stress area [18].

It is an anisotropic and uneven bone composite material, and its mechanical properties are evidently different individually and by parts, as is the hardness of bones in different parts. As one of the most important properties of bone, bone hardness includes elastic deformation and plastic deformation. The nanoindentation method was used to measure the hardness of human bones, which provided valuable data for the preparation of bone repair materials (Table 1) [19–23].

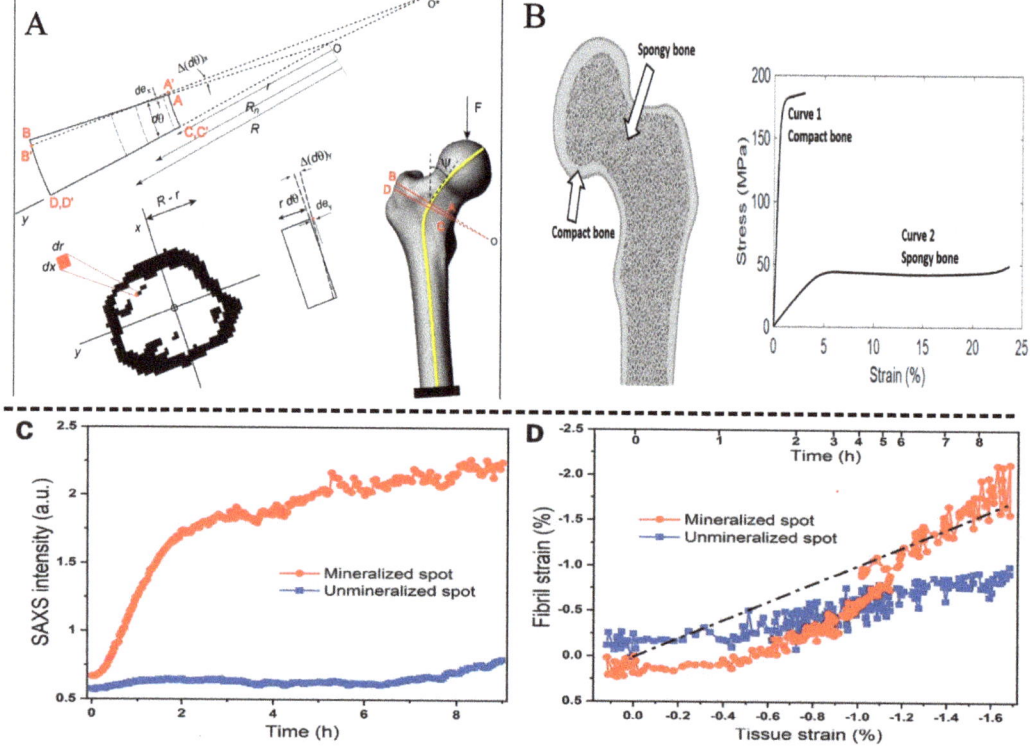

Figure 2. *Cont.*

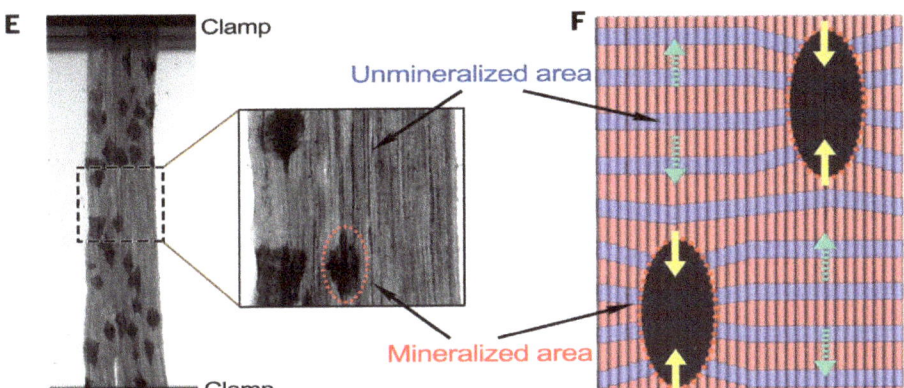

Figure 2. (**A**) Representation of the curve beam model of bone, Reprinted with permission from Ref. [17] (2023, Springer Nature). (**B**) Bone can be divided into two types: cortical bone and cancellous bone. Right panel: Typical stress-strain curves of cortical and cancellous bone, Reprinted with permission from Ref [24] (2023, JMNI). (**C**) Integrated small angle X-ray scattering (SAXS) intensity of mineralized and unmineralized regions. (**D**) Correlation between tissue and fibril strain in the mineralized and unmineralized spots. (**E**) Image of tendon after 4 h of mineralization. (**F**) Schematic of the evolution of strains during tendon mineralization. In the mineralized area, collagen fibers shrink Reprinted with permission from Ref. [25] (2023, AAAS).

The bone has several irregular marrow cavities due to its structure. Bone is classified into two types based on its size and density: cortical (dense) and cancellous (spongy) (Figure 2B). The proportion of each bone varies; however, on average, cortical and cancellous bones account for approximately 80% and 20% of the bone, respectively. These two skeletal components are identical, but macroscopic and microscopic structures differ [24]. The cortical bone serves as the shell of the entire skeleton. The gap within cortical bone is much smaller. The cortical bone has a porosity of 5–10% and an apparent density of 1.5–1.8 g/cm^3 [26]. Cancellous bone is found at the end of the bone or within it, surrounded by outer cortical bone. Cancellous bone consists of thin columns called trabeculae that are loose and dense, with porosity of 50–90% and an apparent density of 0.5–1.0 g/cm^3 [27]. Porosity is one of the most crucial factors that affect the mechanical properties of bone. As a result of significant differences in porosity, the mechanical properties of cortical and cancellous bones are significant. Cortical bone can be tolerant of higher stress (approximately 150 MPa) and lower strain (approximately 3%) before failure, and cancellous bone can be tolerant of lower stress (approximately 50 MPa) and higher strain (approximately 50%) before failure [24]. Furthermore, the distribution of cortical and cancellous bones in the body varies. Cancellous bone is commonly found in the long bone metaphysis, vertebral body, and the interior of the pelvis. By contrast, cortical bone is lamellar and commonly found on the surface of the long bone diaphysis and cancellous bone (such as the vertebral body and pelvis). Furthermore, collagen fibrils are mineralized with HA during bone formation. Mineral precipitation has been shown in studies to cause collagen fibril contraction of collagen fibrils at stress levels of several megapascals. The dimension of the stress depends on the type and quantity of mineral [25].

2. Biomechanical Properties of Biomimetic Bone Materials

In recent decades, bone tissue engineering has received great attention because of its potential to repair the bone matrix of traumatic or nontraumatic destruction. However, because of the different contents of cortical bone and cancellous bone, the biomechanical properties of different parts and shapes of bone are also quite different [24]. In this study, we briefly describe the common types of bone repair and the scaffolds for bone repair.

2.1. Common Types of Bone Repair

Bone differentiates into various shapes and structures based on its roles and functions. Correspondingly, the contents of cortical and cancellous bones vary depending on their location and shape [24]. This indicates that the biomechanical properties of these bones are quite different because of their different porosities [28,29]. There are several classification methods for bone in the academic world. In this study, we divided human bone into load-bearing bone and non-weight-bearing bones according to the location and load size of the bone. Load-bearing bones bear most of the load of the human body, including mainly the spine, limb bones, and joints, whereas non-weight-bearing bones include mainly the skull, maxillofacial, orbital bones, and ear ossicles.

The knee joint bears the maximum joint pressure in daily life, which is approximately 4–4.5 times the body weight [29]. Furthermore, when the body walks, this multiplies further [30]. The hip joint, ankle joint, wrist joint, and other load-bearing parts are subjected to a great deal of stress [31]. When the skeleton is damaged, such as a common bone disease osteoporosis, bone with this condition will become very fragile and prone to fracture, especially in weight-bearing areas such as the pelvis, hips, wrists, and spine [32]. This implies that scaffolds with a similar strength to the original bone need to be designed and that when considering biomimetic alternatives for these parts, materials with good mechanical properties must be chosen.

Compared with load-bearing bone, non-weight-bearing bone is subjected to less mechanical load and has more roles in filling and protection [29]. In recent years, because of the development of medical aesthetics and dentistry, some non-weight-bearing bone replacement materials have received extensive attention (Table 1). Its application can be roughly divided into two parts: non-weight-bearing bone orthopedic implants and bone defect filling. Non-weight-bearing bone orthopedic implants are mainly used in orbital implantation, ossicular replacement, and nasal bone injury. In addition, the main methods of bone defect filling are alveolar ridge elevation, tooth replacement, and maxillofacial reconstruction. For bone defect repair, we summarized different reference repair materials for different sites of the bone defect (Table 1). According to the different application scenarios of restorative materials, orthopedic implants can be divided into non-weight-bearing and load-bearing implants.

Table 1. Application of bone repair materials in common sites.

	Repair Site	Vickers Hardness [19–23]	Material Properties	Application Features	Example
Load bearing bone	Limb bones Joints Spine Ribs Skull	40.39–44.59 HV 38.55 HV 25.47–32.80 HV 37.35 HV 39.86 HV	Metals and Alloys	Weight-bearing, Correction, Immobilization	Nano-titanium and Ti-6Al-4V alloy [29]
Non-weight-bearing bone	Maxillofacial Orbital Dental Middle ear bone	43.54 HV 42.95 HV 278–285 HV 54.11 HV	Bioceramics	Fill, Support, Protect	Calcium Phosphate, HA [33–35]
	Cartilage	0.317 HB	Polymer	Fill, patch, join, join	Collagen and PLA nanofibers [36,37]
	Maxillofacial Dental	43.54 HV 278–285 HV	Composite material	Fill, repair	HA-Collagen [38–41]

2.2. Load Bearing Implant

Load-bearing implants are artificial knee joint and hip joint prostheses and intervertebral fusion, which are used for the limbs and trunk of the implants. These implants do not only have the effect of filling defects but also need to bear the weight and load in the process of movement of patients; therefore, they need higher mechanical properties [42].

Metal (Figure 3) has become the preferred material for load-bearing implants because of its excellent mechanical characteristic and ability to withstand physiological loads. Typically, these materials are stainless steel, cobalt-chromium (Co-Cr) alloy, titanium (Ti), and Ti alloy [43,44]. Although Co-Cr alloy has excellent corrosion resistance, its friction property is poor, which limits its application in joint prostheses. Of all these metals, Ti and its alloys are the most resistant to corrosion [45,46]. Several Ti alloys, such as Ti-6Al-4V and Ti-6Al-7Nb, have sufficient strength and corrosion resistance [47]. However, its main drawbacks are its high cost, poor wear performance, oxygen diffusion to Ti during manufacturing and heat treatment, and dissolved oxygen, which causes Ti embrittlement [48]. In addition, some problems are inevitable. The difference in Young's modulus between metallic materials and bone induces changes in the mechanical stress field, leading to adaptive remodeling and a decrease in local bone density [49,50]. Moreover, the adverse effects of metal materials implanted in the human body need to be reduced caused by fatigue fracture, corrosion, and metal corrosion [51].

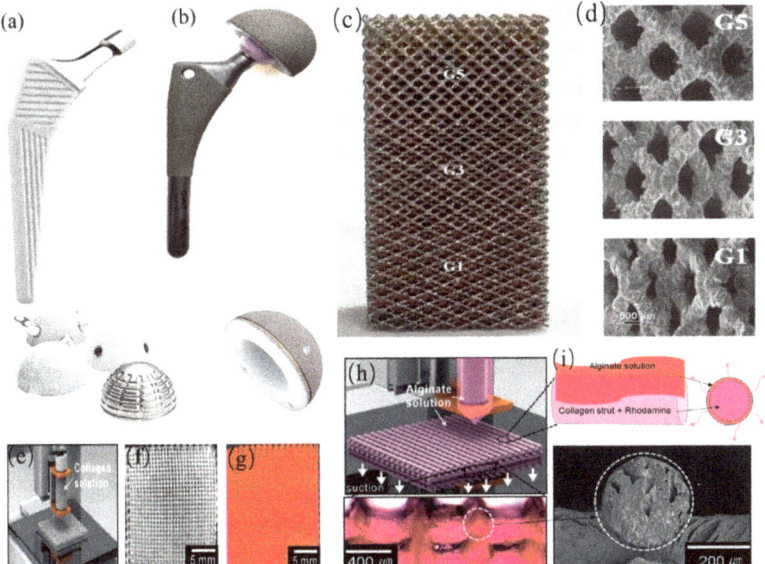

Figure 3. Common bone material scaffolds: (**a**) Femoral stem and articular concave scaffold; (**b**) HA-coated metal scaffold, Reprinted with permission from Ref. [46] (2023, Elsevier). (**c**,**d**) the 3D-printed vertebral body, Reprinted with permission from Ref. [52] (2023, Elsevier). (**e**–**i**) 3D alginate/collagen scaffold preparation process, Reprinted with permission from Ref. [53] (2023, ACS).

Researchers have created porous scaffolds to improve scaffold performance by reducing the influence of stress shielding in metals and alloys [54]. The final density, pore size, material type, and preparation parameters all significantly impact the mechanical properties of porous scaffolds [45]. For example, when the porosity increased from 55% to 75%, the strength of the spongiform bone-like biomimetic Ti scaffold decreased from 120 MPa to 35 MPa [55]. In general, with increasing porosity, the effect of stress shielding is gradually weakened, and it is more conducive to the growth of cells between tissues. However, although high porosity can provide space for bone growth, which is conducive to implant fixation, with increasing porosity, the strength and extensibility of porous structure will decrease; therefore, the porosity also needs to be controlled within a certain range [56]. It is necessary to control the porosity and pore size of the scaffold accurately. Among the several methods, 3D printing has attracted much attention because of its excellent properties, which designs scaffolds with not only different shapes and sizes but also different pore percentages and mechanical strengths [57].

2.3. Non-Weight-Bearing Bone Implants

Non-weight-bearing bone implants are mainly internal fixation implants such as bone plates and bone screws, and filling implants are applied to repair bone defects in non-weight-bearing areas [58]. These implants are used for structural fixation and filling but generally are not used for load bearing. In addition to metal materials for some internal fixation implants, degradable polymer and ceramic materials with similar inorganic composition to bone are preferred materials for non-weight-bearing bone implants [38,39]. Polymers, bioceramics, and composite materials can be classified based on their chemical structures.

Polymers can control the shape, structure, and chemical composition of materials and can be used to fabricate bioscaffolds as artificial biomaterials. Polymeric biomaterials are typically implanted into the human body in various forms, such as tissue scaffolds, gels, particles, or films and degraded into non-toxic products that are absorbed or excreted by the human body through enzymatic reactions [40]. Synthetic polymers, such as poly (a-hydroxy acid), are degraded in vivo to non-toxic lactic acid and glycolic acid, which can be eliminated from the human body by through normal excretion [41,59]. Although synthetic polymer materials are relatively easy to process into a pore scaffold, they may also produce acidic degradation products and change the pH around the tissue. This change in pH affects cell behavior and survival and causes tissue inflammation [60]. Natural biological materials do not have the problems that polymers do, and they have excellent biological activity, biocompatibility, and controllable degradation, all of which are crucial components of tissue engineering materials [61]. Naturally derived biomaterials are typically divided into two categories: protein-based biomaterials, such as collagen and sericin, and glycosyl biomaterials, such as hyaluronic acid and cellulose [62]. However, the degradation rate of naturally derived biomaterials in vivo is not only difficult to control and anticipate, but the mechanical properties are also weak, and the uniformity of composition cannot be regulated [53].

Ceramic material is a type of biological material with a crystal or partially crystal structure, which is stiff but fragile [63]. Moreover, its mechanical properties are associated with chemical elements. Generally, the chemical elements used to make bioceramics are only a small part of the periodic table, indicating that bioceramics can only be made of alumina, zirconia, carbon, and silicon- and calcium-containing compounds [64]. Therefore, bioceramics have excellent biological functions and biocompatibility. For example, after implantation, the formation of apatite on the surface of bioceramics makes the combination of internal tissues and implants stronger [65]. However, its mechanical properties are influenced by its elements and structure, making it stiff and fragile correspondingly [64]. Researchers frequently consider incorporating biological active agents as composite materials.

There are many kinds of chemical elements in the human body. They interact with each other and maintain life homeostasis together. A variety of metal elements, such as magnesium (Mg), zinc (Zn), manganese (Mn), strontium (Sr), copper (Cu), cobalt (Co), ferrum (Fe), aluminium (Al), nickel (Ni), and chromium (Cr), have been found to induce proliferation during tissue regeneration [66]. They also play an important role in promoting bone biomineralization [67]. Therefore, the introduction of metal elements can not only improve the mechanical properties of HA bioceramics but also promote the proliferation, differentiation, and migration of active cells in the bone to regulate bone mineralization. The incorporation of magnesium into bioceramics can promote bone proliferation [68]. The introduction of Hap into Fe can increase biocompatibility and blood compatibility [69]. In the biological experiment of β-SiAlON ceramics, the cells cultured on the surface of β-SiAlON were observed. The increase in AlO_2 concentration had no effect on cell adhesion and spreading, but it may slightly inhibit cell proliferation at high concentrations. Low AlO_2 concentration helps to promote osteogenic differentiation and mineralized nodule formation [70]. Zn-containing bioceramic scaffolds in craniofacial bone repair experiments show that soluble Zn^{2+} can promote osteogenic differentiation of adipose stem cells [71]. Boron silicate nanoparticles merged with Cu and Mn can be used for fusion bone repair and anti-tumor therapy. It can enhance bone regeneration through the osteogenesis of Cu^{2+} and Mn^{3+} and induce tumor cell apoptosis through Cu^{2+} and Mn^{3+} [72]. Sr is a trace element in the human body that is beneficial to bone formation [73]. Sr has a strong affinity for bone. Due to the physical and chemical similarity with Ca, the interaction of Sr in bone tissue is similar to that of Ca [74]. Sr can inhibit the osteoclast differentiation of pre-osteoclast cells and promote the expression of outcome cells and protein secretion. The subsequent rabbit bone scaffold implantation experiment also proved this. The Sr-doped scaffold provides a suitable environment for cell proliferation and differentiation during degradation [75,76]. In addition, the doping of some rare metals, such as praseodymium (Pr), erbium (Er), and yttrium (Y), can also promote cell proliferation and differentiation [77–79].

Composites are typically made up of polymers and mineral salts, with the mineral phase primarily consisting of phosphate, silicate, and other minerals [80]. Composite materials combine polymer toughness and mineral hardness of minerals and become the first choice for future biological materials. For example, Bogdan Conrad and Fan Yang prepared scaffolds from HA-mineralized gelatin, whereas Chen et al. used HA-mineralized silk fibroin (SF)/cellulose [81,82].

Bioglass incorporation into collagen scaffold as a relatively broad composite material has attracted much attention due to its excellent degradability and stability. Bioglass is an excellent biomaterial which is often used in bone defects. Before this, people often combined polylactide-co-glycolide with organic glass to improve the mechanical properties of composite scaffolds [83,84]. Collagen has become the preferred matrix for bioglass doping among many biodegradable materials due to its excellent biocompatibility and biological temperature [85]. In contrast, the incorporation of inorganic bioactive glass has been shown to increase biological activity and mineralization, control scaffold degradation, and improve the mechanical properties of collagen scaffolds [86,87]. Existing studies show that in vivo, mineralized scaffolds doped with bioglass can promote the mineralization of collagen. Nijsure et al. successfully prepared bioglass-incorporated electrochemically aligned collagen [88]. The incorporation of bioglass-incorporated electrochemically aligned collagen will enhance the mechanical properties and cell-mediated mineralization [88]. The dissolution product of the bioglass collagen composite scaffold stimulates osteoblast differentiation and extracellular matrix mineralization in vitro without any osteogenic supplement [89].

As an emerging option for composite materials, biomimetic mineralized collagen is a highly mineralized composite material composed of collagen and HA. It has attracted much attention because it has the same composition as the bone matrix. In addition, their mechanical properties and microstructure are similar to the extracellular matrix of native tissues [90]. Because of its exceptional biological activity, osseointegration, and biological induction ability, it is widely favored by researchers. Wang Xiumei's team developed high-strength bone materials mimicking compact bone and completed the skull defect experiment of adult ovis aries. The results showed excellent osseointegration and osteogenic induction abilities [91,92]. In recent years, mineralized collagen-guided bone tissue regeneration has gradually been used in oral clinical treatment, primarily for the treatment of bone and soft tissue defects caused by periodontal disease or cysts [93–95]. Moreover, while mineralized collagen made good progress in repairing other bone defects, it has the following limitations: the implant material lacks structural strength and requires external fixation increasing patients' pain. Notably, biomimetic mineralized collagen materials, like other traditional bone repair materials, have not been widely used in clinical practice due to insufficient mechanical strength [96]. How to improve mechanical strength is also a research focus.

3. Study to Improve the Mechanical Strength of Mineralized Collagen

Among the various methods for improving the mechanical properties of collagen, in vitro biomimetic mineralization is an effective method for achieving the most accurate simulation of natural bone tissue structure. According to the principle of biomimetics and the metabolism law of human tissue, HA/collagen composite material construct not only has the macroscopic structure of natural bone but also simulates its microscopic characteristics, revealing the benefits of HA and collagen materials complementing each other and significantly improving the compression modulus of the composite scaffold added with HA [97]. It has excellent biocompatibility, bone conductivity, and osteoinductive ability [98]. Numerous studies have been conducted to promote the collagen mineralization process more effectively and enhance the degree of collagen mineralization. At present, the factors affecting the in vitro biomimetic mineralization of collagen include mechanical, biological, chemical, and collagen structure.

3.1. Force to Promote the Mineralization of Collagen

The process of bone healing is affected by several factors, such as biological, chemical, and mechanical factors. Several studies have demonstrated that force acts directly on bone matrix and then on cells. Bone remodeling requires the participation of osteoblasts and osteoclasts. Osteoblasts and osteoclasts respond to force stimulation and show different proliferation abilities and activities. During bone remodeling, mechanical force stimulates the fracture site, accelerating bone formation and inhibiting bone resorption [99]. In contrast, when the body is not stimulated by force for a long time (such as bed rest, joint fixation after surgery, or exposure to a microgravity environment), the body will lose more skeleton. In severe cases, osteoporosis will occur [100–102].

Various methods for simulating the force environment of osteoclasts and osteoblasts in the bone matrix, including fluid shear stress, cyclic stretching, continuous compression force, and mechanical stress from liquid perfusion or compressed air, have been developed to investigate cellular responses to mechanical stimuli [103–110]. Physiological mechanical loading enhances the antiapoptotic effect and promotes osteoblast proliferation and differentiation, resulting in extracellular matrix formation and bone remodeling [106–108].

In addition, stress can induce the mineralization process of collagen. On the one hand, stress can affect the self-assembly of collagen, and on the other hand, stress can also induce collagen mineralization. The experiments were performed within a microfluidic channel, and the size of the channel affects the angular size of the collagen alignment with the axis of the microfluidic channel. Collagen fiber alignment decreases with increasing channel size [109]. Du et al. used a cone-and-plate viscometer to provide fluid shear stress [110]. The results showed that the formation of amorphous calcium phosphate (ACP) was associated with its rate. Fluid shear stress can significantly affect the ACP by somatic transformation and the crystal structure of HA transformed from precursors. Subsequently, periodic shear stress was used again to induce collagen mineralization. The results showed that periodic fluid shear stress could control the size of ACP, such as polyacrylic acid (PAA), avoid aggregation, and contribute to the formation of intrafibrillar mineralization (Figure 4A,B) [111]. Cyclic tensile experiments on demineralized bone demonstrated that cyclic strain increased the migration of mineralized fluid with mineralized precursors to the matrix, resulting in the formation of more calcium phosphate nanocrystals and an increase in the elastic modulus of the collagen matrix (Figure 4C,D) [112]. However, when a constant tensile force is added to the demineralized bone, the mineralization of the demineralized bone is also inhibited. The experiments of Clinical Dentistry showed that collagen mineralization could be more effectively induced by the flowing mineralization solution under focused high-intensity ultrasound. In addition, the amount of mineral formation is proportional to the exposure time [113].

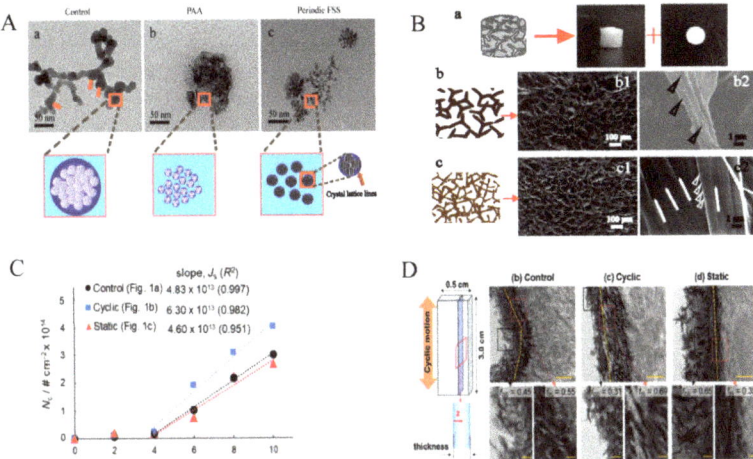

Figure 4. (**A**) Both PAA and periodic fluid shear stress (FSS) can control the generation of ACP Reprinted with permission from Ref. [111] (2023, RSC). (**B**) SEM images of different mineralized collagens. Figure b shows the PAA-induced normal IM collagen and figure c shows the periodic FSS and TPP-co-induced oriented HIM collagen. The SEM image (b1–b2) shows that the FSS group (Bc) is more neatly arranged than the PAA group (Bb), Reprinted with permission from Ref. [114] (2023, MDPI). (**C**,**D**) The loading cycle relative to the control variable group and the static group. The nucleation rate of mineralized collagen was higher in the stress group, Reprinted with permission from Ref. [112] (2023, RSC).

3.2. Collagen Fiber Arrangement Affects Mineralization

The bone structure is constantly regulated by the mechanical environment during the reorganization, thereby maintaining the mechanical strength. Bone is the basic structural unit of cortical bone, which is composed of a concentric lamellar structure around the central Haversian tube. Although the direction of the bone process is mainly parallel to the long axis of the bone, the direction of collagen fibers in a single layer may vary greatly, resulting in many models [10,115] proposed over the years. Over the years, researchers have also studied the effect of collagen fiber orientation on the mechanical properties of bone lamellar structure. The results also confirmed that collagen fiber orientation allows the bone to withstand greater stress without breaking [116–118]. Similarly, in the in vitro biomimetic mineralization experiment of collagen, it was also found that the arrangement of collagen fibers had a great influence on the formation and deposition of HA. As we mentioned earlier, type I collagen has a special amino acid sequence and triple helix structure. Cross-linking generates new chemical bonds through amino acids on adjacent peptide chains, which can improve the stability of collagen conception. In the body, cross-linking is an enzymatic or non-enzymatically mediated enzymatic process mediated by lysyl oxidase to produce trivalent collagen cross-linked pyridinoline (PYD) and deoxypyridino-line (DPD) [119]. In the body, in in vitro experiments, researchers often change the structure of mineralized collagen through physical or chemical cross-linking and then change the arrangement of collagen fibers [120], just as the existence of cross-linking makes the collagen structure different so that the mineralization of collagen is also different. Collagen with a different cross-linking degree was prepared by gamma-ray irradiation. With the increase of cross-linking degree, the pore structure of collagen became denser. The compact structure of collagen enables HA to adhere to collagen fibers [121]. However, using glutaraldehyde as a cross-linking agent to prepare mineralized collagen scaffolds, it was found that with the increase of cross-linking agent, the arrangement of HAP crystals in collagen fibers decreased, and improper use of cross-linking agent would inhibit the mineralization of collagen [122].

3.3. Other Methods to Promote the Mineralization of Collagen

In addition to in vitro biomineralization, other factors can promote collagen mineralization [111,123]. The use of a polymer-induced liquid precursor to mineralize collagen fibers can result in nanostructures that are extremely similar to the bone tissue matrix, according to nonclassical crystallization theory. Calcium ions gradually aggregate with phosphate in this process to form ACP, which is distributed inside and outside collagen fibers and converted to HA [124–126]. Polymers like polyvinyl phosphoric acid and PAA help in the formation of nanosized ACP [127]. As a biological small molecule, poly-aspartic acid can also control the formation of ACP, thereby achieving mineralization in collagen fibers [112]. In recent years, studies have also focused on the phosphorylation of collagen. Compared with the regulation of only orthophosphate, which can only form spherical mineralized crystals, needle-shaped mineralized crystals will be formed in the solution with the addition of alkaline phosphatase (ALP), thereby forming petal-shaped crystals on collagen (Figure 5A–F) [128]. In addition, subsequent experiments proved that compared with fluid shear stress alone, the pore density, hydrophilicity, enzymatic stability, and thermal stability of mineralized collagen were significantly improved after the addition of sodium tripolyphosphate [119]. Several experimental factors influence the size and distribution of HA nanocrystals in the pore region of collagen fibrils and among the fibrils by affecting the formation and transportation of ACP [129,130]. According to previous studies, collagen as the template for biomineralization, its fiber diameter, orientation, degree of cross-linking, and degree of phosphorylation can all affect mineralization. For example, the diameter of collagen affects the mineralization degree inside and outside of the fiber. A thicker collagen fiber is not conducive for HA to enter the fiber and mineralization inside the fiber [131]. With increasing cross-linking, the collagen will become denser, which is conducive to the deposition of HA and the production of highly mineralized collagen (Figure 5G–I) [121].

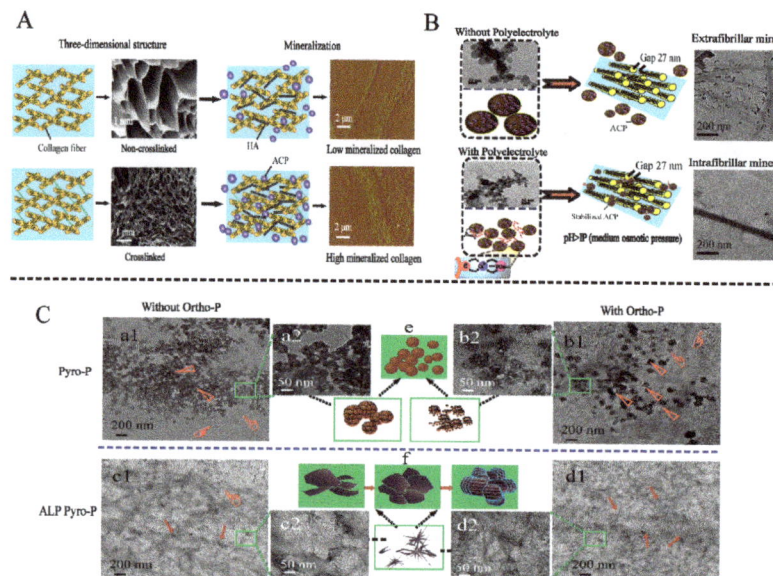

Figure 5. (**A**) Effect of different cross-linking degrees on collagen mineralization. With the increase of collagen cross-linking, more and more HA is attached to the surface of collagen, and collagen fibers are covered by HA (arrows), Reprinted with permission from Ref. [121] (2023, Elsevier). (**B**) The effect of polyelectrolytes on the intrafibrillar mineralization [128]. (**C**) The effect of ALP promotion on mineral crystal shape and crystallinity, c1–d2: the particles were rod-like (ALP), and a1–b2: granular particles were formed, Reprinted with permission from Ref. [128] (2023, Elsevier).

4. Method for Detecting Mechanical Properties of Mineralized Collagen

Although the mineralized collagen scaffolds have the same composition as bone, achieving similar structure and mechanical properties to that of natural bone has always been the focus and a challenge for researchers. In the preparation of mineralized collagen, the detection methods and standards are particularly crucial. Currently, researchers test the mechanical properties of collagen fibers, primarily from the macroscopic and microscopic perspectives, to analyze the material's mechanical properties. This study concentrated on microscopic testing methods because macroscopic mechanical testing, or traditional mechanical testing, is relatively well-developed (Table 2).

4.1. Macroscopic Mechanics Analysis Methods

There are numerous methods to analyze the mechanical properties of materials. The traditional mechanical property testing techniques include stretching, bending, and torsion. Various testing methods improve material performance parameter acquisition methods and provide a broad avenue for material performance testing [132]. These traditional material testing techniques were performed earlier, and we summarized the research status of several typical testing methods in this study. The tensile test of materials can be divided into ex situ and in situ stretching based on real-time observation. Ex situ stretching is traditional stretching (Figure 6A,B). In addition, the branches are more complex and have several directions for development [133]. The studies conducted by people using the extensometer can measure not only the plastic deformation, elastic recovery, and tensile strength of the material but also the test temperature, load frequency, holding load, amplitude, and other parameters, and the total deformation of the specimen [134].

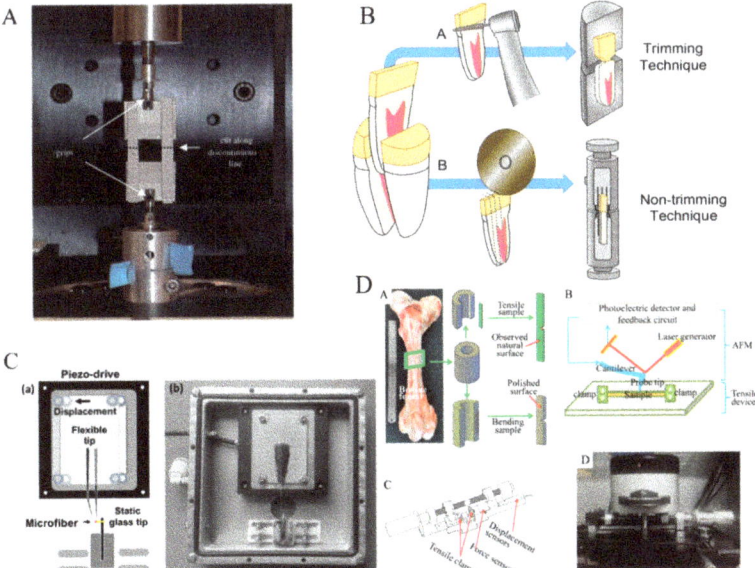

Figure 6. (**A**) Sample from the nano-tensile testing machine and ex situ observation, Reprinted with permission from Ref. [133] (2023, Elsevier). (**B**) In situ observation of the sample, Reprinted with permission from Ref. [132] (2023, Elsevier). (**C**) In situ stretching device for collagen fibers, Reprinted with permission from Ref. [135] (2023, Elsevier). (**D**) In situ micro-mechanical tensile mechanical test experiment by cow bone with AFM, Reprinted with permission from Ref. [136] (2023, Elsevier).

4.2. Microscopic Mechanics Analysis Methods

The continuous maturation of surface topography observation and internal structure flaw detection technology benefits the development of in situ stretching. Surface topography can be observed in situ using charged-coupled device cameras, optical microscopes, atomic force microscopy (AFM), scanning electron microscopy (SEM), and other instruments. The crystal structure of the material was studied using an X-ray diffractometer and a Raman spectrometer (Figure 6C,D) [132,135,136]. Micro-tensile, nanoindentation, and scratch tests are the main methods for detecting the microscopic morphology of materials in materials science. They can accurately measure the hardness and elastic modulus of materials. At the same time, with microscopic imaging, morphology changes can be observed and widely used. This part is further explained below.

4.2.1. Micro Stretching

Micro stretching is an in situ stretching method based on the rapid development of optical microscopies, such as AFM, SEM, and other microscopic observation methods. Typically, micro-stretching can reflect the mechanical changes of the material on a micro- and nano-scale level. Among them, the combination of SEM and the tensile mechanical testing device is an earlier in situ method used in material studies [137]. The imaging speed of SEM is fast, and the micro- and nanoscale topography can be observed clearly. It can provide detailed information on the behavior of materials during mechanical testing that static observation cannot. Some studies performed in situ SEM mechanical tests on transverse and longitudinal bone specimens to further verify the anisotropy of bone mechanical properties and proposed that the mechanical properties of the longitudinal and transverse orientations of the bone were different, which could be attributed to differences in the direction of microcracks [138]. Furthermore, a study reported a novel device with a confocal Raman microscope that enables uniaxial stretching of microfibers ranging in diameter from 10 to 100 microns in length [135].

4.2.2. Nanoindentation and Scratch Experiments

Nanoindentation, known as depth-sensitive indentation technology, is a new type of mechanical property testing method developed on the basis of the traditional Brinell hardness test and Vickers hardness test [125]. Initially, nanoindentation was used to research the mechanical properties of nanomaterials, and it was often used to detect the mechanical properties of thin films and other nanostructured materials [126]. The researchers developed a bone nanoindentation protocol to measure elastic properties consistent with macroscopic level measurement behavior. It is recommended to test large indentations with a diameter of 10 μm and depth of 500–1000 nm [139,140], leading to measured elastic moduli on the order of 10–20 GPa. Anisotropic analysis of the indentation results in two orthogonal planes showed that the moduli were consistent with the micro-tensile specimens [139]. When the same loading protocol was used for trabecular tissue, the measured elastic moduli were similar to cortical bone tissue. Low depths indentation has been used to measure the properties of individual flakes having alternating high and low moduli [141]. The bone indentation protocol is typically held under constant load for a period of time; during this time, the bone exhibits creep and stress relaxation behavior [142,143].

Similarly, nanoindentation can be used to test the mechanical properties of mineralized collagen [144–146]. Stanishevsky et al. prepared HA nanoparticle-collagen composites using solution deposition and electrostatic or spinning collector electrospinning and measured Young's modulus using nanoindentation technique from 0.2 to 20 GPa and hardness from 25 to 500 MPa, depending on the composite preparation process, composition, and microstructure. When the HA content is 45–60%, the nanoindentation of Young's modulus and hardness of the HA/collagen composite are the largest [145]. As mineralized collagens have a unique feature of composite materials in the indentation load-displacement curve and creep, and the appearance of this feature is associated with viscoelasticity, it is necessary to measure and change it to improve the mechanical properties of mineralized collagen [146].

Furthermore, the scratch test method is a high-resolution test and detection method which can observe the surface structure and morphology of materials at the microscopic scale. The test results can reveal critical surface information and mechanical parameters such as the material's friction coefficient, hardness, and surface roughness, and combine the groove and residual morphology of the specimen surface to evaluate the friction and wear resistance of the specimen surface and the bonding ability of the film, revealing the intrinsic relationship between the material's deep structure and its surface properties [147]. Furthermore, the scratch was used for the mechanical testing of collagen. Zhao et al. successfully calculated the critical load value of the mineralized collagen deposition coating in the scratch test [148]. The experimental results demonstrated that the critical load is proportional to the collagen concentration in the electrolyte. At high collagen concentration (500 mg/L), the critical loading of the coating was approximately twice as high as that obtained without collagen addition [148].

4.2.3. AFM

In measuring the elastic modulus of collagen fibers, AFM has more sensitive detection and is less prone to make an error. AFM is an extremely versatile nanotechnology belonging to the scanning probe microscope family, and it can be used as a surface imaging tool and force sensor and actuator technology. AFM is a type of true nanoscale method where forces and deformations are on the nanometer scale. Typically, AFM is used in conjunction with other mechanical loading tools. Colin A performed the dynamic mechanical analysis of individual type I collagen fibers at low frequencies (0.1–2 Hz) using AFM (Figure 7) [133]. Different regions of procollagen have different elastic moduli. The elastic modulus of the overlap area with the highest density (approximately 5 GPa) was 160% of that of the gap area [149]. Later, AFM was used to measure and determine the flexural and shear modulus of electrospun collagen fibers. A triangular silicon nitride cantilever beam was used for vertical bending experiments. Flexural modulus dropped from 7.5716 GPa to 1.4702 GPa up to 250 nm and remained constant at 1.4 GPa for larger diameter fibers [150]. Qian et al. used AFM to record and image the nanomechanical behavior of the medullary surface of the bovine femur in situ [136].

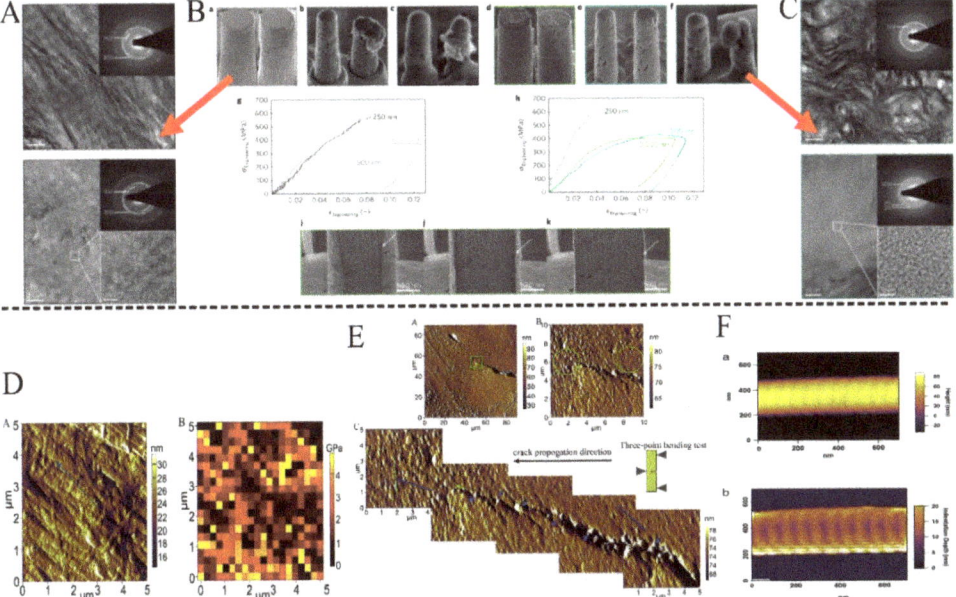

Figure 7. (**A**–**C**) The failure mechanism of the micron bone column under stress was observed under SEM and TEM, Reprinted with permission from Ref. [12] (2023, Springer Nature). (**B**): the failure

mechanism of the ordered and disordered micron bone pillars under stress observed under SEM, (A), and (C): TEM analysis of the ordered and disordered nanostructures. (D) Surface nanoindentation test of compact bone under AFM and figure of stress field distribution. (a) is the AFM image of the natural surface, and the elastic modulus shown by nanoindentation in (b) [136]. (E) The destruction of the bone surface under stress was observed under AFM, Reprinted with permission from Ref. [136] (2023, Elsevier). (F) Collagen image under AFM: an image of collagen in tapping mode; b, indentation data of collagen, Reprinted with permission from Ref. [149] (2023, Elsevier).

4.3. Simulation Analysis Method

In response to the above experimental methods and a large number of experimental data, researchers have also established models to predict the data results. Several models to predict the mechanical properties of mineralized collagen have also been proposed. Computational models involving mostly a finite element method (FEM) and molecular dynamics (MD) atoms are briefly described in this study.

The FEM can consider the geometric details of mineralized collagen in both two-dimensional (2D) and three-dimensional (3D) space. The model can include the shape, orientation, and arrangement of various stages (Figure 8A–C). At the microscopic level, Jager proposed a geometric model for the staggered arrangement of collagen fibers and HA platelets and investigated the increase in elastic modulus and fracture stress with an increasing mineral content in the fiber [151]. Wang proposed a 2D shear lag model to explore stress concentration fields around an initial crack in a mineral-collagen composite [152]. Subsequently, some researchers began to use the cohesive FEM to analyze mineralized collagen [119]. Ana developed a 3D finite element model of staggered mineral distribution within mineralized collagen fibers to characterize the elastic behavior of lamellar bone at the submicron scale [153]. Meanwhile, a multiscale finite element framework was proposed to investigate the effect of intra–and extra-fibrillar mineralization on the elastic properties of bone tissue by considering the structural hierarchy at the nano- and micrometer scales (Figure 8D,E) [154]. The material properties and fiber network of the mineralized collagen fibers have an effect on the mechanical properties of the sub-microscale bone, according to a 3D real model of the mineralized collagen network [118]. In addition, the mechanical response of mineralized collagen at the sub-microscale is associated with the loading direction based on the different arrangements of collagen fibers.

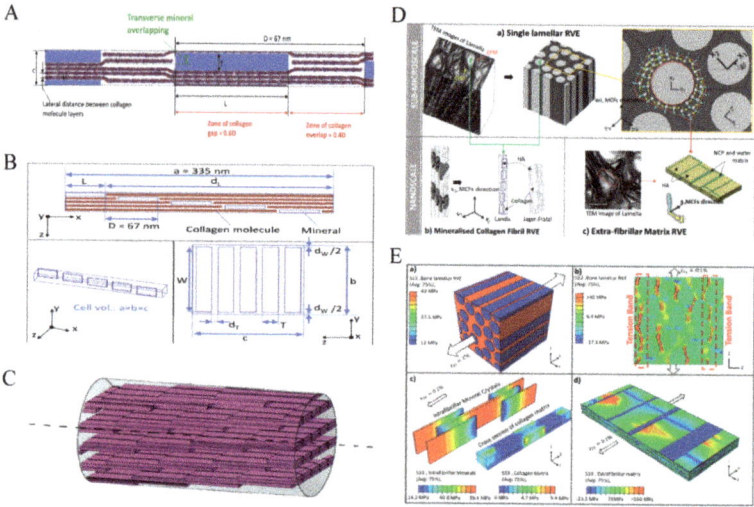

Figure 8. Finite element simulation. (A–C) Collagen and minerals are interlaced within the fibers, Reprinted with permission from Ref. [153] (2023, Elsevier). (D,E) (D): Structural modeling of bone

at the micro and nano-scales; (**E**): Stress cloud diagram of the model at 0.1% strain. a: Axial stress distribution of MCFs at 0.1% strain; b, c: Stress distribution in the lamellar structure of bone; d: Stress distribution of extra-fibrillar matrix, Reprinted with permission from Ref. [154] (2023, Elsevier).

MD simulation obtains the information and behavior of materials at the nanometer scale by studying the interaction between molecules. MD simulation can predict the overall mechanical properties of materials at the microscopic level by simulating the chemical composition and intermolecular forces of materials, and then it can also be used as the input of micromechanics or FEM. By investigating the molecular fiber toughening mechanism of mineralized collagen fibers, it was found that in a multifaceted increase in energy dissipation compared to fibers without a mineral phase [155]. Arun K. Nair investigated the mechanical properties of mineralized collagen with different mineral densities under tensile and compressive loads. Both the tensile and compressive moduli of the network increase monotonically with increasing mineral density (Figure 9B–D) [4,156]. He also investigated the effect of hydration on collagen fibers. With increasing hydration, the stress-strain behavior became more nonlinear, and the Young's modulus of collagen fibers decreased [157]. Furthermore, the mineralized collagen fibers' hydration has an effect on viscoelasticity. The presence of water in the fibers increases their viscosity and the energy dissipation capacity (Figure 9E–G) [158]. MD can similarly be modeled for smaller units of collagen. Computational simulations to study collagen molecular damage during cyclic fatigue loading of tendons showed that the triple-helix degeneration of collagen was positively associated with fatigue and the number of loading cycles, and the damage was associated with creep strain (Figure 9A) [159].

Table 2. Method for detection research mechanical properties of mineralized collagen.

Reference	Subject	Method	Detecting Parameter
Tan [133]	polycaprolactone electrospun ultrafine fiber	fiber stretching and ex situ observation	tensile malleability
Sano [132]	dentin	fiber stretching and in situ observation	bond strength
Koester [138]	bone	In situ mechanical test with SEM	mechanical properties of the longitudinal and transverse orientations of the bone
Hengsberger [139]	cortical bone of cow	nanoindentation	elastic modulus
Isaksson [142]	cortical bone of rabbit	nanoindentation	elastic modulus and viscoelastic parameters
Stanishevsky [145]	HA nanoparticle-collagen composites electrospinning	nanoindentation	Young's modulus and hardness
Grant [149]	collagen fibrils	AFM	elastic (static) and viscous (dynamic) responses
Qian [136]	Bovine Cortical Bone	AFM	crack propagation
Jäger [151]	Bone (submicron)	FEM staggered arrangement model	elastic modulus and fracture stress
Wang [152]	Bone (submicron)	FEM 2D shear lag model	an initial crack
Vercher [153]	lamellar bone	FEM a 3D finite element model	elastic properties
Alijani [154]	lamellar bone	FEM intra and extra-fibrillar mineralization model	elastic properties
Buehler [155]	collagen microfibril	MD simulation	Young's modulus fracture stress (mineral)
Nair [4]	collagen microfibril	MD simulation	modulus of tension (mineral)
Nair [156]	collagen microfibril	MD simulation	modulus of compression (mineral)
Milazzo [158]	collagen microfibril	MD simulation	Viscoelasticity (mineral and water content)

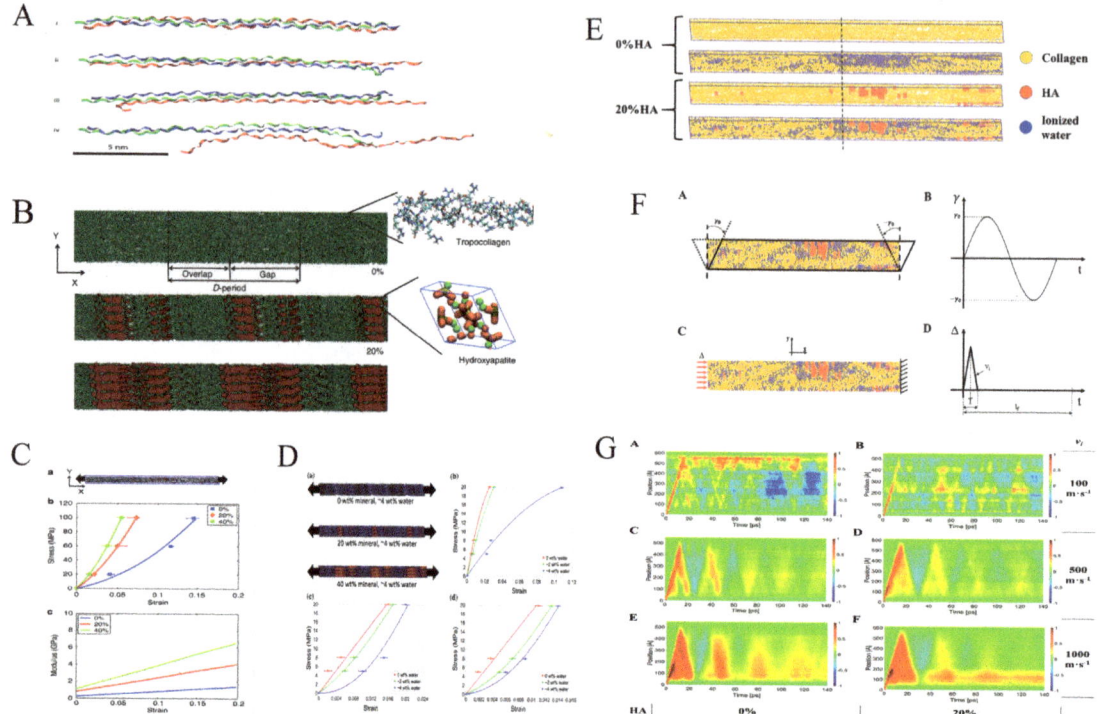

Figure 9. Molecular dynamics simulation. (**A**) The triple helical peptide chain undergoes unwinding under cyclic fatigue loading and thus breaks, Reprinted with permission from Ref. [159] (2023, AAAS). (**B**) Collagen microfibril model with 0%, 20%, and 40% mineralization simulation modeling of mineralized collagen, Reprinted with permission from Ref. [4] (2023, Springer Nature). (**C,D**) Stretching and compression models of collagen microfibrils with different mineralization rates Reprinted with permission from Refs. [4,156] (2023, Springer Nature). (**E–G**) Study on the viscoelastic behavior of mineralized collagen by mineral and water content, Reprinted with permission from Ref. [158] (2023, RSC).

5. Conclusions

Although collagen has excellent biodegradability, low antigenicity, and biological stability, its mechanical strength is unsatisfactory. How to improve the mechanical strength of mineralized collagen is a focus of research. The combination of HA and collagen, which are also biological materials, has given great development to the bionic materials of bone. However, how to prepare bionic bone materials more in line with the natural structure of bone has become the consensus of the research community.

Admittedly, there are many maturely prepared mineralized collagens on the market, which are obtained by the following methods: 1. Precipitate the prepared collagen scaffold in the biomimetic mineralization solution. 2. Codeposition of collagen and HAp. Although this bionic bone also has the pore size and porosity of natural bone, these large pores allow cells and capillaries to grow, showing excellent bone conductivity; however, in the same proportion of components as natural bone, this bionic bone is not as good as the mechanical properties of real bone. In addition, these biomimetic bones are also quite different from natural bones at the collagen fiber level. Under the transmission electron microscope, the collagen fibers of natural bone showed obvious periodic banding (intra-fiber mineralization), while the mineralized collagen of bionic bone was only the attachment of collagen fibers and HAp (extra-fiber mineralization). In Kim's experiment, it also showed that there were debonding particles in the artificially prepared mineralized collagen, which had a negative effect on the elastic modulus of mineralized collagen [112]. More precisely, the realization of in-fiber mineralization is also one of the effective ways to improve the mechanical properties of mineralized collagen.

In this regard, in the first section, starting from the structure of bone, we explained the multi-scale and multi-level structure of bone. From the most basic amino acids and HA crystals to mineralized collagen fibers and even to the whole bone, each layer is extremely complex, which also gives us a great challenge to understand the microstructure and mechanism of bone and to prepare artificial bionic bone. We analyzed the mechanical properties at the level of the bone unit (compact bone and cancellous bone). At the same time, stress is the main stimulus for people's daily activities. Whether the force acts on the cells or directly on the matrix, the stress acts on the bone matrix inevitably. Mechanical stimulation is one of the influencing factors in enhancing collagen mineralization. Mineralized collagen materials are particularly important for bone, but the current research progress is far from satisfactory. Although the study confirmed at the microscopic level that fluid shear forces can regulate the rate, size, and distribution of the mineralized precursor, the implants are subjected to complex and diverse forces in the human body, and it is unknown how the complex and diverse forces regulate collagen mineralization. Bone is a piezoelectric material. Studies have shown that bone has a significant inverse piezoelectric effect on the microscopic surface. This is exciting, which helps to understand the mineralization process, and mechanical stimulation through the piezoelectric effect to produce charge affects the adhesion of HA.

In conclusion, as a bioactive material, collagen can produce different responses to different mechanical stimuli. The goal of research on bionic bone repair materials has always been to change their macroscopic and microscopic structures to have good mechanical properties similar to natural bone. The constitutive and structure-activity relationship between mechanical stimulation, mineralized structure, and mechanical properties are also worthy of further study. Mineralized collagen is expected to be better developed as a new bone repair material.

Author Contributions: Conceptualization, writing—original draft preparation and visualization, Y.N.; conceptualization, writing—review, editing, project administration, and Funding acquisition. T.D. and Y.L. All authors have read and agreed to the published version of the manuscript.

Funding: This work was financially supported by the National Key Research and Development Program of China (2021YFA1000202), National Natural Science Foundation of China (Nos. 11832003 and 12202023), China Postdoctoral Science Foundation (No. 2021TQ0020), Beijing Postdoctoral Research Foundation (No. 2022ZZ041).

Data Availability Statement: There are no additional data available for this study other than what is reported in the manuscript.

Conflicts of Interest: There is no conflict to declare.

References

1. Hamed, E.; Jasiuk, I. Multiscale damage and strength of lamellar bone modeled by cohesive finite elements. *J. Mech. Behav. Biomed. Mater.* **2013**, *28*, 94–110. [CrossRef] [PubMed]
2. Wingender, B.; Bradley, P.; Saxena, N.; Ruberti, J.W.; Gower, L. Biomimetic organization of collagen matrices to template bone-like microstructures. *Matrix Biol.* **2016**, *52–54*, 384–396. [CrossRef]
3. Kikuchi, M.; Ikoma, T.; Itoh, S.; Matsumoto, H.N.; Koyama, Y.; Takakuda, K.; Shinomiya, K.; Tanaka, J. Biomimetic synthesis of bone-like nanocomposites using the self-organization mechanism of hydroxyapatite and collagen. *Compos. Sci. Technol.* **2004**, *64*, 819–825. [CrossRef]
4. Nair, A.K.; Gautieri, A.; Chang, S.W.; Buehler, M.J. Molecular mechanics of mineralized collagen fibrils in bone. *Nat. Commun.* **2013**, *4*, 1724–1729. [CrossRef]
5. Noitup, P.; Garnjanagoonchorn, W.; Morrissey, M.T. Fish skin type I collagen: Characteristic comparison of albacore tuna (Thunnus alalunga) and silver-line grunt (Pomadasys kaakan). *J. Aquat. Food Prod. Technol.* **2005**, *14*, 17–28. [CrossRef]
6. Rho, J.; Kuhn-Spearing, L.; Zioupos, P. Mechanical properties and the hierarchical structure of bone. *Med. Eng. Phys.* **1998**, *20*, 92–102. [CrossRef]

7. Fang, W.J.; Ping, H.; Huang, Y.; Xie, H.; Wang, H.; Wang, W.M.; Fu, Z.Y. Growth of mineralized collagen films by oriented calcium fluoride nanocrystal assembly with enhanced cell proliferation. *J. Mat. Chem. B* **2021**, *9*, 6668–6677. [CrossRef] [PubMed]
8. Cui, F.; Li, Y.; Ge, J. Self-assembly of mineralized collagen composites. *Mat. Sci. Eng. R* **2007**, *57*, 1–27. [CrossRef]
9. Weiner, S.; Wagner, H.D. The material bone: Structure mechanical function relations. *Annu. Rev. Mater. Sci.* **1998**, *28*, 271–298. [CrossRef]
10. Reznikov, N.; Shahar, R.; Weiner, S. Bone hierarchical structure in three dimensions. *Acta Biomater.* **2014**, *10*, 3815–3826. [CrossRef]
11. Reznikov, N.; Bilton, M.; Lari, L.; Stevens, M.M.; Kroger, R. Fractal-like hierarchical organization of bone begins at the nanoscale. *Science* **2018**, *360*, 507. [CrossRef] [PubMed]
12. Tertuliano, O.A.; Greer, J.R. The nanocomposite nature of bone drives its strength and damage resistance. *Nat. Mater.* **2016**, *15*, 1195–1202. [CrossRef]
13. Skedros, J.G.; Dayton, M.R.; Sybrowsky, C.L.; Bloebaum, R.D.; Bachus, K.N. The influence of collagen fiber orientation and other histocompositional characteristics on the mechanical properties of equine cortical bone. *J. Exp. Biol.* **2006**, *209*, 3025–3042. [CrossRef]
14. Oftadeh, R.; Perez-Viloria, M.; Villa-Camacho, J.C.; Vaziri, A.; Nazarian, A. Biomechanics and Mechanobiology of Trabecular Bone: A Review. *J. Biomech. Eng.-Trans. ASME* **2015**, *137*, 010802–01080215. [CrossRef]
15. Launey, M.E.; Buehler, M.J.; Ritchie, R.O. On the Mechanistic Origins of Toughness in Bone. *Annu. Rev. Mater. Rev.* **2010**, *40*, 25–53. [CrossRef]
16. Lu, S.Y.; Jiang, D.J.; Liu, S.H.; Liang, H.F.; Lu, J.R.; Xu, H.; Li, J.; Xiao, J.; Zhang, J.; Fei, Q.M. Effect of different structures fabricated by additive manufacturing on bone ingrowth. *J. Biomater. Appl.* **2022**, *36*, 1863–1872. [CrossRef] [PubMed]
17. Oftadeh, R.; Karimi, Z.; Villa-Camacho, J.; Tanck, E.; Verdonschot, N.; Goebel, R.; Snyder, B.D.; Hashemi, H.N.; Vaziri, A.; Nazarian, A. Curved Beam Computed Tomography based Structural Rigidity Analysis of Bones with Simulated Lytic Defect: A Comparative Study with Finite Element Analysis. *Sci. Rep.* **2016**, *6*, 32397. [CrossRef] [PubMed]
18. Lin, C.Y.; Kang, J.H. Mechanical Properties of Compact Bone Defined by the Stress-Strain Curve Measured Using Uniaxial Tensile Test: A Concise Review and Practical Guide. *Materials* **2021**, *14*, 4224. [CrossRef]
19. Li, S.; Wang, J.Z.; Yin, B.; Hu, Z.S.; Zhang, X.J.; Wu, W.; Liu, G.B.; Liu, Y.K.; Fu, L.; Zhang, Y.Z. Atlas of Human Skeleton Hardness Obtained Using the Micro-indentation Technique. *Orthop. Surg.* **2021**, *13*, 1417–1422. [CrossRef]
20. Duboeuf, F.; Burt-Pichat, B.; Farlay, D.; Suy, P.; Truy, E.; Boivin, G. Bone quality and biomechanical function: A lesson from human ossicles. *Bone* **2015**, *73*, 105–110. [CrossRef]
21. Mieloch, A.A.; Richter, M.; Trzeciak, T.; Giersig, M.; Rybka, J.D. Osteoarthritis Severely Decreases the Elasticity and Hardness of Knee Joint Cartilage: A Nanoindentation Study. *J. Clin. Med.* **2019**, *8*, 1865. [CrossRef]
22. Franke, O.; Durst, K.; Maier, V.; Göken, M.; Birkholz, T.; Schneider, H.; Hennig, F.; Gelse, K. Mechanical properties of hyaline and repair cartilage studied by nanoindentation. *Acta Biomater.* **2007**, *3*, 873–881. [CrossRef]
23. Gnjato, S. Addition to the methodology of research into permanent teeth hardness. *Arch. Biol. Sci.* **2010**, *62*, 739–746. [CrossRef]
24. Hart, N.H.; Nimphius, S.; Rantalainen, T.; Ireland, A.; Siafarikas, A.; Newtonet, R.U. Mechanical basis of bone strength: Influence of bone material, bone structure and muscle action. *J. Musculoskelet. Neuronal Interact.* **2017**, *17*, 114. [PubMed]
25. Ping, H.; Wagermaier, W.; Horbelt, N.; Scoppola, E.; Li, C.; Werner, P.; Fu, Z.; Fratzl, P. Mineralization generates megapascal contractile stresses in collagen fibrils. *Sci. (Am. Assoc. Adv. Sci.)* **2022**, *376*, 188–192. [CrossRef] [PubMed]
26. Levingstone, T.J.; Matsiko, A.; Dickson, G.R.; Brien, F.J.O.; Gleeson, J.P. A biomimetic multi-layered collagen-based scaffold for osteochondral repair. *Acta Biomater.* **2014**, *10*, 1996–2004. [CrossRef] [PubMed]
27. Parkinson, I.H.; Fazzalari, N.L. Interrelationships between structural parameters of cancellous bone reveal accelerated structural change at low bone volume. *J. Bone Miner. Res.* **2003**, *18*, 2200–2205. [CrossRef]
28. Currey, J.D. The design of mineralised hard tissues for their mechanical functions. *J. Exp. Biol.* **1999**, *202*, 3285–3294. [CrossRef]
29. Michael, F.M.; Khalid, M.; Walvekar, R.; Ratnam, C.T.; Ramarad, S.; Siddiqui, H.; Hoque, M.E. Effect of nanofillers on the physico-mechanical properties of load bearing bone implants. *Mat. Sci. Eng.* **2016**, *67*, 792–806. [CrossRef]
30. Stevens, M.M. Biomaterials for bone tissue engineering. *Mater. Today* **2008**, *11*, 18–25. [CrossRef]
31. Paital, S.R.; Dahotre, N.B. Review of laser based biomimetic and bioactive Ca-P coatings. *Mater. Sci. Technol.* **2008**, *24*, 1144–1161. [CrossRef]
32. Sozen, T.; Ozisik, L.; Calik Basaran, N. An overview and management of osteoporosis. *Eur. J. Rheumatol.* **2017**, *4*, 46–56. [CrossRef]
33. Ma, H.; Feng, C.; Chang, J.; Wu, C. 3D-printed bioceramic scaffolds: From bone tissue engineering to tumor therapy. *Acta Biomater.* **2018**, *79*, 37–59. [CrossRef]
34. Ben-Nissan, B. Natural bioceramics: From coral to bone and beyond. *Curr. Opin. Solid. State Mat. Sci.* **2003**, *7*, 283–288. [CrossRef]
35. El-Ghannam, A. Bone reconstruction: From bioceramics to tissue engineering. *Expert Rev. Med. Devices* **2005**, *2*, 87–101. [CrossRef] [PubMed]
36. Liu, X.; Ma, P.X. Polymeric scaffolds for bone tissue engineering. *Ann. Biomed. Eng.* **2004**, *32*, 477–486. [CrossRef] [PubMed]
37. Bharadwaz, A.; Jayasuriya, A.C. Recent trends in the application of widely used natural and synthetic polymer nanocomposites in bone tissue regeneration. *Mat. Sci. Eng. C-Mater.* **2020**, *110*, 110698. [CrossRef] [PubMed]
38. Deshpande, H.; Schindler, C.; Dean, D.; Clem, W.; Bellis, S.L.; Nyairo, E.; Mishra, M.; Thomas, V. Nanocomposite Scaffolds Based on Electrospun Pollycaprolactone/Modified CNF/Nanohydroxyapatite by Electrophoretic Deposition. *J. Biomater. Tissue Eng.* **2011**, *1*, 177–184. [CrossRef]

29. Siddiqui, H.A.; Pickering, K.L.; Mucalo, M.R. A Review on the Use of Hydroxyapatite-Carbonaceous Structure Composites in Bone Replacement Materials for Strengthening Purposes. *Materials* **2018**, *11*, 1813. [CrossRef]
30. Anandagoda, N.; Ezra, D.G.; Cheema, U.; Bailly, M.; Brown, R.A. Hyaluronan hydration generates three-dimensional meso-scale structure in engineered collagen tissues. *J. R. Soc. Interface* **2012**, *9*, 2680–2687. [CrossRef]
31. Lee, S.S.; Hughes, P.; Ross, A.D.; Robinson, M.R. Biodegradable implants for sustained drug release in the eye. *Pharm. Res.* **2010**, *27*, 2043–2053. [CrossRef]
32. Vamsi Krishna, B.; Xue, W.; Bose, S.; Bandyopadhyay, A. Engineered porous metals for implants. *JOM* **2008**, *60*, 45–48. [CrossRef]
33. Weisgerber, D.W.; Erning, K.; Flanagan, C.L.; Hollister, S.J.; Harley, B.A.C. Evaluation of multi-scale mineralized collagen–polycaprolactone composites for bone tissue engineering. *J. Mech. Behav. Biomed. Mater.* **2016**, *61*, 318–327. [CrossRef]
34. Matthews, J.A.; Wnek, G.E.; Simpson, D.G.; Bowlin, G.L. Electrospinning of Collagen Nanofibers. *Biomacromolecules* **2002**, *3*, 232–238. [CrossRef]
35. Nazari, K.A.; Nouri, A.; Hilditch, T. Effects of milling time on powder packing characteristics and compressive mechanical properties of sintered Ti-10Nb-3Mo alloy. *Mater. Lett.* **2015**, *140*, 55–58. [CrossRef]
36. Rony, L.; Lancigu, R.; Hubert, L. Intraosseous metal implants in orthopedics: A review. *Morphologie* **2018**, *102*, 231–242. [CrossRef] [PubMed]
37. Elias, C.N.; Lima, J.; Valiev, R.; Meyers, M.A. Biomedical applications of titanium and its alloys. *JOM* **2008**, *60*, 46–49. [CrossRef]
38. Khorasani, A.M.; Goldberg, M.; Doeven, E.H.; Littlefair, G. Titanium in Biomedical Applications-Properties and Fabrication: A Review. *J. Biomater. Tissue Eng.* **2015**, *5*, 593–619. [CrossRef]
39. Glassman, A.H.; Bobyn, J.D.; Tanzer, M. New femoral designs—Do they influence stress shielding? *Clin. Orthop. Rel. Res.* **2006**, *453*, 64–74. [CrossRef]
40. Zhang, M.; Gregory, T.; Hansen, U.; Cheng, C.K. Effect of stress-shielding-induced bone resorption on glenoid loosening in reverse total shoulder arthroplasty. *J. Orthop. Res.* **2020**, *38*, 1566–1574. [CrossRef]
41. Asri, R.; Harun, W.; Samykano, M.; Lah, N.; Ghani, S.; Tarlochan, F.; Raza, M.R. Corrosion and surface modification on biocompatible metals: A review. *Mater. Sci. Eng. C-Mater. Biol. Appl.* **2017**, *77*, 1261–1274. [CrossRef] [PubMed]
42. Zhang, S.; Li, C.; Hou, W.; Zhao, S.; Li, S. Longitudinal compression behavior of functionally graded Ti–6Al–4V meshes. *J. Mater. Sci. Technol.* **2016**, *32*, 1098–1104. [CrossRef]
43. Lee, H.; Ahn, S.H.; Kim, G.H. Three-dimensional collagen/alginate hybrid scaffolds functionalized with a drug delivery system (DDS) for bone tissue regeneration. *Chem. Mater.* **2012**, *24*, 881–891. [CrossRef]
44. Thompson, M.K.; Moroni, G.; Vaneker, T.; Fadel, G.; Campbell, R.I.; Gibson, I.; Bernard, A.; Schulz, J.; Graf, P.; Ahuja, B.; et al. Design for Additive Manufacturing: Trends, opportunities, considerations, and constraints. *Cirp Ann.-Manuf. Technol.* **2016**, *65*, 737–760. [CrossRef]
45. Pattanayak, D.K.; Fukuda, A.; Matsushita, T.; Takemoto, M.; Fujibayashi, S.; Sasaki, K.; Nishida, N.; Nakamura, T.; Kokubo, T. Bioactive Ti metal analogous to human cancellous bone: Fabrication by selective laser melting and chemical treatments. *Acta Biomater.* **2011**, *7*, 1398–1406. [CrossRef]
46. Liu, Y.; Rath, B.; Tingart, M.; Eschweiler, J. Role of implants surface modification in osseointegration: A systematic review. *J. Biomed. Mater. Res. Part A* **2020**, *108*, 470–484. [CrossRef]
47. Polo-Corrales, L.; Latorre-Esteves, M.; Ramirez-Vick, J.E. Scaffold Design for Bone Regeneration. *J. Nanosci. Nanotechnol.* **2014**, *14*, 15–56. [CrossRef]
48. Yang, H.T.; Qu, X.H.; Wang, M.Q.; Cheng, H.W.; Jia, B.; Nie, J.F.; Dai, K.R.; Zheng, Y.F. Zn-0.4Li alloy shows great potential for the fixation and healing of bone fractures at load-bearing sites. *Chem. Eng. J.* **2021**, *417*, 129317. [CrossRef]
49. McBane, J.E.; Sharifpoor, S.; Cai, K.H.; Labow, R.S.; Santerre, J.P. Biodegradation and in vivo biocompatibility of a degradable, polar/hydrophobic/ionic polyurethane for tissue engineering applications. *Biomaterials* **2011**, *32*, 6034–6044. [CrossRef]
50. Wisniewska, K.; Rybak, Z.; Watrobinski, M.; Struszczyk, M.H.; Filipiak, J.; Zywicka, B.; Szymonowicz, M. Bioresorbable polymeric materials—Current state of knowledge. *Polimery* **2021**, *66*, 3–10. [CrossRef]
51. Ramesh, N.; Moratti, S.C.; Dias, G.J. Hydroxyapatite-polymer biocomposites for bone regeneration: A review of current trends. *J. Biomed. Mater. Res. Part B* **2018**, *106*, 2046–2057. [CrossRef] [PubMed]
52. Park, S.; Gwon, Y.; Kim, W.; Kim, J. Rebirth of the Eggshell Membrane as a Bioactive Nanoscaffold for Tissue Engineering. *ACS Biomater. Sci. Eng.* **2021**, *7*, 2219–2224. [CrossRef]
53. Dorozhkin, S.V. Calcium orthophosphate bioceramics. *Ceram. Int.* **2015**, *41*, 13913–13966. [CrossRef]
54. Dorozhkin, S.V. Calcium Orthophosphates as Bioceramics: State of the Art. *J. Funct. Biomater.* **2010**, *1*, 22–107. [CrossRef] [PubMed]
55. Jakus, A.E.; Rutz, A.L.; Jordan, S.W.; Kannan, A.; Mitchell, S.M.; Yun, C.; Koube, K.D.; Yoo, S.C.; Whiteley, H.E.; Richter, C.; et al. Hyperelastic "bone": A highly versatile, growth factor-free, osteoregenerative, scalable, and surgically friendly biomaterial. *Sci. Transl. Med.* **2016**, *8*, 127r–358r. [CrossRef] [PubMed]
56. Priyadarshini, B.; Rama, M.; Chetan; Vijayalakshmi, U. Bioactive coating as a surface modification technique for biocompatible metallic implants: A review. *J. Asian Ceram. Soc.* **2019**, *7*, 397–406. [CrossRef]
57. Du, T.; Niu, X.; Cao, P.; Zhang, Y.; Liu, Y.; Yang, H.; Qiao, A. Multifarious roles of metal elements in bone mineralization. *Appl. Mater. Today* **2023**, *32*, 101810. [CrossRef]

68. Devi, K.B.; Tripathy, B.; Kumta, P.N.; Nandi, S.K.; Roy, M. In vivo biocompatibility of zinc-doped magnesium silicate bio-ceramics. *ACS Biomater. Sci. Eng.* **2018**, *4*, 62126–62133. [CrossRef] [PubMed]
69. Balakrishnan, S.; Padmanabhan, V.P.; Kulandaivelu, R.; Nellaiappan, T.S.N.; Sagadevan, S.; Paiman, S.; Mohammad, F.; Al-Lohedan, H.A.; Obulapuram, P.K.; Oh, W.C. Influence of iron doping towards the physicochemical and biological characteristics of hydroxyapatite. *Ceram. Int.* **2021**, *47*, 5061–5070. [CrossRef]
70. Zhang, L.; Zhang, C.; Ji, Y.; Xu, E.; Liu, X.; Zhao, F.; Yuan, J.; Cui, J.; Cui, J. Effects of Z-value on physicochemical and biological properties of β-SiAlONs ceramics. *Ceram. Int.* **2021**, *47*, 34810–34819. [CrossRef]
71. Qian, J.; Zhang, W.; Chen, Y.; Zeng, P.; Wang, J.; Zhou, C.; Zeng, H.; Sang, H.; Huang, N.; Zhang, H.; et al. Osteogenic and angiogenic bioactive collagen entrapped calcium/zinc phosphates coating on biodegradable Zn for orthopedic implant applications. *Biomater. Adv.* **2022**, *136*, 212792. [CrossRef] [PubMed]
72. Pang, L.; Zhao, R.; Chen, J.; Ding, J.; Chen, X.; Chai, W.; Cui, X.; Li, X.; Wang, D.; Pan, H. Osteogenic and anti-tumor Cu and Mn-doped borosilicate nanoparticles for syncretic bone repair and chemodynamic therapy in bone tumor treatment. *Bioact. Mater.* **2022**, *12*, 1–15. [CrossRef]
73. Wu, C.; Ramaswamy, Y.; Kwik, D.; Zreiqat, H. The effect of strontium incorporation into CaSiO3 ceramics on their physical and biological properties. *Biomaterials* **2007**, *28*, 3171–3181. [CrossRef] [PubMed]
74. Borciani, G.; Ciapetti, G.; Vitale-Brovarone, C.; Baldini, N. Strontium functionalization of biomaterials for bone tissue engineering purposes: A biological point of view. *Materials* **2022**, *15*, 1724. [CrossRef] [PubMed]
75. Peng, S.; Liu, X.S.; Huang, S.; Li, Z.; Pan, H.; Zhen, W.; Luk, K.D.K.; Guo, X.E.; Lu, W.W. The cross-talk between osteoclasts and osteoblasts in response to strontium treatment: Involvement of osteoprotegerin. *Bone* **2011**, *49*, 1290–1298. [CrossRef]
76. Devi, K.B.; Tripathy, B.; Roy, A.; Lee, B.; Kumta, P.N.; Nandi, S.K.; Roy, M. In vitro biodegradation and in vivo biocompatibility of forsterite bio-ceramics: Effects of strontium substitution. *ACS Biomater. Sci. Eng.* **2018**, *5*, 530–543. [CrossRef]
77. Ibrahimzade, L.; Kaygili, O.; Dundar, S.; Ates, T.; Dorozhkin, S.V.; Bulut, N.; Koytepe, S.; Ercan, F.; Gürses, C.; Hssain, A.H. Theoretical and experimental characterization of Pr/Ce co-doped hydroxyapatites. *J. Mol. Struct.* **2021**, *1240*, 30557. [CrossRef]
78. Shekhawat, D.; Singh, A.; Banerjee, M.K.; Singh, T.; Patnaik, A. Bioceramic composites for orthopaedic applications: A comprehensive review of mechanical, biological, and microstructural properties. *Ceram. Int.* **2021**, *47*, 3013–3030. [CrossRef]
79. Albalwi, H.A.; El-Naggar, M.E.; Abou Taleb, M.; Kalam, A.; Alghamdi, N.A.; Mostafa, M.S.; Salem, S.; Afifi, M. Medical applications of ternary nanocomposites based on hydroxyapatite/ytterbium oxide/graphene oxide: Potential bone tissue engineering and antibacterial properties. *J. Mater. Res. Technol.* **2022**, *18*, 4834–4845. [CrossRef]
80. Jazayeri, H.E.; Rodriguez-Romero, M.; Razavi, M.; Tahriri, M.; Ganjawalla, K.; Rasoulianboroujeni, M.; Malekoshoaraie, M.H.; Khoshroo, K.; Tayebi, L. The cross-disciplinary emergence of 3D printed bioceramic scaffolds in orthopedic bioengineering. *Ceram. Int.* **2018**, *44*, 1–9.
81. Conrad, B.; Yang, F. Hydroxyapatite-coated gelatin microribbon scaffolds induce rapid endogenous cranial bone regeneration in vivo. *Biomater. Adv.* **2022**, *140*, 213050. [CrossRef]
82. Chen, Z.J.; Zhang, Y.; Zheng, L.; Zhang, H.; Shi, H.H.; Zhang, X.C.; Liu, B. Mineralized self-assembled silk fibroin/cellulose interpenetrating network aerogel for bone tissue engineering. *Biomater. Adv.* **2022**, *134*, 112549. [CrossRef]
83. Lu, H.H.; El-Amin, S.F.; Scott, K.D.; Laurencin, C.T. Three-dimensional, bioactive, biodegradable, polymer–bioactive glass composite scaffolds with improved mechanical properties support collagen synthesis and mineralization of human osteoblast-like cells in vitro. *J. Biomed. Mater. Res. A* **2003**, *64*, 465–474. [CrossRef] [PubMed]
84. Tsigkou, O.; Hench, L.L.; Boccaccini, A.R.; Polak, J.M. Enhanced differentiation and mineralization of human fetal osteoblasts on PDLLA containing Bioglass® composite films in the absence of osteogenic supplements. *J. Biomed. Mater. Res. A* **2007**, *80*, 837–851. [CrossRef] [PubMed]
85. Yang, C.R.; Wang, Y.J.; Chen, X.F. Mineralization regulation and biological influence of bioactive glass-collagen-phosphatidylserine composite scaffolds. *Sci. China Life. Sci.* **2012**, *55*, 236–240. [CrossRef]
86. Miri, A.K.; Muja, N.; Kamranpour, N.O.; Lepry, W.C.; Boccaccini, A.R.; Clarke, S.A.; Nazhat, S.N. Ectopic bone formation in rapidly fabricated acellular injectable dense collagen-Bioglass hybrid scaffolds via gel aspiration-ejection. *Biomaterials* **2016**, *85*, 128–141. [CrossRef] [PubMed]
87. Vuornos, K.; Ojansivu, M.; Koivisto, J.T.; Häkkänen, H.; Belay, B.; Montonen, T.; Huhtala, H.; Kääriäinen, M.; Hupa, L.; Kellomäki, M.; et al. Bioactive glass ions induce efficient osteogenic differentiation of human adipose stem cells encapsulated in gellan gum and collagen type I hydrogels. *Mat. Sci. Eng. C Mater.* **2019**, *99*, 905–918. [CrossRef]
88. Gurumurthy, B.; Pal, P.; Griggs, J.A.; Janorkar, A.V. Optimization of collagen-elastin-like polypeptide-bioglass scaffold composition for osteogenic differentiation of adipose-derived stem cells. *Materialia* **2020**, *9*, 100572. [CrossRef]
89. Ferreira, S.A.; Young, G.; Jones, J.R.; Rankin, S. Bioglass/carbonate apatite/collagen composite scaffold dissolution products promote human osteoblast differentiation. *Mat. Sci. Eng. C Mater.* **2021**, *118*, 111393. [CrossRef]
90. Brovold, M.; Almeida, J.I.; Pla-Palacin, I.; Sainz-Arnal, P.; Sanchez-Romero, N.; Rivas, J.J.; Almeida, H.; Dachary, P.R.; Serrano-Aullo, T.; Soker, S.; et al. Naturally-derived biomaterials for tissue engineering applications. *Adv. Exp. Med. Biol.* **2018**, *1077*, 421–449.
91. Wang, S.; Zhao, Z.J.; Yang, Y.D.; Mikos, A.G.; Qiu, Z.Y.; Song, T.X.; Cui, F.Z.; Wang, X.M.; Zhang, C.Y. A high-strength mineralized collagen bone scaffold for large-sized cranial bone defect repair in sheep. *Regen. Biomater.* **2018**, *5*, 283–292. [CrossRef]

92. Wang, S.; Yang, Y.D.; Koons, G.L.; Mikos, A.G.; Qiu, Z.Y.; Song, T.X.; Cui, F.Z.; Wang, X.M. Tuning pore features of mineralized collagen/PCL scaffolds for cranial bone regeneration in a rat model. *Mat. Sci. Eng. C Mater.* **2020**, *106*, 110186. [CrossRef]
93. Ding, Q.F.; Cui, J.J.; Shen, H.Q.; He, C.L.; Wang, X.D.; Shen, S.; Lin, K.L. Advances of nanomaterial applications in oral and maxillofacial tissue regeneration and disease treatment. *Wiley Interdiscip. Rev.-Nanomed. Nanobiotechnol.* **2021**, *13*, e1669. [CrossRef]
94. Boda, S.K.; Almoshari, Y.; Wang, H.; Wang, X.; Reinhardt, R.A.; Duan, B.; Wang, D.; Xie, J. Mineralized nanofiber segments coupled with calcium-binding BMP-2 peptides for alveolar bone regeneration. *Acta Biomater.* **2019**, *85*, 282–293. [CrossRef]
95. Jain, G.; Blaauw, D.; Chang, S. A Comparative Study of Two Bone Graft Substitutes-InterOss((R)) Collagen and OCS-B Collagen((R)). *J. Funct. Biomater.* **2022**, *13*, 28. [CrossRef] [PubMed]
96. Sionkowska, A.; Kozlowska, J. Properties and modification of porous 3-D collagen/hydroxyapatite composites. *Int. J. Biol. Macromol.* **2013**, *52*, 250–259. [CrossRef] [PubMed]
97. Calabrese, G.; Giuffrida, R.; Fabbi, C.; Figallo, E.; Lo Furno, D.; Gulino, R.; Colarossi, C.; Fullone, F.; Giuffrida, R.; Parenti, R.; et al. Collagen-Hydroxyapatite Scaffolds Induce Human Adipose Derived Stem Cells Osteogenic Differentiation In Vitro. *PLoS ONE* **2016**, *11*, e0151181. [CrossRef]
98. Cunniffe, G.M.; Dickson, G.R.; Partap, S.; Stanton, K.T.; O'Brien, F.J. Development and characterisation of a collagen nano-hydroxyapatite composite scaffold for bone tissue engineering. *J. Mater. Sci.-Mater. Med.* **2010**, *21*, 2293–2298. [CrossRef]
99. Wang, L.; You, X.; Zhang, L.; Zhang, C.; Zou, W. Mechanical regulation of bone remodeling. *Bone Res.* **2022**, *10*, 16. [CrossRef] [PubMed]
100. Almeida, M.; Han, L.; Martin-Millan, M.; O'Brien, C.A.; Manolagas, S.C. Oxidative Stress Antagonizes Wnt Signaling in Osteoblast Precursors by Diverting β-Catenin from T Cell Factor- to Forkhead Box O-mediated Transcription. *J. Biol. Chem.* **2007**, *282*, 27298–27305. [CrossRef]
101. Aguirre, J.I.; Plotkin, L.I.; Stewart, S.A.; Weinstein, R.S.; Parfitt, A.M.; Manolagas, S.C.; Bellido, T. Osteocyte apoptosis is induced by weightlessness in mice and precedes osteoclast recruitment and bone loss. *J. Bone Miner. Res.* **2006**, *21*, 605–615. [CrossRef]
102. Ziros, P.G.; Gil, A.P.; Georgakopoulos, T.; Habeos, I.; Kletsas, D.; Basdra, E.K.; Papavassiliou, A.G. The bone-specific transcriptional regulator Cbfa1 is a target of mechanical signals in osteoblastic cells. *J. Biol. Chem.* **2002**, *277*, 23934–23941. [CrossRef] [PubMed]
103. Schaffler, M.B.; Cheung, W.Y.; Majeska, R.; Kennedy, O. Osteocytes: Master orchestrators of bone. *Calcif. Tissue Int.* **2014**, *94*, 5–24. [CrossRef] [PubMed]
104. Bonewald, L.F. The amazing osteocyte. *J. Bone Miner. Res.* **2011**, *26*, 229–238. [CrossRef] [PubMed]
105. Kurata, K.; Uemura, T.; Nemoto, A.; Tateishi, T.; Murakami, T.; Higaki, H.; Miura, H.; Iwamoto, Y. Mechanical strain effect on bone-resorbing activity and messenger RNA expressions of marker enzymes in isolated osteoclast culture. *J. Bone Miner. Res.* **2001**, *16*, 722–730. [CrossRef] [PubMed]
106. Huiskes, R.; Ruimerman, R.; van Lenthe, G.H.; Janssen, J.D. Effects of mechanical forces on maintenance and adaptation of form in trabecular bone. *Nature* **2000**, *405*, 704–706. [CrossRef]
107. Kivell, T.L. A review of trabecular bone functional adaptation: What have we learned from trabecular analyses in extant hominoids and what can we apply to fossils? *J. Anat.* **2016**, *228*, 569–594. [CrossRef] [PubMed]
108. Peyroteo, M.; Belinha, J.; Vinga, S.; Dinis, L.; Natal Jorge, R. A Model for Bone Remodeling: Cellular Dynamics and Mechanical Loading. In Proceedings of the 2017 IEEE 5th Portuguese Meeting on Bioengineering, Coimbra, Portugal, 16–18 February 2017; pp. 1–4.
109. Lee, P.; Lin, R.; Moon, J.; Lee, L.P. Microfluidic alignment of collagen fibers for in vitro cell culture. *Biomed. Microdevices* **2006**, *8*, 35–41. [CrossRef]
110. Niu, X.; Fan, R.; Guo, X.; Du, T.; Yang, Z.; Feng, Q.; Fan, Y. Shear-mediated orientational mineralization of bone apatite on collagen fibrils. *J. Mat. Chem. B* **2017**, *5*, 9141–9147. [CrossRef]
111. Du, T.; Niu, X.; Hou, S.; Xu, M.; Li, Z.; Li, P.; Fan, Y. Highly aligned hierarchical intrafibrillar mineralization of collagen induced by periodic fluid shear-stress. *J. Mat. Chem. B* **2020**, *8*, 2562–2572. [CrossRef]
112. Kim, D.; Lee, B.; Marshall, B.; Thomopoulos, S.; Jun, Y.S. Cyclic strain enhances the early stage mineral nucleation and the modulus of demineralized bone matrix. *Biomater. Sci.* **2021**, *9*, 5907–5916. [CrossRef] [PubMed]
113. Daood, U.; Fawzy, A.S. Minimally invasive high-intensity focused ultrasound (HIFU) improves dentine remineralization with hydroxyapatite nanorods. *Dent. Mater.* **2020**, *36*, 456–467. [CrossRef]
114. Du, T.M.; Niu, Y.M.; Liu, Y.J.; Yang, H.S.; Qiao, A.K.; Niu, X.F. Physical and Chemical Characterization of Biomineralized Collagen with Different Microstructures. *J. Funct. Biomater.* **2022**, *13*, 57. [CrossRef]
115. Giraud-Guille, M.M. Twisted Plywood Architecture of Collagen Fibrils in Human Compact Bone Osteons. *Calcif. Tissue Int.* **1988**, *42*, 167–180. [CrossRef]
116. Stockhausen, K.E.; Qwamizadeh, M.; Wölfel, E.M.; Hemmatian, H.; Fiedler, I.A.K.; Flenner, S.; Longo, E.; Amling, M.; Greving, I.; Ritchie, R.O.; et al. Collagen fiber orientation is coupled with specific nano-compositional patterns in dark and bright osteons modulating their biomechanical properties. *ACS Nano* **2021**, *15*, 455–467. [CrossRef]
117. Wang, Y.; Ural, A. A finite element study evaluating the influence of mineralization distribution and content on the tensile mechanical response of mineralized collagen fibril networks. *J. Biomed. Mater. Res. A* **2019**, *100*, 103361. [CrossRef]
118. Wang, Y.; Ural, A. Effect of modifications in mineralized collagen fibril and extra-fibrillar matrix material properties on submicroscale mechanical behavior of cortical bone. *J. Mech. Behav. Biomed. Mater.* **2018**, *82*, 18–26. [CrossRef] [PubMed]

119. Siegmund, T.; Allen, M.R.; Burr, D.B. Failure of mineralized collagen fibrils: Modeling the role of collagen cross-linking. *J. biomech.* **2008**, *41*, 1427–1435. [CrossRef]
120. Dhand, C.; Ong, S.T.; Dwivedi, N.; Diaz, S.M.; Venugopal, J.R.; Navaneethan, B.; Fazil, M.H.; Liu, S.; Seitz, V.; Wintermantel, E.; et al. Bio-inspired in situ crosslinking and mineralization of electrospun collagen scaffolds for bone tissue engineering. *Biomaterials* **2016**, *104*, 323–338. [CrossRef]
121. Du, T.M.; Niu, X.F.; Li, Z.W.; Li, P.; Feng, Q.L.; Fan, Y.B. Crosslinking induces high mineralization of apatite minerals on collagen fibers. *Int. J. Biol. Macromol.* **2018**, *113*, 450–457. [CrossRef] [PubMed]
122. Kikuchi, M.; Matsumoto, H.N.; Yamada, T.; Koyama, Y.; Takakuda, K.; Tanaka, J. Glutaraldehyde cross-linked hydroxyapatite/collagen self-organized nanocomposites. *Biomaterials* **2004**, *25*, 63–69. [CrossRef] [PubMed]
123. Du, T.; Niu, X.; Hou, S.; Li, Z.; Li, P.; Fan, Y. Apatite minerals derived from collagen phosphorylation modification induce the hierarchical intrafibrillar mineralization of collagen fibers. *J. Biomed. Mater. Res. Part A* **2019**, *107*, 2403–2413. [CrossRef] [PubMed]
124. Nassif, N.; Martineau, F.; Syzgantseva, O.; Gobeaux, F.; Willinger, M.; Coradin, T.; Cassaignon, S.; Azaïs, T.; Giraud-Guille, M.M. In Vivo Inspired Conditions to Synthesize Biomimetic Hydroxyapatite. *Chem. Mat.* **2010**, *22*, 3653–3663. [CrossRef]
125. Niederberger, M.; Colfen, H. Oriented attachment and mesocrystals: Non-classical crystallization mechanisms based on nanoparticle assembly. *Phys. Chem. Chem. Phys.* **2006**, *8*, 3271–3287. [CrossRef] [PubMed]
126. Liu, Y.; Luo, D.; Wang, T. Hierarchical structures of bone and bioinspired bone tissue engineering. *Small* **2016**, *12*, 4611–4632. [CrossRef]
127. Lima, D.B.; de Souza, M.; de Lima, G.G.; Souto, E.; Oliveira, H.; Fook, M.; de Sa, M. Injectable bone substitute based on chitosan with polyethylene glycol polymeric solution and biphasic calcium phosphate microspheres. *Carbohydr. Polym.* **2020**, *245*, 116575. [CrossRef] [PubMed]
128. Du, T.; Niu, Y.; Jia, Z.; Liu, Y.; Qiao, A.; Yang, H.; Niu, X. Orthophosphate and alkaline phosphatase induced the formation of apatite with different multilayered structures and mineralization balance. *Nanoscale* **2022**, *14*, 1814–1825. [CrossRef]
129. Thula, T.T.; Rodriguez, D.E.; Lee, M.H.; Pendi, L.; Podschun, J.; Gower, L.B. In vitro mineralization of dense collagen substrates: A biomimetic approach toward the development of bone-graft materials. *Acta Biomater.* **2011**, *7*, 3158–3169. [CrossRef]
130. Thula, T.T.; Svedlund, F.; Rodriguez, D.E.; Podschun, J.; Pendi, L.; Gower, L.B. Mimicking the Nanostructure of Bone: Comparison of Polymeric Process-Directing Agents. *Polymers* **2011**, *3*, 10–35. [CrossRef]
131. Höhling, H.J.; Barckhaus, R.H.; Krefting, E.-R.; Althoff, J.; Quint, P. Collagen mineralization: Aspects of the structural relationship between collagen and the apatitic crystallites. *Ultrastruct. Skelet. Tissues* **1990**, *7*, 41–62.
132. Sano, H.; Chowdhury, A.F.M.A.; Saikaew, P.; Matsumoto, M.; Hoshika, S.; Yamauti, M. The microtensile bond strength test: Its historical background and application to bond testing. *Jpn. Dent. Sci. Rev.* **2020**, *56*, 24–31. [CrossRef]
133. Tan, E.P.; Ng, S.Y.; Lim, C.T. Tensile testing of a single ultrafine polymeric fiber. *Biomaterials* **2005**, *26*, 1453–1456. [CrossRef]
134. Xiang, D.H.; Zhang, Z.M.; Wu, B.F.; Feng, H.R.; Shi, Z.L.; Zhao, B. Effect of Ultrasonic Vibration Tensile on the Mechanical Properties of High-Volume Fraction SiCp/Al Composite. *Int. J. Precis. Eng. Manuf.* **2020**, *21*, 2051–2066. [CrossRef]
135. Chatzipanagis, K.; Baumann, C.G.; Sandri, M.; Sprio, S.; Tampieri, A.; Kröger, R. In situ mechanical and molecular investigations of collagen/apatite biomimetic composites combining Raman spectroscopy and stress-strain analysis. *Acta Biomater.* **2016**, *46*, 278–285. [CrossRef] [PubMed]
136. Qian, T.; Chen, X.; Hang, F. Investigation of nanoscale failure behaviour of cortical bone under stress by AFM. *J. Mech. Behav. Biomed. Mater.* **2020**, *112*, 103989. [CrossRef]
137. Ritchie, R.O.; Nalla, R.K.; Kruzic, J.J.; Ager, J.W.; Balooch, G.; Kinney, J.H. Fracture and ageing in bone: Toughness and structural characterization. *Strain* **2006**, *42*, 225–232. [CrossRef]
138. Koester, K.J.; Ager, J.W.; Ritchie, R.O. The true toughness of human cortical bone measured with realistically short cracks. *Nat. Mater.* **2008**, *7*, 672–677. [CrossRef]
139. Hengsberger, S.; Enstroem, J.; Peyrin, F.; Zysset, P. How is the indentation modulus of bone tissue related to its macroscopic elastic response? A validation study. *J. Biomech.* **2003**, *36*, 1503–1509. [CrossRef]
140. Hoffler, C.E.; Guo, X.E.; Zysset, P.K.; Goldstein, S.A. An application of nanoindentation technique to measure bone tissue lamellae properties. *J. Biomech. Eng.-Trans. ASME* **2005**, *127*, 1046–1053. [CrossRef]
141. Hengsberger, S.; Kulik, A.; Zysset, P. Nanoindentation discriminates the elastic properties of individual human bone lamellae under dry and physiological conditions. *Bone* **2002**, *30*, 178–184. [CrossRef] [PubMed]
142. Isaksson, H.; Malkiewicz, M.; Nowak, R.; Helminen, H.J.; Jurvelin, J.S. Rabbit cortical bone tissue increases its elastic stiffness but becomes less viscoelastic with age. *Bone* **2010**, *47*, 1030–1038. [CrossRef] [PubMed]
143. Wu, Z.H.; Baker, T.A.; Ovaert, T.C.; Niebur, G.L. The effect of holding time on nanoindentation measurements of creep in bone. *J. Biomech.* **2011**, *44*, 1066–1072. [CrossRef] [PubMed]
144. Ibrahim, A.; Magliulo, N.; Groben, J.; Padilla, A.; Akbik, F.; Abdel, H.Z. Hardness, an Important Indicator of Bone Quality, and the Role of Collagen in Bone Hardness. *J. Funct. Biomater.* **2020**, *11*, 85. [CrossRef]
145. Stanishevsky, A.; Chowdhury, S.; Chinoda, P.; Thomas, V. Hydroxyapatite nanoparticle loaded collagen fiber composites: Microarchitecture and nanoindentation study. *J. Biomed. Mater. Res. Part A* **2008**, *86A*, 873–882. [CrossRef] [PubMed]
146. Rodriguez-Florez, N.; Oyen, M.L.; Shefelbine, S.J. Insight into differences in nanoindentation properties of bone. *J. Mech. Behav. Biomed. Mater.* **2013**, *18*, 90–99. [CrossRef]

47. Johnson, T.B.; Siderits, B.; Nye, S.; Jeong, Y.H.; Han, S.H.; Rhyu, I.C.; Han, J.S.; Deguchi, T.; Beck, F.M.; Kim, D.G. Effect of guided bone regeneration on bone quality surrounding dental implants. *J. Biomech.* **2018**, *80*, 166–170. [CrossRef]
48. Zhao, X.; Hu, T.; Li, H.; Chen, M.; Cao, S.; Zhang, L.; Hou, X. Electrochemically assisted co-deposition of calcium phosphate/collagen coatings on carbon/carbon composites. *Appl. Surf. Sci.* **2011**, *257*, 3612–3619. [CrossRef]
49. Grant, C.A.; Phillips, M.A.; Thomson, N.H. Dynamic mechanical analysis of collagen fibrils at the nanoscale. *J. Mech. Behav. Biomed. Mater.* **2012**, *5*, 165–170. [CrossRef] [PubMed]
50. Yang, L.; Fitié, C.F.C.; Van Der Werf, K.O. Mechanical properties of single electrospun collagen type I fibers. *Biomaterials* **2008**, *29*, 955–962. [CrossRef] [PubMed]
51. Jäger, I.; Fratzl, P. Mineralized Collagen Fibrils: A Mechanical Model with a Staggered Arrangement of Mineral Particles. *Biophys. J.* **2000**, *79*, 1737–1746. [CrossRef]
52. Wang, X.; Qian, C. Prediction of microdamage formation using a mineral-collagen composite model of bone. *J. Biomech.* **2006**, *39*, 595–602. [CrossRef] [PubMed]
53. Vercher-Martínez, A.; Giner, E.; Arango, C.; Javier Fuenmayor, F. Influence of the mineral staggering on the elastic properties of the mineralized collagen fibril in lamellar bone. *J. Mech. Behav. Biomed. Mater.* **2015**, *42*, 243–256. [CrossRef] [PubMed]
54. Alijani, H.; Vaughan, T.J. A multiscale finite element investigation on the role of intra- and extra-fibrillar mineralisation on the elastic properties of bone tissue. *J. Mech. Behav. Biomed. Mater.* **2022**, *129*, 105139. [CrossRef] [PubMed]
55. Buehler, M.J. Molecular nanomechanics of nascent bone: Fibrillar toughening by mineralization. *Nanotechnology* **2007**, *18*, 295102. [CrossRef]
56. Nair, A.K.; Gautieri, A.; Buehler, M.J. Role of Intrafibrillar Collagen Mineralization in Defining the Compressive Properties of Nascent Bone. *Biomacromolecules* **2014**, *15*, 2494–2500. [CrossRef]
57. Fielder, M.; Nair, A.K. Effects of hydration and mineralization on the deformation mechanisms of collagen fibrils in bone at the nanoscale. *Biomech. Model. Mechanobiol.* **2019**, *18*, 57–68. [CrossRef]
58. Milazzo, M.; David, A.; Jung, G.S.; Danti, S.; Buehler, M.J. Molecular origin of viscoelasticity in mineralized collagen fibrils. *Biomater. Sci.* **2021**, *9*, 3390–3400. [CrossRef]
59. Zitnay, J.L.; Jung, G.S.; Lin, A.H.; Qin, Z.; Li, Y.; Yu, S.M.; Buehler, M.J.; Weiss, J.A. Accumulation of collagen molecular unfolding is the mechanism of cyclic fatigue damage and failure in collagenous tissues. *Sci. Adv.* **2020**, *6*, a2795. [CrossRef] [PubMed]

Disclaimer/Publisher's Note: The statements, opinions and data contained in all publications are solely those of the individual author(s) and contributor(s) and not of MDPI and/or the editor(s). MDPI and/or the editor(s) disclaim responsibility for any injury to people or property resulting from any ideas, methods, instructions or products referred to in the content.

Review

Zinc-Based Biodegradable Materials for Orthopaedic Internal Fixation

Yang Liu [1], Tianming Du [1], Aike Qiao [1], Yongliang Mu [2] and Haisheng Yang [1,*]

[1] Department of Biomedical Engineering, Faculty of Environment and Life, Beijing University of Technology, Beijing 100124, China
[2] School of Metallurgy, Northeastern University, Shenyang 110819, China
* Correspondence: haisheng.yang@bjut.edu.cn; Tel.: +86-(010)-6739-6657

Abstract: Traditional inert materials used in internal fixation have caused many complications and generally require removal with secondary surgeries. Biodegradable materials, such as magnesium (Mg)-, iron (Fe)- and zinc (Zn)-based alloys, open up a new pathway to address those issues. During the last decades, Mg-based alloys have attracted much attention by researchers. However, the issues with an over-fast degradation rate and release of hydrogen still need to be overcome. Zn alloys have comparable mechanical properties with traditional metal materials, e.g., titanium (Ti), and have a moderate degradation rate, potentially serving as a good candidate for internal fixation materials, especially at load-bearing sites of the skeleton. Emerging Zn-based alloys and composites have been developed in recent years and in vitro and in vivo studies have been performed to explore their biodegradability, mechanical property, and biocompatibility in order to move towards the ultimate goal of clinical application in fracture fixation. This article seeks to offer a review of related research progress on Zn-based biodegradable materials, which may provide a useful reference for future studies on Zn-based biodegradable materials targeting applications in orthopedic internal fixation.

Keywords: zinc-based biodegradable materials; orthopedic implant; biodegradability; mechanical property; biocompatibility

1. Introduction

Bone fractures are becoming increasingly common with the rapid increases in aging population, traffic accidents, sports injuries and metabolic diseases [1–3]. Fractures have a lifetime prevalence of ~40% and an annual incidence of 3.6% [4]. The most common and burdensome fractures are lower leg fractures of the patella, tibia or fibula, or ankle [5]. Fracture healing is the process of reconstructing bone and restoring its biological and biomechanical functions [6,7]. As one of the common surgical treatments, internal fixation using screws, pins, plates, etc., provides mechanical stability for a fractured bone, allowing weight bearing, early use of the limb, and bone healing [8]. Success in fracture healing is closely related to the internal fixation implants used.

Implants used for internal fixation can be divided into several categories: wires, pins and screws, plates, and intramedullary nails or rods [9]. Staples and clamps are also used occasionally for osteotomy or fracture fixation [9]. Traditional fixation materials are generally nondegradable, including inert stainless steel (SS), titanium (Ti) and its alloys, and cobalt-chromium (Co-Cr) alloys [10]. They possess satisfactory biocompatibility, high wear resistance, and adequate mechanical strength (Table 1) [11–13]. However, they have notable shortcomings when being applied in fracture fixation. For example, metallic materials have much higher elastic modulus values (190–200 GPa for 316L SS, 210–240 GPa for Co-Cr alloys, and 90–110 GPa for Ti alloys) compared with bone tissues (3–30 GPa). Although a rigid fixation is required at the beginning of the healing process to provide a sufficient mechanical stability, a large discrepancy in stiffness between bone and the implant can lead to stress shielding and therefore can delay healing [14]. Even for a successful bone healing, a secondary surgery is often required to remove the implant [15].

Biodegradable materials are well suited to solve the issues above. Fixation implants made of biodegradable materials can provide a strong mechanical support of the fracture site at earlier stages of the healing process, and later on degrade naturally as the healed bone takes over the mechanical loading and their by-products can be absorbed and metabolized [16]. Degradable polymers are intended for applications in soft tissue graft fixation and meniscus repair due to their low strength [17–19]. Compared to polymers, Mg-based biodegradable materials have higher strength and modulus that are close to cortical bone (Table 1). Also, Mg ions released from Mg-based biodegradable implants have beneficial effects on bone regeneration [20]. Due to their appropriate mechanical property, biocompatibility and biodegradability, Mg-based metals have attracted a great deal of attention of in vitro and in vivo research during the last decades. Several Mg-based implants (bone screws, pins, plates) have been available in clinic or undergoing clinical trials [15,21]. However, the issues with an over-fast degradation rate and generation of hydrogen still need to be overcome. Additionally, current Mg alloys (ultimate tensile strength (UTS) < 350 MPa) have relatively low mechanical strength and are only limited to non- or low-load-bearing applications, such as fixation of small bones and cancellous fragments, meniscus repair and soft tissue fixation [21]. Clearly, there remains a critical need for development of biodegradable materials for fixation of fractures at heavy load-bearing skeletal sites where fractures occur most frequently.

The mechanical strength of zinc (Zn) alloys falls in a wide range, from the value of pure Mg to the value of commercial pure Ti and 316 stainless steel (Figure 1). For bone repair, it has been reported that the degradation rates of fixation implants should be between 0.2 and 0.5 mm y^{-1} to match bone healing [1]. Mg-based alloys have degradation rates ranging from 0.8 to 2.7 mm y^{-1} [1,15,22,23], which are above the desired degradation rates of bone implants. The degradation rates of Zn-based alloys are mainly between 0.1 and 0.3 mm y^{-1} [1,24,25]. Moderate corrosion rates and excellent mechanical properties make Zn-based biodegradable metals potential candidates for biomaterial for internal fracture fixation, particularly at heavy load-bearing sites [26–28]. In terms of biocompatibility, Zn is the second most abundant transition metal in humans, serving as a structural or enzymatic cofactor for approximately 10% of the proteome [29]. Consequently, perturbations in Zn homeostasis may lead to various disorders, including growth deficiencies, immune defects, neurological disorders, and cancers [30]. Studies also found that Zn ions (Zn^{2+}) play an important role in promoting fracture healing [30,31].

Zn-based alloys have shown a great potential of application in orthopaedics, particularly for internal fixation of fractures at heavy load-bearing bone [24,25,32,33]. There has been a growing body of in vitro studies on the development of new Zn-based biodegradable materials and testing of their biodegradability, mechanical property and biocompatibility, with fewer in vivo animal studies and no clinical application as yet [34–46] (Figure 2). Although there are several review articles that have elaborated on some aspects of those properties [14,22,26,27,47], it remains unclear if current Zn-based biodegradable material are sufficient to meet clinical needs for orthopaedic internal fixation and what research gap needs to be filled next. In the following sections, we first point out the clinical requirements of implant biomaterials for orthopaedic internal fixation primarily at the heavy load-bearing skeletal sites in terms of their biodegradability, mechanical property and biocompatibility, and then summarize various typical Zn-based biodegradable materials (pure Zn, Zn-based alloys and composites) that have been developed so far and examined in vitro and in vivo for each of these properties. Lastly, unaddressed questions or future research directions are discussed with the aim of moving towards clinical applications of Zn-based biodegradable materials for orthopaedic internal fixation.

Table 1. Characteristics of different typical metallic biomaterials.

Classification	Materials	Biodegradability	Mechanical Properties	Biocompatibility	Applications or Potential Applications	Ref.
Non-biodegradable metallic materials	316L SS	Non-biodegradable	High elastic modulus, low wear and corrosion resistance, high tensile strength	High biocompatibility	Acetabular cup, bone screws, bone plates, pins, etc.	[11]
	Co–Cr alloys	Non-biodegradable	High elastic modulus, high wear and corrosion resistance	Low biocompatibility	Bone screws, bone plates, femoral stems, total hip replacements, etc.	[12]
	Ti alloys	Non-biodegradable	Poor fatigue strength, light weight	High biocompatibility	Dental implants, bone screws, bone plates, etc.	[11,13]
Biodegradable metallic materials	Mg-based alloys	Biodegradable, high degradation rate	Poor mechanical properties, elastic modulus are close to cortical bone	High biocompatibility, H_2 evolution	Bone screws, bone plates (non-load bearing parts), etc.	[2,9]
	Fe-based alloys	Biodegradable, low degradation rate	High elastic modulus, poor mechanical properties	Low biocompatibility	Bone screws, bone plates, etc.	[9]
	Zn-based alloys	Biodegradable, moderate corrosion rate	High elastic modulus, high mechanical properties, low creep resistance	Cytotoxicity, no gas production, high biocompatibility	Bone screws, bone plates (load-bearing parts (potential applications)), etc.	[3,9,10]

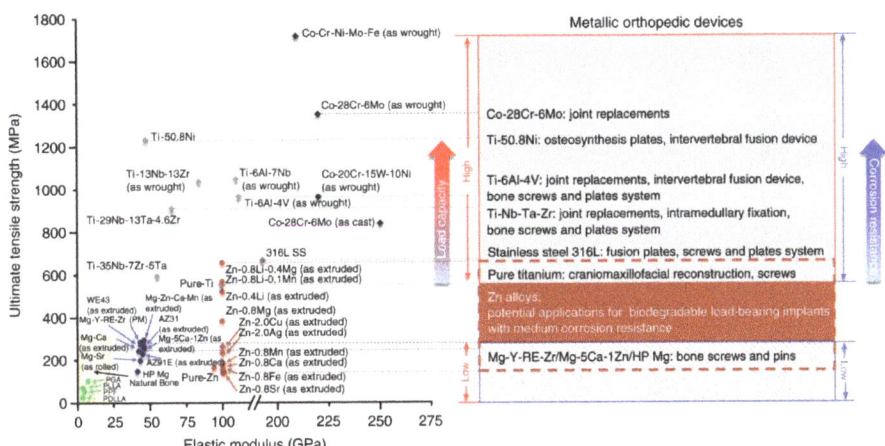

Figure 1. Mechanical properties of biodegradable and non-biodegradable materials for orthopaedic devices and their clinical applications [24].

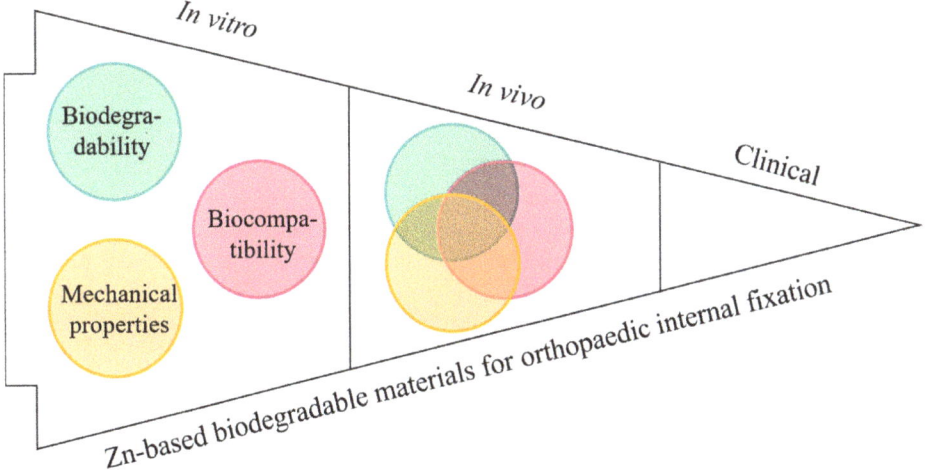

Figure 2. A schematic indicating the state of the art of research on Zn-based biodegradable materials for orthopaedic internal fixation.

2. Biodegradability of Zn-Based Biodegradable Materials

It is well known that human bodies are full of fluid solutions and bear mechanical loading, generating corrosive and mechanical environments for biodegradable materials. It is expected that after biodegradable metals are implanted in the human body, they can gradually degrade at a suitable rate that matches the healing rate of bone tissues (Figure 3). However, different types of fractures at different skeletal sites require different fixation implants (as well as the amounts of degradable materials). Therefore, considering designs of suitable Zn-based devices with a proper degradation rate to meet clinical fixation requirements of different fractures, it is necessary to understand the degradation mechanisms, regulation of the degradation rate, and mechanical factors influencing the degradation of biodegradable metals. It is evident that pure Zn has a relatively low degradation rate. Adding alloy elements or reinforcement materials is commonly used to tune the biodegradability of Zn.

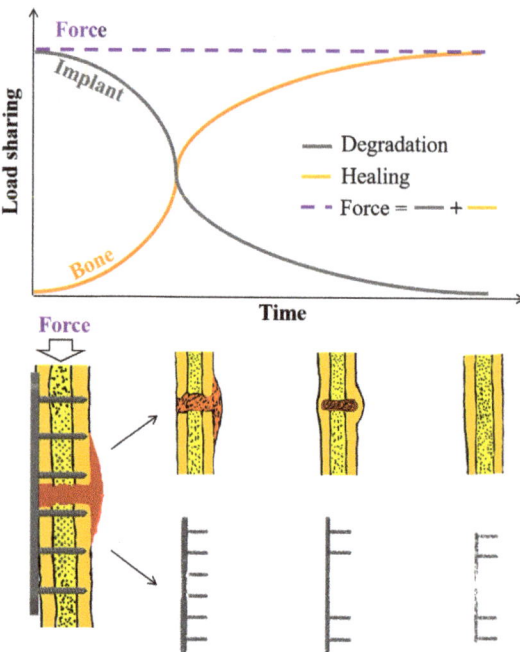

Figure 3. Schematics illustrating the processes of bone healing and implant degradation under a perfect matching scene.

2.1. Biodegradability of Pure Zn

Zn is a relatively low active metal. The standard electrode potential of Zn is −0.76 V, which lies between those of Mg (−2.37 V) and Fe (−0.44 V) [14]. It is prone to corrode in various fluid environments within the human body [22]. Studies have been extensively focused on the in vitro corrosion behavior of pure Zn in different corrosion media, such as Hank's balanced salt solution (HBSS), phosphate buffer saline (PBS), and stimulated body fluid (SBF) [16,28,48]. The medium electrode potential of Zn is associated with a moderate corrosion rate of approximately 0.1 mm y^{-1} [28]. The corrosion mechanism of Zn is regulated by the following reactions:

$$Zn \rightarrow Zn^{2+} + 2e^-, \qquad (1)$$

$$2H_2O + O_2 + 4e^- \rightarrow 4OH^-, \qquad (2)$$

$$2Zn + 2H_2O + O_2 \rightarrow 2Zn(OH)_2, \qquad (3)$$

$$Zn(OH)_2 \rightarrow ZnO + H_2O, \qquad (4)$$

Following the anode reaction, the Zn loses two electrons to generate the Zn^{2+}. Cathode reaction is a process where the electrons of hydrogen dissolve oxygen reduction in the electrolyte to produce hydroxide (OH^-). The simultaneous increases of Zn^{2+} and OH^- in the solution facilitate the precipitation of $Zn(OH)_2$, but the $Zn(OH)_2$ is unstable and may subsequently transform into a thermodynamically more stable ZnO [34,49]. It is evident from these series of reactions above that Zn does not release hydrogen gas during biodegradation like Mg, indicating one of the major benefits of Zn.

2.2. Biodegradability of Zn-Based Alloys

It has been reported that the degradation rate of bone implants should be somewhere between 0.2 and 0.5 mm y^{-1} to match bone healing [1]. Hence, pure Zn clearly does not meet the requirements of biodegradable orthopaedic implants. Adding other alloying elements is one way to alter the corrosion rate of biodegradable metals. To establish binary Zn alloy systems, some studies added alloying elements that are beneficial for bone health (e.g., Mg, Ca, Sr, Li, Mn, Fe, Cu, and Ag) into Zn-based alloys (Table 2) [24,32,40,42,44,50–54]. They used the same melting and extrusion process to prepare a variety of binary Zn-based alloys. Various alloying elements affect the corrosion rate of Zn-based alloys to different degrees [24]. Alloying can lead to an accelerated degradation, and Fe, Ag and Cu had the most significant roles in accelerating corrosion, followed by Li, Sr, Ca and Mg (Table 2) [24]. Studies have shown that corrosion rates of current Zn-based alloys are mainly between 0.1 and 0.3 mm y^{-1} [1]. The choice of Zn-based alloys generally depends on clinical demands, considering the fracture site, and the shape and size of fixators. In general, the degradation rate of the currently developed Zn alloy system is relatively slow. More ternary and quaternary Zn alloys may need to be developed to match the rate of fracture healing.

2.3. Biodegradability of the Zn-Based Composites

Other than alloying, adding reinforcement to form Zn matrix composites is another way to regulate the degradation rate of Zn metals. Composites including pure Zn as a matrix and hydroxyapatite (HA) as reinforcements were prepared by spark plasma sintering (SPS) [55]. A wide range of degradation rates (0.3–0.85 mm y^{-1}) can be achieved by changing the concentration of HA. In addition, the immersion experiment of another beta-tricalcium phosphate (β-TCP)/Zn-Mg composite showed that the corrosion resistance of the composite is slightly decreased with the increase in β-TCP content [56].

2.4. Biodegradability of Zn-Based Biomaterials under Mechanical Loading

The response of biodegradable materials to the combined effect of physiological loading and corrosion environment is an important issue in vivo since stress-induced degradation and cracking are common [57]. Particularly, for load-bearing fracture fixation where biodegradable implant undertakes loading, it is critical to understand how mechanical stress affects the biodegradation behavior of the implant.

The combination of mechanical loading and a specific corrosive medium environment can lead to sudden cracking and failure of degradable metals. This phenomenon is called stress corrosion cracking [57]. In vivo animal experiments and clinical studies have indicated the role of mechanical stress in the early failure of biodegradable implants [19,57–59]. Li et al. conducted slow-strain rate testing (SSRT) and constant-load immersion tests on a promising Zn-0.8 wt%Li alloy [60]. They investigated its stress corrosion cracking susceptibility and examined its feasibility as biodegradable metals with pure Zn serving as a control group. They observed that the Zn-0.8 wt%Li alloy exhibited a low stress corrosion cracking susceptibility. This was attributed to variations in microstructure and deformation mechanism after alloying with Li. In addition, compared to the "no stress" condition (0.124 mm y^{-1}), the corrosion rate of the Zn-Li alloy only increased slightly under tensile stress of 11.1 MPa (0.129 mm y^{-1}) and compressive stress of 17.7 MPa (0.125 mm y^{-1}). Both pure Zn and Zn-0.8 wt%Li alloy did not fracture over a period of 28 days during the constant-load immersion test. The magnitude of the applied stress was close to the physiological loading condition and thus the authors proved the feasibility of both materials as biodegradable metals. So far, there are only a few experimental studies on the stress corrosion of Zn-based biodegradable materials. Since previous experimental studies have shown in degradable polymers or Mg-based alloys that the corrosion rate is affected by the loading mode (tension or compression) and magnitude [18,61–64], it is assumed that these effects also exist in Zn-based biodegradable materials. Identification of the quantitative relationships between various forms of applied loading and degradation behaviors of Zn-based materials is important for the design of load-bearing fixation implants. However, the corrosion behaviors of Zn-based biomaterials under different loading conditions need to be further explored.

Table 2. In vitro experiments of binary Zn-based alloys.

Composition (wt%)	Mechanical Properties			Corrosion Test		Cytocompatibility		Ref.
	σYS (MPa)	σUTS (MPa)	ε (%)	Corrosion Medium	Corrosion Rate (mm y^{-1})	Cell Type	Key Findings	
Zn-0.8Mg	203	301	13	MEM	0.071	U-2OS, L-929	Zn is less biocompatible than magnesium and the maximum safe concentrations of Zn^{2+} for the U-2OS and L929 cells are 120 μM and 80 μM.	[50]
Zn-1.0Ca	206	252	12.7	HBSS	0.09	MG63	Adding the alloying elements Ca into Zn can significantly increase the viability of MG63 and can promote the MG63 cell proliferation compared with the pure Zn and negative control groups.	[51]
Zn-1.1Sr	220	250	22	SBF	0.4	HOBs, hMSCs	The proliferation ability of the two kinds of cells did not decrease in the zinc alloy leaching solution. When the concentration of the leaching solution was low, the growth of the two kinds of cells was promoted.	[32]
Zn-0.4Li	387	520	5.0	SBF	0.019	MC3T3-E1	Zn-0.4Li alloy extract can significantly promote the proliferation of MC3T3-E1 cells.	[24]
Zn-5.0Ge	175	237	22	HBSS	0.051	MC3T3-E1	The diluted extracts at a concentration <12.5% of both the as-cast Zn-5Ge alloy and pure Zn showed grade 0 cytotoxicity; the diluted extracts at the concentrations of 50% and 25% of Zn-5Ge alloy showed a significantly higher cell viability than those of pure Zn.	[52]
Zn-6.0Ag	-	290	-	SBF	0.114	-	-	[44]
Zn-0.8Fe	127	163	28.1	SBF	0.022	MC3T3-E1	MC3T3-E1 cells had unhealthy morphology and low cell viability.	[24]
Zn-4Cu	327	393	44.6	HBSS	0.13	L-929, TAG, SAOS-2	Zn-4Cu alloy had no obvious cytotoxic effect on L929, TAG and Saos-2 cells.	[53]
Zn-0.8Mn	98.4	104.7	1.0	-	-	L-929	Zn-0.8Mn alloy showed 29% to 44% cell viability in 100% extract, indicating moderate cytotoxicity.	[40]
Zn-2Al	142	192	12	SBF	0.13	MG63	Cell viability decreased to 67.5 ± 5.3% in 100% extract cultured for one day, indicating that high concentrations of ions have a negative effect on cell growth. With the extension of culture time, the number of cells increased significantly.	[42]
Zn-0.0.5Zr	104	157	22	-	-	-	-	[54]

YS: yield Strength; UTS: ultimate tensile strength; SBF: stimulated body fluid; MEM: minimum essential medium; HBSS: Hank's balanced salt solution; L-929: mouse fibroblasts; MG63: human osteosarcoma cells; HOBs: human osteoblasts; MSCs: human bone marrow mesenchymal stem cells; MC3T3-E1: mouse preosteoblasts; TAG: human immortalized periosteal cells; SAOS-2: human osteosarcoma cells; U-2OS: human osteosarcoma cells.

Cyclic loading-induced fatigue fractures are very common in engineered metals, where the fatigue strength is further reduced in a corrosive environment [65]. It was reported that under the combined effects of stress and corrosive media, fatigue cracks propagate faster [57]. Corrosion fatigue is of primary concern for metallic internal fixation which commonly bear cyclic dynamic loads in vivo. The corrosion pit propagation rate is influenced by the magnitude of stress, frequency, and cycle number [66,67]. Previous studies have compared the compression-induced fatigue behavior of additively manufactured porous Zn in air and in revised simulated body fluid (r-SBF) [68]. The fatigue strength of the additively-manufactured porous Zn was high in air (i.e., 70% of its yield strength) and even higher in r-SBF (i.e., 80% of its yield strength). The high value of the relative fatigue strength in air could be attributed to the high ductility of pure Zn itself. The formation of corrosion products around the strut junctions might explain the higher fatigue strength of additive manufacturing Zn in r-SBF. The favorable fatigue behavior of additive manufacturing porous Zn further highlights its potential as a promising bone-substituting biomaterial. Another study found in their fatigue testing of Zn-0.5Mg-WC nanocomposites that the material survived after 10 million cycles of tensile loading when the maximum stress was 80% of the yield stress [69]. These results suggest that the Zn-0.5Mg-WC nanocomposite is a promising candidate for biodegradable materials. So far, there has been no report on the fatigue corrosion behavior of Zn-based alloys in vivo. Since the resistance of a material to fatigue and corrosion is an important consideration for designing implants, future relevant studies on Zn-based biomaterials may be required.

3. Mechanical Properties of Zn-Based Biodegradable Materials

In addition to the biodegradable properties, the mechanical properties of the biodegradable metals are also important considerations for designing orthopaedic implants for internal fixation. Yield strength (YS), ultimate tensile strength (UTS), elongation (ε) and elastic modulus (E) are common parameters which are used to indicate the mechanical properties of biomedical materials [11,37,70–73]. Extensive studies have determined those mechanical parameters of Zn-based biodegradable materials. The reported mechanical criteria for degradable metals (e.g., Mg-based) are UTS > 300 MPa and ε > 20% [22]. On the other hand, the current gold standard for medical metal materials, such as Ti and its alloys, has a tensile strength of over 600 MPa [13]. To certain extent, these criteria could provide guides of mechanical properties for development of Zn-based degradable materials. However, the requirement may vary with different load-bearing sites.

3.1. Mechanical Properties of Pure Zn

Pure Zn has extremely low yield strength (29.3 MPa) and elongation (1.2%) in its as-cast condition [74]. The Young's modulus of pure Zn is around 94 GPa [16]. Obviously, it is difficult to meet the mechanical criteria as biodegradable metals [22]. On the other hand, owing to the low melting point of Zn, several additional uncertainties exist with regard to the mechanical properties of biodegradable Zn and Zn-based alloys. Low creep resistance, high susceptibility due to natural aging, and static recrystallization may lead to the failure of Zn-based biodegradable materials during storage at a room temperature and usage at a body temperature [26]. Studies showed that Zn-based alloys underwent appreciable creep deformation under human body temperature (37°) [75]. In addition, recrystallization of Zn-based alloys under stress can reduce their resistance to creep [42]. Thus, creep deformation is an important factor that should be considered in the studies of pure Zn.

3.2. Mechanical Properties of Zn-Based Alloys

Alloying is a common approach to change the mechanical properties of metals, where alloy ratio is essential for studies of Zn-based alloys. Attempts have been made to optimizing the Zn-based alloys by changing the alloy ratio, in order to obtain better mechanical performance in vitro and then move to in vivo conditions [24,32,40,42,44,50–54]. Zn-based alloys have Young's modulus values ranging from 100 to 110 GPa depending on alloying conditions [16]. As summarized in Table 2, the Zn-based alloys with improved mechanical properties to various degrees are generated by adding elements of Mg [50], Ca [51], Sr [32], Li [24], Ge [52], Ag [44], Fe [24], Cu [53], Mn [40], Al [42], Zr [54]. The improvement of adding Li elements is particularly obvious, but the elongation of Zn-Li is only 5%. Following addition of the Cu element, the elongation of the Zn-based alloys reaches 44.6% Binary Zn-based alloys have poor mechanical properties and may not be applicable in load-bearing sites of the skeleton. Table 3 summarizes the mechanical properties of ternary Zn-based alloys on the basis of binary Zn-based alloys.

Different mechanical processing methods have great influences on the mechanical properties of the same Zn-based alloys. Among the three common mechanical processing operations (hot extrusion, hot rolling, and casting), the hot extrusion can produce the greatest improvement in mechanical properties of Zn-based alloys. Compared with binary Zn-based alloys, ternary Zn-based alloys have largely improved mechanical properties. For example, the tensile strength of Zn-0.8Li-0.4Mg is 646 MPa, which is greater than those of pure titanium or 316L SS (Figure 1) [24]. In addition, reasonable mechanical integrity of Zn-0.8Li-0.4Mg was maintained in vitro, and is expected to be used for bone repair at load-bearing sites.

3.3. Mechanical Properties of Zn-Based Composites

Apart from the addition of alloying elements, adding reinforcement matrix as composite could also regulate the mechanical properties of Zn metals. The biocompatibility and the mechanical properties were improved by controlling the type and content of the second phase to form a composite material. In a previous study, Zn-HA composites were prepared with pure Zn as matrix and hydroxyapatite (HA) as reinforcement by spark plasma sintering [55]. In vivo tests showed that the addition of HA resulted in a better performance in osteogenesis with prolonged fixation time. In another study, Zn-Mg-β-TCP composites were prepared with Zn-Mg as matrix and β-TCP as enhancer by the mechanical stirring combined with ultrasonic assisted casting and hot extrusion technology [56]. This material had an ultimate tensile strength of 330.5 MPa and showed better biocompatibility than Zn-Mg alloys in cellular experiments. A barrier layer of ZrO_2 nanofilm was constructed on the surface of Zn-0.1 wt%Li alloy via atomic layer deposition (ALD) [76]. Their results indicated that the addition of ZrO_2 could effectively improve cell adhesion and vitality, and promote osseointegration, but the non-degradation of ZrO_2 brought new challenges. Composites often have advantages over alloys due to the addition of second-phase enhancers. Compared with pure Zn, the addition of a second-phase material largely enhances its mechanical strength and biocompatibility. However, it was reported that the ductility of Zn-based composite materials is only 10% or even lower, with a greater brittleness [22], bringing difficulties to the processing of orthopaedic devices (such as bone screws and bone plates). In addition, the complex manufacturing process, high cost of composite materials, and a lack of sufficient basic theoretical supports in the field of preparation and processing still limit their developments.

Table 3. In vitro experiments of ternary Zn-based alloys.

Composition (wt%) and Manufacturing Process	Mechanical Properties			Corrosion Test		Cytocompatibility		Ref.
	σYS (MPa)	σUTS (MPa)	ε (%)	Corrosion Medium	Corrosion Rate (mm/y)	Cell Type	Key Findings	
Zn-1.5Mg-0.5Zr HE	350	425	12	-	-	L-929	Overall, the L-929 cells exhibit polygonal or spindle shape, and well spread and proliferated in the extracts of pure Zn and Zn alloys.	[39]
Zn-1.0Ca-1Sr Cast	86	140	1.2	SBF	-	MG63	Adding the alloying elements Mg, Ca and Sr into Zn can significantly increase the viability of MG63 cell proliferation compared with the pure Zn and negative control groups.	[77]
Zn-1.0Ca-1Sr HE	212	260	6.7	SBF	0.11			
Zn-1.0Ca-1Sr HR	144	203	8.8	SBF	-			
Zn-0.8Li-0.4Mg HE	438	646	3.68	-	-	-	-	[24]
Zn-3Ge-0.5Mg Cast	66.9	88.3	1.4	HBSS	0.062	MG63	The extract with a concentration of 100% had obvious cytotoxicity to MG63 cells. When the concentration of the extract was diluted to 12.5% or lower, the survival rate of MG-63 cells was all above 90%.	[78]
Zn-3Ge-0.5Mg HR	253	208	9.2	HBSS	0.075			
Zn-4Ag-0.6Mn HE	-	302	35	HBSS	0.012	-	-	[79]
Zn-1Fe-1Mg Cast	146	157	2.3	SBF	0.027	-	-	[80]
Zn-0.8Mn-0.4 Cast	112	120	0.3	-	-	-	-	[68]
Zn-0.8Mn-0.4 HE	253	343	8	-	-			
Zn-0.8Mn-0.4 HR	245	323	12	-	-			

HE: Hot extrusion; HR: Hot rolling.

4. Biocompatibility of Zn-Based Biodegradable Materials

Biocompatibility is the ability of a material to conform to the host response, cell response, and living systems, and it is a vital property of metallic internal fixation for bone repair [1]. The metallic fixation implants directly release ions into the human body, affecting the surrounding cells, tissues, and blood (Figure 4) and leading to either positive or negative results [47,81,82]. Additionally, biomaterial-induced infections are one of the leading causes of implant failure in orthopaedic surgery [25]. Postoperative wound infection may cause an increase in the cost of pain treatment and even sequelae such as limb malformation and dysfunction of the implants [26]. Thus, exploring the biocompatibility of Zn and its alloys is important considering their ultimate implanting in the human body.

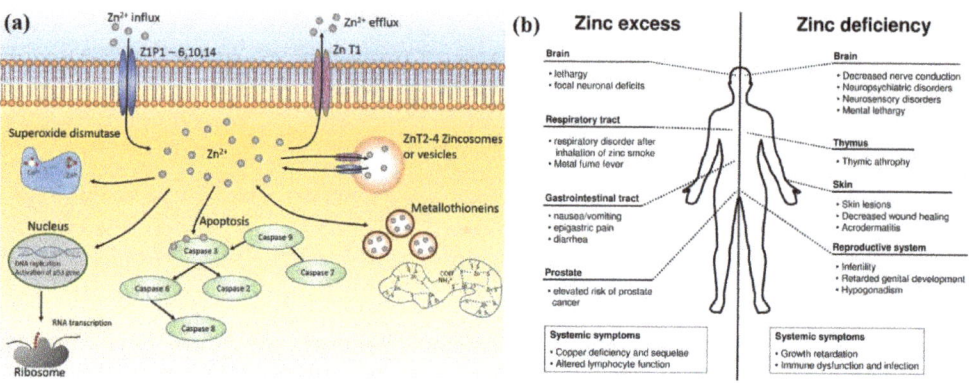

Figure 4. (a) Biological roles of Zn. Reprinted with permission from Ref. [81], Copyright 2016 Wiley. (b) Comparison of the influence of zinc excess versus deficiency [82].

4.1. Biocompatibility of Pure Zn

Zn plays a fundamental role in multiple biochemical functions of the human body, including cell division, cell growth, wound healing, and the breakdown of carbohydrates [83]. Dietary Zn^{2+} deficiency has been linked to impaired skeletal development and bone growth in humans and animals (Figure 5) [83]. Specifically, 85% of Zn in the human body is found in muscle and bone, 11% in the skin and liver, and the rest in other tissues [84]. Zn is located at sites of soft tissue calcification, including osteons and calcified cartilage. Zn levels in bone tissue increase as bone mineralization increases. The skeletal growth was reduced during Zn deficiency. Zn plays a key biological role in the development, differentiation and growth of various tissues in the human body [85], including nucleic acid metabolism, stimulation of new bone formation, signal transduction, protection of bone mass, regulation of apoptosis, and gene expression [14]. Zn not only inhibits related diseases such as bone loss and inflammation, but also plays an important role in cartilage matrix metabolism and cartilage II gene expression [86]. The following symptoms are associated with Zn deficiency, including impaired physical growth and development in infants and young adults, the increased risk of infection, the loss of cognitive function, the problems of memory and behavioral, and learning disability. However, excessive Zn may cause neurotoxicity problems [87]. Based on the RDI (Reference Daily Intake) values reported for mature adults, the biocompatibility of Zn (RDI: 8–20 mg/day) is not as good as that of Mg (RDI: 240–420 mg/day), but very similar to that of Fe (RDI: 8–18 mg/day) [88]. Excessive Zn can cause symptoms such as nausea, vomiting, abdominal pain, diarrhea, fatigue, and can weaken immune function and delay bone development [87]. Therefore, when Zn-based biodegradable materials are implanted into the body as bone implant materials, the toxicity of their degradation products should be considered.

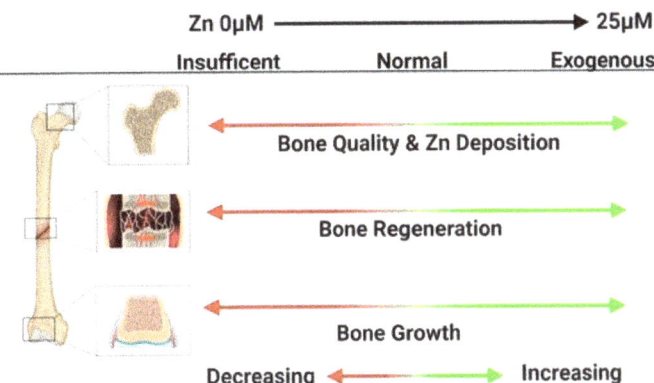

Figure 5. Zn^{2+} deficiency has been linked to impaired skeletal development and bone growth in humans and animals [83].

4.2. Biocompatibility of Zn-Based Alloys

The results of cytotoxicity tests can reflect the biological safety of the material to some extent. Tables 2 and 3 summarize the results of cytocompatibility testing of alloying elements. Specifically, according to the cytocompatibility testing, additions of Mg, Ca and Li do not produce cytotoxicity, but can promote cell proliferation. However, Cu, Al, and Fe show varying degrees of toxic effects on bone cells [34,37,40,42,43,73,74,89].

Regarding the effect of metal ions on antibacterial activity, Sukhodub et al. [90] systematically examined the antibacterial abilities of metal ions and reported that the sterilization rate of the metal ions from high to low was as follows: Ag^+, Cu^{2+}, Zn^{2+}, and Mg^{2+}. Among these metal ions, Zn ions have a good antibacterial ability when they reach a certain concentration and can kill various bacteria and fungi. Zinc is an essential element with intrinsic antibacterial and osteoinductive capacity [91]. Zn-based antimicrobial materials generally consist of zinc complexes and ZnO nano-particles. Complexes such as zinc pyrithione and its derivatives are well known antifungal compounds and have been broadly applied in medicines [92]. Lima et al. [93] prepared Zn-doped mesoporous hydroxyapatites (HAps) with various Zn contents by co-precipitation using a phosphoprotein as the porous template. They found that the antibacterial activity of the HAps samples depended strongly on their Zn^{2+} contents. Tong et al. [94] examined the bacterial distributions of the Zn-Cu foams pre- and post-heat treatment after co-culturing with staphylococcus aureus for 24 h, and observed good antibacterial properties of the Zn-Cu foams. Lin et al. [74] observed better antibacterial properties of Zn-1Cu-0.1Ti than pure Zn. Ren et al. [95] systematically investigated a variety of Cu-containing medical metals including stainless steels, Ti alloys, and Co-based alloys, and demonstrated good antibacterial abilities of those materials stemming from the durable and broad-spectrum antibacterial characteristics of Cu ions. Therefore, Cu-containing Zn alloys may be expected to be promising implant materials with intrinsic antibacterial ability.

4.3. Biocompatibility of Zn-Based Composites

HA is a well-known bioceramic with bioactivity that supports cell proliferation, bone ingrowth and osseointegration. HA has similar chemical and crystallographic structures to bone, which can form a chemical bond with osseous tissue, and act like nucleation for new bone [17]. Yang et al. [55] fabricated Zn-(1, 5, 10 wt%) HA composites using the SPS technique and investigated their in vitro degradation behaviors. Zn-HA composites showed significantly improved cell viability of osteoblastic MC3T3-E1 cells compared with pure Zn. An effective antibacterial property was observed as well. As a bioactive ceramic, β-TCP has good biocompatibility, osteoconductivity and biodegradability [96]. In a study by Pan et al. [56], the biocompatibility of Zn-1Mg-xβ-TCP (x = 0, 1, 3, 5 vol%) composites were investigated. When L-929 and MC3T3 cells were cultured in different concentrations for one day, the relative proliferation rate of the cells is above 80%, and the cytotoxicity is 0–1. Moreover, the addition of β-TCP makes the compatibility of the composite material to MC3T3 cells significantly higher than that of the Zn-Mg alloy.

5. In Vivo Evaluation of Zn-Based Biodegradable Materials with Animal Models

In addition to in vitro testing, in vivo animal experiments are a necessary step in assessing the performances of Zn-based biodegradable materials prior to translation into clinical applications. Different from in vitro experiments where the biodegradability, mechanical property, and biocompatibility of a material are often tested separately, animal models can be used to examine all these properties together in an in vivo condition. Although the in vivo animal experiments may not be able to fully mimic the mechanical, biological and chemical environments in the human body, they are currently the best way to evaluate the interactions between Zn-based biodegradable materials and host [15,24,25,36,96]. There are far fewer in vivo studies on Zn-based biomaterials than in vitro studies. Several representative in vivo studies on Zn-based biodegradable materials were summarized in Table 4.

Yang et al. [24] implanted the pure Zn into the rat femur condyle. A serious fibrous tissue encapsulation was found for pure Zn, resulting in the lack of direct bonding between bone and implant (Figure 6a). The delayed osseointegration of pure Zn is claimed to be attributed to the local high Zn ion concentration. Consistent with the observations in vitro, the in vivo results confirmed that alloying with appropriate elements such as Mg, Ca and Sr can effectively improve the biocompatibility. Yang et al. [55] implanted the pure Zn into the rat femur condyle. A serious fibrous tissue encapsulation was found for pure Zn, resulting in the lack of direct bonding between bone and implant (Figure 6b). Meanwhile, Jia et al. [70] implanted the Zn-0.8 wt.%Mn alloy into the rat femoral condyle for repairing bone defects with pure Ti as control. Their results showed that the new bone tissue at the bone defect site in both groups gradually increased with time, but a large amount of new bone tissue was observed around the Zn-0.8Mn alloy scaffold (Figure 6c). More importantly, in a heavy load-bearing rabbit shaft fracture model, the Zn-0.4Li-based bone plates and screws showed comparable performance in bone fracture fixation compared to the Ti-6Al-4 V counterpart whereas the cortical bone in the Zn-0.4Li alloy group was much thicker (Figure 6d). The results suggest the great potential of Zn-Li based alloys for degradable biomaterials in heavy load-bearing applications [25].

It can be seen from those in vivo studies above (Table 4) that the Zn-based biodegradable materials play an important role in promoting osteogenesis. The corrosion rate of Zn-based biodegradable materials is relatively slow in vivo and can provide a long-term mechanical support in the period of fracture healing [24,25,36]. No incomplete fracture healing and structural collapse of the implant were reported during the animal experiments on load-bearing parts such as the femur [25]. However, the long-term results of the Zn-based implants remain unknown since those animal studies generally lasted for 8–24 weeks [24,25,46]. Additionally, the in vivo studies testing performances of Zn-based implants in fracture fixation are limited [24].

Currently, only small animal models, such as mice, rats and rabbits have been used to examine primarily the biodegradability and biocompatibility of Zn^{2+} metals on bone defect sites (Table 4) [24,25,33,36,46,51,55,56,70,97]. Although mammals have many similarities, differences across small animals, large animals and humans should be recognized [98]. For example, the difference in skeletal size across various species affects the amount of Zn materials that needs to be degraded or absorbed as well as the mechanical environment, which may lead to varied results between preclinical studies and clinical applications. Therefore, with clinical translations in mind, future studies may be warranted with large animal models. In addition, as Zn-based implants are expected to be used at heavy load-bearing sites for internal fracture fixation, proper site-specific in vivo animal models should be used to test their biodegradability, mechanical properties and biocompatibility (Figure 7).

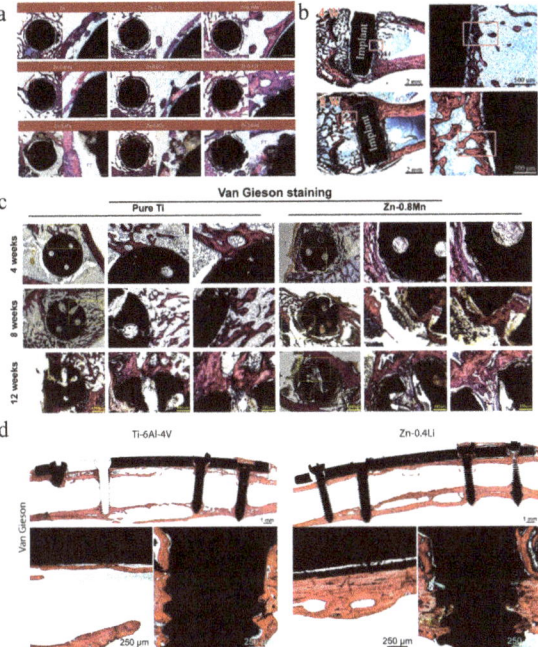

Figure 6. (**a**) Hard tissue sections of pure Zn, Zn-0.4Li, Zn-0.1Mn, Zn-0.8 Mg, Zn-0.8Ca, Zn-0.1Sr, Zn-0.4Fe, Zn-0.4Cu and Zn-2Ag in metaphysis. The magnified region is marked by red rectangle. NB, new bone; DP, degradation products; FT, fibrous tissue. Scale bar, 0.5 mm in low magnification, 500 μm in high magnification [24]. (**b**) Histological characterization of hard tissue sections at implant sites. Van Gieson staining of pure Zn. Reprinted with permission from Ref. [55], Copyright 2018, Elsevier. (**c**) The Van Gieson staining results of specimens 4 weeks, 8 weeks, and 12 weeks postoperatively. Within each row, full-view images of bone defect areas (20×), medium magnification images (50×), and higher magnification images (100×) arranged from left to right. Reprinted with permission from Ref. [70], Copyright 2020, Elsevier. (**d**) Van Gieson staining of representative histological images of femoral fracture healing at 6 months. The fracture healing and fixation screws are magnified and marked by red rectangles. Reprinted with permission from Ref. [25], Copyright 2021, Elsevier.

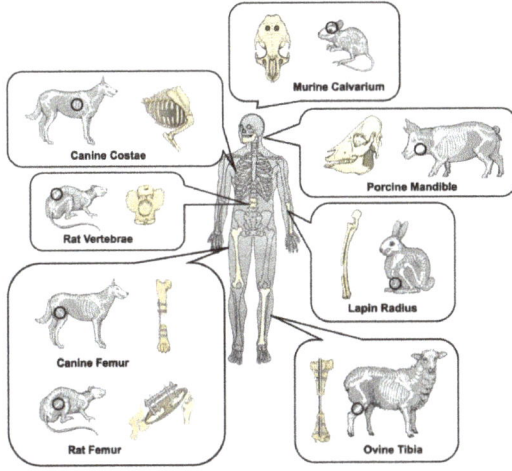

Figure 7. Schematic representation of common animal bone defect models [98].

Table 4. Relevant animal studies of Zn-based biodegradable materials as potential orthopaedic implants.

Zn-Based Metals	Designed Implants	Control	Surgeries	Animal Species	Major Findings	Ref.
Zn-Mn	Scaffold	Pure Ti	Insertion into femoral condyle	Rats	The new bone tissues at the bone defect sites gradually increased with time in both groups, and numerous new bone tissues were observed around the Zn-0.8Mn alloy scaffold	[70]
Zn-1Mg, Zn-1Ca, Zn-1Sr	Intramedullary nails	NA	Insertion into femoral marrow medullary cavity	Mice	There was no inflammation observed around the implantation site and no mouse died after operation. The new bone thickness of Zn-1Mg, Zn-1Ca and Zn-1Sr pin groups are significantly larger than the sham control group.	[51]
Zn-HA	Pin	Pure Zn	Insertion into femoral condyle	Rats	There was new bone formation around the Zn-HA composite, and the bone mass increased over time. With prolonged implantation time, the Zn-HA composite was more effective than pure Zn in promoting new bone formation.	[55]
Zn-0.05Mg	Pin	Pure Zn	Insertion into femoral condyle	Rabbits	No inflammatory cells were found at the fracture site, and new bone tissue formation was confirmed at the bone/implant interface, proving that the Zn-0.05Mg alloy promoted the formation of new bone tissue.	[46]
Zn-(0.001% < Mg < 2.5%, 0.01% < Fe < 2.5%)	Screw and plate	PLLA, Ti-based alloys	Mandible fracture	Beagles	The new bone formation in the Zn alloy group and the titanium alloy group was significantly higher than that in the PLLA group. In addition, the new bone formation in the Zn-based alloys group was slightly higher than that in the Ti-based alloys group. The degradation of Zn implants in vivo would not increase the concentration of Zn^{2+}.	[97]
Zn-X (Fe, Cu, Ag, Mg, Ca, Sr, Mn, Li)	Intramedullary nails	Pure Zn	Insertion into femoral marrow medullary cavity	Rats	Pure Zn, Zn-0.4Fe, Zn-0.4Cu and Zn-2.0Ag alloy implants showed localized degradation patterns with local accumulation of products. In contrast, the degradation of Zn-0.8Mg, Zn-0.8Ca, Zn-0.1Sr, Zn-0.4Li and Zn-0.1Mn was more uniform on the macroscopic scale.	[24]
Zn-0.8Sr	Scaffold	Pure Ti	Insertion into femoral condyle	Rats	Zn-based alloys promote bone regeneration by promoting the proliferation and differentiation of MC3T3-E1 cells, upregulating the expression of osteogenesis-related genes and proteins, and stimulating angiogenesis.	[36]
Zn-0.8Li-0.1Ca	Scaffold	Pure Ti	Insertion into radial defect	Rabbits	The Zn-0.8Li-0.1Ca alloy has a similar level of biocompatibility to pure titanium, but it promotes regeneration significantly faster than pure Ti.	[33]
Zn-0.4Li	Screw and plate	Ti-6Al-4V	Femoral shaft fracture	Rabbits	Plates and screws made of Zn-0.4Li alloy showed comparable performance to Ti-6Al-4V in fracture fixation, and the fractured bone healed completely six months after surgery.	[25]
Zn-1Mg-nvol%β-TCP (n = 0, 1)	Columnar samples	Zn-1Mg	Specimens in lateral thighs.	Rats	Zn-1Mg alloy and Zn-1Mg-β-TCP composites had no significant tissue inflammation and showed good biocompatibility.	[56]

6. Summary and Future Directions

A growing number of new Zn-based biodegradable materials have been developed and their biodegradability, mechanical properties, and biocompatibility were tested mostly in vitro and partially in vivo. An ideal biodegradable material for orthopaedic internal fixation should have a suitable combination of biocompatibility, biodegradability, and mechanical properties (YS, UTS, and ε). Although the mechanical properties of pure Zn are difficult to meet the requirements of orthopaedic fixation, Zn-based alloys can achieve the mechanical properties of traditional implants used in internal fixation at load-bearing sites. Zn-based materials have a moderate corrosion rate and good biocompatibility. Their degradation by-product Zn^{2+} can promote bone growth and mineralization. These properties support Zn-based biodegradable materials as an alternative for internal fixation implants at heavy load-bearing skeletal sites. However, many questions still need to be addressed before Zn-based biodegradable materials can be used for fracture fixation in clinics (Figure 8).

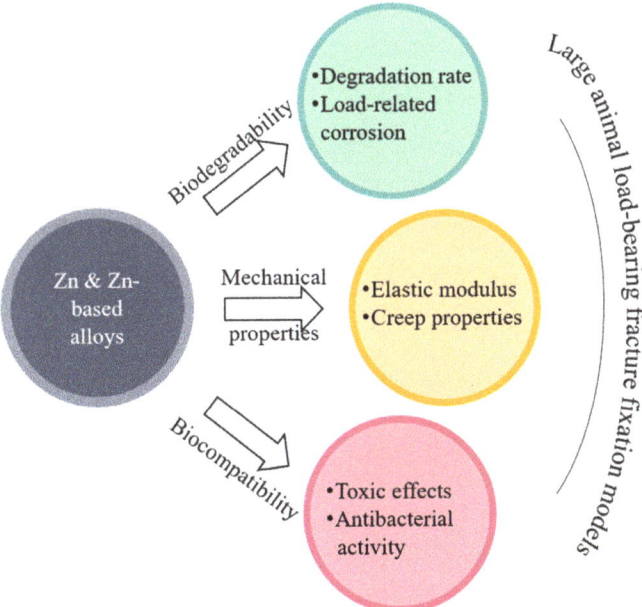

Figure 8. Future directions of Zn-based biodegradable materials.

In terms of biodegradability, (1) the current degradation rate of Zn-based biodegradable materials remains relatively slow, and it needs to be further tuned according to the target skeletal site to match its healing rate; (2) due to existence of static and dynamic loads on the skeleton, the stress corrosion and fatigue corrosion of the materials need to be better understood. In terms of the mechanical properties, (1) the elastic modulus (94–110 GPa) of current Zn-based biodegradable materials is higher than that of bone and should be reduced to avoid stress shielding when being used in internal fixation; (2) the creep effect of Zn-based alloys on the failure of the internal fixation implant at a physiological temperature of the human body should be explored further. In terms of biocompatibility, (1) since high content of Zn^{2+} has toxic effects on cells, attempts should be made to regulate the degradation rate of internal fixation to ensure that the concentration of degradation products does not exceed the safe concentration range of the implant site; (2) antibacterial properties may be explored further. Additionally, in vivo experiments should move from small animal models to large animal models for heavy load-bearing fracture fixation.

Author Contributions: Conceptualization, Y.L. and H.Y.; validation, T.D., A.Q., Y.M. and H.Y.; formal analysis, Y.M.; investigation, Y.L.; writing—original draft preparation, Y.L.; writing—review and editing, H.Y. All authors have read and agreed to the published version of the manuscript.

Funding: This research was funded by National Natural Science Foundation of China, grant numbers 11702008 and 12272017; Beijing Natural Science Foundation, grant number 7202003; Beijing Municipal Education Commission Research Program, grant number KM202010005035; China Postdoctoral Science Foundation funded project, grant number 2021TQ0020; Cultivation fund of Faculty of Enviroment and Life.

Data Availability Statement: Not applicable.

Conflicts of Interest: The authors declare no conflict of interest. The funders had no role in the design of the study; in the collection, analyses, or interpretation of data; in the writing of the manuscript; or in the decision to publish the results.

References

1. Shuai, C.; Li, S.; Peng, S.; Feng, P.; Lai, Y.; Gao, C. Biodegradable metallic bone implants. *Mater. Chem. Front.* **2019**, *3*, 544–562 [CrossRef]
2. Fu, R.; Feng, Y.; Liu, Y.; Willie, B.M.; Yang, H. The combined effects of dynamization time and degree on bone healing. *J. Orthop. Res.* **2022**, *40*, 634–643. [CrossRef] [PubMed]
3. Fu, R.; Feng, Y.; Bertrand, D.; Du, T.; Liu, Y.; Willie, B.M.; Yang, H. Enhancing the efficiency of distraction osteogenesis through rate-varying distraction: A computational study. *Int. J. Mol. Sci.* **2021**, *22*, 11734. [CrossRef] [PubMed]
4. Tucker, A. Management of common upper limb fractures in adults. *Surgery* **2022**, *40*, 184–191. [CrossRef]
5. Wu, A.M.; Bisignano, C.; James, S.L.; Abady, G.G.; Abedi, A.; Abu-Gharbieh, E.; Alhassan, R.K.; Alipour, V.; Arabloo, J.; Asaad, M.; et al. Global, regional, and national burden of bone fractures in 204 countries and territories, 1990–2019: A systematic analysis from the Global Burden of Disease Study 2019. *Lancet Healthy Longev.* **2021**, *2*, e580–e592. [CrossRef]
6. Fu, R.; Feng, Y.; Liu, Y.; Yang, H. Mechanical regulation of bone regeneration during distraction osteogenesis. *Med. Nov. Technol. Devices* **2021**, *11*, 100077. [CrossRef]
7. Fu, R.; Bertrand, D.; Wang, J.; Kavaseri, K.; Feng, Y.; Du, T.; Liu, Y.; Willie, B.M.; Yang, H. In vivo and in silico monitoring bone regeneration during distraction osteogenesis of the mouse femur. *Comput. Methods Programs Biomed.* **2022**, *216*, 106679. [CrossRef]
8. Stiffler, K.S. Internal fracture fixation. *Clin. Tech. Small Anim. Pract.* **2004**, *19*, 105–113. [CrossRef]
9. Taljanovic, M.S.; Jones, M.D.; Ruth, J.T.; Benjamin, J.B.; Sheppard, J.E.; Hunter, T.B. Fracture Fixation. *Radiographics* **2003**, *23*, 1569–1590. [CrossRef]
10. Manam, N.S.; Harun, W.S.W.; Shri, D.N.A.; Ghani, S.A.C.; Kurniawan, T.; Ismail, M.H.; Ibrahim, M.H.I. Study of corrosion in biocompatible metals for implants: A review. *J. Alloys Compd.* **2017**, *701*, 698–715. [CrossRef]
11. Hasiak, M.; Sobieszczanska, B.; Laszcz, A.; Bialy, M.; Checmanowski, J.; Zatonski, T.; Bozemska, E.; Wawrzynska, M. Production, mechanical properties and biomedical characterization of ZrTi-based bulk metallic glasses in comparison with 316L stainless steel and Ti6Al4V alloy. *Materials* **2021**, *15*, 252. [CrossRef] [PubMed]
12. Iatecola, A.; Longhitano, G.A.; Antunes, L.H.M.; Jardini, A.L.; Miguel, E.C.; Beres, M.; Lambert, C.S.; Andrade, T.N.; Buchaim, R.L.; Buchaim, D.V.; et al. Osseointegration improvement of Co-Cr-Mo alloy produced by additive manufacturing. *Pharmaceutics* **2021**, *13*, 724. [CrossRef] [PubMed]
13. Okazaki, Y.; Katsuda, S.I. Biological safety evaluation and surface modification of biocompatible Ti-15Zr-4Nb alloy. *Materials* **2021**, *14*, 731. [CrossRef] [PubMed]
14. Kabir, H.; Munir, K.; Wen, C.; Li, Y. Recent research and progress of biodegradable zinc alloys and composites for biomedical applications: Biomechanical and biocorrosion perspectives. *Bioact. Mater.* **2021**, *6*, 836–879. [CrossRef]
15. Wang, J.L.; Xu, J.K.; Hopkins, C.; Chow, D.H.; Qin, L. Biodegradable magnesium-based implants in orthopedics-A general review and perspectives. *Adv. Sci.* **2020**, *7*, 1902443. [CrossRef]
16. Witte, F.; Hort, N.; Vogt, C.; Cohen, S.; Kainer, K.U.; Willumeit, R.; Feyerabend, F. Degradable biomaterials based on magnesium corrosion. *Curr. Opin. Solid State Mater. Sci.* **2008**, *12*, 63–72. [CrossRef]
17. Jang, H.Y.; Shin, J.Y.; Oh, S.H.; Byun, J.H.; Lee, J.H. PCL/HA hybrid microspheres for effective osteogenic differentiation and bone regeneration. *ACS Biomater. Sci. Eng.* **2020**, *6*, 5172–5180. [CrossRef]
18. Yang, Y.; Zhao, Y.; Tang, G.; Li, H.; Yuan, X.; Fan, Y. In vitro degradation of porous poly(l-lactide-co-glycolide)/β-tricalcium phosphate (PLGA/β-TCP) scaffolds under dynamic and static conditions. *Polym. Degrad. Stab.* **2008**, *93*, 1838–1845. [CrossRef]
19. Zhou, H.; Lawrence, J.G.; Bhaduri, S.B. Fabrication aspects of PLA-CaP/PLGA-CaP composites for orthopedic applications: A review. *Acta Biomater.* **2012**, *8*, 1999–2016. [CrossRef]
20. Zhang, Y.; Xu, J.; Ruan, Y.C.; Yu, M.K.; O'Laughlin, M.; Wise, H.; Chen, D.; Tian, L.; Shi, D.; Wang, J.; et al. Implant-derived magnesium induces local neuronal production of CGRP to improve bone-fracture healing in rats. *Nat. Med.* **2016**, *22*, 1160–1169. [CrossRef]

1. Lee, J.W.; Han, H.S.; Han, K.J.; Park, J.; Jeon, H.; Ok, M.R.; Seok, H.K.; Ahn, J.P.; Lee, K.E.; Lee, D.H.; et al. Long-term clinical study and multiscale analysis of in vivo biodegradation mechanism of Mg alloy. *Proc. Natl. Acad. Sci. USA* **2016**, *113*, 716–721. [CrossRef] [PubMed]
2. Zheng, Y.F.; Gu, X.N.; Witte, F. Biodegradable metals. *Mater. Sci. Eng. R Rep.* **2014**, *77*, 1–34. [CrossRef]
3. Kumar, K.; Gill, R.S.; Batra, U. Challenges and opportunities for biodegradable magnesium alloy implants. *Mater. Technol.* **2017**, *33*, 153–172. [CrossRef]
4. Yang, H.; Jia, B.; Zhang, Z.; Qu, X.; Li, G.; Lin, W.; Zhu, D.; Dai, K.; Zheng, Y. Alloying design of biodegradable zinc as promising bone implants for load-bearing applications. *Nat. Commun.* **2020**, *11*, 401. [CrossRef] [PubMed]
5. Yang, H.; Qu, X.; Wang, M.; Cheng, H.; Jia, B.; Nie, J.; Dai, K.; Zheng, Y. Zn-0.4Li alloy shows great potential for the fixation and healing of bone fractures at load-bearing sites. *Chem. Eng. J.* **2021**, *417*, 129317. [CrossRef]
6. Li, G.; Yang, H.; Zheng, Y.; Chen, X.H.; Yang, J.A.; Zhu, D.; Ruan, L.; Takashima, K. Challenges in the use of zinc and its alloys as biodegradable metals: Perspective from biomechanical compatibility. *Acta Biomater.* **2019**, *97*, 23–45. [CrossRef] [PubMed]
7. Li, H.F.; Shi, Z.Z.; Wang, L.N. Opportunities and challenges of biodegradable Zn-based alloys. *J. Mater. Sci. Technol.* **2020**, *46*, 136–138. [CrossRef]
8. Torne, K.; Larsson, M.; Norlin, A.; Weissenrieder, J. Degradation of Zinc in saline solutions, plasma, and whole blood. *J. Biomed. Mater. Res. B Appl. Biomater.* **2016**, *104*, 1141–1151. [CrossRef]
9. Andreini, C.; Banci, L.; Bertini, I.; Rosato, A. Counting the Zinc-proteins encoded in the human genome. *J. Proteome Res.* **2006**, *5*, 196–201. [CrossRef]
10. Weiss, A.; Murdoch, C.C.; Edmonds, K.A.; Jordan, M.R.; Monteith, A.J.; Perera, Y.R.; Rodriguez Nassif, A.M.; Petoletti, A.M.; Beavers, W.N.; Munneke, M.J.; et al. Zn-regulated GTPase metalloprotein activator 1 modulates vertebrate zinc homeostasis. *Cell* **2022**, *185*, 2148–2163. [CrossRef]
11. Qiao, W.; Pan, D.; Zheng, Y.; Wu, S.; Liu, X.; Chen, Z.; Wan, M.; Feng, S.; Cheung, K.M.C.; Yeung, K.W.K.; et al. Divalent metal cations stimulate skeleton interoception for new bone formation in mouse injury models. *Nat. Commun.* **2022**, *13*, 535. [CrossRef]
12. Zhu, D.; Cockerill, I.; Su, Y.; Zhang, Z.; Fu, J.; Lee, K.W.; Ma, J.; Okpokwasili, C.; Tang, L.; Zheng, Y.; et al. Mechanical strength, biodegradation, and in vitro and in vivo biocompatibility of Zn biomaterials. *ACS Appl. Mater. Interfaces* **2019**, *11*, 6809–6819. [CrossRef] [PubMed]
13. Zhang, Z.; Jia, B.; Yang, H.; Han, Y.; Wu, Q.; Dai, K.; Zheng, Y. Biodegradable ZnLiCa ternary alloys for critical-sized bone defect regeneration at load-bearing sites: In vitro and in vivo studies. *Bioact. Mater.* **2021**, *6*, 3999–4013. [CrossRef]
14. Bowen, P.K.; Seitz, J.M.; Guillory, R.J.; Braykovich, J.P.; Zhao, S.; Goldman, J.; Drelich, J.W. Evaluation of wrought Zn-Al alloys (1; 3; and 5 wt % Al) through mechanical and in vivo testing for stent applications. *J. Biomed. Mater. Res. B Appl. Biomater.* **2018**, *106*, 245–258. [CrossRef] [PubMed]
15. Dambatta, M.S.; Izman, S.; Kurniawan, D.; Hermawan, H. Processing of Zn-3Mg alloy by equal channel angular pressing for biodegradable metal implants. *J. King Saud. Univ. Sci.* **2017**, *29*, 455–461. [CrossRef]
16. Jia, B.; Yang, H.; Zhang, Z.; Qu, X.; Jia, X.; Wu, Q.; Han, Y.; Zheng, Y.; Dai, K. Biodegradable Zn-Sr alloy for bone regeneration in rat femoral condyle defect model: In vitro and in vivo studies. *Bioact. Mater.* **2021**, *6*, 1588–1604. [CrossRef]
17. Kafri, A.; Ovadia, S.; Goldman, J.; Drelich, J.; Aghion, E. The suitability of Zn-1.3%Fe alloy as a biodegradable implant material. *Metals* **2018**, *8*, 153. [CrossRef]
18. Krezel, A.; Maret, W. The biological inorganic chemistry of zinc ions. *Arch. Biochem. Biophys.* **2016**, *611*, 3–19. [CrossRef] [PubMed]
19. Ren, T.; Gao, X.; Xu, C.; Yang, L.; Guo, P.; Liu, H.; Chen, Y.; Sun, W.; Song, Z. Evaluation of as-extruded ternary Zn–Mg–Zr alloys for biomedical implantation material: In vitro and in vivo behavior. *Mater. Corros.* **2019**, *70*, 1056–1070. [CrossRef]
20. Shi, Z.Z.; Yu, J.; Liu, X.F.; Zhang, H.J.; Zhang, D.W.; Yin, Y.X.; Wang, L.N. Effects of Ag, Cu or Ca addition on microstructure and comprehensive properties of biodegradable Zn-0.8Mn alloy. *Mater. Sci. Eng. C Mater. Biol. Appl.* **2019**, *99*, 969–978. [CrossRef]
21. Shi, Z.; Yu, J.; Liu, X. Microalloyed Zn-Mn alloys: From extremely brittle to extraordinarily ductile at room temperature. *Mater. Des.* **2018**, *144*, 343–352. [CrossRef]
22. Shuai, C.; Cheng, Y.; Yang, S.; Yang, W.; Qi, F. Laser additive manufacturing of Zn-2Al part for bone repair: Formability; microstructure and properties. *J. Alloys Compd.* **2019**, *798*, 606–615. [CrossRef]
23. Shuai, C.; Xue, L.; Gao, C.; Peng, S.; Zhao, Z. Rod-like eutectic structure in biodegradable Zn-Al-Sn alloy exhibiting enhanced mechanical strength. *ACS Biomater. Sci. Eng.* **2020**, *6*, 3821–3831. [CrossRef]
24. Shuai, C.; Xue, L.; Gao, C.; Yang, Y.; Peng, S.; Zhang, Y. Selective laser melting of Zn–Ag alloys for bone repair: Microstructure; mechanical properties and degradation behavior. *Virtual Phys. Prototyp.* **2018**, *13*, 146–154. [CrossRef]
25. Shi, Z.; Yu, J.; Liu, X.; Wang, L. Fabrication and characterization of novel biodegradable Zn-Mn-Cu alloys. *J. Mater. Sci. Technol.* **2018**, *34*, 1008–1015. [CrossRef]
26. Xiao, C.; Wang, L.; Ren, Y.; Sun, S.; Zhang, E.; Yan, C.; Liu, Q.; Sun, X.; Shou, F.; Duan, J.; et al. Indirectly extruded biodegradable Zn-0.05wt%Mg alloy with improved strength and ductility: In vitro and in vivo studies. *J. Mater. Sci. Technol.* **2018**, *34*, 1618–1627. [CrossRef]
27. Yuan, W.; Xia, D.; Wu, S.; Zheng, Y.; Guan, Z.; Rau, J.V. A review on current research status of the surface modification of Zn-based biodegradable metals. *Bioact. Mater.* **2022**, *7*, 192–216. [CrossRef] [PubMed]
28. Chen, K.; Lu, Y.; Tang, H.; Gao, Y.; Zhao, F.; Gu, X.; Fan, Y. Effect of strain on degradation behaviors of WE43, Fe and Zn wires. *Acta Biomater.* **2020**, *113*, 627–645. [CrossRef]

49. Yang, J.; Yim, C.D.; You, B.S. Effects of solute Zn on corrosion film of Mg–Sn–Zn alloy formed in NaCl solution. *J. Electrochem. Soc.* **2016**, *163*, C839–C844. [CrossRef]
50. Kubasek, J.; Vojtech, D.; Jablonska, E.; Pospisilova, I.; Lipov, J.; Ruml, T. Structure, mechanical characteristics and in vitro degradation; cytotoxicity; genotoxicity and mutagenicity of novel biodegradable Zn-Mg alloys. *Mater. Sci. Eng. C Mater. Biol. Appl.* **2016**, *58*, 24–35. [CrossRef]
51. Li, H.F.; Xie, X.H.; Zheng, Y.F.; Cong, Y.; Zhou, F.Y.; Qiu, K.J.; Wang, X.; Chen, S.H.; Huang, L.; Tian, L.; et al. Development of biodegradable Zn-1X binary alloys with nutrient alloying elements Mg, Ca and Sr. *Sci. Rep.* **2015**, *5*, 10719. [CrossRef] [PubMed]
52. Tong, X.; Zhang, D.; Zhang, X.; Su, Y.; Shi, Z.; Wang, K.; Lin, J.; Li, Y.; Lin, J.; Wen, C. Microstructure; mechanical properties; biocompatibility; and in vitro corrosion and degradation behavior of a new Zn-5Ge alloy for biodegradable implant materials. *Acta Biomater.* **2018**, *82*, 197–204. [CrossRef] [PubMed]
53. Li, P.; Zhang, W.; Dai, J.; Xepapadeas, A.B.; Schweizer, E.; Alexander, D.; Scheideler, L.; Zhou, C.; Zhang, H.; Wan, G.; et al. Investigation of zinccopper alloys as potential materials for craniomaxillofacial osteosynthesis implants. *Mater. Sci. Eng. C Mater. Biol. Appl.* **2019**, *103*, 109826. [CrossRef]
54. Wątroba, M.; Bednarczyk, W.; Kawałko, J.; Bała, P. Effect of zirconium microaddition on the microstructure and mechanical properties of Zn-Zr alloys. *Mater. Charact.* **2018**, *142*, 187–194. [CrossRef]
55. Yang, H.; Qu, X.; Lin, W.; Wang, C.; Zhu, D.; Dai, K.; Zheng, Y. In vitro and in vivo studies on zinc-hydroxyapatite composites as novel biodegradable metal matrix composite for orthopedic applications. *Acta Biomater.* **2018**, *71*, 200–214. [CrossRef]
56. Pan, C.; Sun, X.; Xu, G.; Su, Y.; Liu, D. The effects of beta-TCP on mechanical properties; corrosion behavior and biocompatibility of beta-TCP/Zn-Mg composites. *Mater. Sci. Eng. C Mater. Biol. Appl.* **2020**, *108*, 110397. [CrossRef]
57. Li, X.; Chu, C.; Chu, P.K. Effects of external stress on biodegradable orthopedic materials: A review. *Bioact. Mater.* **2016**, *1*, 77–84. [CrossRef]
58. Li, N.; Zheng, Y. Novel magnesium alloys developed for biomedical application: A review. *J. Mater. Sci. Technol.* **2013**, *29*, 489–502. [CrossRef]
59. Kirkland, N.T.; Birbilis, N.; Staiger, M.P. Assessing the corrosion of biodegradable magnesium implants: A critical review of current methodologies and their limitations. *Acta Biomater.* **2012**, *8*, 925–936. [CrossRef]
60. Li, G.N.; Zhu, S.M.; Nie, J.F.; Zheng, Y.; Sun, Z. Investigating the stress corrosion cracking of a biodegradable Zn-0.8 wt%Li alloy in simulated body fluid. *Bioact. Mater.* **2021**, *6*, 1468–1478. [CrossRef]
61. Li, P.; Feng, X.; Jia, X.; Fan, Y. Influences of tensile load on in vitro degradation of an electrospun poly(L-lactide-co-glycolide) scaffold. *Acta Biomater.* **2010**, *6*, 2991–2996. [CrossRef] [PubMed]
62. Guo, M.; Chu, Z.; Yao, J.; Feng, W.; Wang, Y.; Wang, L.; Fan, Y. The effects of tensile stress on degradation of biodegradable PLGA membranes: A quantitative study. *Polym. Degrad. Stab.* **2016**, *124*, 95–100. [CrossRef]
63. Gao, Y.; Wang, L.; Li, L.; Gu, X.; Zhang, K.; Xia, J.; Fan, Y. Effect of stress on corrosion of high-purity magnesium in vitro and in vivo. *Acta Biomater.* **2019**, *83*, 477–486. [CrossRef] [PubMed]
64. Gao, Y.; Wang, L.; Gu, X.; Chu, Z.; Guo, M.; Fan, Y. A quantitative study on magnesium alloy stent biodegradation. *J. Biomech.* **2018**, *74*, 98–105. [CrossRef]
65. Vasudevan, A.K.; Sadananda, K. Classification of environmentally assisted fatigue crack growth behavior. *Int. J. Fatigue* **2009**, *31*, 1696–1708. [CrossRef]
66. Jafari, S.; Singh Raman, R.K.; Davies, C.H.J. Corrosion fatigue of a magnesium alloy in modified simulated body fluid. *Eng. Fract. Mech.* **2015**, *137*, 2–11. [CrossRef]
67. Zhao, J.; Gao, L.L.; Gao, H.; Yuan, X.; Chen, X. Biodegradable behaviour and fatigue life of ZEK100 magnesium alloy in simulated physiological environment. *Fatigue Fract. Eng. Mater. Struct.* **2015**, *38*, 904–913. [CrossRef]
68. Shi, Z.; Li, H.; Xu, J.; Gao, X.; Liu, X. Microstructure evolution of a high-strength low-alloy Zn–Mn–Ca alloy through casting; hot extrusion and warm caliber rolling. *Mater. Sci. Eng. A* **2020**, *771*, 138626. [CrossRef]
69. Guan, Z.; Linsley, C.S.; Pan, S.; Yao, G.; Wu, B.M.; Levi, D.S.; Li, X. Zn-Mg-WC nanocomposites for bioresorbable cardiovascular stents: Microstructure, mechanical properties, fatigue, shelf life, and corrosion. *ACS Biomater. Sci. Eng.* **2022**, *8*, 328–339. [CrossRef]
70. Jia, B.; Yang, H.; Han, Y.; Zhang, Z.; Qu, X.; Zhuang, Y.; Wu, Q.; Zheng, Y.; Dai, K. In vitro and in vivo studies of Zn-Mn biodegradable metals designed for orthopedic applications. *Acta Biomater.* **2020**, *108*, 358–372. [CrossRef]
71. Katarivas Levy, G.; Leon, A.; Kafri, A.; Ventura, Y.; Drelich, J.W.; Goldman, J.; Vago, R.; Aghion, E. Evaluation of biodegradable Zn-1%Mg and Zn-1%Mg-0.5%Ca alloys for biomedical applications. *J. Mater. Sci. Mater. Med.* **2017**, *28*, 174. [CrossRef] [PubMed]
72. Liu, C.; Li, Y.; Ge, Q.; Liu, Z.; Qiao, A.; Mu, Y. Mechanical characteristics and in vitro degradation of biodegradable Zn-Al alloy. *Mater. Lett.* **2021**, *300*, 130181. [CrossRef]
73. Tang, Z.; Niu, J.; Huang, H.; Zhang, H.; Pei, J.; Ou, J.; Yuan, G. Potential biodegradable Zn-Cu binary alloys developed for cardiovascular implant applications. *J. Mech. Behav. Biomed. Mater.* **2017**, *72*, 182–191. [CrossRef]
74. Lin, J.; Tong, X.; Shi, Z.; Zhang, D.; Zhang, L.; Wang, K.; Wei, A.; Jin, L.; Lin, J.; Li, Y.; et al. A biodegradable Zn-1Cu-0.1Ti alloy with antibacterial properties for orthopedic applications. *Acta Biomater.* **2020**, *106*, 410–427. [PubMed]
75. Zhu, S.; Wu, C.; Li, G.; Zheng, Y.; Nie, J. Creep properties of biodegradable Zn-0.1Li alloy at human body temperature: Implications for its durability as stents. *Mater. Res. Lett.* **2019**, *7*, 347–353. [CrossRef]

36. Yuan, W.; Xia, D.; Zheng, Y.; Liu, X.; Wu, S.; Li, B.; Han, Y.; Jia, Z.; Zhu, D.; Ruan, L.; et al. Controllable biodegradation and enhanced osseointegration of ZrO2-nanofilm coated Zn-Li alloy: In vitro and in vivo studies. *Acta Biomater.* **2020**, *105*, 290–303. [CrossRef]
37. Li, H.; Yang, H.; Zheng, Y.; Zhou, F.; Qiu, K.; Wang, X. Design and characterizations of novel biodegradable ternary Zn-based alloys with IIA nutrient alloying elements Mg, Ca and Sr. *Mater. Des.* **2015**, *83*, 95–102. [CrossRef]
38. Lin, J.; Tong, X.; Sun, Q.; Luan, Y.; Zhang, D.; Shi, Z.; Wang, K.; Lin, J.; Li, Y.; Dargusch, M.; et al. Biodegradable ternary Zn-3Ge-0.5X (X = Cu; Mg; and Fe) alloys for orthopedic applications. *Acta Biomater.* **2020**, *115*, 432–446. [CrossRef]
39. Mostaed, E.; Sikora-Jasinska, M.; Ardakani, M.S.; Mostaed, A.; Reaney, I.M.; Goldman, J.; Drelich, J.W. Towards revealing key factors in mechanical instability of bioabsorbable Zn-based alloys for intended vascular stenting. *Acta Biomater.* **2020**, *105*, 319–335. [CrossRef]
40. Xue, P.; Ma, M.; Li, Y.; Li, X.; Yuan, J.; Shi, G.; Wang, K.; Zhang, K. Microstructure, mechanical properties, and in vitro corrosion behavior of biodegradable Zn-1Fe-xMg alloy. *Materials* **2020**, *13*, 4835. [CrossRef]
41. Bowen, P.K.; Shearier, E.R.; Zhao, S.; Guillory II, R.J.; Zhao, F.; Goldman, J.; Drelich, J.W. Biodegradable metals for cardiovascular stents: From clinical concerns to recent Zn-alloys. *Adv. Healthc. Mater.* **2016**, *5*, 1121–1140. [CrossRef]
42. Plum, L.M.; Rink, L.; Haase, H. The essential toxin: Impact of zinc on human health. *Int. J. Environ. Res. Publ. Health.* **2010**, *7*, 1342–1365. [CrossRef] [PubMed]
43. O'Connor, J.P.; Kanjilal, D.; Teitelbaum, M.; Lin, S.S.; Cottrell, J.A. Zinc as a therapeutic agent in bone regeneration. *Materials* **2020**, *13*, 2211. [CrossRef] [PubMed]
44. Tapiero, H.; Tew, K.D. Trace elements in human physiology and pathology: Zinc and metallothioneins. *Biomed. Pharmacother.* **2003**, *57*, 399–411. [CrossRef]
45. Glutsch, V.; Hamm, H.; Goebeler, M. Zinc and skin: An update. *J. Dtsch. Dermatol. Ges.* **2019**, *17*, 589–596. [CrossRef]
46. Jimenez, M.; Abradelo, C.; San Roman, J.; Rojo, L. Bibliographic review on the state of the art of strontium and zinc based regenerative therapies. Recent developments and clinical applications. *J. Mater. Chem. B* **2019**, *7*, 1974–1985. [CrossRef] [PubMed]
47. Hernandez-Escobar, D.; Champagne, S.; Yilmazer, H.; Dikici, B.; Boehlert, C.J.; Hermawan, H. Current status and perspectives of zinc-based absorbable alloys for biomedical applications. *Acta Biomater.* **2019**, *97*, 1–22. [CrossRef]
48. Venezuela, J.; Dargusch, M.S. The influence of alloying and fabrication techniques on the mechanical properties, biodegradability and biocompatibility of zinc: A comprehensive review. *Acta Biomater.* **2019**, *87*, 1–40. [CrossRef] [PubMed]
49. Bakhsheshi-Rad, H.R.; Hamzah, E.; Low, H.T.; Kasiri-Asgarani, M.; Farahany, S.; Akbari, E.; Cho, M.H. Fabrication of biodegradable Zn-Al-Mg alloy: Mechanical properties, corrosion behavior, cytotoxicity and antibacterial activities. *Mater. Sci. Eng. C Mater. Biol. Appl.* **2017**, *73*, 215–219. [CrossRef] [PubMed]
50. Sukhodub, L.B. Antimicrobial activity of Ag^+, Cu^{2+}, Zn^{2+}, Mg^{2+} ions doped chitosan nanoparticles. *Ann. Mechnikov's Inst.* **2015**, *1*, 39–43.
51. Serrano-Aroca, A.; Cano-Vicent, A.; Sabater, I.S.R.; El-Tanani, M.; Aljabali, A.; Tambuwala, M.M.; Mishra, Y.K. Scaffolds in the microbial resistant era: Fabrication, materials, properties and tissue engineering applications. *Mater Today Bio* **2022**, *16*, 100412. [CrossRef]
52. Riduan, S.N.; Zhang, Y. Recent Advances of Zinc-based Antimicrobial Materials. *Chem. Asian J.* **2021**, *16*, 2588–2595. [CrossRef]
53. de Lima, C.O.; de Oliveira, A.L.M.; Chantelle, L.; Silva Filho, E.C.; Jaber, M.; Fonseca, M.G. Zn-doped mesoporous hydroxyapatites and their antimicrobial properties. *Colloids Surf.* **2021**, *198*, 111471. [CrossRef] [PubMed]
54. Tong, X.; Shi, Z.; Xu, L.; Lin, J.; Zhang, D.; Wang, K.; Li, Y.; Wen, C. Degradation behavior, cytotoxicity, hemolysis, and antibacterial properties of electro-deposited Zn-Cu metal foams as potential biodegradable bone implants. *Acta Biomater.* **2020**, *102*, 481–492. [CrossRef] [PubMed]
55. Ren, L.; Yang, K. Antibacterial design for metal implants. In *Metallic Foam Bone*; Woodhead Publishing: Sawston, UK, 2017; pp. 203–216.
56. Cui, Z.; Zhang, Y.; Cheng, Y.; Gong, D.; Wang, W. Microstructure; mechanical, corrosion properties and cytotoxicity of betacalcium polyphosphate reinforced ZK61 magnesium alloy composite by spark plasma sintering. *Mater. Sci. Eng. C Mater. Biol. Appl.* **2019**, *99*, 1035–1047. [CrossRef] [PubMed]
57. Wang, X.; Shao, X.; Dai, T.; Xu, F.; Zhou, J.G.; Qu, G.; Tian, L.; Liu, B.; Liu, Y. In vivo study of the efficacy, biosafety, and degradation of a zinc alloy osteosynthesis system. *Acta Biomater.* **2019**, *92*, 351–361. [CrossRef]
58. Taguchi, T.; Lopez, M.J. An overview of de novo bone generation in animal models. *J. Orthop. Res.* **2021**, *39*, 7–21. [CrossRef]

Article

Compressive Properties and Degradable Behavior of Biodegradable Porous Zinc Fabricated with the Protein Foaming Method

Qiqi Ge [1], Xiaoqian Liu [1], Aike Qiao [2] and Yongliang Mu [1,*]

1 School of Metallurgy, Northeastern University, Shenyang 110819, China
2 Faculty of Environment and Life, Beijing University of Technology, Beijing 100124, China
* Correspondence: myledu@sina.com; Tel.: +86-18802440693

Citation: Ge, Q.; Liu, X.; Qiao, A.; Mu, Y. Compressive Properties and Degradable Behavior of Biodegradable Porous Zinc Fabricated with the Protein Foaming Method. *J. Funct. Biomater.* **2022**, *13*, 151. https://doi.org/10.3390/jfb13030151

Academic Editor: Seung-Kyun Kang

Received: 29 July 2022
Accepted: 9 September 2022
Published: 13 September 2022

Publisher's Note: MDPI stays neutral with regard to jurisdictional claims in published maps and institutional affiliations.

Copyright: © 2022 by the authors. Licensee MDPI, Basel, Switzerland. This article is an open access article distributed under the terms and conditions of the Creative Commons Attribution (CC BY) license (https://creativecommons.org/licenses/by/4.0/).

Abstract: A new protein foaming–consolidation method for preparing porous zinc was developed using three proteins (egg white protein (EWP), bovine bone collagen protein (BBCP), and fish bone collagen protein (FBCP)) as both consolidating and foaming agents. The preparation route utilized powder mixing and sintering processing, which could be divided into three steps: slurry preparation, low-temperature foaming, and high-temperature sintering. The morphological characteristics of the pore structures revealed that the porous zinc had an interconnected open-cell structure. Compared to the porous zinc prepared with EWP or BBCP, the porous zinc prepared with FBCP possessed the largest average pore size and the highest compressive properties. The porosity of the porous zinc increased with the stirring time, the content of protein and sucrose, and higher sintering temperatures. Moreover, a compression test and immersion test were performed to investigate the stress–strain behavior and corrosion properties of the resulting porous zinc. A fluctuated stress plateau could be found due to the brittle fracture of the porous cells. The porous zinc prepared with FBCP showed the highest compressive strength and elastic modulus. The corrosion rate of the porous zinc obtained through an immersion test in vitro using simulated bodily fluids on the thirty-second day was close to 0.02 mm/year. The corresponding corrosion mechanism of porous zinc was also discussed.

Keywords: medical degradation; porous zinc; protein foaming; elasticity modulus; compressive strength

1. Introduction

Medically degradable metal can be metabolized and absorbed into the human body, and the degradation products are harmless for it. The demand for such metals has growing over recent decades due to their excellent biocompatibility and biodegradability and their adequate mechanical properties [1,2]. Porous material is a potential material for implantation for bone tissue regeneration and substitution. Voids and holes in porous materials can ensure that fresh fluid is easily sent into porous implantation materials, allowing new bone or vascular tissue to grow into the material [3]. With the proliferation and differentiation of new bone or vascular cells, the medically degradable porous material will gradually disappear [4,5]. In addition, as medically degradable porous materials must also possess adequate mechanical properties for supporting enough strength for surgery and in vivo tissue recovery processes [6–8].

Over recent decades, bone implantation materials have been made with glass–ceramic [9], polymer [10], and poly-hydroxyapatite composites [11]. However, the mechanical properties of these implantations cannot be satisfactory for the whole process of implantation and healing [12]. On the other hand, though permanent implantation materials such as tantalum [13], titanium [14], and their alloys have adequate mechanical properties, a stress-shielding influence would emerge due to their mechanical properties far more than it would with natural bone [15]. Therefore, degradable metal materials have become a research focus in recent years.

Currently, the most noticeable medically degradable porous metals include zinc (Zn) [16], iron (Fe) [2], magnesium (Mg), and their alloys [17]. Mg possesses suitable mechanical properties and good biocompatibility [18]; however, it releases hydrogen when it degrades in the body, and the degradation rate is so fast that the integrity of the scaffolding function of the implantation cannot be promised [19–21]. Fe possesses a higher elastic modulus, which leads to stress shielding [22]. Furthermore, the corrosion rate of Fe is so slow that it would induce the occurrence of inflammation in the human body [23–26]. Out of these, zinc possesses the most suitable degradation rate [27,28], and, as an essential trace element in the human body, it is integral for nucleic acid metabolism and in the induction of bone cell growth [29–31]. Some parameters of various porous materials are summed up in Table 1. Several fabrication methods for porous zinc have been investigated, including infiltration casting [32], additive manufacturing [33], and selective laser melting [33]. Compared with these technologies, the protein foaming method is one of the most promising approaches due to its easy attainability, friendly interaction with the environment, and harmless effect on the human body [34].

Table 1. Performance parameters of porous materials.

Material and Method	Compressive Strength/MPa	Application	Porosity	Ref.
Fe/3DP	16.7	Bone tissue engineering	80~80.6%	[24]
Fe/PU	0.382 ± 0.024	Bone tissue engineering	96~97%	[25]
Fe/TAED	3.5	Tissue engineering	>90%	[26]
Mg/FDHP	11.1~30.3	Bone substitute applications	33~54%	[18]
Mg/PM	4.4~38	Orthopedic applications	12~38%	[21]
Zn/AMC	6~11	Orthopedic applications	22~65%	[27]
Zn/AM	10.8~13.9	Bone substitution	60~67%	[28]
(PF/HAP) PS	0.3~1.1	Bone tissue engineering	79~89%	[11]
Glass-ceramic/PF	2.6~6.2	Biomaterials scaffold	68~78%	[9]
Polymer/3DP	2.6~6.2	Engineering architected foams	68~78%	[10]
Zn/Protein foaming	1.19~9.20	Cancellous bone substitution	50~85.8%	This work

Various cell structures for porous ceramics and metals can be produced by means of the protein foaming method [35]. Pore structures rely on the diversity of amino acids of proteins to determine the good binder properties of proteins [36–39]. In particular, forming a steady two-dimensional network structure with a protein molecule would cause the bubbles obtained through mechanical stirring in a slurry to be stable [35,40]. Currently, the protein foaming method is adopted to produce porous materials of Ti_4Al_6V [41] and porous ceramic [42], which cannot degrade within the human body. Porous zinc, however, is promising for use as a substitution for cancellous bone.

In this work, various proteins were used as foaming agents to prepare porous zinc [43,44]. The pore structures and microstructures were characterized, and the compressive properties and corrosion behaviors in simulated body fluid (SBF) were also investigated.

2. Materials and Methods

2.1. Raw Materials

Porous zinc was produced from zinc powder (granularity: 103 μm; density of zinc powder; 7.14 g/cm^3; 99.99%), egg white protein (EWP) purchased from a local market, bovine bone collagen protein (BBCP; Shaanxi Chenming biological Co., Ltd., Xi'an, China), and fish bone collagen protein (FBCP; Dongsheng Biotechnology Co., Ltd., Guangzhou, China). Sucrose (AR) and polyvinyl alcohol (AR) were added as a binder and a dispersant. The distribution of particle sizes for zinc powder is illustrated in Figure 1; it was measured by using a laser particle size analyzer (Master Size 2000, Malvern Panalytical, Malvern, UK).

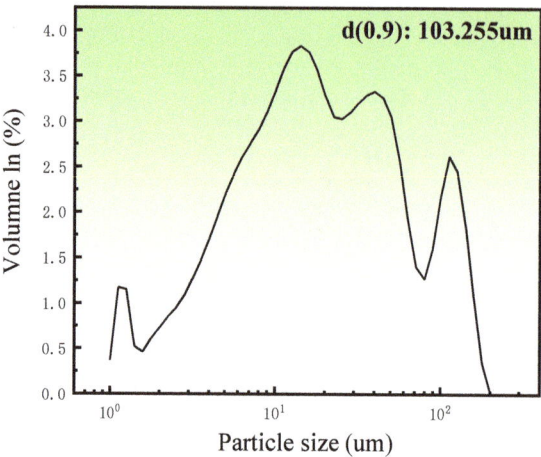

Figure 1. Zinc particle size distribution.

2.2. Fabrication of Porous Zinc

In order to determine the decomposition temperatures of sucrose, proteins, and a mixture of zinc powder and additives, a sintering process, differential thermal analysis (DTA, NETZSCH Gerätebau GmbH, Selb, Germany), and thermogravimetric analysis (TGA, NETZSCH Gerätebau GmbH, Selb, Germany) were used, as shown in Figure 2. of the endothermic and exothermic peaks of the three types of proteins used in this study almost overlapped; Figure 2b represent all of the proteins used in this study.

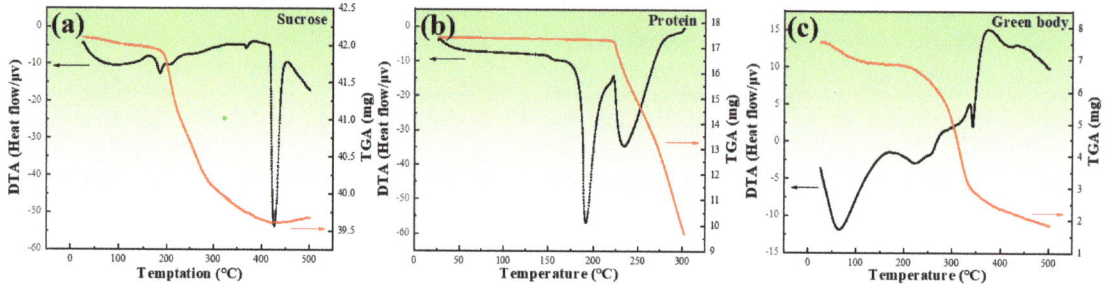

Figure 2. DTA and TGA for the raw materials: (**a**) sucrose; (**b**) proteins; (**c**) green body of porous zinc. The red line represents the TGA, and the black line represents the DTA.

Figure 2a shows the result of the DTA and TGA for sucrose, which exhibited endothermic peaks at 190 and 236 °C, respectively. The endothermic peak in DTA was attributed to the phase transition that occurred during heating. The sudden drop in the TGA was due to the gases released during thermal decomposition. In combination with the sudden drop in the TGA curve at around 213 °C, this indicated the melting and decomposition of sucrose. Figure 2b presents the results of the DTA and TGA for the proteins, where three endothermic peaks can be observed, and the mass loss appeared from the beginning of the heating. It was obvious that the proteins were decomposing. The decomposition rate increased from about 200 °C. Figure 2c exhibits the results of the DTA and TGA for the mixture. The TGA curve sharply decreased at about 200 °C. Four endothermic peaks at around 188, 207, 373, and 425 °C could be observed, corresponding to the melting and decomposition of sucrose, the decomposition of the proteins, and the melting of the zinc powder, respectively.

Figure 3 shows the fabrication process of porous zinc, which could be divided into three stages. (a) The slurry preparation stage: The zinc powder, protein (EWP, BBCP, or FBCP), sucrose, polyvinyl alcohol, and dilute HCl were added into deionized water. Table 2 shows the composition and content of the slurry. By carrying out substantial experiments, optimal parameters were obtained and porous zinc was successfully fabricated. Proteins were chosen according to the relevant literature [43,44]. A stirring time of 5–20 min at a speed of 80–100 rpm was necessary to introduce numerous air bubbles. The slurry was then allowed to stand at room temperature for 0–20 min. (b) The low-temperature foaming stage: The slurry was placed in a drying oven for foaming at a temperature of 70–100 °C for a period of 2–4 h, during which the moisture was removed and a preliminary porous structure formed. It can be inferred from Figure 2 that, at this temperature, the proteins and sucrose did not decompose. (c) High-temperature sintering stage: A preform covered with graphite was sintered at a temperature of 435–490 °C for 7–10 h in order to prevent the samples from cracking. The samples were then taken out to cool down when the sintering was completed. It can be observed from Figure 2 that the proteins and sucrose decomposed at this temperature and released gases, which affected the morphology of the porous zinc.

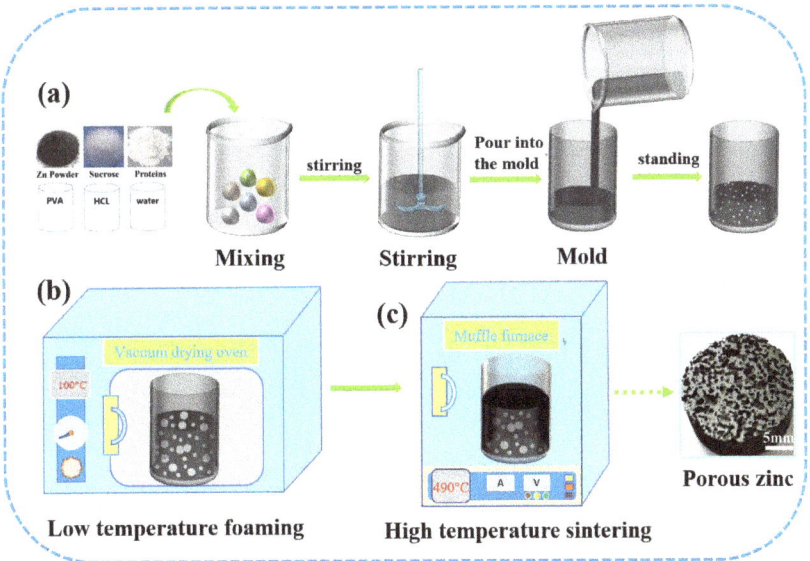

Figure 3. Fabrication process of porous zinc.

Table 2. Slurries consisting of different compositions and contents for the fabrication of porous zinc.

Types of Protein	Content (wt.%)				
	Protein	Sucrose	Polyvinyl Alcohol	Deionized Water	1M HCL
FBCP	0.24	5	5	9.53	0
FBCP	1.2	10	0	8.8	0.67
BBCP	0.24	15	3	8.96	0.45
BBCP	0.80	5	0	3.9	0.22
EWP	3	5	5	4.7	0
EWP	15	10	3	2.92	0.67

2.3. Property Characterization

The porosity of the porous zinc was measured with the following equation.

$$P = \left(1 - \frac{\rho^*}{\rho_s}\right) \times 100\% \tag{1}$$

where ρ^* and ρ_s are the densities of the porous zinc and the cell wall material, respectively.

The pore size was measured by using Image J on SEM images, and the distribution of pore sizes was obtained by using statistics. Field-emission scanning electron microscopy (Quanta250FEG, FEI Co., Ltd., Hillsboro, OR, USA) was used to observe the pore morphology, pore size, and distribution. X-ray diffraction patterns of porous zinc were obtained using an X-ray diffractometer (D8 ADVANCE, Bruker-AXS Co., Ltd., Karlsruhe, Germany) with Cu–Kα radiation at 40 kV with a scanning rate of 10°/min in the range of 10–90°.

Quasi-static compression tests were conducted in a CMT5105 material testing system (MTS Co., Ltd., Suncheon-si, Korea) with a strain rate of 10^{-3} on specimens with dimensions of 15 mm × 18 mm [45]. At least five tests were conducted for each specimen to guarantee the reliability of the results.

The corrosion behavior of porous zinc produced with FBCP was investigated with an in vitro immersion test in SBF [46] for up to 32 d based on ASTM G31-12. The chemical composition of the solution is indicated in Table 3. The pH of the SBF solution was between 7.4 and 7.45 and was adjusted with 1 M HCl. The diameter and height of the specimen for immersion are 10 mm and 10 mm respectively. The volume of SBF added was 20 mL/cm^2 [47] according to the surface area of the porous zinc, supposing that it was made up of many spheres. The size and number of spheres were related to the porosity and average pore size of porous zinc. The surface areas (cm^2) of the samples were calculated with Equation (2).

$$A = 4\pi \left(\frac{d^2}{2}\right) \times \left(\frac{V \times P}{\frac{4}{3} \times \pi \left(\frac{d}{2}\right)^3}\right) = \frac{6\,V\,P}{d} \tag{2}$$

where d is the average diameter (cm), V is the volume of the samples (cm^3), and P is the porosity of the samples (%).

Table 3. Regents of preparing SBF.

Rank	Reagent	Content	Purity	Molecular Weight
1	NaCl	8.035 g/L	99.5%	58.4430 g/mol
2	NaHCO$_3$	0.355 g/L	99.5%	84.0068 g/mol
3	KCl	0.225 g/L	99.5%	74.5515 g/mol
4	K$_2$HPO$_4$ 3H$_2$O	0.231 g/L	99.0%	228.2220 g/mol
5	MgCl$_2$ 6H$_2$O	0.311 g/L	99.0%	203.3034 g/mol
6	1.0M HCl	39 mL/L	-	-
7	CaCl$_2$	0.292 g/L	95.0%	110.9848 g/mol
8	NaSO$_4$	0.072 g/L	99.0%	142.0428 g/mol
9	Tris	6.118 g/L	99.0%	121.1356 g/mol
10	1.0M HCl	0~5 mL/L	-	-

The immersion temperature was 37 ± 0.5 °C, and the solution was renewed every two days. During immersion, the amounts of zinc ions in the SBF were evaluated with a spectrophotometer (722N, Shang Hai Jing Hua Instrument, Shanghai, China). The degradation byproducts of the porous zinc were removed according to the method [48] using chromic acid solution (200 g/L CrO$_3$). The process of cleaning the degradation byproducts was carried out at 80 °C for 1 min, and then alcohol and water were used to clean again at room temperature. The corrosion rate (mm/year) was calculated with the following equation [47].

$$CR = 8.76 \times 10^4 \frac{M}{A\,t\,\rho} \tag{3}$$

where M is the mass loss weight (g), A is the surface area of the samples (cm^2), t is the test duration (h), and ρ is the density of zinc (g/cm^3).

Field-emission scanning electron microscopy (Quanta250FEG, FEI Co., Ltd., Hillsboro, America) was used to observe the pore morphology and corrosion products of porous Zn. X-ray diffraction patterns of porous zinc were obtained using an X-ray diffractometer with Cu–Kα radiation at 40 KV with a scanning rate of 10°/min in the range of 10–90°.

3. Results and Discussion
3.1. Pore Structure of Porous Zinc

Figure 4 shows the macrostructures of the porous zinc prepared by using different proteins. The samples prepared with EWP (Figure 4a,b), BBCP (Figure 4c,d), and FBCP (Figure 4e,f) exhibited similar structures.

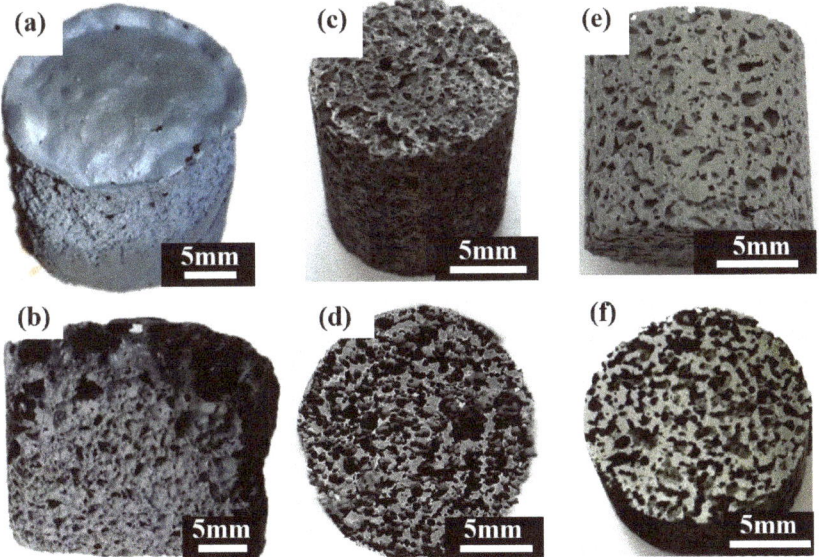

Figure 4. Macrostructures of porous zinc using different proteins: (**a,b**) EWP; (**c,d**) BBCP; (**e,f**) FBCP.

The porosity of porous zinc increased if the stirring time was longer and more protein and sucrose were added. A maximum porosity of nearly 85% was achieved with a longer stirring time and more sucrose and protein. A minimum porosity of nearly 50% was achieved with shorter agitation time and less protein and sucrose. The porosity could even be further decreased below the minimum value by further reducing the stirring time and the content of protein and sucrose.

Figure 5a–f show the microstructures of porous zinc produced with different proteins. An interconnected open-cell structure could be observed in three samples. The pores of the porous zinc foamed with EWP (Figure 5a,b) with a porosity of 76% ± 2% were irregular and dense. However, the sample foamed with BBCP (Figure 5c,d) with a porosity of 76% ± 2% showed circular and regular pores. The pore walls contained numerous hollow spheres (Figure 5d). These could have been contributed by the gas produced through the decomposition of the organic additives during sintering, which was then wrapped by the partly molten zinc. Compared to the samples foamed with EWP and BBCP, that foamed with FBCP with a porosity of 53% ± 2% displayed a larger pore size and thicker pore walls, as shown in Figure 5e,f. A large number of small and roughly circular pores could be found. The results suggested that the porous zinc produced with BBCP/FBCP had a suitable pore structure for implants.

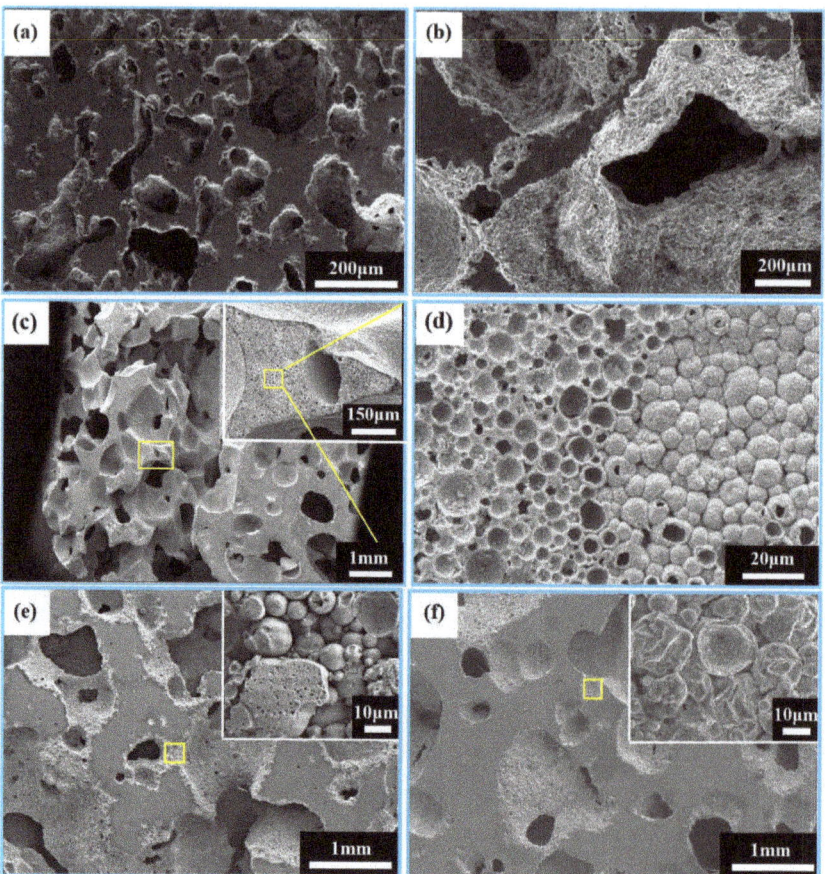

Figure 5. SEM images of porous zinc using different proteins: (**a,b**) EWP with a porosity of 76% ± 2%; (**c,d**) BBCP with a porosity of 52% ± 2% (stirring time: 20 min, sintering temperature: 490 °C); (**e,f**) FBCP with a porosity of 53% ± 2%.

Figure 6 shows the pore size distributions of the porous zinc prepared with EWP, BBCP, and FBCP; the porosity of the various samples of porous Zn was 76% ± 2%, 52% ± 2%, and 53% ± 2%, respectively. It can be seen that the pore sizes of the samples foamed with BBCP and FBCP were much larger than that of the sample foamed with EWP. Figure 7 shows the distribution of pore sizes of the micropores in the pore walls of the sample produced with BBCP that was shown in Figure 5c. The average diameter of the hollow spheres was 7–8 μm. It could be observed that the porous zinc produced with FBCP possessed the highest average pore size, making it more suitable as an implant for the growth of bone tissue.

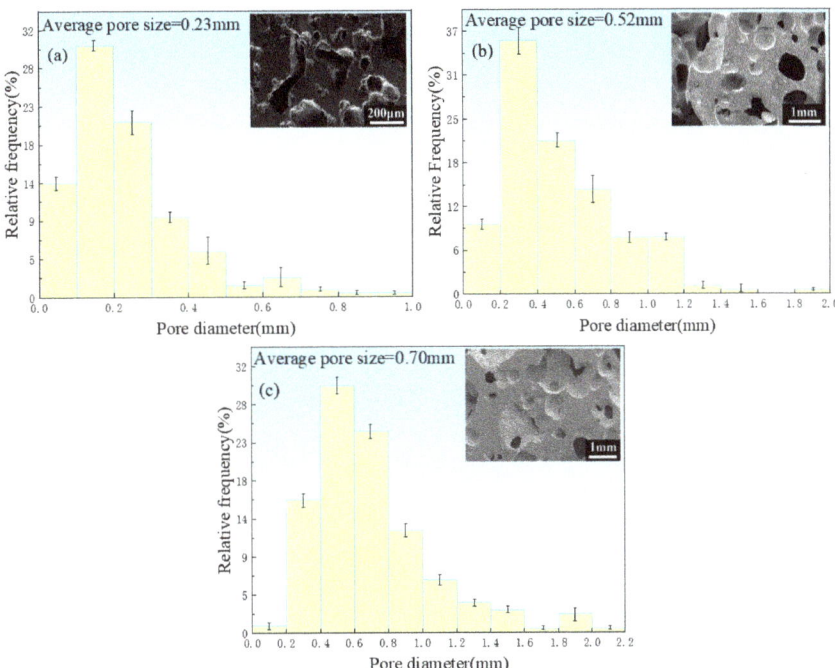

Figure 6. Diameter distribution in porous zinc produced with (**a**) EWP with a porosity of 76% ± 2%, (**b**) BBCP with a porosity of 52% ± 2%, and (**c**) FBCP with a porosity of 53% ± 2%.

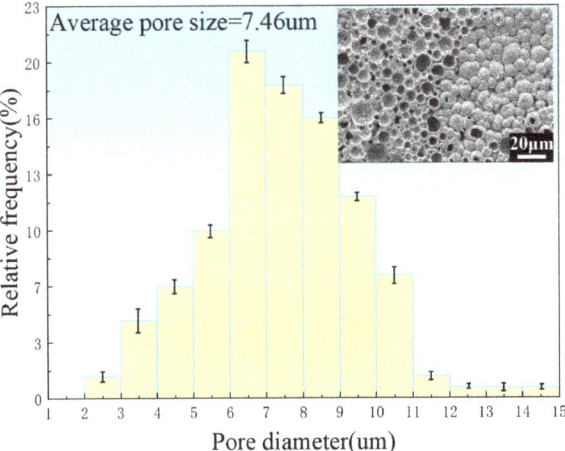

Figure 7. Size distribution of micropores in the pore walls of the porous zinc produced with BBCP with a porosity of 52% ± 2 %.

The XRD patterns of the porous zinc samples sintered at different temperatures are shown in Figure 8. It can be seen that the sintered samples were mainly composed of Zn, ZnO, and $C_{12}H_{10}N_4O_4Zn$. The reason for the formation of $C_{12}H_{10}N_4O_4Zn$ can be attributed to the combination reaction that occurred during the decomposition of proteins and sucrose and the melting of zinc. The peak intensity of $C_{12}H_{10}N_4O_4Zn$ decreased with increasing sintering temperature, demonstrating that a higher sintering temperature was beneficial for the removal of organics.

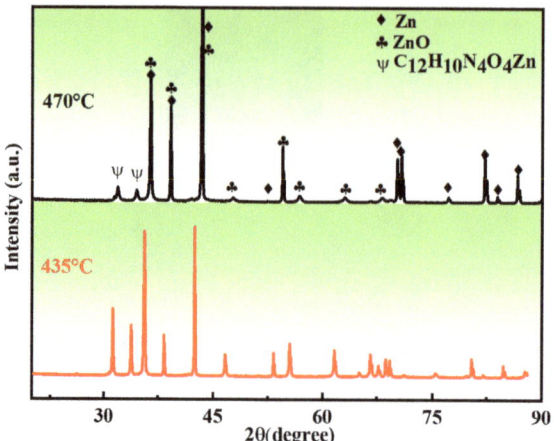

Figure 8. XRD patterns of the porous zinc produced with FBCP with a porosity of 53% ± 2% at different sintering temperatures: (**top**) 470 and (**bottom**) 435 °C.

3.2. Compression Properties

The compressive stress–strain curves of porous zinc are shown in Figure 9. The curves could be divided into three typical stages. The first was the elastic stage. The FBCP samples showed a higher peak stress and elastic modulus compared to those of the BBCP and EWP samples. The EWP samples showed the lowest peak stress and elastic modulus. The second was the stress fluctuation stage, during which the pore walls collapsed with increasing stress. A large stress drop occurred at this stage. It is worth noting that the stress plateau decreased with strain at the second stage when the porosity of sample was lower than 60%, which can be attributed to the brittle fracture of the pore walls. Finally, a densification stage was exhibited, in which the subtended pore walls came into contact, causing an abrupt increase in stress.

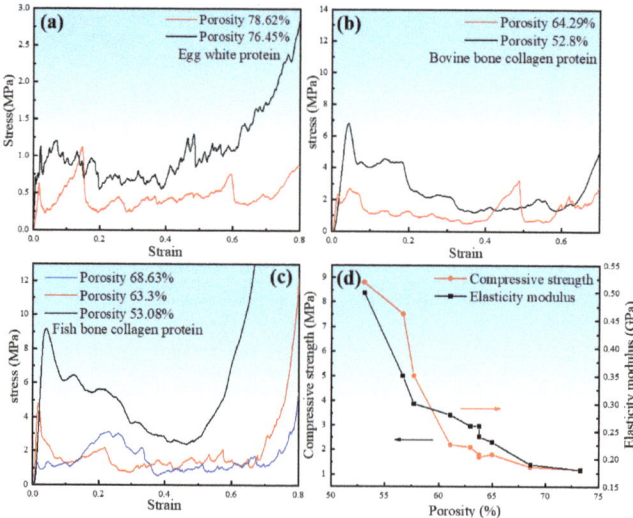

Figure 9. The effect of porosity on compression stress–strain curves and the relationship of porosity with the compressive strength and modulus of elasticity: (**a**) EWP, (**b**) BBCP, and (**c**) FBCP. (**d**) The relationship of porosity with compressive strength and the modulus of elasticity in porous zinc produced with FBCP.

The porous zinc produced with EWP with porosities of 78.62% and 76.45% presented the lowest compressive strengths of 0.63 and 0.72 MPa (Figure 9a). The BBCP porous zinc samples with porosities of 62.29% and 52.82% presented the compressive strengths of 2.33 and 6.80 MPa (Figure 9b). Figure 9c presents the stress–strain curves of the porous zinc produced with FBCP with porosities of 68.63%, 63.30%, and 53.08%. The corresponding compressive strengths were 1.30, 4.8, and 9.2 MPa, respectively. The fabricated porous zinc produced with FBCP/BBCP/EWP exhibited porosities and mechanical properties close to those of a human cancellous bone, with porosities from 30% to 95%, a compressive strength from 2 to 12 MPa, and a modulus of elasticity from 0.1 to 5 GPa [24]. The results exhibited that the porous zinc produced with FBCP (porosity of 53.08%) had the highest compressive strength compared to the porous zinc produced with EWP/BBCP. As described above, the porous zinc produced with FBCP possessed the highest compressive properties and a suitable pore structure, which was what we needed.

The effect of porosity on the compressive strength and elastic modulus of porous zinc presented in Figure 9d indicated that the compressive strength and elastic modulus of porous zinc decreased with the porosity.

3.3. Corrosion Behavior

Figure 10 shows the corrosion rates and content of Zn^{2+} of the porous zinc produced with FBCP with a porosity of 53–58%. The corrosion rates of the porous zinc decreased with the increase in immersion time, indicating that the formation of corrosion products had a protective effect on the surface of the porous zinc. In two days, a quick improvement in the corrosion rate and content of Zn^{2+} was observed because bare porous zinc was in contact with the SBF solution. The occurrence of a sudden decrease in the corrosion rate on the third day was mainly caused by the production of corrosion products on the surface, which impeded the corrosion process. As the pore structure was not absolutely the same for the experiments, the corrosion rate of the porous zinc underwent minor fluctuations. Moreover, with the increase in the corrosion products, the corrosion rate decreased, and consequently, the corrosion rate was close to 0.02 mm/year [16], which satisfied the requirements for corrosion rates in bone implants of less than 0.5 mm/year [16], indicating that porous zinc could be applied for cancellous bone implantations.

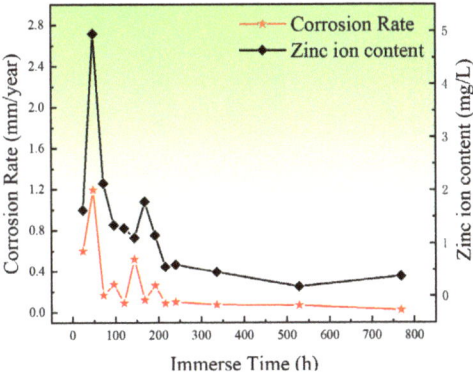

Figure 10. Corrosion rates of the porous zinc prepared with FBCP with a porosity of 53–58% and the content of zinc ions.

The corrosion behavior of the porous zinc produced with FBCP was evaluated through the characterization of the SEM surface morphologies of the degraded porous zinc after 32 days of immersion, as shown in Figure 11a–f. The pore sizes of the porous zinc (Figure 11a–f) produced with FBCP ranged from 0.01 to 2.08 mm, with a porosity of 53–58%. Figure 11 shows that the majority of the corrosion occurred in the bonding of particles, and corrosion pits occurred on the surfaces of the particles. It appeared that the bonding of the particles was corroded more easily than their surfaces. This may be attributed to the oxidation and cathodic reactions.

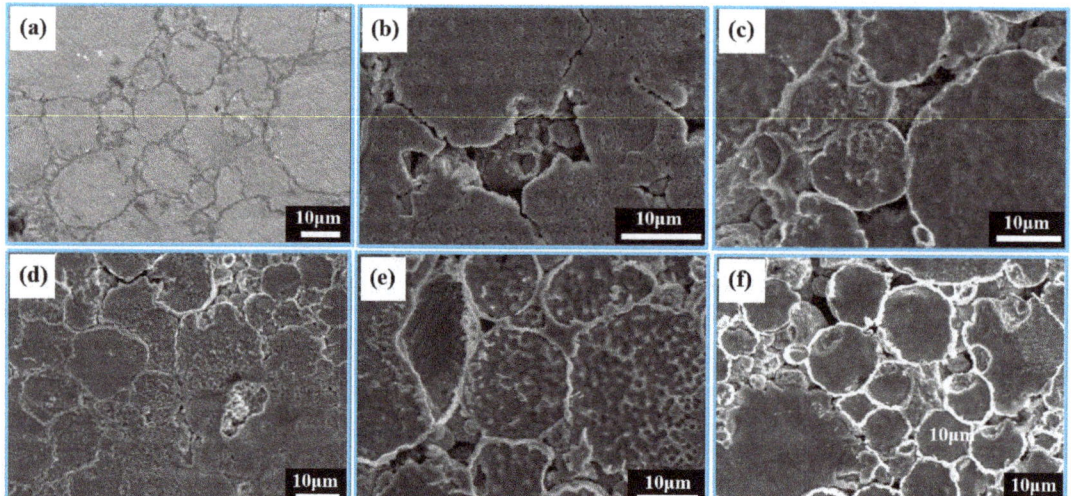

Figure 11. SEM images of porous zinc produced with FBCP with a porosity of 53–58% after immersion for different periods of time: (**a**) 0, (**b**) 5, (**c**) 10, (**d**) 15, (**e**) 25, and (**f**) 32 d.

It can be observed from Figure 12 that the EDS analysis after immersion of the porous zinc produced with FBCP indicated that the surface covered with corrosion products consisted of C, O, Zn, Na, P, Ca, S, and K (Figure 12a), while the surface after cleaning consisted of C, O, Zn, and Na (Figure 12b), indicating that the corrosion product mainly consisted of P, Ca, S, and K, and that P-based and Ca-based corrosion products were located on the top of the corrosion layer.

Figure 12. EDS analysis of the porous zinc produced with FBCP after immersion: (**a**) porous zinc produced with FBCP covered with corrosion products before cleaning; (**b**) porous zinc produced with FBCP after cleaning.

The corresponding EDS mapping results for the porous zinc produced with FBCP covered with the corrosion product are shown in Figure 13; the whole region mainly contained Na, O, P, Zn, Ca, Mg, Cl, and K. Many corrosion products were distributed on the surface of the porous zinc. There were more corrosion products attached to the surface in the bright portion with micro-cracks.

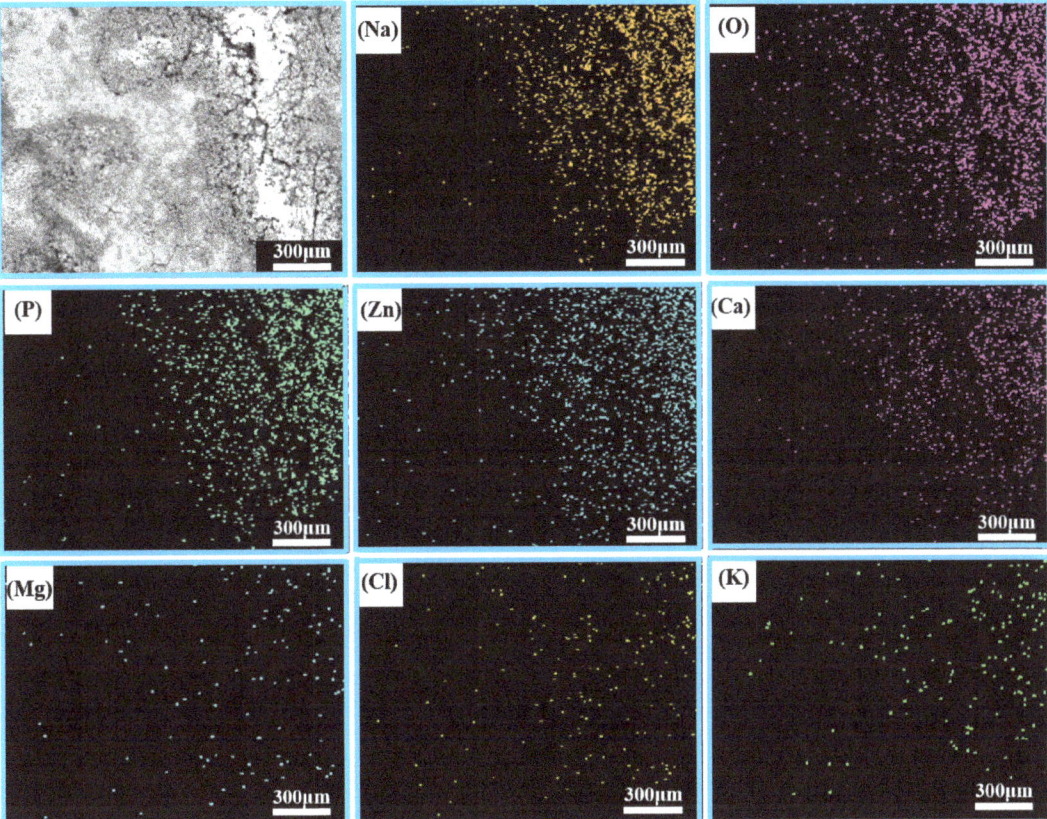

Figure 13. EDS spectra of the porous zinc produced with FBCP after immersion.

To analyze the phase compositions of the corrosion products of the porous zinc, EDS and XRD (Figures 12–14) were performed; the corrosion morphology was also analyzed (Figures 11–13). The XRD patterns of the corroded porous zinc produced with FBCP after immersion in SBF for 32 d are given in Figure 14. The corrosion products of porous zinc may include ZnO, $ZnCO_3$, $Zn_3(PO_4)_2$, $Zn(OH)_2$, $CaCO_3$, $Ca_3(PO_4)_2$, and $Ca_3Zn_2(PO_4)_2CO_3(OH)_2$ [31,49–52], which can be degraded in the human body through metabolism, and they have no toxicity. These corrosion products were in agreement with the loose and porous surface morphology of the porous zinc after corrosion (Figures 11–13). A passive film, such as $Zn_3(PO_4)_2$, can protect the underlying porous zinc and delay degradation. However, ZnO enhanced the degradation by forming a galvanic couple with Zn. The degradation of zinc in the nearly neutral environment in the SBF can be described as follows [53].

$$Zn \rightarrow Zn^{2+} + 2e^- \qquad (4)$$

$$O_2 + 2H_2O + 4e^- \rightarrow 4OH^- \qquad (5)$$

$$Zn^{2+} + 2OH^- \rightarrow Zn(OH)_2 \qquad (6)$$

$$2Zn^{2+} + 2OH^- \rightarrow 2ZnO + H_2 \qquad (7)$$

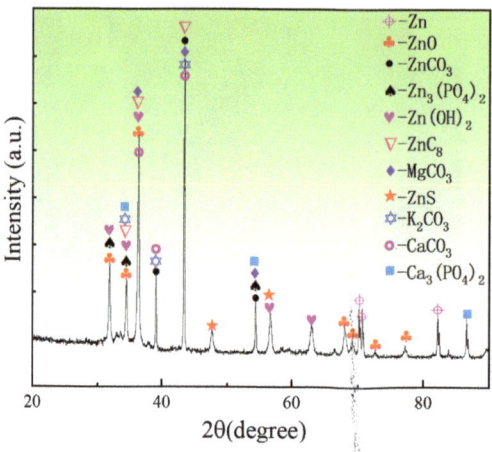

Figure 14. XRD patterns of porous zinc produced with FBCP after immersion in SBF for 32 d.

Zinc hydroxide ($Zn(OH)_2$) can redissolve due to chloride attack in physiological conditions; thus, zincite (ZnO) was the more dominant corrosion product, as it had more thermodynamically stable oxidation, and the released Zn ions could react with phosphate ions to form insoluble Zn phosphate ($Zn_3(PO4)_2$) [30].

$$Zn(OH)_2 + 2Cl^- \rightarrow Zn^{2+} + 2OH^- + 2Cl^- \tag{8}$$

$$3Zn^{2+} + 2HPO_4^{2-} + 2OH^- + 2H_2O \rightarrow Zn_3(PO_4)_2 \cdot 4H_2O \tag{9}$$

4. Conclusions

In the present study, a new protein foaming–consolidation porous zinc was successfully fabricated with three proteins as consolidating and foaming agents. The effects of stirring time, content of protein and sucrose, and sintering temperature on the microstructure, compressive properties, and corrosion properties of porous zinc were investigated. The main conclusions drawn from the current work are as follows:

1. Porous zinc was produced using three simple steps of slurry preparation, low-temperature foaming, and high-temperature sintering. The processing method is applicable to the preparation of porous zinc with porosities in the range of 50–85.8% and pore sizes in the range of 0.012 to 2.08 mm. The porous zinc produced with FBCP exhibited the highest compressive strength and elastic modulus.
2. The macrostructure of porous zinc can be changed with different protein types, the content of protein and sucrose, the stirring time, and the sintering temperature.
3. The porosity increased with the stirring time, as well as the content of protein and sucrose. The porous zinc produced with FBCP exhibited more circular and regular pores and the largest pore size.
4. The compressive properties of porous zinc were highly dependent on the porosity and types of proteins. Porosity and compressive strength were inversely proportional. Porous zinc prepared with FBCP exhibited a superior compressive strength and elastic modulus. The compressive strength of the porous zinc produced with FBCP was about eight times higher than that of the porous zinc produced with EWP.
5. The main corrosion mechanisms of porous zinc showed that Zn ions would react with hydroxyl ions, carbonate ions, phosphate ions, etc. The corrosion products were determined to be ZnO, $ZnCO_3$, $Zn_3(PO_4)_2$, $Zn(OH)_2$, $CaCO_3$, $Ca_3(PO_4)_2$, and $Ca_3Zn_2(PO_4)_2CO_3(OH)_2$, which can be degraded in the human body through metabolism. The corrosion rate of porous zinc obtained through an in vitro immersion test using simulated body fluid on the thirty-second day was close to 0.02 mm/year.

6. Overall, porous zinc shows an optimal combination of compressive and corrosion properties and is considered as highly promising for the requirements of cancellous bone implantation.

Author Contributions: Q.G.: Data curation, Investigation, Writing—original draft; X.L.: Data curation, Writing—original draft; A.Q.: Supervision, Resources, Funding acquisition, Writing—review and editing; Y.M.: Methodology, Project administration, Supervision, Resources, Writing—review and editing. All authors have read and agreed to the published version of the manuscript.

Funding: This research was funded by the Joint Program of Beijing Municipality Beijing Natural Science Foundation grant number (No. KZ202110005004) and National Natural Science Foundation of China grant number (No. 12172018).

Institutional Review Board Statement: Not applicable.

Informed Consent Statement: Not applicable.

Data Availability Statement: Not applicable.

Conflicts of Interest: The authors declare that they have no known competing financial interests or personal relationships that could have appeared to influence the work reported in this paper.

References

1. Yusop, A.H.; Bakir, A.A.; Shaharom, N.A.; Abdul Kadir, M.R.; Hermawan, H. Porous Biodegradable Metals for Hard Tissue Scaffolds: A Review. *Int. J. Biomater.* **2012**, *2012*, 1–10. [CrossRef] [PubMed]
2. He, J.; Fang, J.; Wei, P.; Li, Y.; Guo, H.; Mei, Q.; Ren, F. Cancellous bone-like porous Fe@Zn scaffolds with core-shell-structured skeletons for biodegradable bone implants. *Acta Biomater.* **2021**, *121*, 665–681. [CrossRef] [PubMed]
3. Wang, Y.C.; Huang, H.; Jia, G.Z.; Zeng, H.; Yuan, G.Y. Fatigue and dynamic biodegradation behavior of additively manu-factured Mg scaffolds. *Acta Biomater.* **2021**, *135*, 705–722. [CrossRef]
4. Chang, C.-H.; Kuo, T.-F.; Lin, F.-H.; Wang, J.-H.; Hsu, Y.-M.; Huang, H.-T.; Loo, S.-T.; Fang, H.-W.; Liu, H.-C.; Wang, W.-C. Tissue engineering-based cartilage repair with mesenchymal stem cells in a porcine model. *J. Orthop. Res.* **2011**, *29*, 1874–1880. [CrossRef]
5. Risbud, M.V.; Sittinger, M. Tissue engineering: Advances in in vitro cartilage generation. *Trends Biotechnol.* **2002**, *20*, 351–356. [CrossRef]
6. Bonithon, R.; Kao, A.P.; Fernández, M.P.; Dunlop, J.N.; Blunn, G.W.; Witte, F.; Tozzi, G. Multi-scale mechanical and mor-phological characterisation of sintered porous magnesium-based scaffolds for bone regeneration in critical-sized defects. *Acta Biomater.* **2021**, *127*, 338–352. [CrossRef]
7. Bonadio, J.; Smiley, E.; Patil, P.; Goldstein, S. Localized, direct plasmid gene delivery in vivo: Prolonged therapy results in reproducible tissue regeneration. *Nat. Med.* **1999**, *5*, 753–759. [CrossRef]
8. Cheung, H.Y.; Lau, K.T.; Lu, T.P.; Hui, D. A critical review on polymer-based bio-engineered materials for scaffold development. *Compos. Part B Eng.* **2007**, *38*, 291–300. [CrossRef]
9. Elsayed, H.; Romero, A.R.; Picicco, M.; Kraxnerd, J.; Galusekd, D.; Colomboa, P.; Bernardo, E. Glass-ceramic foams and retic-ulated scaffolds by sinte-r-crystallization of a hardystonite glass. *J. Non-Cryst. Solids* **2020**, *528*, 119744. [CrossRef]
10. Jiang, H.; Ziegler, H.; Zhang, Z.; Meng, H.; Chronopoulos, D.; Chen, Y. Mechanical properties of 3D printed architected polymer foams under large deformation. *Mater. Des.* **2020**, *194*, 108946. [CrossRef]
11. Xi, J.; Zhang, L.; Zheng, Z.H.; Chen, G.Q.; Gong, Y.D.; Zhao, N.M.; Zhang, X.F. Preparation and evaluation of porous poly(3-hydroxybutyrate-co-3-hy-droxyhexanoate) hydroxyapatite composite scaffolds. *J. Biomater. Appl.* **2008**, *22*, 293–307.
12. Yang, S.; Wang, J.; Tang, L.; Ao, H.; Tan, H.; Tang, T.; Liu, C. Mesoporous bioactive glass doped-poly (3-hydroxybutyrate-co-3-hydroxyhexanoate) composite scaffolds with 3-dimensionally hierarchical pore networks for bone regeneration. *Colloids Surf. B Biointerfaces* **2014**, *116*, 72–80. [CrossRef] [PubMed]
13. Bobyn, J.D.; Stackpool, G.J.; Hacking, S.A.; Tanzer, M.; Krygier, J.J. Characteristics of bone ingrowth and interface mechanics of a new porous tantalum biomaterial. *J. Bone Joint Surg. Br. Vol.* **1999**, *81*, 907–914. [CrossRef]
14. Van Bael, S.; Chai, Y.C.; Truscello, S.; Moesen, M.; Kerckhofs, G.; Van Oosterwyck, H.; Kruth, J.-P.; Schrooten, J. The effect of pore geometry on the in vitro biological behavior of human periosteum-derived cells seeded on selective laser-melted Ti6Al4V bone scaffolds. *Acta Biomater.* **2012**, *8*, 2824–2834. [CrossRef] [PubMed]
15. Geetha, M.; Singh, A.K.; Asokamani, R.; Gogia, A.K. Ti based biomaterials, the ultimate choice for orthopaedic implants-A review. *Prog. Mater. Sci.* **2009**, *54*, 397–425. [CrossRef]
16. Tong, X.; Shi, Z.M.; Xu, Z.M.; Lin, J.X.; Zhang, D.C.; Wang, K.; Li, Y.C.; Wen, C.E. Degradation behavior, cytotoxicity, he-molysis, and antibacterial properties of electro-deposited Zn-Cu metal foams as potential biodegradable bone implants. *Acta Biomater.* **2020**, *102*, 481–492. [CrossRef] [PubMed]

17. Tian, Y.; Miao, H.-W.; Niu, J.-L.; Huang, H.; Kang, B.; Zeng, H.; Ding, W.-J.; Yuan, G.-Y. Effects of annealing on mechanical properties and degradation behavior of biodegradable JDBM magnesium alloy wires. *Trans. Nonferrous Met. Soc. China* **2021**, *31*, 2615–2625. [CrossRef]
18. Zhang, X.; Li, X.-W.; Li, J.-G.; Sun, X.-D. Preparation and mechanical property of a novel 3D porous magnesium scaffold for bone tissue engineering. *Mater. Sci. Eng. C* **2014**, *42*, 362–367. [CrossRef]
19. Sharma, P.; Pandey, P.M. Corrosion behaviour of the porous iron scaffold in simulated body fluid for biodegradable implant application. *Mater. Sci. Eng. C* **2019**, *99*, 838–852. [CrossRef]
20. Lai, Y.; Li, Y.; Cao, H.; Long, J.; Wang, X.; Li, L.; Li, C.; Jia, Q.; Teng, B.; Tang, T.; et al. Osteogenic magnesium incorporated into PLGA/TCP porous scaffold by 3D printing for repairing challenging bone defect. *Biomaterials* **2019**, *197*, 207–219. [CrossRef]
21. Čapek, J.; Vojtěch, D. Properties of porous magnesium prepared by powder metallurgy. *Mat. Sci. Eng. C* **2013**, *33*, 564–569. [CrossRef] [PubMed]
22. Carluccio, D.; Xu, C.; Venezuela, J.; Cao, Y.; Kent, D.; Bermingham, M.; Demir, A.G.; Previtali, B.; Ye, Q.; Dargusch, M. Additively manufactured iron-manganese for biodegradable porous load-bearing bone scaffold applications. *Acta Biomater.* **2020**, *103*, 346–360. [CrossRef] [PubMed]
23. Sharma, P.; Jain, K.G.; Pandey, P.M.; Mohanty, S. In vitro degradation behaviour, cytocompatibility and hemocompatibility of topologically ordered porous iron scaffold prepared using 3D printing and pressureless microwave sintering. *Mater. Sci. Eng. C* **2020**, *106*, 110247. [CrossRef] [PubMed]
24. Sharma, P.; Pandey, P.M. A novel manufacturing route for the fabrication of topologically-ordered open-cell porous iron scaffold. *Mater. Lett.* **2018**, *222*, 160–163. [CrossRef]
25. Alavi, R.; Akbarzadeh, A.; Hermawan, H. Post-corrosion mechanical properties of absorbable open cell iron foams with hollow struts. *J. Mech. Behav. Biomed. Mater.* **2021**, *117*, 104413. [CrossRef]
26. He, J.; Ye, H.; Li, Y.; Fang, J.; Mei, Q.; Lu, X.; Ren, F. Cancellous-Bone-like Porous Iron Scaffold Coated with Strontium Incorporated Octacalcium Phosphate Nanowhiskers for Bone Regeneration. *ACS Biomater. Sci. Eng.* **2019**, *5*, 509–518. [CrossRef]
27. Cockerill, I.; Su, Y.; Sinha, S.; Qin, Y.-X.; Zheng, Y.; Young, M.L.; Zhu, D. Porous zinc scaffolds for bone tissue engineering applications: A novel additive manufacturing and casting approach. *Mater. Sci. Eng. C* **2020**, *110*, 110738. [CrossRef]
28. Li, Y.; Pavanram, P.; Zhou, J.; Lietaert, K.; Taheri, P.; Li, W.; San, H.; Leeflang, M.; Mol, J.; Jahr, H.; et al. Additively manufactured biodegradable porous zinc. *Acta Biomater.* **2020**, *101*, 609–623. [CrossRef]
29. Ke, G.Z.; Yue, R.; Huang, H.; Kang, B.; Zeng, H.; Yuan, G.Y. Effects of Sr addition on microstructure, mechanical properties and in vitro degradation behavior of as-extruded Zn-Sr binary alloys. *Trans. Nonferrous Met. Soc. China* **2020**, *30*, 1873–1883. [CrossRef]
30. Guan, X.M.; Xiong, M.P.; Zeng, F.Y.; Xu, B.; Yang, L.D.; Guo, H.; Niu, J.L.; Zhang, J.; Chen, C.X.; Pei, J.; et al. Enhancement of Osteogenesis and Biodegradation Control by Brushite Coating on Mg-Nd-Zn-Zr Alloy for Mandibular Bone Repair. *Acs Appl. Mater Inter.* **2014**, *6*, 21525–21533. [CrossRef]
31. Venezuela, J.; Dargusch, M. The influence of alloying and fabrication techniques on the mechanical properties, biodegradability and biocompatibility of zinc: A comprehensive review. *Acta Biomater.* **2019**, *87*, 1–40. [CrossRef] [PubMed]
32. Hou, Y.; Jia, G.; Yue, R.; Chen, C.; Pei, J.; Zhang, H.; Huang, H.; Xiong, M.; Yuan, G. Synthesis of biodegradable Zn-based scaffolds using NaCl templates: Relationship between porosity, compressive properties and degradation behavior. *Mater. Charact.* **2018**, *137*, 162–169. [CrossRef]
33. Yuan, L.; Ding, S.; Wen, C. Additive manufacturing technology for porous metal implant applications and triple minimal surface structures: A review. *Bioact. Mater.* **2019**, *4*, 56–70. [CrossRef] [PubMed]
34. Lyckfeldt, O.; Brandt, J.; Lesca, S. Protein forming—A novel shaping technique for ceramics. *J. Eur. Ceram. Soc.* **2000**, *20*, 2551–2559. [CrossRef]
35. Pottathara, Y.B.; Vuherer, T.; Maver, U.; Kokol, V. Morphological, mechanical, and in-vitro bioactivity of gela-tine/collagen/hydroxyapatite based scaffolds prepared by unidirectional freeze-casting. *Polym Test* **2021**, *102*, 107308. [CrossRef]
36. He, X.; Su, B.; Zhou, X.G.; Yang, J.H.; Zhao, B.; Wang, X.Y.; Yang, G.Z.; Tang, Z.H.; Qiu, H.X. Gelcasting of Alumina ceramics using an egg white protein binder system. *Ceram. Silikáty* **2011**, *55*, 1–7.
37. Dhara, S.; Pradhan, M.; Ghosh, D.; Bhargava, P. Nature Inspired Novel Pro-cessing Routes to Ceramic Foams. *Adv. Appl. Ceram.* **2005**, *104*, 9–21. [CrossRef]
38. Kinsella, J. Functional properties of proteins: Possible relationships between structure and function in foams. *Food Chem.* **1981**, *7*, 273–288. [CrossRef]
39. Ahmad, F.; Sopyan, L. Porous ceramics with controllable properties prepared by protein foaming-consolidation method. *J. Porous Mat.* **2011**, *18*, 195–203.
40. Ahmad, F.; Sopyan, L.; Mel, M.; Ahmad, Z. Porous alumina through protein foaming–consolidation method effect of dispersant concentration on the physical properties. *Asia Pac. J. Chem. Eng.* **2011**, *6*, 863–869.
41. Kapat, K.; Srivas, P.K.; Rameshbabu, A.P.; Maity, P.P.; Jana, S.; Dutta, J.; Majumdar, P.; Chakrabarti, D.; Dhara, S. Influence of porosity and pore-size distribution in Ti6Al4V foam on physicomechanical properties, osteogenesis, and quantitative validation of bone ingrowth by micro-computed tomography. *ACS Appl. Mater. Interfaces* **2017**, *9*, 39235–39248. [CrossRef] [PubMed]
42. Stochero, N.P.; Moraes, E.G.D.; Moreira, A.C.; Fernandes, C.P.; Innocentini, M.D.M.; Oliveira, A.P.N.D. Ceramic shell foams produced by direct foaming and gelcasting of proteins: Permeability and microstructural characterization by X-ray microtomography. *J. Eur. Ceram. Soc.* **2020**, *40*, 4224–4231. [CrossRef]

3. Kapat, K.; Srivas, P.K.; Dhara, S. Coagulant assisted foaming—A method for cellular Ti6Al4V: Influence of microstructure on mechanical properties. *Mater. Sci. Eng. C* **2017**, *689*, 63–71. [CrossRef]
4. Fadli, A.; Sopyan, I. Preparation of porous alumina for biomedical applications through protein foaming–consolidation method. *Mater. Res. Innov.* **2009**, *13*, 327–329. [CrossRef]
5. *ISO 13314-2011*; Mechanical testing of metals- Ductility testing- Compression test for porous and cellular metals. International Organization for Standardization: Genève, Switzerland, 2011.
6. Kokubo, T.; Takadama, H. How useful is SBF in predicting in vivo bone bioactivity? *Biomaterials* **2006**, *27*, 2907–2915. [CrossRef]
7. *ASTM G31-21*; Standard Guide for Laboratory Immersion Corrosion Testing of Metals. ASTM International: West Conshohocken, PA, USA, 2021.
8. *ASTM G1-03*; Standard Practice for Preparing, Cleaning, and Evaluating Corrosion Test Specimens. ASTM International: West Conshohocken, PA, USA, 2011.
9. Bagha, P.S.; Khakbiz, M.; Sheibani, S.; Hermawan, H. Design and characterization of nano and bimodal structured biode-gradable Fe-Mn-Ag alloy with accelerated corrosion rate. *J. Alloys Compd.* **2018**, *767*, 955–965. [CrossRef]
10. Hernández-Escobar, D.; Champagne, S.; Yilmazer, H.; Dikici, B.; Boehlert, C.J.; Hermawan, H. Current status and perspectives of zinc-based absorbable alloys for biomedical applications. *Acta Biomater.* **2019**, *97*, 1–22. [CrossRef]
11. Yang, H.; Wang, C.; Liu, C.; Chen, H.; Wu, Y.; Han, J.; Jia, Z.; Lin, W.; Zhang, D.; Li, W.; et al. Evolution of the degradation mechanism of pure zinc stent in the one-year study of rabbit abdominal aorta model. *Biomaterials* **2017**, *145*, 92–105. [CrossRef]
12. Törne, K.; Örnberg, A.; Weissenrieder, J. Influence of strain on the corrosion of magnesium alloys and zinc in physiological environments. *Acta Biomater.* **2017**, *48*, 541–550. [CrossRef]
13. Alves, M.M.; Proek, T.; Santos, C.F.; Montemora, M.F. Evolution of the in vitro degradation of Zn-Mg alloys under simulated physiological conditions. *Rsc. Adv.* **2017**, *7*, 28224–28233. [CrossRef]

Article

Collarless Polished Tapered Stems of Identical Shape Provide Differing Outcomes for Stainless Steel and Cobalt Chrome: A Biomechanical Study

Ayumi Kaneuji [1,*], Mingliang Chen [1,2], Eiji Takahashi [2], Noriyuki Takano [3], Makoto Fukui [1], Daisuke Soma [1], Yoshiyuki Tachi [1], Yugo Orita [1], Toru Ichiseki [1] and Norio Kawahara [1]

[1] Department of Orthopaedic Surgery, Kanazawa Medical University, Kahoku-gun 920-0293, Japan
[2] Department of Orthopaedics, Affiliated Renhe Hospital of China Three Gorges University, Yichang 443000, China
[3] Department of Mechanical Engineering, Kanazawa Institution of Technology, Nonoichi 921-8501, Japan
* Correspondence: kaneuji@kanazawa-med.ac.jp; Tel.: +81-076-286-2211

Abstract: Cemented polished tapered femoral stems (PTS) made of cobalt–chrome alloy (CoCr) are a known risk factor for periprosthetic fracture (PPF). The mechanical differences between CoCr-PTS and stainless-steel (SUS) PTS were investigated. CoCr stems having the same shape and surface roughness as the SUS Exeter® stem were manufactured and dynamic loading tests were performed on three each. Stem subsidence and the compressive force at the bone–cement interface were recorded. Tantalum balls were injected into the cement, and their movement was tracked to indicate cement movement. Stem motions in the cement were greater for the CoCr stems than for the SUS stems. In addition, although we found a significant positive correlation between stem subsidence and compressive force in all stems, CoCr stems generated a compressive force over three times higher than SUS stems at the bone–cement interface with the same stem subsidence ($p < 0.01$). The final stem subsidence amount and final force were greater in the CoCr group ($p < 0.01$), and the ratio of tantalum ball vertical distance to stem subsidence was significantly smaller for CoCr than for SUS ($p < 0.01$). CoCr stems appear to move more easily in cement than SUS stems, which might contribute to the increased occurrence of PPF with the use of CoCr-PTS.

Keywords: bone cement; cobalt–chrome alloy; Exeter stem; periprosthetic femoral fractures; polished tapered stem

1. Introduction

For cementless total hip arthroplasty (THA), the femoral stem is typically made of a titanium alloy due to its ease of processing, good biocompatibility, and minimal stress shielding. In contrast, cemented stems have been fabricated using various metal materials, such as titanium alloy, stainless steel (SUS), cobalt–chromium alloy (CoCr), among others. In the cemented stems, collarless polished tapered stems (PTS) having a surface roughness of less than 0.1 μm provide long-term results superior to stems with a surface roughness of 2.5 μm to 12 μm [1–4]. However, several studies have shown a greater frequency of postoperative periprosthetic fractures (PPF) with PTS than with composite beam-type stems [5–8]. Importantly, national joint registries have reported a higher PPF revision rate for PTS made of cobalt–chromium alloy (CoCr) than for PTS of stainless steel (SUS) [9,10].

Differences in stem design and surface roughness are thought to affect the frequency of PPFs, but insufficient data exist to confirm uniform in-cement stem behavior across stems fabricated from differing materials.

In a biomechanical study on differences between CoCr and SUS using cone rods, the researcher prepared cylindrical polished tapered rods of the same shape and having the same coefficient of friction. When the rods were fixed into cement and placed under load, the CoCr rods slipped significantly farther than the SUS rods, and the strain gauge on the cement surface around the CoCr rods demonstrated more cement creep than around the SUS rods [11]. That experiment pointed to differences in rod subsidence and cement stress between polished SUS and CoCr rods, but provided no information about differences in biomechanical behavior among stem materials in cemented stems. This experiment suggests that there may be differences in subsidence and stress on the cement between polished SUS and CoCr stems. If the CoCr-PTS exhibits greater subsidence than the SUS-PTS, this could explain the difference in PPF fracture frequency.

The Exeter® stem (Stryker, Allendale, NJ, USA) is made of SUS and is the most widely used PTS in the world, providing excellent clinical results [12]. Therefore, a materials comparison between a CoCr product and the Exeter stem could be highly meaningful.

For this study, original Exeter stems made of SUS were compared to copied Exeter stems made of CoCr with the exact same shape and surface roughness, which exhibited different mechanical behaviors.

2. Materials and Methods

2.1. Implant Preparation

The original Exeter® stem was made of SUS, but no CoCr counterpart was available. At our request, Stryker company approved this study and provided original computer-aided design (CAD) data on the Exeter stem, size 1 with 44 mm offset, for experimental purposes limited to this study. A CoCr stem identical in shape to the Exeter stem was then manufactured (Nagumo Manufacturing, Jyoetsu, Niigata, Japan) using the CAD data provided by Stryker. The manufacturing error was ±1 µ [13]. The surface of each CoCr stem was ground to a polished surface roughness of less than 0.1 µm (Yoshida Seiko, Kanazawa, Ishikawa, Japan). For this study, three Exeter (SUS) stems were prepared, and three CoCr stems with the same shape and surface finish as the Exeter stems were also prepared (Figure 1). Prior to the loading experiment, the surface roughness of the stems (one original Exeter stem and all three CoCr-Exeter-shaped stems) was measured (Figure 2).

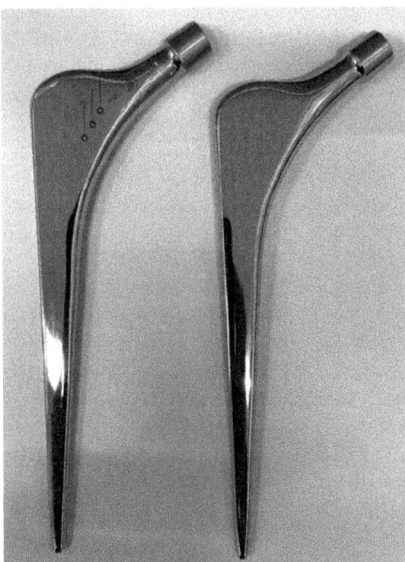

Figure 1. Exeter stem (SUS) and Exeter-shaped CoCr stem. On the left is an original Exeter stem. On the right is our Exeter-shaped stem of CoCr. The Exeter-shaped CoCr stem was made using the exact same design and surface roughness as the original stem.

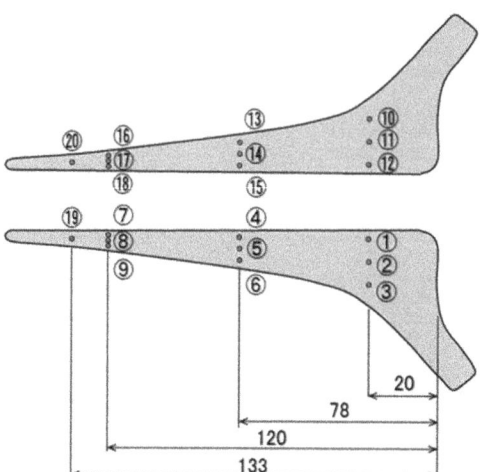

Figure 2. Measurement of stem surface roughness. Prior to the loading experiment, the surface roughness of the stem was analyzed. Using a laser microscope (VK-8700, Keyence, Osaka, Japan), surface roughness was assessed at 18 sites on the front and back surfaces of the stem: three positions at 20 mm (proximal), 78 mm (middle), and 120 mm (distal) from the top of the stem.

2.2. An Experimental Device

The equipment for load testing and the data measurement methods used in this study were consistent with those previously reported [14–16]. Composite femurs were used instead of cadaveric femurs to avoid variations based on individual differences in bone strength and canal shape. The composite femurs (#3403®, Pacific Research Laboratories, Vashon, WA, USA) had mechanical properties comparable to the human femur. We cut the composite femur neck obliquely at 20 mm distal to the top of the greater trochanter, the same procedure as for THA, and we cut the distal portion of the femur 230 mm from the top of the greater trochanter. To create the composite femoral canal, we used an Exeter stem broach that was attached to a test fixator of S45C structural carbon steel and epoxy resin (Devcon B, ITW Industry, Osaka, Japan). Both the fixator and the composite femur were penetrated by the rod that protected the measuring equipment. This allowed us to conduct measurements at a constant site. The rod passing through the inner tube was secured on the face of the medullary canal of the composite femur (Figure 3A). To mimic the wet conditions of the in vivo femoral environment, the composite femurs were soaked in blended vegetable oil for 24 h before attaching them to the test equipment. An 80 g sample of bone cement (Simplex p, Stryker, NJ, USA) was mixed in a vacuum cement mixing system (ACM vacuum mixing ball, Stryker, Tokyo, Japan), and a cement injection gun was used to insert the mixture into the femoral canal. Based on reports that the creep and mechanical behavior of bone cement can differ significantly between room temperature and body temperature [17], the experiment was performed to mimic in vivo environmental conditions by maintaining the test equipment at 37 °C with a temperature sensor (T-35®, Takigen MFG Co., Ltd., Tokyo, Japan) and a heater (G6A92® [240 V, 250 W], Takigen MFG Co., Ltd., Tokyo, Japan).

2.3. Stem and Tantalum Ball Insertion

After the cement was delivered, each stem was inserted into the composite femur. Before the cement hardened, we used an indwelling needle to inject tantalum balls 0.6 mm in diameter into the cement around the stem. Before and after the loading test, micro-computed tomography (CT) was taken from the area of the neck cut to 30 mm below the lesser trochanter, to measure the three-dimensional movement of each tantalum ball. The movement of the tantalum ball was assumed to represent the movement of the cement around the stem. Measurements were performed according to the method described by Takahashi et al. [15]. The movement of the balls was measured both horizontally and vertically, independent of the stem, and the final amount of subsidence for each stem and the ratios of ball subsidence/stem subsidence were calculated.

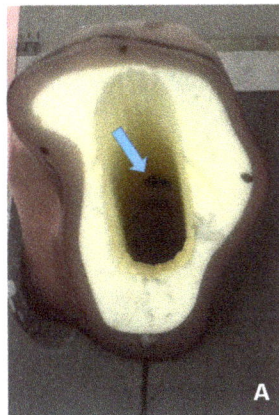
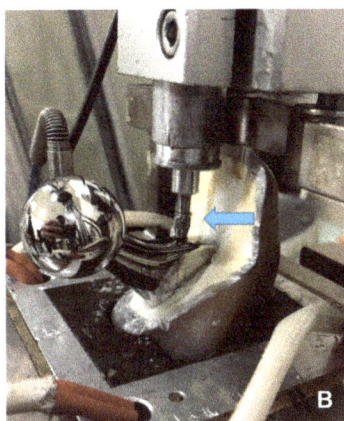
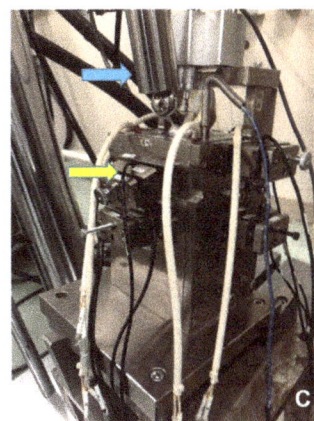

Figure 3. The experimental device. A unit consisting of the stem, cement, and composite femur is placed in the tester to simulate the internal human body conditions as closely as possible. (**A**) A rod (blue arrow) passed through the composite femur from the outside of the fixator. A load cell sensor to sense the pressure was placed at the bone–cement interface through this hole. (**B**). After a stem was fixed with cement into the composite femur, the digital gauge (blue arrow) was in contact with the shoulder of the stem. (**C**). The loading element was inclined at 15° with the front portion raised, so that the load was applied 15° medially to the femoral head at the top of the stem (blue arrow). The yellow arrow indicates the load cell on the proximal medial side.

2.4. Loading Method and Loading Period

Cyclical loading (1 Hz, 30 to 3000 N) was applied 500,000 times to a metal femoral head (cobalt–chromium alloy, 26 mm in diameter) attached to the stem neck. A fatigue testing system (EHF-UM 300KN-70L; Shimadzu Corporation, Kyoto, Japan) was used at an angle of 15° medially to the coronal plane of the model to mimic normal in vivo loading in a 70 kg adult [18]. In this experiment, an eight-hour non-loading period was incorporated between the 16 h loading periods to simulate the actual sleep period in a clinical setting and allow for stress relaxation of the cement.

2.5. Measurement of Stem Subsidence

A digital displacement gauge (DTH-A-5, 5 mm; Kyowa Electronic Instruments Co., Ltd., Tokyo, Japan, Figure 3B) was placed on the proximal lateral portion of the stem to measure stem subsidence over time. Data were transmitted automatically to a computer through data collection and analysis software (Sensor Interface PCD-300A, Kyowa Electronic Instruments Co., Tokyo, Japan). Due to the vast amount of collected data, a single file containing 8 min of measurement data every hour was used, and 216 files (24 h a day for 9 days) were generated for each loading test. The 16 h loading period for each day was divided into early, middle, and late phases, and the first two files from each phase (a total of 20,000 values) were collected for use. Subsidence was defined as the mean of the maximum values for sine waves from the 20,000 values during each phase of the loading period. Additionally, the mean values of the 20,000 data sets were collected for the duration of non-loading. On the final day of the experiment, the final compressive force and stress relaxation values were collected as the mean and standard deviation (SD) values. Graphs were made for continuous data on stem subsidence for all models throughout the experiment.

2.6. Measurement of Compressive Force at the Bone–Cement Interface

After the cement had hardened, a rod in the tube from outside the fixator was connected to a load cell (TR20I 500N/fs®, TR20I 200N/fs®, Kyowa Electronic Instruments Co., Ltd., Tokyo, Japan, Figure 3C) and the pressure transducer. The rod was placed at eight sites on the medial, lateral, anterior, and posterior sides of the proximal and distal femur at the interface with the bone cement in the femoral canal. We measured compressive forces at the cement–bone interface over time after calibration and input those measured values automatically into a computer. The loading and non-loading periods each day were classified into 3 periods (early, middle, and late). The compressive force in each period was defined as the mean of the collected 960 maximum values of sine waves in the 2 consecutive files after the start of that period (60 values/min × 8 min × 2 times). In total, 27 averaged values (3 × 9 days) were used to analyze stem subsidence and compressive force, respectively. Continuous data of compressive force were graphed throughout the experiment.

2.7. Measurement of Cement Thickness

After the loading tests were completed, all stems were removed from the cement. CT images were then taken with a slice spacing of 1 mm to measure the thickness of the cement mantle. The average thickness of the cement was calculated using four scans of horizontal sections aligned with the insertion holes of the proximal rods. The thickness of the cement layer was measured in the middle of each of the anterior, posterior, medial, and lateral areas (Figure 4).

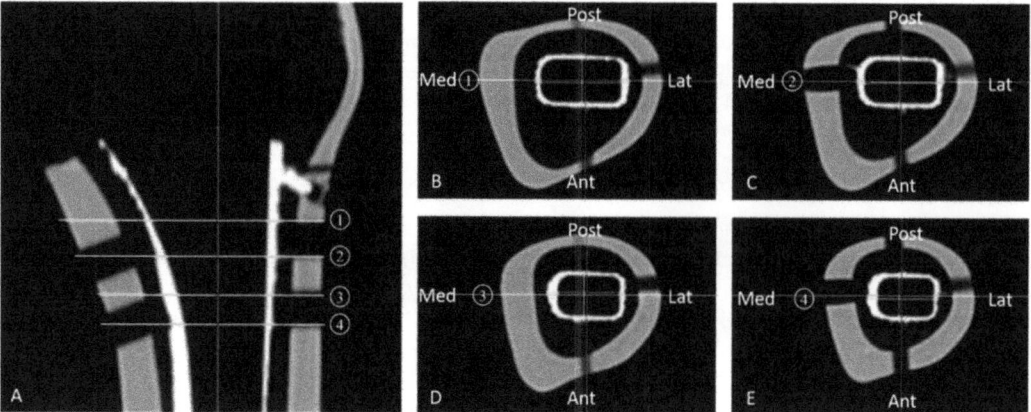

Figure 4. CT images, taken after stem removal. CT images were taken at 1 mm slice spacing in four horizontal sections aligned with the insertion holes of the proximal lateral femoral rods to measure the cement thickness. (**A**) Coronal slice of femoral canal center. 1: Upper edge of proximal lateral hole for rod. 2: Lower edge of proximal lateral hole for rod. 3: Upper edge of distal lateral hole for rod. 4: Lower edge of distal lateral hole for rod. (**B**) Axial slice at the level of 1; (**C**) Axial slice at the level of 2; (**D**) Axial slice at the level of 3; (**E**) Axial slice at the level of 4. The thickness of the cement layer was measured at the midpoint of each area.

2.8. Statistical Analysis

We compared the amount of stem subsidence, compressive force at the bone–cement interface, tantalum ball behavior, and surrounding bone cement thickness between the two groups using the Mann–Whitney U test. The relationship between stem subsidence and compressive force at the bone–cement interface was analyzed using Pearson's correlation coefficient. Statistical significance was indicated for a probability (alpha) of 0.01. All statistical analyses were performed using the Social Science Package version 18.0 (SPSS Inc., Chicago, IL, USA).

3. Results

3.1. Surface Roughness of the Stem before the Experiment

Stem surface roughness was measured before the load test was initiated. The surface roughness of the Exeter stem and the CoCr stems was Ra = 49.61 (SD 13.59) nm and Ra = 52.56 (SD 9.32) nm, respectively. The surface roughness for all areas of all polished stems was less than 100 nm (=0.1 μm). There were no obvious differences in surface roughness across all areas and all stems (Table 1).

Table 1. Surface roughness in all stem areas.

		SUS Group			CoCr Group		
		No.1	No.2	No.3	No.1	No.2	No.3
Front	proximal	50.67 (22.19)	35.00 (9.64)	49.33 (22.12)	54.00 (1.00)	38.67 (0.58)	46.33 (1.15)
	middle	41.33 (3.21)	40.00 (3.61)	43.00 (3.46)	48.33 (3.06)	49.00 (5.29)	46.33 (4.04)
	distal	58.67 (10.40)	66.33 (6.03)	54.33 (10.26)	54.33 (8.50)	55.33 (14.01)	57.33 (8.14)
Back	proximal	47.00 (1.73)	32.00 (2.65)	43.67 (2.31)	60.00 (7.21)	51.33 (6.43)	42.33 (1.52)
	middle	57.33 (10.12)	37.00 (1.00)	59.00 (10.39)	56.33 (3.21)	53.00 (5.57)	42.00 (5.29)
	distal	67.00 (8.89)	45.67 (9.02)	65.67 (9.45)	64.33 (10.50)	68.33 (7.64)	58.67 (1.53)

Surface roughness was assessed on the front and back surfaces of the stem: proximal, middle, and distal from the top of the stem as shown in Figure 2. Numbers indicate mean thickness in each group. Numbers within ( ) are standard deviation. The notation unit is nm.

3.2. The Amount of Stem Subsidence (Figure 5)

Subsidence was greatest on the first day, followed subsequently by gradual subsidence in both stem models. Mean final subsidence was significantly greater in the CoCr group, at 0.478 (SD 0.063) mm vs. 0.257 (SD 0.094) mm in the SUS group ($p < 0.001$). During the loading period, first-day subsidence was 0.123 (SD 0.046) in the SUS group and 0.138 (SD 0.056) in the CoCr group ($p = 0.355$), and first-day rise was 0.114 (SD 0.035) and 0.131 (SD 0.050), respectively ($p = 0.323$), with no significant difference between the two groups in either subsidence or rise during the first day. Excluding the first day, stem subsidence was significantly greater in the CoCr group, at 0.1028 (SD 0.0207) mm vs. 0.0437 (SD 0.0203) mm in the SUS group ($p < 0.001$). Stem rise throughout the non-loading period was greater for CoCr, at 0.0674 (SD 0.0153) mm than 0.0295 (SD 0.0148) mm for SUS ($p < 0.001$).

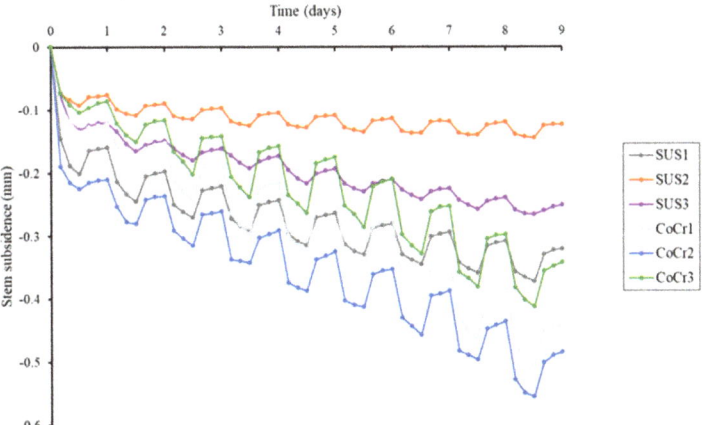

Figure 5. Diagram showing the subsidence of stems. All stems subsided rapidly on the first day and then gradually continued to subside. All stems subsided during loading and rose during non-loading. Except for the first day, daily stem subsidence and stem rise was significantly greater for the CoCr stems than for the SUS stems. Final subsidence of the three CoCr stems was significantly greater than that of the three SUS stems.

3.3. Compressive Force at the Bone–Cement Interface

The final maximum compressive force at the bone–cement interface for both stems was measured for four stems (SUS1, SUS3, CoCr1, CoCr2) on the proximal medial side and for two on the proximal lateral side (SUS2, CoCr3) (Figure 6). The final compressive force was greater in the CoCr group than in the SUS group at the proximal medial ($p < 0.001$) and distal lateral ($p = 0.004$) region significantly (Figure 6).

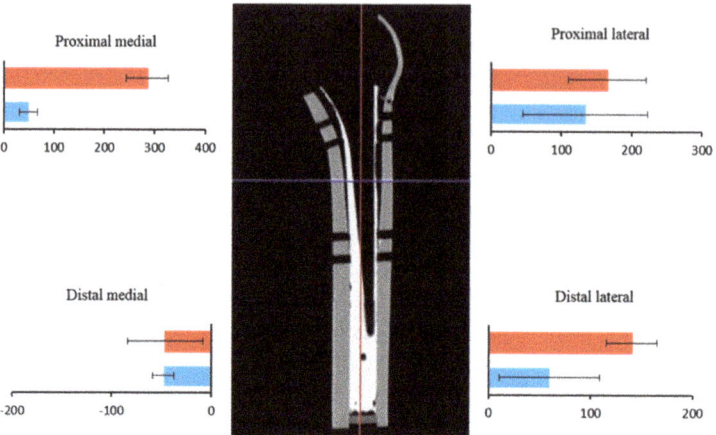

Figure 6. Final forces at the cement–bone interface. Final compressive forces are shown for the SUS (blue) and CoCr (red) stems. The three major regions of compressive force were the proximal medial and lateral regions and the distal lateral region. The bars show mean force in each group, and the lines show standard deviation. The forces differed significantly between the two groups in the proximal medial and distal lateral regions. (Proximal medial p value < 0.001; distal medial p value = 0.258; proximal lateral p value = 0.258; distal lateral p value = 0.004).

The forces in the anterior and posterior regions were small values compared to the medial and lateral regions. The compressive forces at the cement–bone interface in the proximal medial region gradually increased over the course of the experiment. The compressive forces generated by all CoCr stems were significantly greater than those generated by all SUS stems throughout the experiment, and the rate of increase in compressive forces was greater for the CoCr stems than for the SUS stems (Figure 7).

A strongly significant positive correlation was demonstrated between subsidence values and compressive forces in all stems, indicating that the stem subsidence generated compressive force at the cement–bone interface (Figure 8). Between 0.05 mm and 0.1 mm of subsidence, the compressive force was more than four times greater for CoCr than for SUS, a significant difference (37.42 (SD 7.41) N for SUS and 166.34 (SD 23.14) N for CoCr, $p < 0.001$) and consistent with our findings for subsidence between 0.1 and 0.15 mm (53.68 (SD 5.65) N in the SUS group and 186.62 (SD 31.64) N in the CoCr group, $p < 0.001$). In other words, even with the same amount of subsidence, CoCr stems generated a larger compressive force, more than three times that for SUS stems at the bone–cement interface.

A strongly significant positive correlation was observed between stem subsidence and compressive force at the cement–bone interface in all stems. Correlation equations, p values, and coefficient of determination (r^2) are shown below, from the left:

SUS1: $y = 329x + 8.7903$, $p < 0.01$ ($r^2 = 0.9787$)
SUS2: $y = 93.348x + 19.413$, $p < 0.01$ ($r^2 = 0.5432$)
SUS3: $y = 394.85x + 9.5199$, $p < 0.01$ ($r^2 = 0.9626$)
CoCr1: $y = 433.82x + 150.43$, $p < 0.01$ ($r^2 = 0.9668$)
CoCr2: $y = 567.14x + 135.41$, $p < 0.01$ ($r^2 = 0.9881$)
CoCr3: $y = 407.12x + 104.86$, $p < 0.01$ ($r^2 = 0.9775$).

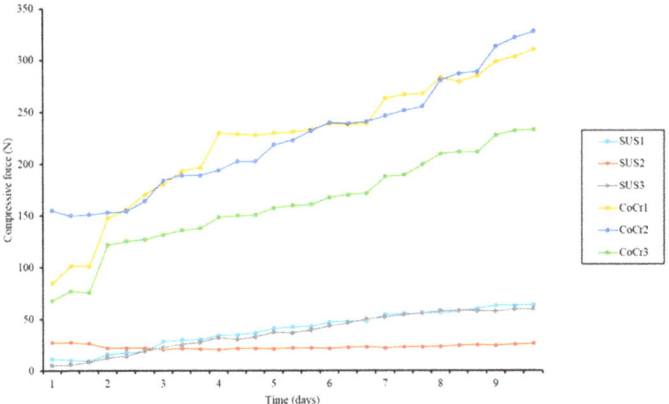

Figure 7. Compressive forces at cement–bone interface in proximal medial region throughout the experiment. Compressive forces from CoCr stems were much greater than from SUS stems throughout the experiment. The mean compressive force associated with CoCr was significantly greater than with SUS on the final day.

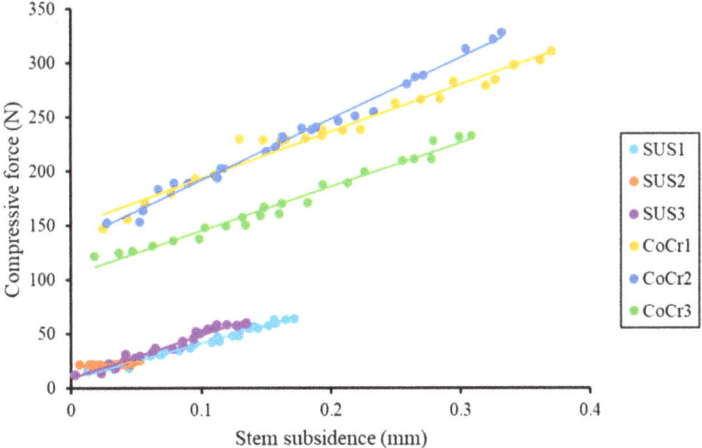

Figure 8. Correlation of stem subsidence and compressive force in proximal medial region. Dots represent the values of each point of stem subsidence and compressive force since the second day. An approximate straight line for each dot was calculated for each stem, and a straight line was generated.

3.4. Movement of Tantalum Balls

The tantalum balls were attempted to be positioned around the stem on the medial, lateral, anterior, and posterior sides, but in some cases, it was not possible to successfully place the balls on the posterior side of the stems. As a result, the tantalum balls in the posterior cement were excluded from the analysis.

The actual distance of tantalum ball movement may vary for each composite femur due to calibration differences; therefore, the distances cannot be compared simply. For this reason, the ratio of horizontal movement to vertical movement was compared for each ball used with each femur. Although there was no difference between the SUS and CoCr stems in the ratio of horizontal/vertical ball movement ($p = 0.863$), the ratio of ball vertical movement to stem subsidence was significantly smaller for CoCr than for SUS ($p = 0.008$).

This result indicated that CoCr stems slipped more against the cement surface than did the SUS stems, even with identical ratios for cement creep (Table 2).

Table 2. There were no significant differences in the horizontal/vertical distance ratio of balls between the SUS and CoCr groups. However, the ratio of vertical distance/stem subsidence was significantly smaller in the CoCr group than in the SUS group.

Position of Ball	SUS1			SUS2			SUS3			CoCr1			CoCr2			CoCr3			p Value
	Medial	Lateral	Anterior	Medial	Lateral	Anterior	Medial	Lateral	Anterior	Medial	Lateral	Anterior	Medial	Lateral	Anterior	Medial	Lateral	Anterior	
Number of balls	1	1	3	1	1	3	1	1	3	1	2	2	1	1	2	1	1	2	
Stem subsidence (mm)		0.36527			0.14203			0.26417			0.49105			0.54482			0.39862		
Horizontal/Vertical movement of ball (A)	0.0926	0.7883	0.5541	0.0811	0.3944	0.2513	0.0244	0.6087	0.1162	0.1077	2.0332	0.1863	0.1667	0.3774	0.1130	0.0515	5.5714	0.3562	
Vertical movement of ball/Stem subsidence (B)	0.4435	0.4184	0.2378	1.0420	0.5622	0.7456	0.1552	0.0871	0.1313	0.3971	0.0697	0.1445	0.0661	0.0973	0.0631	0.2433	0.0176	0.0609	
Mean ± SD (A)				0.3235 (0.2751)									0.9959 (1.8237)						0.863
Mean ± SD (B)				0.4248 (0.3184)									0.1288 (0.1199)						0.008

3.5. Cement Thickness

Due to the hard cancellous bone in this composite bone model, there was minimal cement integration into the bone. As a result, the cement mantles were highly detectable and measurable with accuracy on all slices of CT in all composite femurs.

Between the two groups, there was no significant difference in cement thickness around the lesser trochanter between the SUS and CoCr groups (mean thickness 1.813 mm for SUS and 1.842 mm for CoCr) (Table 3). Results also showed no difference in stem alignment between the two groups.

Table 3. Thickness of cement in each CT slice. Numbers indicate mean thickness in each group. Numbers within ( ) are standard deviation. The notation unit is nm.

	SUS Group				CoCr Group			
	Medial	Lateral	Anterior	Posterior	Medial	Lateral	Anterior	Posterior
No.1	2.23 (0.334)	1.95 (0.250)	1.73 (0.179)	1.38 (0.083)	2.28 (0.130)	2.1 (0.187)	1.63 (0.148)	1.33 (0.083)
No.2	1.98 (0.192)	2.03 (0.148)	1.5 (0.071)	1.28 (0.083)	2.38 (0.083)	2.23 (0.109)	1.53 (0.109)	1.6 (0.071)
No.3	2.38 (0.130)	2.28 (0.130)	1.55 (0.112)	1.50 (0.122)	2.10 (0.158)	2.13 (0.311)	1.53 (0.148)	1.30 (0.071)

4. Discussion

Different alloy materials exhibiting different mechanical behaviors have been investigated by researchers. Takegami et al. [19] performed biomechanical tests on the CPT, VerSys Advocate, and CMK stems, and noted that the median fracture torque for the CPT stem was significantly lower than for the CMK stem. They suggested that variations in the mechanical behavior of these prosthetic materials might contribute to differences in the incidence of PPF.

To replicate a physiologic environment as closely as possible, our experiment employed 1 Hz sine curve loading to simulate a normal human walking pace, maintained a wet environment for the composite bone, kept the model at a temperature of 37 °C, and provided an 8 h period of stress relaxation per day. To ensure a reliable comparison between CoCr and SUS materials, stems were confirmed to have identical design and surface roughness, and were placed in cement of identical thickness. Therefore, our results were considered highly reliable.

Under these reliable conditions, significant differences were observed between CoCr and SUS stems. CoCr stems exhibited greater daily subsidence and rise, as well as greater final subsidence, suggesting that the CoCr stems were more prone to slippage within the cement compared to the SUS stems.

By investigating the movement of tantalum balls within the cement, we confirmed the movement of the cement around the stem. As a result, CoCr stems were shown to be more susceptible to slippage on the cement surface compared to SUS stems.

Previous reports have suggested that, with a polished surface of Ra = 0.06 μm and bone cement of moderate viscosity, the frictional coefficient was significantly lower for CoCr than for SUS [20], consistent with findings that CoCr had lower wettability than SUS and provided relatively poor adhesion to bone cement [11,20]. This would explain the greater mobility of CoCr stems in the cement.

Grammatopoulos et al. [21] investigated the double taper PTS fracture type and reported that 16 of 21 clinical cases were of the Vancouver B2 type, involving spiral fractures, stem subsidence, and cement rupture. Morishima et al. [22] fixed the Exeter stem with cement in composite bone and applied internal rotation while loading the prosthesis, resulting in type B2 spiral fractures in all cases. The Vancouver type is considered to be the most common clinical PPF morphology associated with PTSs, and these findings indicate that stem subsidence with rotational force is an important factor in the occurrence of PPF when using a PTS [21–24].

Stem subsidence is thought to generate compressive force at the cement–bone interface because of the strong correlation between stem subsidence and these compressive forces [14–16]. In this study, mean final subsidence was significantly greater for CoCr stems than for SUS (0.491 mm and 0.261 mm, respectively). Compressive forces were also significantly higher for CoCr than for SUS. Furthermore, even with equal levels of subsidence, compressive forces were found to be more than three times higher for CoCr stems than for SUS stems at the cement–bone interface. This was believed to be due to the fact that CoCr stems were more prone to slip on the cement surface compared to SUS stems, as evidenced by the results of the tantalum balls. Our findings suggest that CoCr stems exert greater pressure on the femur compared to SUS stems. This conclusion aligns with previous research indicating that significant stem subsidence may induce great hoop stress [14].

Some reports also show that, when inserted into a thicker cement layer, a PTS stem of CoCr will subside more, causing the cement surrounding the stem to be dragged in the direction of subsidence, which is toward the shear force [15,16]. The significant subsidence of the PTS in a thicker cement mantle may likewise generate high stress in the cement and predispose to PPF [15,25]. However, there was no difference in the thickness of the cement between the CoCr stem and SUS stem groups in this study. Therefore, our results, which excluded factors other than material differences, suggest that the CoCr stem, which is prone to large movements and sinking, may be more likely to cause PPF than the SUS stem.

Our study has some limitations. (1) Only a limited number of stems (three SUS and three CoCr) were used in the experiment. (2) Loading was only applied from one direction, and no real clinical rotation was added; other results might be obtained with the addition of stem rotation. (3) The composite femur used in this experiment might differ from the actual behavior of a human femur. (4) The use of different stem designs and stem cements could yield different results. (5) In contrast to cadaveric femurs, composite bone has the benefit of being highly standardized. However, composite femurs lack cancellous bone in the diaphysis area, resulting in a thick mantle (more than 3 mm) in the distal region.

Strengths of our study include the use of composite bone, which in contrast to cadaveric femurs, has the benefit of being highly standardized even though only three pairs of stems were used. Although the differences between stem brands cannot be excluded from consideration, the differences in mechanical behavior between materials are practical and quantifiable. Our findings would be expected to differ if applied to cadaver bones or to individual living organisms. However, we consider the data in this study to be as reliable as previous similar biomechanical experiments using composite bones [14–16].

5. Conclusions

Our findings clearly demonstrated that the Exeter-shaped CoCr stem showed greater subsidence in the cement and higher bone–cement interface forces than the original Exeter stem made of SUS. Our study suggested that the CoCr stem was more prone to slipping on the cement surface than SUS and generated over three times the hoop stress of SUS with the same amount of subsidence. These differences might appear to contribute to the increased risk of PPF when using CoCr-PTS.

Author Contributions: Conceptualization, A.K. and M.C.; methodology, A.K., M.C. and E.T.; software, A.K. and N.T.; validation, A.K., T.I. and N.K.; formal analysis, M.F. and D.S.; investigation, A.K., M.C., Y.T. and Y.O.; resources, N.K.; data curation, M.C. and E.T.; writing—original draft preparation, A.K. and M.C.; writing—review and editing, A.K., T.I. and N.K.; visualization, N.K.; supervision, N.K.; project administration, A.K. and N.K.; funding acquisition, A.K. and N.K. All authors have read and agreed to the published version of the manuscript.

Funding: This study was performed using implants provided free of charge from Stryker Co. A. Kaneuji and N. Kawahara declare the receipt of research grants from Zimmer Biomet Co. as chief members of their university research department.

Data Availability Statement: The data presented in this study are available on request from the corresponding author.

Conflicts of Interest: Kaneuji is a paid consultant for Johnson and Johnson Co.

References

1. Collis, D.K.; Mohler, C.G. Comparison of clinical outcomes in total hip arthroplasty using rough and polished cemented stems with essentially the same geometry. *J. Bone Jt. Surg. Am.* **2002**, *84*, 586–592. [CrossRef]
2. Della Valle, A.G.; Zoppi, A.; Peterson, M.G.E.; Salvati, E.A. A Rough Surface Finish Adversely Affects the Survivorship of a Cemented Femoral Stem. *Clin. Orthop. Relat. Res.* **2005**, *436*, 158–163. [CrossRef]
3. Hamadouche, M.; Baqué, F.; Lefevre, N.; Kerboull, M. Minimum 10-year survival of Kerboull cemented stems according to surface finish. *Clin. Orthop. Relat. Res.* **2008**, *466*, 332–339. [CrossRef]
4. Howie, D.; Middleton, R.G.; Costi, K. Loosening of matt and polished cemented femoral stems. *J. Bone Jt. Surg. Br.* **1998**, *80*, 573–576. [CrossRef]
5. Carli, A.V.; Negus, J.J.; Haddad, F.S. Periprosthetic femoral fractures and trying to avoid them what is the contribution of femoral component design to the increased risk of periprosthetic femoral fracture? *Bone Jt. J.* **2017**, *99-B* (Suppl. A1), 50–59. [CrossRef]
6. Raut, S.; Parker, M.J. Medium to long term follow up of a consecutive series of 604 Exeter Trauma Stem Hemiarthroplasties (ETS) for the treatment of displaced intracapsular femoral neck fractures. *Injury* **2016**, *47*, 721–724. [CrossRef]
7. Mellner, C.; Mohammed, J.; Larsson, M.; Esberg, S.; Szymanski, M.; Hellström, N.; Chang, C.; Berg, H.E.; Sköldenberg, O.; Knutsson, B.; et al. Increased risk for postoperative periprosthetic fracture in hip fracture patients with the Exeter stem than the anatomic SP2 Lubinus stem. *Eur. J. Trauma Emerg. Surg.* **2021**, *47*, 803–809. [CrossRef]
8. Scott, T.; Salvatore, A.; Woo, P.; Lee, Y.-Y.; Salvati, E.A.; Della Valle, A.G. Polished, Collarless, Tapered, Cemented Stems for Primary Hip Arthroplasty May Exhibit High Rate of Periprosthetic Fracture at Short-Term Follow-Up. *J. Arthroplast.* **2018**, *33*, 1120–1125. [CrossRef]
9. Palan, J.; Smith, M.C.; Gregg, P.; Mellon, S.; Kulkarni, A.; Tucker, K.; Blom, A.W.; Murray, D.W.; Pandit, H. The influence of cemented femoral stem choice on the incidence of revision for periprosthetic fracture after primary total hip arthroplasty: An analysis of national joint registry data. *Bone Jt. J.* **2016**, *98-B*, 1347–1354. [CrossRef]
10. Lamb, J.; Jain, S.; King, S.; West, R.; Pandit, H. Risk Factors for Revision of Polished Taper-Slip Cemented Stems for Periprosthetic Femoral Fracture After Primary Total Hip Replacement: A Registry-Based Cohort Study from the National Joint Registry for England, Wales, Northern Ireland and the Isle of Man. *J. Bone Jt. Surg.* **2020**, *102*, 1600–1608. [CrossRef]
11. Tsuda, R. Differences in mechanical behavior between Cobalt-chrome alloy and Stainless-steel alloy in polished tapered femoral stems fixed with bone cement. *J. Kanazawa Med. Univ.* **2016**, *41*, 1–9. (In Japanese)
12. Keeling, P.; Howell, J.R.; Kassam, A.-A.M.; Sathu, A.; Timperley, A.J.; Hubble, M.J.; Wilson, M.J.; Whitehouse, S.L. Long-Term Survival of the Cemented Exeter Universal Stem in Patients 50 Years and Younger: An Update on 130 Hips. *J. Arthroplast.* **2020**, *35*, 1042–1047. [CrossRef] [PubMed]
13. Available online: https://nagumo-ss.com/works/ (accessed on 20 March 2023). (In Japanese).
14. Kaneuji, A.; Yamada, K.; Hirosaki, K.; Takano, M.; Matsumoto, T. Stem subsidence of polished and rough double-taper stems: In vitro mechanical effects on the cement-bone interface. *Acta Orthop.* **2009**, *80*, 270–276. [CrossRef] [PubMed]
15. Takahashi, E.; Kaneuji, A.; Tsuda, R.; Numata, Y.; Ichiseki, T.; Fukui, K.; Kawahara, N. The influence of cement thickness on stem subsidence and cement creep in a collarless polished tapered stem: When are thick cement mantles detrimental? *Bone Jt. Res.* **2017**, *6*, 351–357. [CrossRef]
16. Numata, Y.; Kaneuji, A.; Kerboull, L.; Takahashi, E.; Ichiseki, T.; Fukui, K.; Tsujioka, J.; Kawahara, N. Biomechanical behaviour of a French femoral component with thin cement mantle: The "French paradox" may not be a paradox after all. *Bone Jt. Res.* **2018**, *7*, 485–493. [CrossRef]
17. Lee, A.J.C.; Ling, R.S.M.; Gheduzzi, S.; Simon, J.-P.; Renfro, R.J. Factors affecting the mechanical and viscoelastic properties of acrylic bone cement. *J. Mater. Sci. Mater. Med.* **2002**, *13*, 723–733. [CrossRef]
18. Bergmann, G.; Graichen, F.; Rohlmann, A. Hip joint loading during walking and running, measured in two patients. *J. Biomech.* **1993**, *26*, 969–990. [CrossRef]
19. Takegami, Y.; Seki, T.; Osawa, Y.; Imagama, S. Comparison of periprosthetic femoral fracture torque and strain pattern of three types of femoral components in experimental model. *Bone Jt. Res.* **2022**, *11*, 270–277. [CrossRef]
20. Hirata, M.; Oe, K.; Kaneuji, A.; Uozu, R.; Shintani, K.; Saito, T. Relationship between the surface roughness of material and bone cement: An increased "polished" stem may result in the excessive taper-slip. *Materials* **2021**, *14*, 3702. [CrossRef]
21. Grammatopoulos, G.; Pandit, H.; Kambouroglou, G.; Deakin, M.; Gundle, R.; McLardy-Smith, P.; Taylor, A.; Murray, D. A unique peri-prosthetic fracture pattern in well fixed femoral stems with polished, tapered, collarless design of total hip replacement. *Injury* **2011**, *42*, 1271–1276. [CrossRef]
22. Morishima, T.; Ginsel, B.L.; Choy, G.G.; Wilson, L.J.; Whitehouse, S.L.; Crawford, R.W. Periprosthetic fracture torque for short versus standard cemented hip stems: An experimental in vitro study. *J. Arthroplast.* **2014**, *29*, 1067–1071. [CrossRef] [PubMed]
23. Brodén, C.; Mukka, S.; Muren, O.; Eisler, T.; Boden, H.; Stark, A.; Sköldenberg, O. High risk of early periprosthetic fractures after primary hip arthroplasty in elderly patients using a cemented, tapered, polished stem. *Acta Orthop.* **2015**, *86*, 169–174. [CrossRef]

24. Baryeh, K.; Mendis, J.; Sochart, D.H. Temporal Subsidence Patterns of Cemented Polished Taper-Slip Stems: A Systematic Review. *EFORT Open Rev.* **2021**, *6*, 331–342. [CrossRef]
25. Korsnes, L.; Gottvall, A.; Buttazzoni, C.; Mints, M. Undersizing the Exeter stem in hip hemiarthroplasty increases the risk of periprosthetic fracture. *HIP Int.* **2019**, *30*, 469–473. [CrossRef] [PubMed]

Disclaimer/Publisher's Note: The statements, opinions and data contained in all publications are solely those of the individual author(s) and contributor(s) and not of MDPI and/or the editor(s). MDPI and/or the editor(s) disclaim responsibility for any injury to people or property resulting from any ideas, methods, instructions or products referred to in the content.

MDPI AG
Grosspeteranlage 5
4052 Basel
Switzerland
Tel.: +41 61 683 77 34

Journal of Functional Biomaterials Editorial Office
E-mail: jfb@mdpi.com
www.mdpi.com/journal/jfb

Disclaimer/Publisher's Note: The statements, opinions and data contained in all publications are solely those of the individual author(s) and contributor(s) and not of MDPI and/or the editor(s). MDPI and/or the editor(s) disclaim responsibility for any injury to people or property resulting from any ideas, methods, instructions or products referred to in the content.

www.ingramcontent.com/pod-product-compliance
Lightning Source LLC
LaVergne TN
LVHW070240100526
838202LV00015B/2160